MICROWAVE ENGINEERING
Passive Circuits

Peter A. Rizzi

SOUTHEASTERN MASSACHUSETTS UNIVERSITY

Prentice-Hall International, Inc.

ISBN 0-13-581711-0

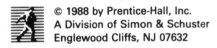 © 1988 by Prentice-Hall, Inc.
A Division of Simon & Schuster
Englewood Cliffs, NJ 07632

Printed in the United States of America

10 9 8 9 6 5 4 3 2 1

ISBN   0-13-581711-0

Prentice-Hall International (UK) Limited, *London*
Prentice-Hall of Australia Pty. Limited, *Sydney*
Prentice-Hall Canada Inc., *Toronto*
Prentice-Hall Hispanoamericana, S.A., *Mexico*
Prentice-Hall of India Private Limited, *New Delhi*
Prentice-Hall of Japan, Inc., *Tokyo*
Simon & Schuster Asia Pte. Ltd., *Singapore*
Editora Prentice-Hall do Brasil, Ltda., *Rio de Janeiro*
Prentice-Hall, Inc., *Englewood Cliffs, New Jersey*

TO
ALL MY STUDENTS, PAST AND
PRESENT, WHO HAVE MADE MY
WORK MEANINGFUL AND ENJOYABLE

AND TO
BOB COMMERFORD
MY FRIEND AND COLLEAGUE

Contents

Preface

This book is intended as an introductory text on microwave theory and techniques. The topics are presented at a level appropriate for use in undergraduate electrical engineering and technology programs or for self-study by practicing engineers. The main objectives are to provide the reader with an understanding of microwave devices and circuits, as well as some of the methods used in their analysis and design.

In 1970, after 18 years of experience in the microwave industry, I joined the electrical engineering faculty at Southeastern Massachusetts University. One of my first tasks was to develop a two-semester elective sequence in microwave engineering. In searching for a suitable text, it was apparent that the available microwave books were either for use in training technicians or for students engaged in engineering study at the graduate level. An appropriate text for undergraduate study was simply not available. As a result, it became necessary to develop a comprehensive set of lecture notes for the senior-level course. Those notes are the basis for much of the material included here. A series of minicourses presented to practicing microwave engineers has also provided a valuable source of content.

The text requires that the student have a working knowledge of electricity and magnetism, ac circuits, and basic transmission-line theory. A solid background in mathematics including calculus and elementary differential equations is also assumed. Certain sections, indicated by a star ($\star$), include derivations that require a knowledge of the differential operators used in vector calculus. Students unfamiliar with their mathematical properties may skip these sections. However, the final results of these derivations are provided in subsequent sections as they are needed. With this approach, the text is designed to address the needs of both engineering and technology programs. The effectiveness of this approach has been observed by the writer in over a decade of teaching within both programs.

This writer believes that most concepts are more easily learned by starting with an elementary discussion of the subject complete with appropriate examples before proceeding to a more generalized treatment. The text, for the most part, reflects this pedagogical viewpoint. Numerous examples of design and analysis that utilize transmission-line theory, equivalent circuits, and the vector-phasor method are presented. Occasionally, a particular component is examined from more than one point of view in order to gain some additional insight. In many cases, an introduction to the more complex aspects of the topic is included in order to stimulate the student to consider further study.

Chapters 2 and 3 review the fundamentals of fields and waves and basic transmission-line theory. Despite its review nature, a perusal of the material is suggested since it influences the presentation of topics in later chapters. Chapter 4 describes some impedance matching techniques and the analysis of two-port networks. Transmission and scattering matrix methods are utilized as analytical tools in this and subsequent chapters. A summary of their properties are given in Appendices C and D. Chapter 5 discusses the characteristics of several types of transmission lines used in microwave systems. A wide variety of passive microwave components are described and analyzed in Chapters 6 through 8. The special topic of microwave filters and resonators is covered in Chapter 9.

Following each chapter is a list of useful references. In addition, many practice problems are included so that the student has an opportunity to gain skill in applying the principles and techniques learned. Answers to all problems are given at the end of the text. A detailed solutions manual is available from the publisher.

The text can be used in either a one-or two-semester course. For students having the appropriate background in electromagnetic theory and transmission lines, a one-semester course covering Chapters 4 to 9 is suggested. If the instructor feels that a detailed review of transmission-line theory is necessary, the following course outline is suggested: Chapters 3 through 9, excluding Secs. 4–4, 5–6, 6–7, 7–5, 7–6, 8–4, 8–5, 9–5, and 9–6. A two-semester course covering all chapters is suggested for students whose only exposure to electricity and magnetism was the usual sophomore-level physics course.

Given the constraints of producing a text of reasonable size while simultaneously insuring proper depth of content, certain topics such as microwave measurements, electronics, ferrites, and antennas have been omitted. Appendix F does, however, clarify some of the terminology used in describing antennas that form part of a microwave system.

In summary, I have tried to present the subject of microwaves so that it is readily understood and clearly applicable. The reader can best determine how well I have succeeded in this endeavor. Comments regarding organization, clarity of presentation, usefulness, and any other helpful suggestions will be most welcomed.

To undertake and complete the writing of a textbook requires help and encouragement from many people. First of all, I wish to express my gratitude to Frank Brand, Dan Murphy, and Ted Saad for their kind words of support and encouragement over the years. Next, I wish to thank Arthur Uhlir, Jr. for reading the complete manuscript and providing me with many valuable suggestions. Thanks are also due Ken Carr, Harlan Howe, Don Soorian, and Joe White for their thoughtful comments on various portions of the text. Special thanks are due Sue Fontaine and Liz Moreau for typing the final version on the magical word processor. The staff at Prentice Hall has been very helpful in seeing the text through its various stages. In particular the infinite patience and encouragement of Bernard Goodwin is gratefully acknowledged. Finally, my thanks to Tim Bozik, editor, and to Edie Riker for supervising the production of this book.

I cannot conclude this litany of acknowledgements without mentioning my golfing buddies without whose efforts this book would have been finished *3 years ago*!

Foreword

One of the great pleasures in life is watching a truly great craftsman ply his or her trade. It is an experience that can be appreciated whether the artisan is a woodworker, a painter, a fine photographer, a musician, or a teacher. As a teacher, Peter Rizzi is that kind of craftsman. The loyalty and appreciation of his students is impressive and reflects that craftsmanship. One of my colleagues, who is a former Rizzi student, asserts that he "would swim upstream through hot tar" for the opportunity of taking another course from Peter Rizzi. I am not that strong a swimmer myself, but I have had the pleasure of being taught by Peter Rizzi and I understand what he means.

This book has evolved from the course notes that Dr. Rizzi has put together through the years. However, the book is not a dry academic exercise. Because of Peter's industrial background, it is rich with analogies and examples. Further, he has drawn from a host of skilled colleagues in industry and academia and has produced a text that is smooth and easy to read—a text that, above all, teaches. An essential part of the learning experience involves the use of examples and problems which students themselves solve. Here again, the problems in this book are not tests, they are teaching aids. Every problem illustrates a point and in almost all cases, the solutions are simple. If, in solving one of these problems, a reader gets bogged down in four pages of matrix algebra, then either the main point or a key assumption has been missed.

The introduction of this book describes it as an undergraduate text. I believe Dr. Rizzi is too modest. This book should also be useful at the graduate level and to practicing engineers. As a matter of fact, I don't believe there is a single working engineer in industry who can't learn something from this text.

The book has been a long time coming. I recall some discussions with Peter nearly 10 years ago when he was first considering the possibility of writing such a book. At that time, we observed that there was no book like this in existence. With the exception of what you are holding in your hand, that is still true today. I think it was well worth the wait.

Harlan Howe, Jr.
Boxborough, Massachusetts
December 1986

EDITOR's NOTE
Mr. Howe is a fellow of the IEEE and is a Past President of the Microwave Theory and Techniques Society. He is currently Senior Vice President of the M/A-COM Corporate Component Technology Center in Burlington, Massachusetts.

1

Introduction

This chapter presents a description of microwaves and some of its many applications. Significant historical developments related to the advancement of microwave engineering are also presented. Students of engineering are urged to study the history of science, mathematics, and engineering. I have observed that those possessing a *sense of history* develop a deeper understanding of new concepts and are more creative problem solvers (see Ref. 1–10). Besides, it's fun! The chapter concludes with a review of the units and notation used throughout the text.

1–1 THE MICROWAVE SPECTRUM

Microwaves are electromagnetic waves whose frequencies range from approximately 300 megahertz (MHz) to 1000 gigahertz (GHz). Most applications of microwave technology make use of frequencies in the 1 to 40 GHz range. The letter designations for the most commonly used microwave bands are listed in Table 1–1.

TABLE 1–1 Commonly Used Microwave Frequency Bands

LETTER DESIGNATION	FREQUENCY RANGE
L band	1 to 2 GHz
S band	2 to 4 GHz
C band	4 to 8 GHz
X band	8 to 12 GHz
K_u band	12 to 18 GHz
K band	18 to 26 GHz
K_a band	26 to 40 GHz

The diagram of the electromagnetic spectrum in Fig. 1–1 illustrates that the lower end of the microwave region borders radio and television frequencies, while the upper end is adjacent to the infrared and optical spectrums. As a result, microwave engineers often employ ideas and techniques derived from both of these well-established disciplines. For example, optical techniques are used to design microwave antennas and lenses, while microwave circuit design usually involves concepts associated with ac network theory.

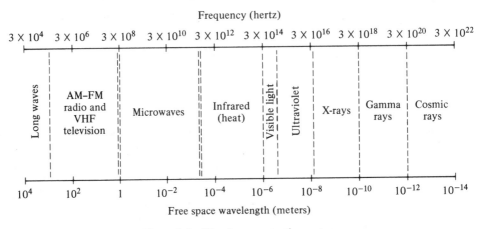

Figure 1–1 The electromagnetic spectrum.

There are three important characteristics that differentiate microwave engineering from its low-frequency and optical counterparts. First, the size of most microwave components and circuits are within an order of magnitude of the operating wavelength.[1] This is not the case for ordinary optical devices and conventional electric circuits. For example, the validity of the ray concept in optics stems from the fact that components, such as lenses and mirrors, are much larger than the operating wavelength. On the other hand, the application of ac network theory requires that all dimensions of the circuit elements be much smaller than the wavelength. To verify this statement, consider the carbon resistor and its model illustrated in Fig. 1–2. A basic assumption of ac theory is that the current entering the resistor exactly equals that leaving it, both in amplitude and phase. However, they cannot be exactly equal since it takes a finite time for the wave associated with the current to travel the length l. Since phase delay is merely a normalized way of expressing the time delay of a sinusoid, the conclusion is that the phase of I_2 will not necessarily be the same as I_1. Thus the statement $I_1 = I_2$ assumes that the time delay t_d is negligible compared to the period T of the ac signal.[2] Expressed mathematically $t_d = l/v \ll T = 1/f$. Since all waves are governed by the relationship

$$f\lambda = v \qquad\qquad (1\text{–}1)$$

---

[1] It is interesting to note that the size of many audio components, such as speakers and microphones, are comparable to the wavelength of audible sound waves. Consequently, audio engineering techniques are also useful to the microwave engineer.

[2] All symbols and their SI units are listed in Appendix A.

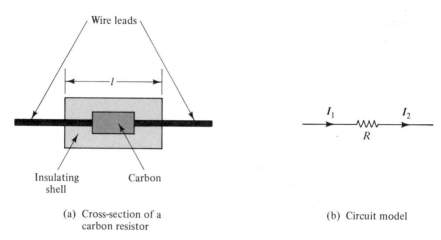

(a) Cross-section of a   (b) Circuit model
      carbon resistor

Figure 1–2   The resistance model.

where, in this case, v is the electromagnetic wave velocity, the inequality may be rewritten as $l \ll \lambda$. In other words, an alternate way of expressing the underlying assumption of ac theory is that the dimensions of the circuit elements must be small compared to the operating wavelength. In Chapter 3, transmission-line methods will be explored which partially circumvent the restrictiveness of this assumption and hence are very useful in the design and analysis of microwave circuits.

It should be noted that even when the actual circuit element is small compared to the wavelength, it may not behave as expected. For example, to the designer of low-frequency networks, an ordinary low-power carbon resistor may be modeled as a simple resistance. For the microwave designer, however, an accurate representation must include the shunt capacitance due to the insulating shell and the series inductance associated with the wire leads. In many cases, the reactances associated with this capacitance and inductance obscure the resistive properties making it unsuitable for microwave applications. Specially designed microwave resistors are commercially available that minimize these unwanted or *parasitic* reactances. The same holds true for diodes, transistors, and other elements.

A second phenomenon that is of particular concern to the microwave engineer is that of skin effect. Skin depth, defined precisely in Chapter 2, is a measure of the degree to which an electromagnetic field penetrates a conducting material. This effect is a function of frequency, the depth of penetration decreasing with increasing frequency. For example, at 1000 MHz, the skin depth for copper is only a few microns. As a result, microwave currents tend to flow along the surface of a conductor, which dramatically increases its resistive effect. At microwave frequencies, the resistance of a copper wire can easily be a thousand times greater than its low-frequency value. Therefore, the use of special techniques for minimizing circuit losses are particularly important at microwaves. In some cases the silver-plating and polishing of metallic surfaces is necessary.

The third unique characteristic of microwave work is associated with the measurement techniques used. At low frequencies, the properties of a circuit or system are determined by measuring voltages and currents. This approach is not applicable to microwave circuits since oftentimes these quantities are not uniquely defined. Even when the direct measurement of microwave voltage is possible, noise fluctuations usually degrade the accuracy of the measurement. As a result, most microwave experimentation involves the accurate measurement of impedance and power rather than voltage and current.

1–2 HISTORICAL BACKGROUND

Michael Faraday was one of the greatest of all scientific researchers. His careful attention to the detail and organization of experiments, as well as his systematic analysis of data resulted in many remarkable contributions to the science of electricity and magnetism. In 1845, for example, he studied the effect of a magnetic field on the propagation of light through glass. He observed that when the magnetic field was aligned with the direction of propagation, the plane of polarization of the light beam rotated slightly. This observation suggested the existence of a relationship between electromagnetism and light. In his "Thoughts on Ray Vibrations," Faraday speculated that light might be a transverse magnetic disturbance with *wavelike* characteristics. Twenty years later, James Clerk Maxwell published "A Dynamical Theory of the Electromagnetic Field." In this remarkable paper, and later in his definitive work, *Electricity and Magnetism,* Maxwell formally developed the electromagnetic theory of light. Starting with four basic relationships, now called Maxwell's equations, he proved mathematically that electromagnetic waves could propagate through a nonconducting medium. Furthermore, he demonstrated that the electric and magnetic fields are transverse to the direction of propagation and travel at the speed of light. In fact, his predicted value of wave velocity was so near the known speed of light, that he stated, "we have strong reason to conclude that light itself is an electromagnetic disturbance in the form of waves . . ." Maxwell's theory of electromagnetic waves was not readily accepted by all physicists. Many scientists subscribed to the Weber theory which held that electric and magnetic fields merely represented "instantaneous action-at-a-distance."

In the early 1880s, Heinrich Hertz set out to determine which of the two theories was correct. Through a brilliant series of experiments, he succeeded in verifying Maxwell's theory of electromagnetic waves. These experiments are described in two papers, published in 1888, entitled "On Electromagnetic Waves in Air, and Their Reflection" and "On Electric Radiation." A detailed description of these experiments may be found in Refs. 1–1 and 1–2. In the Hertz experiments, the electromagnetic generator consisted of a spark gap connected to an induction coil. The interruption of current in the coil produced a high voltage, causing a spark to appear across the gap. This effect was oscillatory with a frequency in the megahertz range. For a detector, Hertz used a circular loop of thick wire with the ends separated by a very tiny gap. He reasoned that the induced voltage in the loop would be sufficient to cause a small observable spark across the tiny gap. This experiment was conducted in a room where one wall was covered with a large sheet of zinc to reflect the waves (if they existed). At the time, the principle that wave reflections caused a standing wave pattern was well known, the spacing between successive nulls being one-half the wavelength. By moving the detector about, he was able to determine the location of the nulls and hence measure the wavelength. The observed peaks and nulls were considered proof of the existence of electromagnetic waves. Furthermore, since Hertz knew the frequency of his generator (about 30 MHz), the measurement of wavelength allowed him to calculate the velocity of these waves. The wavelength was determined to be about 10 m, from which the velocity was calculated to be 3×10^8 m/s. This value of velocity was that predicted by Maxwell (namely, the speed of light), hence validating Maxwell's theory of electromagnetism. Thus, for an electromagnetic wave in free space, Eq. (1–1) becomes

$$f\lambda_0 = c \qquad (1-2)$$

where λ_0 is the free space wavelength and c denotes the velocity of light in free space. The value of c and other physical constants are listed in Table 1–3.

In further experiments, described in his second paper, Hertz used a generator having a frequency of approximately 450 MHz. He employed parabolic reflectors to concentrate the electromagnetic energy which improved the efficiency of transmission. Reflectors of this type continue to be used today as efficient microwave antennas.

The first practical application of electromagnetic waves was in the field of communications. The major figure in this endeavor was the dynamic Guglielmo Marconi of Italy. In 1895, he succeeded in transmitting radio signals over a distance of 1 mile. By using parabolic reflectors, he increased the transmission distance to 4 miles. In this particular demonstration, the operating wavelength was 25 cm. In 1901, with considerably improved equipment, he successfully transmitted the first transatlantic wireless message from England to Newfoundland, a distance of about 3000 miles. Over the next 30 years, the progress in wireless transmission was remarkable. Frequencies from hundreds of kilohertz up to hundreds of megahertz were transmitted through space over thousands of miles.

While all these achievements were taking place, another method of transmitting signals had developed and grown. The sending of telegraph and telephone signals along wires, in fact, predated much of the work in wireless communication. Samuel Morse's telegraph was put into operation between Baltimore and Washington in 1844. In the early 1850s, the possibility of laying a transatlantic submarine cable to provide a direct telegraphic link between England and America was proposed. Sir William Thomson (later Lord Kelvin) was retained as a consultant by the Atlantic Cable Company to study the problem of electrical transmission along the cable. In 1855, his studies led to the first distributed circuit analysis of a transmission line. His model for the line consisted of uniformly distributed series resistances and shunt capacitances. This was the beginning of transmission-line theory. Although Thomson was aware that the line also contained series inductance and shunt conductance, he concluded that they were not significant in this particular application.

With the invention of the telephone in 1876 by Bell and Gray, a more detailed study of electrical signals on transmission lines was required. This project was undertaken by the brilliant mathematician and engineer, Oliver Heaviside of England. His analysis was so thorough that it appears practically unaltered in today's textbooks. Twenty-one years later, Lord Rayleigh presented theoretical proof that electromagnetic waves could propagate through hollow metal pipes (waveguides). At about the same time, Oliver Lodge demonstrated propagation in waveguides at the Royal Institution in London. The operating frequencies for Lodge's experiments were reported to be 1.5 and 4.0 GHz. Due to the lack of dependable microwave generators, the study of waveguide transmission lay dormant for the next 30 years.

Thanks to the pioneering work of Barkhausen and Kurz on positive-grid oscillators (1919) and Hull on the smooth-bore magnetron (1921), reliable microwave sources became a reality. A tube with 20 watts output at 3 GHz was constructed by British scientists in 1936. A year later, the Varian brothers at Stanford conceived the idea of velocity modulating an electron beam. This discovery led to the invention of the klystron which continues to be an excellent source of microwave power. In 1939, a team of British researchers led by Randall and Boot used the cavity resonator principle to develop a new microwave oscillator, the cavity magnetron. With an applied magnetic field of 1000 oersteds and a dc voltage of 16,000 volts, the magnetron produced an amazing 400 watts at 3 GHz. Significant contributions to the development of both the klystron and magnetron were made by the tireless W. W. Hansen of Stanford University, the inventor of the cavity resonator. Microwave tube development in the 1930s was given great impetus by the threatening war clouds over Europe. Britain's concern over the German threat prompted the British to install radars along their coastline. This effort was directed by Robert Watson-Watt. The use of such radars to insure early detection of enemy aircraft was a decisive factor in the defeat of the German Luftwaffe in the Battle of Britain.

With the availability of microwave sources, theoretical and experimental waveguide development proceeded rapidly. Southworth at Bell Labs and Chu and Barrow at MIT not only demonstrated wave propagation through hollow pipes, but also proved that in many cases it was superior to the recently developed coaxial line. During the next several years, Southworth and his colleagues at Bell Labs developed many waveguide components. These and many other microwave devices are described in Southworth's book (Ref. 1–5). In a later book (Ref. 1–6), Southworth gives a historical perspective of the development of microwave techniques. It makes fascinating reading.

Precipitated by events surrounding World War II, the MIT Radiation Laboratory was established in 1940. Many of the greatest scientists in the United States worked long and hard at this facility to further the development of radar systems. The contributions of men like Schwinger, Marcuvitz, Ragan, Montgomery, Dicke, Purcell, Pound, Smullin, Slater, and many others resulted in major advances in the theory and practice of microwaves. The twenty-seven volumes of the MIT Radiation Lab Series is still a reliable source of information for practicing engineers.

Despite the disbanding of the Radiation Lab at the end of World War II, research and development efforts continued at a rapid pace, particularly in microwave magnetics and electronics. For example, the successful development of ferrite materials led to the realization of passive, nonreciprocal components. One of the first, developed by C. L. Hogan, was a microwave gyrator based on the Faraday rotation principle. A major advance in the field of low-noise receivers was achieved with the development of varactor-type parametric amplifiers and upconverters by M. Hines, A. Uhlir, Jr., and their coworkers at Bell Labs and also by H. Hefner and G. Wade at Stanford. Other significant achievements in the field of microwave amplification included the development of the helix-type traveling-wave tube by R. Kompfner, the CW maser described by N. Bloembergen, and the tunnel diode by L. Esaki.

The developments in passive components were equally spectacular, particularly in the area of filter and coupler design. The fine work of the engineering group at the Stanford Research Institute, led by S. Cohn and others, must be acknowledged. The results of their efforts are well known by those in the microwave field.

The early 1960s saw the emergence of two more major development areas, microwave integrated circuits and solid-state microwave sources. The pioneering efforts of J. B. Gunn, W. T. Read, B. C. DeLoach, and many others led to the successful development of Gunn-effect and Impatt-type oscillators. The research in these and other areas, such as microwave transistors, is ongoing and offers exciting possibilities for the future. The special centennial issue of MTT (Ref. 1-11) provides a fascinating historical account of the many advances in microwave technology.

With the development of satellite communication, microwave relay stations, and the further growth in commercial and military radars, microwave technology has become a billion-dollar industry. The recognition of microwaves as a major field in electrical engineering resulted in the creation of the IRE group on Microwave Theory and Techniques (MTT) in 1952.[3] Prime movers during its formative years included B. Warriner, A. G. Clavier, W. W. Mumford, D. D. King, G. C. Southworth, and others. The International MTT Symposium and the Transactions of the MTT are major sources of information on developments in the theory and practice of microwave engineering. These publications are available in most technical libraries.

1-3 MICROWAVE APPLICATIONS

Microwaves are used to satisfy many functions in our modern society. These run the gamut from sending a television signal across continents to cooking an oven roast in minutes. As indicated in the previous section, two of the earliest uses of electromagnetic waves were for point-to-point communication and radar. One might ask, "What is the advantage of microwave frequencies in these applications?" The ability to focus a radio wave is a function of the antenna size and the operating wavelength. For a fixed antenna size, the focusing ability improves as the wavelength decreases. For instance, the width of a radio beam from a parabolic antenna 1 meter in diameter is about 50 degrees at 1 GHz but only 5 degrees at 10 GHz. To communicate efficiently between two points, it is important that the transmitted signal be sharply focused and aimed at the receiving antenna. Since microwave frequencies have this ability, they are ideally suited for wireless type point-to-point communication. An analysis of point-to-point communication systems is presented in Appendix F. It is interesting to note that the function of ordinary radio and television broadcasting is

---

[3] In 1963, the Institute of Radio Engineers (IRE) and the American Institute of Electrical Engineers (AIEE) merged to form the Institute of Electrical and Electronics Engineers (IEEE).

not to focus but to *cast* the radio signal over as *broad* an area as possible. For that reason, AM, FM, and TV broadcast frequencies are much lower than those in the microwave range.

There are many examples of point-to-point communications at microwave frequencies. A sight familiar to highway travelers is the microwave repeater station shown in Fig. 1–3. A series of these stations spaced along line-of-sight paths can provide a communication link between any two cities in the country. The repeater station collects the microwave signal with one antenna, passes it through an amplifier and transmits the amplified signal via a second antenna to the next repeater station. Usual distances between stations are 25 to 75 miles. A link of many such stations can, for example, deliver a TV signal originating in New York to any other city in the United States and Canada. These links are also used by Western Union, the telephone companies, railroads, public utilities, and many large industries.

Figure 1–3   A typical microwave repeater station. (Courtesy of Andrew Corp., Orland Park, Ill.)

The combination of satellites and point-to-point microwave transmission results in the ability to communicate between continents. For example, a TV signal originating in Europe can be transmitted to a satellite via a large ground-based antenna. The satellite receives, amplifies, and retransmits the signal to a large receiving antenna, possibly located in the eastern United States. In effect the satellite simulates a repeater station with a 500 mile high tower! Since atmospheric noise is particularly low in the 3 to 6 GHz range, most satellite communication systems operate within this frequency band.

The microwave spectrum contains a wide band of frequencies, which is advantageous for use in transmitting information. Communication theory holds that the amount of information that can be transmitted is directly proportional to the available bandwidth. Thus the microwave spectrum can accommodate many more communication channels than the radio and television bands. With the ever-expanding need to transmit information, microwave communications has become increasingly common in our society.

Transmitting microwave signals is sometimes accomplished with transmission lines rather than antenna systems.[4] Some typical lines are pictured in Fig. 1–4. The open two-wire line is used mainly at the lower frequencies over short distances, an example being the TV twin-lead used to connect an antenna to a television receiver. Its main disadvantage is that it is unshielded and hence tends to radiate energy. On the other hand, the coaxial and waveguide configurations are inherently shielded and therefore are the more popular choices for low-loss microwave transmission. One of the major uses of coaxial lines is the mushrooming cable TV industry.

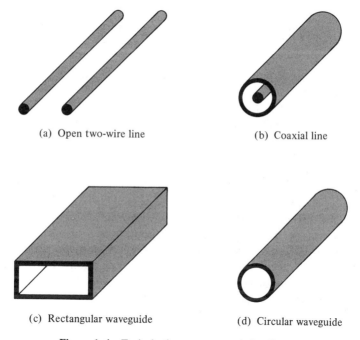

(a) Open two-wire line

(b) Coaxial line

(c) Rectangular waveguide

(d) Circular waveguide

Figure 1–4   Typical microwave transmission lines.

Radar systems represent another major application of microwaves. They are used to detect aircraft, guide supersonic missiles, observe and track weather patterns, and control flight traffic at airports. Radar is also used in burglar alarms, garage-door openers, and in the speed detectors utilized by law enforcement officials. The ability to sharply focus a radiated wave is what makes microwaves so useful in radar applications. For example, an airport radar must be able to discern separate airplanes in the traffic pattern. Consequently, the radar beam must be sufficiently narrow so that when the antenna is aimed at one airplane, the received signal represents a reflected wave from that aircraft and not from another one, say,

---

[4] The two methods of electromagnetic transmission are compared in Appendix F.

15 degrees away. This angular resolution of reflected signals requires a narrow beam and hence the use of microwave frequencies. Sharply focused microwaves from an aircraft radar can also be used to map the terrain of a large ground area. This has both civilian and military application.

The heating properties of microwave power is useful in a wide variety of commercial and industrial applications. The microwave oven is a well-known example. In ordinary ovens, food is heated by infrared. Since the skin depth of foodstuffs is small at infrared frequencies, the heat is absorbed by the surface of the substance. This heat is delivered to the inside of the foodstuff by conduction, a relatively slow process. Hence, it takes a few hours for the center of an average-sized roast to be cooked properly. At the same time, the air in the oven becomes quite hot. In comparison, the air in a microwave oven remains cool since it is a very low-loss medium at microwave frequencies. Moreover, since the skin depth at microwave frequencies for most foods is the same order of magnitude as their dimensions, the entire volume of the food is heated directly and uniformly by the microwave radiation. As a result, microwave cooking is accomplished in about one-tenth the time required by conventional oven methods. Microwave heating properties are also used in drying potato chips, paper, coffee beans, and the like. The use of microwaves for drying clothes has recently been proposed.

Medical applications of microwave technology are being explored in many hospitals and laboratories. The possibility of exposing malignant cells to microwave heat is being investigated as a method for treating cancer. Much more needs to be learned regarding the physiological and psychological effects of microwaves. The digest of the International Microwave Power Institute (IMPI) and the Journal of Microwave Power are excellent sources of information on commercial and industrial applications of microwave power.

Another research area that makes use of microwave signals and techniques is radio astronomy. Karl Jansky of Bell Labs is generally credited with being the pioneer in this exciting field. In the early 1930s he made an extensive study of atmospheric noise at high frequencies. His work led to the discovery and identification of electromagnetic noise from radio stars. This represented the beginning of the science of radio astronomy. Today, sensitive microwave receivers are not only used to study radiation from the sun and the stars, but also as a passive navigational aid to ships at sea.

Two other investigative areas that utilize microwave techniques are material science and high-energy particle physics. Many substances exhibit atomic and molecular resonances in the microwave range. The analysis and interpretation of these resonances, called microwave spectroscopy, is an important vehicle in the scientific effort to understand the fundamental nature of solids, liquids, and gases. In the field of high-energy physics, it is often necessary to accelerate electron beams to velocities within an order of magnitude of the speed of light. Periodically loaded waveguide structures are often employed to slow an electromagnetic wave so that it can efficiently interact with an electron beam. In this manner, electromagnetic energy from a microwave source can be used to increase the kinetic energy of the electron beam.

While current methods of using microwaves are certainly exciting and dynamic, the real challenge and satisfaction for future microwave engineers lies in the applications and techniques yet to be envisioned. I hope you will be a participant in these exciting new developments.

1–4 STANDARD NOTATION, PREFIXES, AND PHYSICAL CONSTANTS

Unless otherwise indicated, the International System of Units (SI) is employed throughout the text. Appendix A contains a compilation of symbols and units for the various quantities that will be encountered. Table 1–2 lists the standard prefixes as adopted by the IEEE and used here. Some of the more common physical constants are given in Table 1–3.

TABLE 1–2   Standard Prefixes and Their Symbols

Prefix	Factor	Symbol
tera	10^{12}	T
giga	10^{9}	G
mega	10^{6}	M
kilo	10^{3}	k
hecto	10^{2}	h
deka	10	da
deci	10^{-1}	d
centi	10^{-2}	c
milli	10^{-3}	m
micro	10^{-6}	μ
nano	10^{-9}	n
pico	10^{-12}	p
femto	10^{-15}	f
atto	10^{-18}	a

TABLE 1–3   Commonly Used Physical Constants

DESCRIPTION	SYMBOL	VALUE
Boltzmann's Constant	k	1.38×10^{-23} J/°K
Electron volt	eV	1.60×10^{-19} J
Electron charge	q_e	1.60×10^{-19} C
Electron mass	m_e	9.11×10^{-31} kg
Permeability of free space	μ_0	$4\pi \times 10^{-7}$ H/m
Permittivity of free space	ϵ_0	8.854×10^{-12} F/m
		$\approx 10^{-9}/36\pi$ F/m
Planck's constant	h	6.626×10^{-34} J–s
Velocity of light in free space	c	2.998×10^{8} m/s
		$\approx 3.00 \times 10^{8}$ m/s

Because of the various ways of expressing electrical quantities, it is essential that a consistent notational system be used. The following format will be utilized:

dc and rms quantities	Italic type	$(V, I, E,$ etc.)
Time-varying quantities	Script letters	$(\mathcal{V}, \mathcal{I}, \mathcal{E},$ etc.)
Phasor quantities	Boldface Italic type	$(\boldsymbol{V}, \boldsymbol{I}, \boldsymbol{E},$ etc.)
Vector quantities	Arrow above the symbol	$(\vec{E}, \vec{H},$ etc.)
Normalized quantities	Bar above the symbol	$(\bar{Z}, \bar{Y},$ etc.)

Phasor notation. It is important to note that the rms-phasor form is used throughout the text. That is, given the general sinusoidal time function

$$\mathcal{E} = \sqrt{2}\,E\,\cos\,(\omega t + \psi) = \mathrm{Re}\,\{\sqrt{2}\,Ee^{j(\omega t + \psi)}\}$$

its phasor representation is

$$\boldsymbol{E} \equiv E\,\underline{/\psi} \equiv Ee^{j\psi}$$

where $\boldsymbol{E}$ is the rms-phasor of $\mathcal{E}$ and E is its rms value. If the phasor value of a quantity is known, its time-varying form may be reconstructed by multiplying the rms phasor by $\sqrt{2}e^{j\omega t}$ and taking the real part thereof.

Decibels and nepers. The ratio of two power levels is often expressed in decibels (dB). The decibel equivalent of a power ratio P_1/P_2 is

$$10\,\log\frac{P_1}{P_2} \quad \mathrm{dB} \tag{1-3}$$

For example, 6 dB is equivalent to $P_1/P_2 = 3.98 \approx 4.0$.

In network and transmission-line analysis, a ratio of voltages appearing across the same or equal impedances is often expressed in nepers (Np). The neper equivalent of a voltage ratio V_1/V_2 is

$$\ln\frac{V_1}{V_2} \quad \mathrm{Np} \tag{1-4}$$

Since power is proportional to the square of voltage, its neper equivalent may be written as

$$\frac{1}{2}\ln\frac{P_1}{P_2} \quad \mathrm{Np} \tag{1-5}$$

The conversion between nepers and decibels is

$$1.00\ \mathrm{Np} = 8.686\ \mathrm{dB}$$

since $\log P_1/P_2 = 0.434 \ln P_1/P_2$.

The decibel-type notation may be used to indicate a power level when P_2 is specified. With the reference level $P_2 \equiv 1$ mW, power may be expressed in dBm. That is,

$$10\,\log P_1 \quad \mathrm{dBm} \tag{1-6}$$

is equivalent to a power level of P_1 milliwatts. Thus 0 dBm is equivalent to 1 mW, +3 dBm to 2 mW, −3 dBm to $\frac{1}{2}$ mW, etc. The dBm method of indicating a power level is widely used in communications work.

REFERENCES

Books

1–1. Holt, C., *Introduction to Electric Fields and Waves,* 4th ed., John Wiley & Sons, New York, 1967.

1–2. Skilling, H., *Exploring Electricity,* Ronald Press Co., New York, 1948.

1–3. Dunlap, O., *The Story of Radio,* Dial Press, New York, 1935.

1–4. Meyer, H., *A History of Electricity and Magnetism,* MIT Press, Cambridge, MA, 1971.

1–5. Southworth, G. C., *Principles and Applications of Waveguide Transmission,* D. Van Nostrand Co., Princeton, NJ, 1950.

1–6. Southworth, G. C., *Forty Years of Radio Research,* Gordon & Breach, New York, 1962.

1–7. Larsen, E., *Telecommunications: A History,* F. Muller, Ltd., London, 1977.

Articles

1–8. Moncrief, F., and M. Bennett, Microwave Milestones. *Microwave System News,* 8, June 1978, pp. 34–49.

1–9. Ramsey, J. F., Microwave Antenna and Waveguide Techniques Before 1900. *Proc. IRE,* 46, February 1958, pp. 405–415.

1–10. Cohn, S. B., Breaking Through the Mental Barrier. *IRE Trans. Microwave Theory and Techniques,* MTT-7, April 1959, pp. 189–191.

1–11. Special Centennial Issue. Historical Perspectives of Microwave Technology. *IEEE Trans. Microwave Theory and Techniques,* MTT-32, September 1984.

2

Elementary Fields and Waves

At low frequencies, electrical phenomena are usually described and measured in terms of charge, voltage, and current; whereas at microwave frequencies, fields and waves are often used. Therefore, it is important for the microwave engineer to appreciate and understand these concepts. This chapter reviews some of the basic principles of electromagnetic fields and waves.

2-1 ELECTRIC AND MAGNETIC FIELDS

The concept of electric and magnetic fields is very useful in analyzing the forces associated with electric charges and their motion.

2-1a The Electric Field

The force between electric charges is governed by Coulomb's law. If the charges are of opposite sign, the force is one of attraction, while for those with like sign, the force is a repulsive one. It is convenient to express this force in terms of an electric field intensity $\overrightarrow{E}$ defined as the force per unit positive charge. That is,

$$\overrightarrow{E} \equiv \frac{\overrightarrow{F}}{q} \tag{2-1}$$

where the SI unit for $\overrightarrow{E}$ is V/m.[1] The electric field due to a single positive charge q_1 is shown in Fig. 2-1. The field lines at any point in the surrounding region indicate

[1] A list of symbols and their SI units is given in Appendix A.

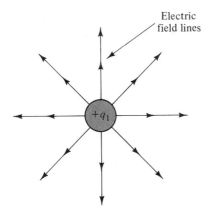

Figure 2-1 The electric field due to a positive point charge.

the direction of the force that would exist on a unit positive charge placed at that point. The magnitude of the field at a distance r from q_1 is given by

$$E = \frac{q_1}{4\pi\epsilon_0\epsilon_R r^2} \tag{2-2}$$

where ϵ_R is the relative permittivity or dielectric constant of the surrounding region. For free space, $\epsilon_R = 1.00$.

Since the density of E lines is the same in all directions, its value is inversely proportional to r^2, as is the magnitude of $\vec{E}$. Therefore, one can say that the density of lines is a relative indication of the electric field strength. This statement is true for any combination of charges, since with E linearly proportional to charge, superposition applies.

The concept of a flux density $\vec{D}$ that is proportional to $\vec{E}$ is useful when dealing with electric fields in a dielectric. It is defined as

$$\vec{D} \equiv \epsilon_R\epsilon_0 \vec{E} \tag{2-3}$$

where $\vec{D}$ is the electric flux density (C/m²). With $\vec{E}$, and hence $\vec{D}$, inversely proportional to r^2, the relationship between flux density and charge can be expressed in terms of Gauss' law. Namely, the net outflow of electric flux through a closed surface S is equal to the net electric charge enclosed. Stated mathematically,

$$\oint \vec{D} \cdot d\vec{S} = \sum_1^m q_n \quad \text{or} \quad \int_v \rho_v dv \tag{2-4}$$

where ρ_v is the volume charge density within the volume v. The integral form on the right side of the equation is used when the charge distribution is continuous rather than an array of point charges.

Another useful concept in the study of electric fields is that of potential or potential difference. The potential V at a point P is defined as the work required to bring a unit positive test charge from infinity to the point P. This work is necessary to overcome the electric field created by one or more fixed charges. That is,

$$V \equiv -\int_\infty^P \vec{E} \cdot d\vec{l} \tag{2-5}$$

the vector $d\vec{l}$ being an element of length along the integration path. In most cases, one is interested in the potential difference V_{AB}, namely the work required to move a unit positive test charge from point A to point B. Thus,

$$V_{AB} \equiv -\int_A^B \vec{E} \cdot d\vec{l} \qquad (2\text{--}6)$$

The concept of stored energy in an electric field is also useful. A free charge placed in a region containing an electric field will be accelerated or decelerated by the field. The change in kinetic energy of the moving charge can be said to have come from the potential energy of the region containing the field. This stored electric energy U_E in a volume v is given by

$$U_E = \frac{1}{2}\int_v \vec{D} \cdot \vec{E} \, dv \qquad (2\text{--}7)$$

Proof of this relationship is given in most texts on electromagnetism (see, for example, Refs. 2–1 and 2–6).

2–1b The Magnetic Field

A method analogous to that described for electric fields may be used to define the magnetic field intensity $\vec{H}$. Namely, $\vec{H}$ represents the force exerted on a unit north pole. Given an isolated magnetic pole m_1, the force exerted by it on a unit north pole is inversely proportional to the square of the distance between them.[2] The magnitude is given by

$$H = \frac{m_1}{4\pi\mu_R\mu_0 r^2} \qquad (2\text{--}8)$$

where μ_R is the relative permeability of the surrounding region. For free space and nonmagnetic materials, $\mu_R = 1.00$. In the SI system, m_1 is in webers and H in A/m.

Again, because of the inverse square relation with distance r, the density of lines is proportional to the magnitude of $\vec{H}$. The concept of a magnetic flux density $\vec{B}$ is useful when describing magnetic forces within a magnetic material. It is defined as

$$\vec{B} \equiv \mu_R\mu_0\vec{H} \qquad (2\text{--}9)$$

where its SI unit is the tesla, which is equivalent to a weber per square meter. One can also write Gauss' law for magnetic fields. Namely

$$\oint \vec{B} \cdot d\vec{S} = 0 \qquad (2\text{--}10)$$

Note that the right-hand side is zero, since the existence of isolated magnetic poles has never been shown. As a result, magnetic flux lines form complete loops. On the

---

[2] Although no such physical entity is known to exist, the concept of an isolated magnetic pole is sometimes useful!

other hand, electric flux lines emerge from positive charge and terminate on negative charge, except for time-varying fields, where they may form complete loops.

It is an experimental fact that a magnetic field is associated with the movement of electric charge. As an example, consider the long wire shown in Fig. 2–2. For a current I, the magnetic field in the region surrounding the wire is given by

$$H = \frac{I}{2\pi r} \qquad (2\text{–}11)$$

where r is the radial distance from the center of the wire to the point in question. This phenomenon can be generalized into Ampere's law, namely, the line integral of $\vec{H}$ around a closed path is equal to the net current passing through a surface $\vec{S}$ defined by the closed path. Stated mathematically,

$$\oint \vec{H} \cdot d\vec{l} = \sum_{1}^{m} I_n \qquad \text{or} \qquad \int \vec{J} \cdot d\vec{S} \qquad (2\text{–}12)$$

The integral form on the right side of the equation is useful when the current passing through $\vec{S}$ has a nonuniform distribution. $\vec{J}$ is the current density (A/m²) at any point on the surface $\vec{S}$. Later, this equation will be modified to include the effect of time-varying electric fields.

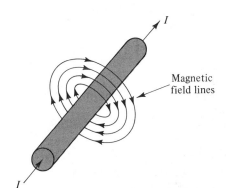

Magnetic field lines

Figure 2–2   The magnetic field due to an infinitely long, current-carrying conductor.

Experimentally, a magnetic field is known to exert a force on moving charge and therefore one can construct an alternate definition of $\vec{B}$ (and hence $\vec{H}$). Referring to Fig. 2–3, the force on a charge q, moving with velocity $\vec{v}$ in a magnetic field of flux density $\vec{B}$ is given by

$$\vec{F} = q(\vec{v} \times \vec{B}) \qquad (2\text{–}13)$$

The cross-product reflects the fact that the force is perpendicular to both $\vec{v}$ and $\vec{B}$. The magnitude of the force is $qvB \sin \theta$. Since force, charge, and velocity are measurable fundamental quantities, Eq. (2–13) may be considered a definition of $\vec{B}$. Most texts, in fact, present this as the primary definition, since it is useful in studying the motion of charged particles. The earlier-stated definition, referring to isolated magnetic poles, is useful in the analysis of ferromagnetic circuits.

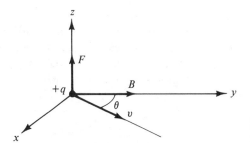

Figure 2–3 The force on a charge moving in a magnetic field.

Since the magnetic field can exert a force on either a magnetic pole or moving electric charge, the concept of stored energy associated with a field can again be utilized. The stored magnetic energy U_M in a volume v is given by

$$U_M = \frac{1}{2} \int_v \overrightarrow{B} \cdot \overrightarrow{H} \, dv \qquad (2\text{--}14)$$

Note the similarity to Eq. (2–7), the expression for stored electric energy.

2–2 FIELDS IN CONDUCTORS AND INSULATORS

The flux density concept is useful in describing the interaction between matter and the electric and magnetic fields. On a macroscopic basis, σ, ϵ_R, and μ_R may be used to describe the properties of the material. For the range of field values normally encountered many materials are linear, homogeneous and isotropic. That is, their properties are independent of the magnitude, location, and direction of the applied field. For these materials, σ, ϵ_R and μ_R are scalar quantities.

2–2a The Electric Field in Conducting Materials

Consider the elemental cylinder of conducting material shown in Fig. 2–4. With a voltage V applied as shown, a current I will flow in accordance with Ohm's law. That is, $I = V/R = GV$, where R and G are respectively the resistance and conductance of the sample. If the current is uniformly distributed over the cross-sectional area A,

$$R = \rho \frac{l}{A} = \frac{l}{\sigma A} \qquad (2\text{--}15)$$

where ρ is the resistivity of the material (ohm-m) and $\sigma \equiv 1/\rho$ is its conductivity (mho/m). Since it may be assumed that the electric field and the current distribution are uniform over the elemental volume, $\overrightarrow{E} = V/l$, $\overrightarrow{J} = I/A$ and hence

$$\overrightarrow{J} = \sigma \overrightarrow{E} = \overrightarrow{E}/\rho \qquad (2\text{--}16)$$

This form of Ohm's law is particularly useful when analyzing nonuniform fields and currents. Values of conductivity for some commonly used metals are given in Appendix B, Table B–1.

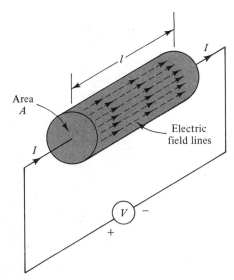

Figure 2–4 An elemental volume of conductive material.

2–2b The Electric Field in Dielectrics

An electric field in a dielectric causes its atoms to become polarized. On a macroscopic basis, this effect can be accounted for by the material's dielectric constant ϵ_R. To illustrate, consider the parallel-plate capacitor shown in Fig. 2–5a. If one plate is charged $+q_f$ and the other $-q_f$, an electric field is created. In the absence of dielectric material between the plates, $E = q_f/A\epsilon_0$. This result is obtained by applying Gauss' law to a closed surface surrounding one of the plates. If the capacitor contains a dielectric, the expression for electric field must be modified to account for the atomic polarization. Referring to part b of the figure, $E = (q_f/A\epsilon_0) - (q_p/A\epsilon_0)$, where q_f is the charge resulting from an external source (free charge) and q_p is the induced polarization charge due to the dielectric material. In most cases, the polarization effect is linearly proportional to the electric field. That is, $q_p/A\epsilon_0 = \chi_e E$, where the constant of proportionality χ_e is known as the *electric susceptibility* of the material. Substitution into the above expression for E yields

$$E = \frac{q_f}{A(1 + \chi_e)\epsilon_0} = \frac{q_f}{A\epsilon_R\epsilon_0} \qquad (2\text{--}17)$$

where $1 + \chi_e \equiv \epsilon_R$ is the relative permittivity or dielectric constant of the material. Note that for a fixed value of free charge, the effect of dielectric polarization is to reduce the electric field by ϵ_R.[3] Conversely, for a fixed voltage between the plates, the electric field is unchanged, and hence the presence of the dielectric increases q_f.

Room temperature values of ϵ_R at 3.0 GHz for various insulators are given in Appendix B, Table B–2. For most dielectrics, ϵ_R is only a slight function of frequency and temperature.

---

[3] With the effect of dielectric polarization accounted for by ϵ_R, the right-hand side of Gauss' law, Eq. (2–4), is interpreted as the net *free* charge enclosed. A detailed discussion of dielectric polarization may be found in Chapter 5 of Ref. 2–1.

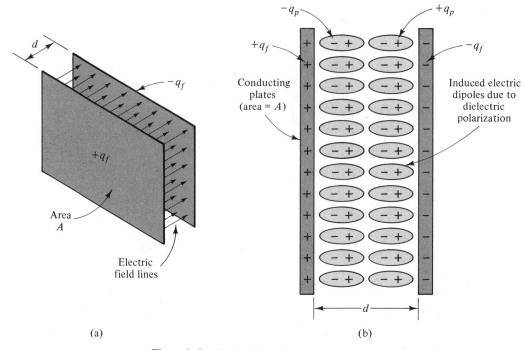

Figure 2–5   Dielectric polarization in a charged capacitor.

Another consequence of dielectric polarization is an increase in the value of capacitance. Capacitance is defined as

$$C \equiv \frac{q_f}{V} \tag{2–18}$$

where V is the voltage between the plates. For the parallel-plate capacitor, the electric field is uniform and hence $E = V/d$. Substitution into Eq. (2–17) yields the following expression for capacitance of a parallel-plate capacitor.

$$C = \frac{A \epsilon_R \epsilon_0}{d} \tag{2–19}$$

where A and d are defined in Fig. 2–5. Thus the capacitance value is directly proportional to the dielectric constant of the material between the capacitor plates. This is true for any configuration of capacitor plates.

The following example illustrates the use of Gauss' law and the capacitance definition.

Example 2–1:

Derive an expression for the capacitance of the coaxial configuration shown in Fig. 2–6. Ignore fringing at the ends by assuming $l \gg b$.

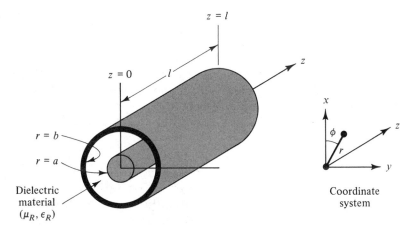

Figure 2–6   A coaxial transmission line. (See Exs. 2–1 and 2–2.)

Solution: Assume that the center conductor has been charged to $+q_f$ and the outer conductor to $-q_f$. Ignoring end effects, the E lines will be directed radially outward. Applying Gauss' law to a cylindrical surface of length l and radius r, where $a < r < b$, yields

$$\int_0^l \int_0^{2\pi} D \, r \, d\phi \, dz = q_f$$

where standard cylindrical coordinates have been used. Because of symmetry, D is independent of ϕ and z. Hence

$$D = \frac{q_f}{2\pi r l} \quad \text{and} \quad E = \frac{q_f}{2\pi \epsilon_0 \epsilon_R r l} \tag{2–20}$$

From Eq. (2–6), the voltage between the plates is

$$V = -\int_a^b \vec{E} \cdot d\vec{r} = \frac{q_f}{2\pi \epsilon_0 \epsilon_R l} \ln \frac{b}{a}$$

and therefore

$$C \equiv \frac{q_f}{V} = \frac{2\pi \epsilon_0 \epsilon_R l}{\ln (b/a)} \tag{2–21}$$

Lossy dielectrics.   All dielectrics have a finite amount of conductivity (σ). At microwave frequencies, this dissipative property is expressed in terms of the material's dielectric loss tangent (tan δ). To illustrate its meaning, consider the parallel-plate capacitor in Fig. 2–5. Its capacitance is given by Eq. (2–19). If the dielectric material is lossy, there will be some conductance G in parallel with C. Therefore, the ac admittance of the lossy capacitor becomes $Y = G + j\omega C$, where $G = \sigma A/d$. Utilizing Eq. (2–19),

$$Y = \omega C \tan \delta + j\omega C \tag{2–22}$$

where

$$\tan \delta \equiv \frac{\sigma}{\omega \epsilon_R \epsilon_0}$$

Note that the conductance due to the lossy dielectric is simply the capacitive susceptance (ωC) multiplied by the material's loss tangent. Although Eq. (2–22) was
derived for the parallel-plate capacitor, it is valid for any configuration. For example, the conductance of the coaxial capacitor in Fig. 2–6 is

$$G = \omega C \tan \delta = \frac{2\pi \epsilon_0 \epsilon_R l}{\ln (b/a)} \omega \tan \delta \qquad (2-23)$$

since C is given by Eq. (2–21). Values of tan δ at 3.0 GHz for some dielectric materials are given in Appendix B, Table B–2. In most cases, the value is only slightly
affected by frequency and temperature variations.[4]

2–2c The Magnetic Field in Magnetic Materials

Most materials exhibit some magnetic properties. The effect is quite pronounced in
materials such as iron, nickel, cobalt, and ferrite. In others, such as the dielectrics
listed in Table B–2, it is practically negligible. On a macroscopic basis, the magnetic
properties of a material may be accounted for by its relative permeability (μ_R). To
illustrate, consider the long solenoid in Fig. 2–7. For $l \gg r$, the magnetic field
within the solenoid is essentially uniform and is given by

$$H = \frac{NI}{l} \qquad (2-24)$$

where I is the current in the coil. When the material within the solenoid is nonmagnetic, the flux density is given by $B = \mu_0 H$. However, when the material is para-
magnetic or ferromagnetic, the flux density will be greater than $\mu_0 H$. This increase

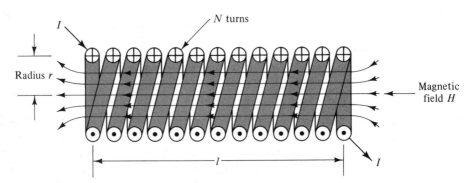

Figure 2–7   The magnetic field in a long, current-carrying solenoid.

[4] Values of ϵ_R and tan δ for many materials may be found in Refs. 2–2 and 2–3. In these references, the term *dissipation factor* is used rather than *loss tangent*.

in B is associated with the spin motion of the electrons in the material. Since an electron possesses charge, the spin creates magnetic flux in a manner similar to a tiny current-carrying coil.[5] With current flowing in the coil, the total flux density is

$$B = \mu_0(H + M) \tag{2–25}$$

where the magnetization M is the net magnetic field contributed by the material. In certain cases, M is linearly proportional to the magnetic field H. That is,

$$M = \chi_m H \tag{2–26}$$

where χ_m is the *magnetic susceptibility* of the material. Substitution into Eq. (2–25) yields

$$B = \mu_0(1 + \chi_m)H = \mu_0 \mu_R H \tag{2–27}$$

where $1 + \chi_m \equiv \mu_R$, the relative permeability of the material. Thus the effect of the material is to increase the flux density by the factor μ_R. Note that for free space and nonmagnetic materials, $\chi_m = 0$ and hence $\mu_R = 1.00$.

The inductance of a solenoid or any other inductor is affected by the presence of magnetic material. To illustrate, consider again the long solenoid in Fig. 2–7. Inductance (L) is defined as the flux linkages per unit of current. That is,

$$L \equiv \frac{N\Phi}{I} \tag{2–28}$$

where N is the number of turns and Φ is the total magnetic flux. For a given flux density, Φ may be determined from

$$\Phi = \int_S \vec{B} \cdot d\vec{S} \tag{2–29}$$

where $\vec{S}$ is the surface through which the flux passes. Since in this case the field is uniform, $\Phi = BA$, where $A = \pi r^2$. Making use of Eqs. (2–24) and (2–27) results in the following expression for the inductance of a long solenoid ($l \gg r$).

$$L = \mu_0 \mu_R \frac{N^2 A}{l} \tag{2–30}$$

This equation is accurate to within ten percent when $l > 3r$. Thus, assuming a linear relation between M and H, the inductance is directly proportional to the relative permeability of the material.

The following example illustrates the use of Ampere's law and the inductance definition in Eq. (2–28).

Example 2–2:

Derive an expression for the inductance of the coaxial configuration in Fig. 2–6. Assume that the current flows along the surface of the inner conductor and returns via the inner surface of the outer conductor, which is the case in microwave applications.

---

[5] A detailed discussion of the magnetic properties of matter may be found in Refs. 2–1, 2–4, and 2–6.

Solution: A current flow I down the center conductor produces a magnetic field around it. Since the conductor is round, the application of Ampere's law to a circle of radius r, where $a \leq r \leq b$, yields

$$\int_0^{2\pi} H \, r \, d\phi = 2\pi r H = I$$

and hence $H = I/2\pi r$ for $a \leq r \leq b$. For $r < a$, $H = 0$ since the current enclosed is zero. For $r > b$, the net current enclosed is also zero, since the return current I flows along the $r = b$ surface. Thus $H = 0$ for $r > b$. From Eq. (2–29), the total magnetic flux passing between the conductors is

$$\Phi = \int_0^l \int_a^b B \, dr \, dz = \mu_R \mu_0 l \int_a^b \frac{I}{2\pi r} \, dr = \frac{\mu_0 \mu_R I l}{2\pi} \ln \frac{b}{a}$$

Substituting into Eq. (2–28) and noting that $N = 1$,

$$L = \frac{\mu_0 \mu_R l}{2\pi} \ln \frac{b}{a} \qquad (2\text{--}31)$$

which is the *high-frequency* inductance of a coaxial line.

2–3 MAXWELL'S EQUATIONS AND BOUNDARY CONDITIONS

The contributions of James Clerk Maxwell to the understanding of electricity and magnetism were discussed in Sec. 1–2. His theoretical studies resulted in the formulation of a general theory of electromagnetism. The four relationships derived by him bear his name. This section reviews the ideas leading to these equations.

2–3a Maxwell's Equations

Maxwell's four equations are mathematical expressions of Gauss' laws for electric and magnetic fields, (Eqs. 2–4 and 2–10), Faraday's law, and Ampere's law. The mathematical form of these laws are developed here.

Faraday's law. Faraday's law of induction states that time-varying magnetic flux passing through a surface produces an emf around the perimeter of the surface. Since emf is the electric force exerted over a path and magnetic flux is the surface integral of $\vec{\mathscr{B}}$, Faraday's law may be written as

$$\oint \vec{\mathscr{E}} \cdot d\vec{l} = -\int_S \frac{\partial \vec{\mathscr{B}}}{\partial t} \cdot d\vec{S} \qquad (2\text{--}32)$$

The minus sign is a result of Lenz's principle which states that changing magnetic flux produces an emf which tends to oppose the flux change that created the emf.

Ampere's law and displacement current. In developing his theory of electromagnetism, Maxwell suggested that a time-varying electric field would produce a magnetic field. In effect, he was saying that Ampere's law, Eq. (2–12), had to be modified to account for time-varying fields. The complete representation of Ampere's law is

$$\oint \vec{\mathcal{H}} \cdot d\vec{l} = \int_S \vec{\mathcal{J}} \cdot d\vec{S} + \int_S \frac{\partial \vec{\mathcal{D}}}{\partial t} \cdot d\vec{S} \qquad (2\text{–}33)$$

The first term on the right side of the equation is the total conduction current through a surface enclosed by the path, while the second term is the total displacement current enclosed. The quantity $\partial \vec{\mathcal{D}} / \partial t$ is the displacement current density (A/m^2). For good conductors, the first term is dominant, while the second term dominates for time-varying fields in an insulator.

To understand the displacement current concept, consider the capacitor in Fig. 2–8. With a dc voltage source V connected to the capacitor, the plates become charged. When the value of q_f reaches CV, the capacitor is fully charged and current flow ceases. The electric field (E) and flux density (D) between the plates are related to the charge q_f by Gauss' law. For a parallel-plate capacitor, $D = q_f/A$, where A is the area of one plate.

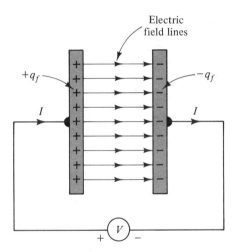

Figure 2–8   The displacement current concept in a capacitor.

Suppose the source voltage V is not constant but increasing with time. This requires that $q_f = CV$ also increases with time. The increase in charge is supplied by the current $I = dq_f/dt$. In this situation, it would appear that Kirchhoff's current law is violated. That is, current entering the plate is I, but current leaving via the insulating space is zero since a perfect insulator cannot support a conduction current. This dilemma is resolved by noting that the current flow that produces a change in q_f also causes a change in flux density, and that the unit for $d\mathcal{D}/dt$ is A/m^2, a current density. By considering this quantity as a current density, Kirchhoff's current law may be applied to time-varying fields. The quantity $\partial \vec{\mathcal{D}} / \partial t$ is known as the *displacement current density* and hence the displacement current through a surface S is

$$I_d = \int_S \frac{\partial \vec{\mathcal{D}}}{\partial t} \cdot d\vec{S} \qquad (2\text{–}34)$$

To verify that the current law is applicable, consider again the parallel-plate capacitor. In this case, $\mathcal{D} = q_f/A$ and hence the displacement current $A(d\mathcal{D}/dt)$ equals dq_f/dt. Thus the displacement current leaving via the insulating region equals the conduction current entering the capacitor plate.

Displacement current only exists when the electric field is time-varying. Since it is a current, one would expect it to create a magnetic field. Displacement current does indeed produce a magnetic field and hence Ampere's law must be restated to reflect this experimental fact. This has been done in Eq. (2–33). As explained in Sec. 2–4, the most important consequence of displacement current is its key role in electromagnetic wave propagation.

With Ampere's law modified to include the effect of displacement currents, all four of Maxwell's equations may now be stated. They are summarized here in both integral and differential forms.

Maxwell's equations in integral form.

$$\oint \vec{\mathcal{D}} \cdot d\vec{S} = \int_v \rho_v \, dv \quad , \quad \oint \vec{\mathcal{B}} \cdot d\vec{S} = 0 \qquad (2\text{–}35)$$

$$\oint \vec{\mathcal{E}} \cdot d\vec{l} = -\int_S \frac{\partial \vec{\mathcal{B}}}{\partial t} \cdot d\vec{S} \qquad (2\text{–}36)$$

$$\oint \vec{\mathcal{H}} \cdot d\vec{l} = \int_S \vec{\mathcal{J}} \cdot d\vec{S} + \int_S \frac{\partial \vec{\mathcal{D}}}{\partial t} \cdot d\vec{S} \qquad (2\text{–}37)$$

where $\quad \vec{\mathcal{D}} = \epsilon_R \epsilon_0 \vec{\mathcal{E}} \quad , \quad \vec{\mathcal{B}} = \mu_R \mu_0 \vec{\mathcal{H}} \quad$ and $\quad \vec{\mathcal{J}} = \sigma \vec{\mathcal{E}}$

Maxwell's equations in differential form.   By applying the rules of vector calculus, Maxwell's equations may be written in differential or point form.[6] Namely,

$$\nabla \cdot \vec{\mathcal{D}} = \rho_v \quad , \quad \nabla \cdot \vec{\mathcal{B}} = 0 \qquad (2\text{–}38)$$

$$\nabla \times \vec{\mathcal{E}} = -\frac{\partial \vec{\mathcal{B}}}{\partial t} \qquad (2\text{–}39)$$

$$\nabla \times \vec{\mathcal{H}} = \vec{\mathcal{J}} + \frac{\partial \vec{\mathcal{D}}}{\partial t} \qquad (2\text{–}40)$$

These relationships represent the fundamental equations of electromagnetism.

2–3b Boundary Conditions for Electric and Magnetic Fields

Most electromagnetic problems involve boundaries between regions having different electric and magnetic properties. The conditions that must exist at these boundaries may be deduced from Maxwell's equations. Table 2–1 lists the conditions for tangential (subscript T) and normal (subscript N) components of the fields.

---

[6] As explained in the preface, the use of the differential form of Maxwell's equations is limited to a few sections in the text. These sections are indicated in the table of contents by a star (✱).

TABLE 2–1   Boundary Conditions for Electromagnetic Fields at the Interface Between Two
Materials

TANGENTIAL COMPONENTS	NORMAL COMPONENTS
$\mathscr{E}_{T_1} = \mathscr{E}_{T_2}$ $\mathscr{H}_{T_1} = \mathscr{H}_{T_2} + \mathscr{K}$ where $\mathscr{K} \equiv$ Surface current density (A/m)	$\mathscr{D}_{N_1} = \mathscr{D}_{N_2} - \rho_s$ $\mathscr{B}_{N_1} = \mathscr{B}_{N_2}$ where $\rho_s \equiv$ Surface charge density (C/m²)

NOTE: 1. If the boundary is charge-free, $\rho_s = 0$.
2. In the absence of conduction currents, $\mathscr{K} = 0$.

These conditions may be verified with the aid of Fig. 2–9. In part *a* of the
figure, the application of Faraday's law to the closed path *a-b-c-d* results in
$\mathscr{E}_{T_1}l - \mathscr{E}_{T_2}l \approx -(\partial\mathscr{B}/\partial t)(l\Delta n)$, where it is assumed $\Delta n \ll l$ and both approach
zero. As $l\Delta n \to 0$, $\mathscr{E}_{T_1} = \mathscr{E}_{T_2}$, which proves the first tangential condition in Table
2–1. The second condition may be verified by applying Ampere's law to the same
closed path *a-b-c-d*, except that $\mathscr{E}_{T_1}$ and $\mathscr{E}_{T_2}$ are replaced by $\mathscr{H}_{T_1}$ and $\mathscr{H}_{T_2}$, respectively. The result is that $\mathscr{H}_{T_1}l - \mathscr{H}_{T_2}l = \mathscr{K}l$, where $\mathscr{K}$ is the conduction current density (A/m) at the boundary surface. A positive value of $\mathscr{K}$ denotes a surface current
directed into the page.

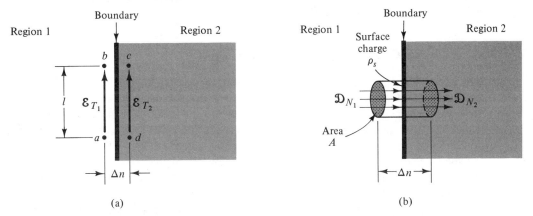

Figure 2–9   Tangential (T) and normal (N) field components at a boundary.

The conditions for the normal components may be verified with the aid of Fig.
2–9b. If both A and Δn are sufficiently small, the application of Gauss' law to the
closed cylinder yields $\mathscr{D}_{N_2}A - \mathscr{D}_{N_1}A \approx \rho_s A$, where ρ_s is the surface charge density
(C/m²). As A and Δn approach zero, $\mathscr{D}_{N_1} = \mathscr{D}_{N_2} - \rho_s$. Replacing $\mathscr{D}_{N_1}$ and $\mathscr{D}_{N_2}$ with
$\mathscr{B}_{N_1}$ and $\mathscr{B}_{N_2}$ and using Gauss' law for magnetic fields results in $\mathscr{B}_{N_1} = \mathscr{B}_{N_2}$, which
means that magnetic flux lines are continuous. Electric flux lines are only continu-
ous when the boundary is charge-free ($\rho_s = 0$).

Special conditions exist when one of the two regions is a perfect conductor
(that is, $\sigma \to \infty$). This requires that $\mathscr{D}$ and $\mathscr{E}$ in that region be identically zero. If, for
example, region 2 has infinite conductivity,

$$\mathscr{E}_{T_1} = 0 \qquad \text{and} \qquad \mathscr{D}_{N_1} = -\rho_s \qquad (2\text{–}41)$$

Thus the direction of $\mathscr{E}$ lines terminating on the surface of a perfect conductor must be perpendicular to that surface.

When the fields are time-varying, $\mathscr{H}$ and $\mathscr{B}$ must also be zero within a perfect conductor. This is a direct consequence of Faraday's law, Eq. (2–36), and the fact that $\mathscr{E} \equiv 0$ in a perfect conductor. Thus for time-varying fields at the boundary of a perfect conductor, the magnetic field conditions become

$$\mathscr{H}_{T_1} = \mathscr{H} \qquad \text{and} \qquad \mathscr{B}_{N_1} = 0 \qquad (2\text{--}42)$$

where again it is assumed that region 2 is the perfect conductor. If μ_{R_1} is a scalar, $\mathscr{H}_{N_1} = 0$, which means that the magnetic field must be tangential to the surface of a perfect conductor. Furthermore, if region 1 is a perfect insulator ($\tan \delta = 0$), the following boundary condition also holds for time-varying fields.

$$\frac{\partial \mathscr{H}_{T_1}}{\partial n} = 0 \qquad (2\text{--}43)$$

where n represents a direction perpendicular to the boundary surface.

The analysis of microwave transmission lines and components make extensive use of the boundary conditions derived here. An appreciation of their meaning is also useful in estimating the electromagnetic field pattern in structures constrained by dielectric and metallic boundaries.

2–4 WAVE PROPAGATION IN PERFECT INSULATORS

The existence of self-propagating electromagnetic waves is the single most important consequence of Maxwell's equations. This propagation results from the fact that a time-varying magnetic field produces a time-varying electric field (Faraday's law) which, in turn, produces a time-varying magnetic field (Ampere's law) . . . and so forth. Since one type field produces the other in a plane normal to it, the electric and magnetic fields are always perpendicular to each other and travel at the same speed. This section reviews the important case of electromagnetic wave propagation in unbounded insulators.

The situation in which the fields are a function of only one coordinate represents the simplest application of Maxwell's equations. The resultant analysis describes the phenomena of uniform plane waves, which is of considerable engineering importance.

Consider the coordinate system shown in Fig. 2–10 and assume a time-varying electric field in the x direction that is independent of x and y. That is, $\mathscr{E}_x = f(z, t)$.

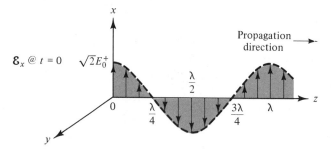

Figure 2–10 $\mathscr{E}_x$ at $t = 0$ as a function of position along the propagation axis.

Also assume that all space consists of a perfect insulator devoid of free charges, which means that both ρ_v and $\vec{J}$ are zero in Eqs. (2–35) to (2–40). Since $\vec{\mathscr{E}}$ is not a function of x and y and only has a component in the x direction, Maxwell's equations reduce to[7]

$$\frac{\partial \mathscr{E}_x}{\partial z} = -\mu_R \mu_0 \frac{\partial \mathscr{H}_y}{\partial t} \quad \text{and} \quad -\frac{\partial \mathscr{H}_y}{\partial z} = \epsilon_R \epsilon_0 \frac{\partial \mathscr{E}_x}{\partial t} \quad (2\text{–}44)$$

By differentiating the first equation with respect to z and the second one with respect to t, $\mathscr{H}_y$ can be eliminated, resulting in

$$\frac{\partial^2 \mathscr{E}_x}{\partial z^2} = \mu_R \mu_0 \epsilon_R \epsilon_0 \frac{\partial^2 \mathscr{E}_x}{\partial t^2} \quad (2\text{–}45)$$

This is the well-known wave equation of mathematical physics. Let us now solve this equation for steady-state sinusoidal excitation using the rms-phasor method.[8]

Rewriting Eqs. (2–44) and (2–45) in phasor form yields

$$\frac{dE_x}{dz} = -j\omega\mu_R \mu_0 H_y \quad , \quad -\frac{dH_y}{dz} = j\omega\epsilon_R \epsilon_0 E_x \quad (2\text{–}46)$$

and

$$\frac{d^2 E_x}{dz^2} = -\omega^2 \mu_R \mu_0 \epsilon_R \epsilon_0 E_x \quad (2\text{–}47)$$

where each differentiation with respect to time introduced a multiplying factor $j\omega$. Since Eq. (2–47) is a second-order differential equation, it has two independent solutions, which may be written as

$$E_x = E_0^+ e^{-j\beta z} + E_0^- e^{+j\beta z} \quad (2\text{–}48)$$

where $\quad \beta \equiv \dfrac{\omega}{v} \quad , \quad v = \dfrac{1}{\sqrt{\mu_R \mu_0 \epsilon_R \epsilon_0}} \quad$ and $\quad \omega = 2\pi f = 2\pi/T.$

Substitution into the first of Eqs. (2–46) yields the two associated solutions for H_y.

$$H_y = H_0^+ e^{-j\beta z} - H_0^- e^{+j\beta z} \quad (2\text{–}49)$$

where $\quad \dfrac{E_0^+}{H_0^+} = \dfrac{E_0^-}{H_0^-} \equiv \eta = \sqrt{\dfrac{\mu_R \mu_0}{\epsilon_R \epsilon_0}}.$

To reconstruct the instantaneous form of the solutions, multiply the rms-phasor solutions by $\sqrt{2}\, e^{j\omega t}$ and take the real parts thereof. Thus,

$$\mathscr{E}_x = \sqrt{2}\, E_0^+ \cos(\omega t - \beta z) + \sqrt{2}\, E_0^- \cos(\omega t + \beta z) \quad (2\text{–}50)$$

and

$$\mathscr{H}_y = \sqrt{2}\, H_0^+ \cos(\omega t - \beta z) - \sqrt{2}\, H_0^- \cos(\omega t + \beta z) \quad (2\text{–}51)$$

[7] The reduction of the differential form of Maxwell's equations is found in most books on electromagnetic theory (for example, Chapter 11 in Ref. 2–4). An excellent treatment of the integral form and its simplification is given in Chapter 12 of Ref. 2–1.

[8] The rms-phasor is described in Sec. 1–4.

To understand the nature of the above equations, consider the first term in Eq. (2–50). Since $\beta = \omega/v$, it can be rewritten as $\sqrt{2} \, E_0^+ \cos \left[\omega(t - z/v)\right]$. At $z = 0$, the electric field is given by $\sqrt{2} \, E_0^+ \cos \omega t$ and is plotted in Fig. 2–11. This curve does *not* represent a wave, but merely the time variation of $\mathcal{E}_x$ at one place, the $z = 0$ plane. A wave implies the movement of a time function from one place to another. The term $\sqrt{2} \, E_0^+ \cos \left[\omega(t - z/v)\right]$ does represent a wave since its cosinusoidal variation at $z = 0$ is repeated at the plane $z = l$ with a time delay l/v. Thus it appears that the $\mathcal{E}_x$ time function has traveled a distance l with a velocity v. The conclusion is that the first term in Eq. (2–50) represents an electric wave traveling in the positive z direction with a velocity v. By a similar argument, the second term describes a wave traveling in the negative z direction. From ac theory, phase delay is merely normalized time delay multiplied by 2π rad (that is, $2\pi t_d/T$). Since $\beta z = (2\pi/T)(z/v)$, it represents a phase delay for the forward traveling wave ($+z$ direction). The quantity β is the phase shift per unit length and is known as the *phase constant* of the wave.

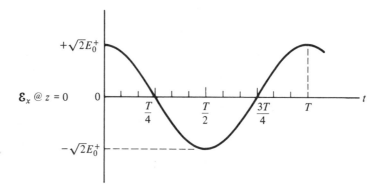

Figure 2–11 $\mathcal{E}_x$ at $z = 0$ as a function of time.

The wavelength λ and period T are related by the wave velocity. Wavelength is defined as the distance one must traverse for a function, periodic in space but fixed in time, to repeat itself. This quantity is described in Fig. 2–10 for the first term in Eq. (2–50) at $t = 0$. Period, on the other hand, is the time required for a function, periodic in time but at a fixed point in space, to repeat itself. It is described in Fig. 2–11. If the field pattern shown in Fig. 2–10 moves in the $+z$ direction with a velocity v, we have a wave. An observer situated at, say, $z = \lambda$ will see a cosinusoidal variation of $\mathcal{E}_x$ with time. When $t = 0$, its value will be a positive maximum. The time required for the next positive maximum, presently located at $z = 0$, to arrive at $z = \lambda$ represents one period. Thus, $T = \lambda/v$. Since $T = 1/f$, this leads to the familiar relationship for all wave propagation

$$f\lambda = v \qquad\qquad (2\text{--}52)$$

For the electromagnetic wave described here, v is given by the expression below Eq. (2–48). With

$$\mu_0 = 4\pi \times 10^{-7} \text{ H/m} \quad \text{and} \quad \epsilon_0 \approx \frac{1}{36\pi} \times 10^{-9} \text{ F/m},$$

$$v = \frac{c}{\sqrt{\mu_R \epsilon_R}} \tag{2-53}$$

where $c \approx 3 \times 10^8$ m/s is the velocity of light in free space. If the wavelength in free space is denoted by λ_0,

$$\lambda_0 = \frac{c}{f} \tag{2-54}$$

and hence for any other insulator,

$$\lambda = \frac{\lambda_0}{\sqrt{\mu_R \epsilon_R}} \tag{2-55}$$

Keep in mind that for nonmagnetic insulators $\mu_R = 1.00$. Equations (2–53) and (2–55) show that both the velocity and wavelength of an electric wave in a dielectric are smaller than their values in free space. This effect is useful in reducing the size of microwave components. Also, since the phase constant $\beta = \omega/v$,

$$\beta = \frac{2\pi}{\lambda} = \frac{2\pi}{\lambda_0} \sqrt{\mu_R \epsilon_R} \qquad \text{rad/length} \tag{2-56}$$

Equation (2–51) reveals that the traveling waves also contain magnetic field components having the same velocity as the electric field. This is understandable because the two fields generate each other. Also, the ratio of their magnitudes is a constant. The relationship is given below Eq. (2–49). The ratio η is called the *intrinsic impedance* of the medium. Note that it is a real number since a lossless insulator has been assumed. This means that the traveling electric and magnetic waves are in phase. Substituting in the MKS values of μ_0 and ϵ_0 yields the equation

$$\eta = 120\pi \sqrt{\frac{\mu_R}{\epsilon_R}} = 377 \sqrt{\frac{\mu_R}{\epsilon_R}} \qquad \text{ohms} \tag{2-57}$$

Note that for free space, the intrinsic impedance is 377 ohms.

Figure 2–12 shows a sketch of the forward traveling electromagnetic wave described by the first terms in Eqs. (2–50) and (2–51). The peak values are $\sqrt{2}\, E_0^+$ and $\sqrt{2}\, H_0^+$ and therefore E_0^+ and H_0^+ are the rms values. Keep in mind that since $\mathscr{E}_x$ and $\mathscr{H}_y$ are independent of x and y, the $\mathscr{E}$-$\mathscr{H}$ pattern shown in the figure is the same along any other line parallel to the z axis. For this reason, the wave is called a *uniform plane wave*.

Figure 2–12 shows the electromagnetic wave at two instances in time, namely, $t = 0$ and $t = T/4$. Since it is traveling in the $+z$ direction, the second wave is merely the wave at $t = 0$ displaced one quarter of a wavelength in the positive z direction. Note that the electric and magnetic fields are perpendicular to each other and both lie in a plane transverse to the direction of propagation. For this reason the uniform plane wave is called a *transverse electromagnetic* (TEM) wave. The fact that

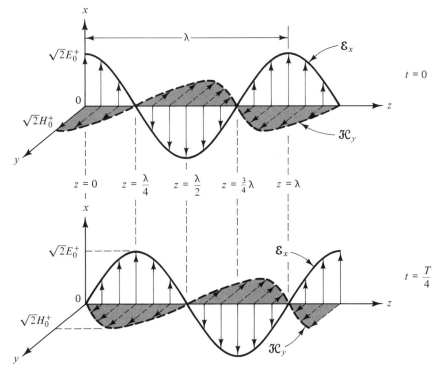

Figure 2–12 Description of an electromagnetic wave traveling in the positive z direction (shown at $t = 0$ and $t = T/4$).

the electric and magnetic fields are perpendicular to each other is a direct consequence of Faraday's and Ampere's laws. To understand this, consider the electric field at $z = \lambda/4$ in Fig. 2–12. Note that when $t = 0$, $\mathcal{E}_x = 0$ at that point. However, as the wave travels to the right, $\mathcal{E}_x$ immediately starts to increase to some finite positive value. This means that its time rate of change is positive and hence the displacement current is in the positive x direction when $t = 0$. In other words, although $\mathcal{E}_x = 0$ at $t = 0$, its time derivative is at a positive maximum. Ampere's law states that a line integral of $\mathcal{H}$ enclosing this displacement current must also be positive. By applying the right-hand rule, the magnetic field when $t = 0$ must be in the positive y direction for z slightly less than $\lambda/4$ and in the negative y direction for z slightly greater than $\lambda/4$. This is indeed the case as shown. If the wave were traveling in the negative z direction, the displacement current would be reversed and hence the direction of the magnetic field would be reversed. A similar argument, utilizing Faraday's law, shows that the movement of the magnetic wave leads to a finite line integral of $\mathcal{E}$ in a plane perpendicular to $\mathcal{H}$. In developing this argument, remember that Faraday's law has a minus sign, while Ampere's law does not.

Let us now consider the power flow associated with the electromagnetic wave propagation we have been describing. Since $\mathcal{E}$ and $\mathcal{H}$ are force fields and contain stored energy, it makes sense that electromagnetic waves involve the propagation of energy. A formal derivation of the energy in an electromagnetic field results in the following vector relationship for power flow.[9]

$$\vec{p} = \vec{\mathscr{E}} \times \vec{\mathscr{H}} \tag{2–58}$$

$\vec{p}$ is the instantaneous vector power density (W/m^2) and is known as the *Poynting vector*. Its magnitude represents the value of instantaneous power density, while its direction indicates the direction of power flow. Because Eq. (2–58) involves the vector cross product, the direction of $\vec{p}$ is always perpendicular to both $\vec{\mathscr{E}}$ and $\vec{\mathscr{H}}$. In our example (Fig. 2–12), $\vec{\mathscr{E}}$ crossed into $\vec{\mathscr{H}}$ is in the positive z direction for any and all values of position and time. To reverse the direction of propagation and hence power flow, either $\vec{\mathscr{E}}$ or $\vec{\mathscr{H}}$ (not both) must be reversed. This agrees with the conclusion arrived at using the displacement current concept in combination with Faraday's and Ampere's laws.

The average power densities for the forward and reverse traveling waves described by Eqs. (2–50) and (2–51) are

$$p_z^+ = E_0^+ H_0^+ \qquad \text{and} \qquad p_z^- = E_0^- H_0^- \tag{2–59}$$

where E_0 and H_0 represent rms values. The average power flow through a surface S perpendicular to the direction of propagation is given by

$$P = \int_S p_z \, dS \qquad \text{where} \qquad p_z = p_z^+ - p_z^- \tag{2–60}$$

For uniform plane waves, p_z is independent of position and hence $P = p_z S$.

The following illustrative example is intended to reinforce the various ideas discussed in this section.

Example 2–3:

A 500 MHz electromagnetic wave is propagating through a perfect nonmagnetic dielectric having $\epsilon_R = 6$.
(a) Calculate the wavelength and the phase constant.
(b) With the wave traveling in the $+z$ direction, the sinusoidal electric field at $z = 80$ cm is delayed relative to the field at $z = 65$ cm. Calculate the time delay (in nanoseconds) and the phase delay (in degrees).
(c) Calculate the average power density in the wave if the peak value of magnetic field is 0.5 A/m.

Solution:
(a) $\lambda_0 = c/f = (3 \times 10^8)/(500 \times 10^6) = 0.6$ m.
Since $\mu_R = 1$, $\lambda = 0.6/\sqrt{6} = 0.245$ m
and $\beta = 2\pi/0.245 = 25.65$ rad/m or 1470°/m.
(b) $t_d = \Delta z/v$, where $v = 3 \times 10^8/\sqrt{6} = 1.22 \times 10^8$ m/s.
Thus, $t_d = (0.80 - 0.65)/v = 1.23 \times 10^{-9}$ s or 1.23 ns.
Phase Delay $= \beta \Delta z = 1470(0.15) = 220°$.
(c) With $\mu_R = 1$, $\eta = 377/\sqrt{6} = 154$ ohms.
$p_z^+ = E_0^+ H_0^+ = \eta (H_0^+)^2$, where $H_0^+ = 0.5/\sqrt{2} = 0.354$ A/m.
Therefore, $p_z^+ = 154(0.354)^2 = 19.3$ W/m^2.

[9] See, for example, Chapter 12 in Ref. 2–1 or Chapter 4 in Ref. 2–6.

2–5 WAVE POLARIZATION

There are four types of wave polarization: linear, circular, elliptic, and random. The wave described in Fig. 2–12 is an example of a linear or plane polarized wave. By convention, the direction of electric field is used to denote the polarization. Therefore the wave in Fig. 2–12 is *vertically polarized* because the electric field is always and everywhere in the vertical direction. A horizontally polarized wave would, of course, be one in which the E lines are horizontal. Linear polarized waves are thus characterized by the fact that the orientation of the field is the same everywhere in space and is independent of time.

It is very useful to describe an electromagnetic wave in terms of phasor quantities. Figure 2–13 shows the phasor representation at five points along the propagation axis for the forward traveling wave depicted in Fig. 2–12. The rms-phasor representation of $\mathscr{E}_x$ at $z = 0$ is given by $E_0^+ \underline{/0}$. With the wave traveling in the $+z$ direction, E at $z = l$ will be phase delayed βl. For instance at $z = \lambda/4$, the phase delay is $(2\pi/\lambda)(\lambda/4) = \pi/2$ rad and hence $E = E_0^+ \underline{/-\pi/2}$. For a given propagation direction, knowledge of E at one point defines its value at all other points along the propagation axis. The magnetic field phasors are also shown in the figure. As explained previously, the electric and magnetic fields are perpendicular to each other, have the same velocity and are related by the intrinsic impedance η. Note that at any point, E and H are in phase, which is the case for η real.

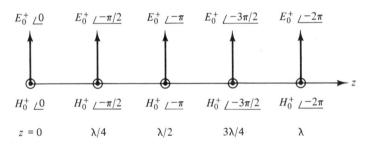

Figure 2–13   Rms-phasor representation of a linear polarized wave.

Since we are dealing with sinusoidal *vector* fields, the quantities in the figure are actually *vector phasors*. They can be expressed mathematically with the aid of unit vectors ($\vec{a}$). The vector-phasor representation for the electric and magnetic fields in Fig. 2–13 are $\vec{E} = \vec{a}_x E$ and $\vec{H} = \vec{a}_y H$, where $\vec{a}_x$ and $\vec{a}_y$ are the unit vectors in the x and y directions. For example, at $z = \lambda/4$

$$\vec{E} = \vec{a}_x E_0^+ \underline{/-\pi/2} = \vec{a}_x E_0^+ e^{-j\pi/2} \quad \text{and} \quad \vec{H} = \vec{a}_y H_0^+ \underline{/-\pi/2} = \vec{a}_y H_0^+ e^{-j\pi/2}$$

This method of describing an electromagnetic wave is a very useful analytical tool. In most cases, it is sufficient to specify the electric field and η, since $H = E/\eta$ and, for a given propagation direction, the magnetic field direction can be deduced from Eq. (2–58).

The rules of vector decomposition may be applied to vector phasors as illustrated by the following example.

Example 2–4:

The metal grating in Fig. 2–14 has the property that a wave with E lines perpendicular to the metal strips propagates through it, while one with E lines parallel to the strips is completely reflected by it. This assumes that the spacing between the strips is much less than the wavelength.

A linearly polarized wave propagating in free space and oriented 30 degrees from vertical ($\theta = 30°$) impinges upon the grating. Determine the rms value and the direction of the electric field at the output side of the grating. The incoming wave has an average power density of 50 W/m².

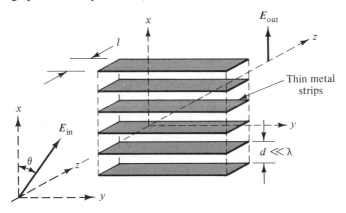

Figure 2–14   Propagation of a linear polarized wave through a metal grating. (See Ex. 2–4.)

Solution: For free space, $\eta = 377$ ohms and hence $p_{in} = (E_{in}^+)^2/377 = 50$ W/m². Therefore, $E_{in}^+ = 137.3$ V/m and its phasor value may be written as $E_{in}^+ = 137.3\underline{/0}$. The incoming vector wave may be decomposed into two components, one parallel and one perpendicular to the metal strips. That is,

$$\overrightarrow{E}_{in} = \vec{a}_x E_{in}^+ \cos 30° + \vec{a}_y E_{in}^+ \sin 30° = \vec{a}_x(119\ \underline{/0}) + \vec{a}_y(68.7\ \underline{/0})$$

Since only the x component passes through the grating, $\overrightarrow{E}_{out} = \vec{a}_x(119\underline{/-\beta l})$, where l is the length of the grating and βl is the phase delay through it. The rms value of the output electric field is 119 V/m and it is oriented in the x direction as shown. Thus the output wave is vertically polarized.

Vector addition and subtraction can also be applied to the vector-phasor representation of linearly polarized waves as long as the waves all have the same frequency and propagation direction. If the individual waves are all in phase (cophasal), a vector diagram may be used to determine the resultant wave. Figure 2–15 describes the addition of two such waves. The individual linear waves and their vector phasors are shown in part a. By utilizing the properties of right triangles, their vector addition yields the resultant phasor $\mathbf{E}_r = 2E_0\underline{/0}$ which is oriented 30° from the vertical. The wave associated with $\mathbf{E}_r$ is shown in part b. It has a peak value of $2\sqrt{2}\,E_0$ and its plane of polarization is 30° from the vertical (the x'-z plane). Although the magnetic field is not shown, it is perpendicular to the electric field and its direction may be deduced from the Poynting vector.

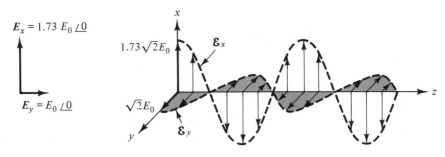

(a) The individual linear polarized waves

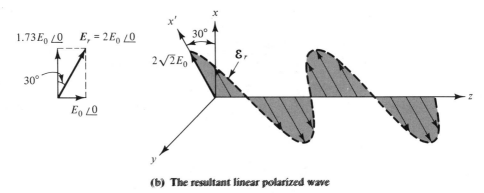

(b) The resultant linear polarized wave

Figure 2–15   The addition of two cophasal linear waves using the vector-phasor representation.

Circular polarized waves.   One example of an electromagnetic wave that is *not* linearly polarized is the circular polarized wave. It is characterized by a constant magnitude electric field vector (and magnetic field vector) whose orientation rotates in a plane transverse to the direction of propagation. The electric field pattern for such a wave is shown in Figs. 2–16 and 2–17. The angular velocity of the rotation ω is equal to $2\pi f$, where f is the frequency of the wave. The rotation in a fixed transverse plane can be either clockwise or counterclockwise.

Let us assume that the pattern shown in Fig. 2–16 represents a wave traveling in the positive z direction at $t = 0$. This pattern with its fixed E vectors, all having the same magnitude, moves in the positive z direction with a velocity v. Therefore, an observer looking down the propagation axis sees the rotating field pattern described in Fig. 2–17. Note that the vectors rotate counterclockwise with time. This is usually referred to as *left-hand circular polarization,* while clockwise rotation is called *right-hand circular polarization.* The solid arrows in the figure represent the electric field vectors at $t = 0$ and the dashed arrows indicate their direction of rotation with time. As in the case of linear polarization, one symbol is sufficient to completely describe the circular polarized wave when the direction of propagation is known. If the first symbol in Fig. 2–17 represents the counterclockwise wave at $z = 0$ and $t = 0$, then the wave at any other time and position is defined. At a fixed

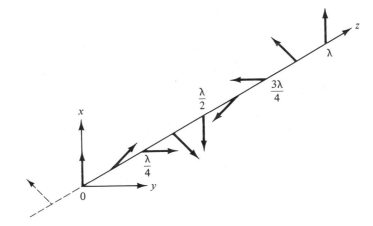

Figure 2–16   Description of a circular polarized wave at one instant in time. (Magnetic vectors not shown.)

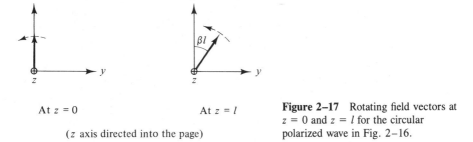

At $z = 0$ At $z = l$

(z axis directed into the page)

Figure 2–17   Rotating field vectors at $z = 0$ and $z = l$ for the circular polarized wave in Fig. 2–16.

time, the vector at $z = l$ is displaced *clockwise* by an angle βl relative to the vector at $z = 0$. Also, at any fixed point, say $z = l$, the vector is defined for any future time t, since it is rotated *counterclockwise* by an angle $\omega t = 2\pi t/T$ relative to its position at $t = 0$.

A circularly polarized wave may be represented by two orthogonal, linearly polarized waves equal in amplitude and 90° out-of-phase. The frequency of the waves must be identical and equal to $\omega/2\pi$, where ω is the angular velocity of the rotating field vectors. Figure 2–18 shows two such waves $\mathscr{E}_x$ and $\mathscr{E}_y$ at $t = 0$. Addition of their $\mathscr{E}$ vectors verifies that they are equivalent to the circular polarized wave shown at the right. Observe that the two linear waves are orthogonal in space and at any point are 90° out-of-phase in time (when one is maximum, the other is zero). If the peak value of each linear wave is $\sqrt{2}\,E_0$, then the amplitude of the circular polarized vector is $\sqrt{2}\,E_0$. The vector phasors for the two linear waves and the equivalent circular polarized representation are also shown, assuming propagation in the $+z$ direction.[10] In this example, $\mathbf{E}_y$ leads $\mathbf{E}_x$ by 90° which results in a counterclockwise circular polarized wave. For a clockwise wave, $\mathbf{E}_y$ would have to lag $\mathbf{E}_x$ by 90°.

[10] Caution must be exercised in adding or subtracting vector phasors by means of a vector diagram. This technique is valid *only* when the waves are cophasal. For example, one might conclude that the sum of the vector-phasors in Fig. 2–18 results in a linear wave polarized 45° from the vertical. This is incorrect! With the waves 90° out-of-phase, the resultant wave is *circularly polarized*.

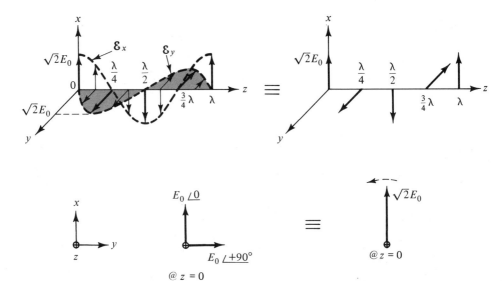

Figure 2–18 The equivalence between a circular polarized wave and two orthogonal, linear polarized waves equal in amplitude and 90° out-of-phase.

The relative time phase of the linear waves can be deduced by merely plotting $\mathscr{E}_x$ and $\mathscr{E}_y$ versus time for a fixed value of z.

The mathematical expression for the counterclockwise (ccw) circular polarized wave in Fig. 2–18 at $z = 0$ is

$$\vec{E}_{ccw} = \vec{a}_x E_0\underline{/0} + \vec{a}_y E_0\underline{/+90°} = E_0(\vec{a}_x + j\vec{a}_y)$$

At any other point, say $z = l$, the waves are delayed βl and therefore

$$\vec{E}_{ccw} = (E_0\underline{/-\beta l})(\vec{a}_x + j\vec{a}_y) \tag{2–61}$$

For a clockwise (cw) wave propagating in the $+z$ direction, the y component lags the x component. Therefore at $z = l$,

$$\vec{E}_{cw} = (E_0\underline{/-\beta l})(\vec{a}_x - j\vec{a}_y) \tag{2–62}$$

A circular polarized wave can be represented by any pair of orthogonal linear waves that are equal in amplitude and 90° out-of-phase. Figure 2–19 shows two equivalent representations of a clockwise wave. The following analysis verifies the equivalence. The pair on the left are given by $E_0(\vec{a}_x - j\vec{a}_y)$. The pair on the right are given by

$$\vec{E}_{cw} = \vec{a}_{x'}(E_0\underline{/-\theta}) + \vec{a}_{y'}(E_0\underline{/-90° - \theta}) = (E_0\underline{/-\theta})(\vec{a}_{x'} - j\vec{a}_{y'})$$

where $\vec{a}_{x'}$ and $\vec{a}_{y'}$ are the unit vectors in the x' and y' directions. These can be written in terms of $\vec{a}_x$ and $\vec{a}_y$ as follows.

$$\vec{a}_{x'} = \vec{a}_x \cos\theta + \vec{a}_y \sin\theta \qquad \text{and} \qquad \vec{a}_{y'} = -\vec{a}_x \sin\theta + \vec{a}_y \cos\theta$$

Therefore,

$$\vec{E}_{cw} = E_0 e^{-j\theta}[\vec{a}_x \cos\theta + \vec{a}_y \sin\theta + j\vec{a}_x \sin\theta - j\vec{a}_y \cos\theta]$$
$$= E_0 e^{-j\theta}[(\cos\theta + j\sin\theta)(\vec{a}_x - j\vec{a}_y)] = E_0(\vec{a}_x - j\vec{a}_y)$$

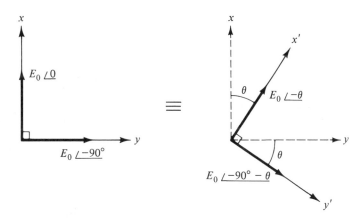

Figure 2-19 Two equivalent representations of a clockwise circular polarized wave.

which verifies that the two representations are equivalent. Thus, for a clockwise circular polarized wave, a *clockwise* rotation θ of the polarization of the orthogonal linear waves requires that phase *delays* equal to θ be added to their phasor values.

In practice, a circular polarized wave can be generated from a single linear wave. The following example illustrates the technique.

Example 2-5:

The array of dielectric strips described in Fig. 2-20 has the property that a wave with E lines parallel to the strips is phase delayed relative to one with E lines perpendicular

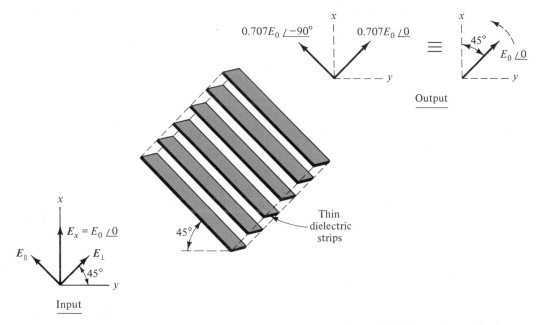

Figure 2-20 A dielectric grating for converting a linear polarized wave into a circular polarized wave. (See Ex. 2-5.)

to the strips. Show that an incoming vertically polarized wave $\overrightarrow{\mathbf{E}}_x$ is converted to a circular polarized wave if the strips are oriented 45° from the vertical and the phase delay is 90°.

Solution: Let $\mathbf{E}_x = E_0\underline{/0}$ represent the incoming linear wave. It can be decomposed into two components, one parallel $(\mathbf{E}_{\parallel})$ and one perpendicular $(\mathbf{E}_{\perp})$ to the dielectric strips. Since the strips are at 45°, the input components are $\mathbf{E}_{\parallel} = 0.707E_0\underline{/0}$ and $\mathbf{E}_{\perp} = 0.707E_0\underline{/0}$. At the output, $\mathbf{E}_{\parallel}$ is delayed 90° relative to $\mathbf{E}_{\perp}$. Thus, as indicated in the figure, there exists at the output two orthogonal waves, equal in amplitude and 90° out-of-phase. This represents a counterclockwise circular polarized wave of amplitude $\sqrt{2}(0.707E_0) = E_0$.

We conclude this section by briefly discussing elliptical and random polarization. An elliptical polarized wave is one in which the tip of the rotating field vector in a fixed transverse plane traces an elliptical path with time. The ellipticity (in dB) of the wave is defined as

$$\text{Ellipticity (dB)} \equiv 20 \log \frac{E_{\max}}{E_{\min}} \qquad (2\text{--}63)$$

where $E_{\max}$ and $E_{\min}$ are, respectively, the electric field amplitudes along the major and minor axes of the ellipse. A circularly polarized wave has 0 dB ellipticity since $E_{\max} = E_{\min}$.

One method of generating an elliptical wave is to combine two orthogonal linearly polarized waves, *unequal* in amplitude and 90° out-of-phase. Actually, circular and linear polarization are special cases of elliptical polarization. In the case of circular polarization, the amplitudes of the two orthogonal waves must be equal. For linear polarization, either one of the amplitudes must be zero or the two orthogonal waves must be cophasal. A more detailed discussion of elliptical polarization may be found in Refs. 2–5 and 2–6.

Random polarization, as the name implies, is electromagnetic energy in which the orientation of the field vector varies randomly with time. Examples of this are radiation from radio stars and the sun.

2–6 WAVE PROPAGATION IN IMPERFECT INSULATORS AND CONDUCTORS

This section reviews uniform plane wave propagation in imperfect media. Propagation through an imperfect dielectric is considered first. An expression for wave attenuation as a function of material parameters is developed. Next, wave propagation in good conductors (such as metals) is analyzed. The concept of skin depth and surface resistance is explained.

2–6a Plane Waves in a Lossy Dielectric

All dielectrics have a finite amount of conductivity. If $\tan \delta < 0.10$, its effect on electromagnetic propagation is mainly some attenuation of the wave. This is verified by the following analysis of uniform plane waves in a lossy dielectric.

For sinusoidal time variations, Faraday and Ampere's laws may be written in phasor form. That is,

$$\frac{d\mathbf{E}_x}{dz} = -j\omega\mu_R\mu_0\mathbf{H}_y \qquad \text{and} \qquad -\frac{d\mathbf{H}_y}{dz} = (\sigma + j\omega\epsilon_R\epsilon_0)\mathbf{E}_x \qquad (2\text{-}64)$$

Note that these expressions are similar to those in Eq. (2-46) except that a conduction current term $\mathbf{J} = \sigma\mathbf{E}_x$ has been added to Ampere's law. Eliminating $\mathbf{H}_y$ results in the following second order differential equation.

$$\frac{d^2\mathbf{E}_x}{dz^2} = -\omega^2\mu_R\mu_0\epsilon_R\epsilon_0(1 - j\tan\delta)\mathbf{E}_x \qquad (2\text{-}65)$$

where, as explained in Sec. 2-2, $\tan\delta \equiv \sigma/\omega\epsilon_R\epsilon_0$ is known as the *loss tangent* of the material.

As before, the two solutions represent traveling waves in the $+$ and $-z$ directions. In rms-phasor form, they are

$$\mathbf{E}_x = E_0^+ e^{-\gamma z} + E_0^- e^{+\gamma z}$$

where $\gamma = (j\omega\sqrt{\mu_R\mu_0\epsilon_R\epsilon_0})(\sqrt{1 - j\tan\delta})$ is known as the *propagation constant*. For dielectric materials with small conductivity ($\tan\delta < 0.10$), the approximation $\sqrt{1 - j\tan\delta} \approx 1 - j\frac{1}{2}\tan\delta$ may be used. Therefore,

$$\gamma \approx \frac{1}{2}\omega\sqrt{\mu_R\mu_0\epsilon_R\epsilon_0}\tan\delta + j\omega\sqrt{\mu_R\mu_0\epsilon_R\epsilon_0}$$

The real part of γ is called the *attenuation constant* (α), while the imaginary part is the previously defined phase constant β. Note that

$$\alpha = \frac{1}{2}\beta\tan\delta = \frac{\pi}{\lambda}\tan\delta \qquad \text{Np/length} \qquad (2\text{-}66)$$

Since $\gamma = \alpha + j\beta$, the rms-phasor solutions may be written as

$$\mathbf{E}_x = E_0^+ e^{-\alpha z}e^{-j\beta z} + E_0^- e^{+\alpha z}e^{+j\beta z} \qquad (2\text{-}67)$$

The magnetic field solutions are

$$\mathbf{H}_y = H_0^+ e^{-\alpha z}e^{-j\beta z} - H_0^- e^{+\alpha z}e^{+j\beta z} \qquad (2\text{-}68)$$

where $H_0^+ = E_0^+/\eta$ and $H_0^- = E_0^-/\eta$. In this case, the intrinsic impedance is

$$\eta = 377\sqrt{\frac{\mu_R}{\epsilon_R}}\left(\frac{1}{\sqrt{1 - j\tan\delta}}\right)$$

$$\approx 377\sqrt{\frac{\mu_R}{\epsilon_R}}(1 + j\frac{1}{2}\tan\delta) \qquad \text{ohms} \qquad (2\text{-}69)$$

What conclusions may be drawn from this analysis? First, since $\beta = \omega/v$, the wave velocity is approximately the same as in the lossless case, namely, $v = c/\sqrt{\mu_R\epsilon_R}$. Also, the fact that η is complex means that $\mathbf{H}_y$ is phase shifted relative to $\mathbf{E}_x$. For $\tan\delta < 0.10$, the phase shift is less than three degrees, a negligible value. Thus for good insulators v, β, λ, and η are unaffected by the material's finite conductivity and hence Eqs. (2–53), (2–55), (2–56), and (2–57) may be used with negligible error.

The only significant effect of finite (but small) conductivity is that the amplitude of the traveling waves are attenuated as they propagate along the z axis. For the forward traveling wave, the ratio of amplitudes at any two points ($z = l_1$ and $z = l_2$) is given by

$$\frac{E_0^+ e^{-\alpha l_1}}{E_0^+ e^{-\alpha l_2}} = \frac{H_0^+ e^{-\alpha l_1}}{H_0^+ e^{-\alpha l_2}} = e^{-\alpha(l_1 - l_2)}$$

Since the unit for α is Np/length, the exponent $\alpha(l_1 - l_2)$ is the total attenuation in nepers. Nepers and decibels are reviewed in Sec. 1–4. The following example gives some indication of the relation between attenuation and the properties of the dielectric.

Example 2–6:

A uniform plane wave at 1000 MHz is propagating through a non-magnetic dielectric whose properties are defined by $\epsilon_R = 4$ and $\tan \delta = 0.02$. The rms value of the forward traveling wave at $z = 10$ cm is 100 V/m. Calculate the loss in power density between $z = 10$ cm and $z = 70$ cm.

Solution: Since $\lambda_0 = 0.3$ m at 1000 MHz and $\mu_R = 1$, $\lambda = \lambda_0/\sqrt{4} = 0.15$ m. Therefore, $\beta = 2\pi/0.15$ rad/m and $\alpha \approx (\pi/0.15) \tan \delta = 0.42$ Np/m. The ratio of rms values is

$$\frac{E^+ \text{ at } z = 70 \text{ cm}}{E^+ \text{ at } z = 10 \text{ cm}} = e^{-0.42(0.70 - 0.10)} = e^{-0.252} = 0.78$$

and thus E^+ at $z = 70$ cm is 78 V/m.

With both E_x and H_y being attenuated at the same rate, the average power density p_z of the traveling wave is attenuated by

$$\frac{p_z \text{ at } z = l_1}{p_z \text{ at } z = l_2} = e^{-2\alpha(l_1 - l_2)}$$

For $l_1 = 0.70$ m and $l_2 = 0.10$ m, the power ratio is $(0.78)^2 \approx 0.60$, which represents an attenuation of 0.252 Np or 2.19 dB. Thus over a distance of 60 cm (four wavelengths), about 40% of the power in the uniform plane wave has been lost due to I^2R losses in the dielectric.

Microwave devices and lines often extend over many wavelengths, which means that the material chosen in this example would not be suitable for low-loss applications. For such applications, microwave engineers usually select materials with loss tangents less than 0.001. Most of the dielectrics listed in Table B–2 of Appendix B satisfy this condition. The conclusions described here apply equally well to a wave traveling in the negative z direction since it will be attenuated at the same rate as a forward traveling wave.

2–6b Plane Waves in a Good Conductor

Let us now consider the case of uniform plane waves in a good conductor (such as a metal). Since these materials are characterized by very high conductivity, the conduction current ($\sigma\mathbf{E}$) is much greater than the displacement current ($j\omega\epsilon_R\,\epsilon_0\,\mathbf{E}$). Therefore Eqs. (2–64) reduce to

$$\frac{d\mathbf{E}_x}{dz} = -j\omega\mu_R\,\mu_0\,\mathbf{H}_y \qquad \text{and} \qquad -\frac{d\mathbf{H}_y}{dz} = \sigma\mathbf{E}_x \qquad (2\text{–}70)$$

Eliminating $\mathbf{H}_y$ results in the following second order differential equation

$$\frac{d^2\mathbf{E}_x}{dz^2} = j\omega\mu_R\,\mu_0\,\sigma\mathbf{E}_x \qquad (2\text{–}71)$$

Although this equation has two solutions, only the one associated with a forward traveling wave will be considered, namely,

$$\mathbf{E}_x = E_0^+ e^{-\gamma z} \qquad \text{where} \qquad \gamma = \sqrt{j\omega\mu_R\,\mu_0\,\sigma}$$

Since $\omega = 2\pi f$ and $\sqrt{j} = 1\underline{/45°} = (1 + j1)/\sqrt{2}$,

$$\gamma = \alpha + j\beta = \sqrt{\pi f\mu_R\,\mu_0\,\sigma} + j\sqrt{\pi f\mu_R\,\mu_0\,\sigma}$$

Thus,

$$\mathbf{E}_x = E_0^+ e^{-z/\delta_s}e^{-jz/\delta_s} \qquad (2\text{–}72)$$

where

$$\delta_s \equiv \frac{1}{\sqrt{\pi f\mu_R\,\mu_0\,\sigma}} \qquad (2\text{–}73)$$

is known as the *skin depth* of the material.

To understand the meaning of Eqs. (2–72) and (2–73), consider the sketch in Fig. 2–21. Assume that an ac source in region 1 is connected to the conductor at the $z = 0$ plane resulting in a rms electric field E_0^+ as shown. Recalling that tangential E must be continuous across a boundary, the electric field just inside the conductor also equals E_0^+. From Eq. (2–72), the rms value of $\mathbf{E}_x$ decreases as the wave attempts to penetrate the conductor. If δ_s is small, the decay of E_x with distance is very rapid. Part b of the figure shows a plot of the rms value of current density as a

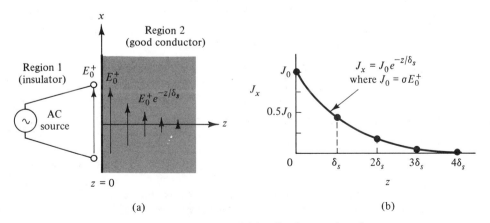

Figure 2–21 Description of skin effect in a good conductor.

function of penetration into the conductor. Its rate of decay with distance is the same as E_x since $J_x = \sigma E_x$.

The magnetic field also decreases at the same rate since $\mathbf{H}_y = \mathbf{E}_x/\eta_s$, where

$$\eta_s = \frac{1 + j}{\sigma \delta_s} = (1 + j)R_s = \sqrt{2}\, R_s \underline{/45°} \quad \text{ohms} \tag{2-74}$$

and R_s is defined in Eq. (2–78). This result is obtained by substituting $\mathbf{E}_x$ into the first of Eqs. (2–70). Its significance is that in a good conductor, the magnetic field lags the electric field by 45°.

At one skin depth ($z = \delta_s$), E_x is e^{-1} or 37 percent of its surface value. This, in fact, may be considered a definition of skin depth. Three skin depths into the conductor, the electric field amplitude is down to e^{-3} or 5 percent of its surface value. These comments also apply to J_x and H_y. Since power density is proportional to E_x^2, its value at three skin depths is e^{-6} or 0.25 percent of the surface power density. Thus, practically speaking, all of the power is located within a few skin depths of the surface.

For copper, $\mu_R = 1$ and $\sigma = 5.8 \times 10^7$ mho/m. Therefore at a frequency f (Hz), the expression for skin depth reduces to

$$\delta_{cu} = \frac{0.066}{\sqrt{f}} \quad \text{meter} \tag{2-75}$$

For example, at 10,000 MHz, the skin depth of copper is 6.6×10^{-7} m or 0.66 micron. Thus a few skin depths represents only 2 microns or *80 millionths of an inch*. In other words, at microwave frequencies, power and current flow in metals is essentially a surface phenomenon. Microwave engineers take advantage of this fact by plating a fairly good conductor (brass or aluminum) with several skin depths of an excellent conductor (silver or gold). In this manner, the electrical properties of the excellent conductor are obtained with minimum cost.

Skin-Effect resistance.   Figure 2–22a shows an ac source connected to a semi-infinite bar of conducting material. The connection is via two strips located at $x = 0$ and $x = l$. Assuming uniform fields in the x-y plane, the applied voltage $\mathbf{V} = E_0^+ l \underline{/0}$ since $\mathbf{E}_x = E_0^+ \underline{/0}$ at $z = 0$. The current density is a function of z since $\mathbf{J}_x = \sigma \mathbf{E}_x = \sigma E_0^+ e^{-\gamma z}$. Therefore, the total current $\mathbf{I}_x$ is

$$\mathbf{I}_x = \int_0^\infty \int_0^b \mathbf{J}_x \, dy \, dz = \frac{\sigma E_0^+ b}{\gamma} = \frac{\delta_s \sigma b E_0^+}{\sqrt{2}} \underline{/-45°}$$

and the ac impedance of the semi-infinite bar is

$$Z = \frac{\mathbf{V}}{\mathbf{I}_x} = \frac{\sqrt{2}\, l}{\sigma b \delta_s} \underline{/+45°} = \frac{l}{\sigma b \delta_s} + j \frac{l}{\sigma b \delta_s} \tag{2-76}$$

The imaginary portion of Z does not concern us since at high frequencies external inductive effects usually dominate. The resistive portion, however, is important. It

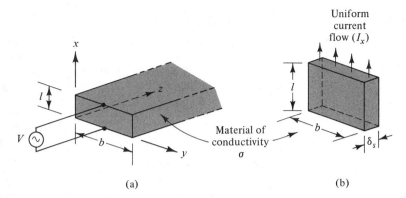

Figure 2–22 The influence of skin effect on the ac resistance of a good conductor.

is given by

$$R_{ac} = \frac{l}{\sigma b \delta_s} \quad \text{ohms} \tag{2–77}$$

Comparing this expression with Eq. (2–15) leads us to an interesting and useful conclusion. Namely, the ac resistance of the semi-infinite conductor is the same as the dc resistance of a rectangular bar of length l and transverse dimensions b and δ_s. In other words, the skin-effect resistance may be computed by assuming that the total current is uniformly distributed over a thickness of *one* skin depth. This concept is illustrated in part b of Fig. 2–22. The same conclusion may be obtained by analyzing the power dissipated in the conductor (Refs. 2–6 and 2–7).

The resistance of a square section ($l = b$), denoted by R_s, is given by

$$R_s = \frac{1}{\sigma \delta_s} = \sqrt{\frac{\pi f \mu_R \mu_0}{\sigma}} \quad \text{ohms/square} \tag{2–78}$$

When the skin depth is small compared to the conductor's thickness (that is, its z dimension), R_s is called the *surface resistivity*.

The phenomenon of skin effect can also be explained using circuit concepts. An excellent treatment is found in Chapter 7 of Ref. 2–7. Also included is a derivation of the skin-effect resistance of round wire. At high frequencies (that is, when $\delta_s \ll a$, the radius of the wire), the concept associated with Eq. (2–77) may be used to calculate the ac resistance. Namely,

$$R_{ac} \approx \frac{l}{\sigma (2\pi a) \delta_s} \quad \text{ohms} \tag{2–79}$$

where l is the length of the wire. In other words, at high frequencies, the wire may be considered as a thin hollow tube of length l, radius a and wall thickness δ_s.

As an example, consider a 100-m length of copper wire having a radius $a = 0.005$ m. At 1000 MHz, $\delta_s = 2.09 \times 10^{-6}$ m and therefore $R_{ac} = 26.3$ ohms. The dc resistance, from Eq. (2–15), is only 0.022 ohms, which means that the 1000 MHz resistance is more than a thousand times greater. Thus because of skin effect, $I^2 R$ losses are significantly greater at the higher frequencies.

2–7 REFLECTIONS AT CONDUCTING AND DIELECTRIC BOUNDARIES

When an electromagnetic wave impinges upon the boundary between two materials, a reflection invariably occurs. The magnitude and form of this reflection is a function of the angle of incidence of the wave as well as the properties of the two materials. This section reviews some of the cases that will be useful in subsequent discussions on guided waves.

2–7a Reflections from a Perfect Conductor

Let us first consider the case of a linearly polarized, uniform plane wave in a perfect insulator propagating toward a perfectly conducting surface (Fig. 2–23). The direction of propagation has been chosen perpendicular to the surface of the conductor (normal incidence). By virtue of Poynting's vector, both the electric and the magnetic vectors are parallel to the plane of the boundary. For the coordinate system shown, the electric and magnetic fields are given by Eqs. (2–50) and (2–51). Since a perfect insulating medium has been assumed, the attenuation in region 1 is zero. It is convenient to choose the $z = 0$ plane at the boundary between the insulating and conducting regions. This choice of coordinates is indicated in Fig. 2–23. Since a perfect conductor has been assumed for region 2, the tangential component of electric field must be zero at the boundary and thus $E_0^- = -E_0^+$. With $E_0^+/H_0^+ = E_0^-/H_0^-$, $H_0^- = -H_0^+$ and hence the magnetic field at $z = 0$ is $2\sqrt{2}H_0^+ \cos \omega t$. As a result, the electric field is zero at the conducting surface, while the magnetic field is double the value of $\mathcal{H}$ associated with the incident wave. For a perfect conductor, $\delta_s = 0$ and therefore ac current flow is restricted to the surface. By virtue of Eq. (2–42), the surface current density $\mathcal{H}$ is in the x direction and equal to $2\sqrt{2}H_0^+ \cos \omega t$.

The following argument may help in understanding the above conclusions. As the E_0^+ wave impinges upon the conductor, a surface current flows in the same direction as the electric field. The resultant charge separation creates an E_0^- equal and opposite to E_0^+. By Ampere's law (right-hand rule), the surface current creates a magnetic field in the y direction that adds to that associated with the incident wave. Thus

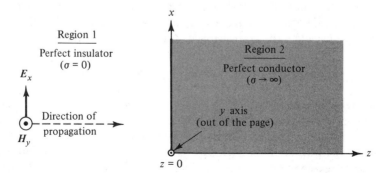

Figure 2–23   Wave propagation from a perfect insulator toward a perfect conductor (normal incidence).

the electric field experiences a reversal but the magnetic field does not, which means that the direction of propagation of E_0^- and H_0^- is opposite to that of the incident wave. Furthermore, since the magnitudes of E_0^- and H_0^- are the same as E_0^+ and H_0^+, the power density in the reflected wave is equal to that in the incident wave. This result is a direct consequence of the fact that a perfect conductor cannot absorb power.

Let us now consider the expressions for $\mathscr{E}_x$ and $\mathscr{H}_y$ at a distance d from the boundary plane. For $z = -d$ (region 1),

$$\mathscr{E}_x = \sqrt{2}\, E_0^+ \cos(\omega t + \beta d) - \sqrt{2}\, E_0^+ \cos(\omega t - \beta d)$$
$$\mathscr{H}_y = \sqrt{2}\, H_0^+ \cos(\omega t + \beta d) + \sqrt{2}\, H_0^+ \cos(\omega t - \beta d)$$
(2-80)

Using the trigonometric identity $\cos(A \pm B) = \cos A \cos B \mp \sin A \sin B$, the above equations become

$$\mathscr{E}_x = -\{2\sqrt{2}\, E_0^+ \sin \beta d\} \sin \omega t$$
$$\mathscr{H}_y = \{2\sqrt{2}\, H_0^+ \cos \beta d\} \cos \omega t$$
(2-81)

These expressions do not represent traveling waves since the arguments of the sin ωt and cos ωt terms do not contain βd terms. What they do represent are sinusoidal time functions whose amplitudes (the bracketed quantities) are a function of position along the z axis. They are called *standing waves*, the implication being that there is no net power flow in the z direction. This is understandable since it has already been shown that the power in the reflected wave is exactly equal to the power in the incident wave. Note that since Eqs. (2-80) and (2-81) are equivalent, two oppositely directed traveling waves and a standing wave are merely different ways of viewing the same phenomenon.

To better understand the nature of these standing waves, consider the expression for $\mathscr{E}_x$ in Eq. (2-81). Since $\beta = 2\pi/\lambda$, then $\mathscr{E}_x = 0$ at $d = 0, \lambda/2, \lambda$, etc. On the other hand, at $d = \lambda/4, 3\lambda/4, 5\lambda/4$, etc., $\mathscr{E}_x = \pm 2\sqrt{2}\, E_0^+ \sin \omega t$, which is an ac electric field having an rms value of $2E_0^+$. At other positions along the z axis, the rms value of $\mathscr{E}_x$ is less than $2E_0^+$. Thus, the largest possible rms value of $\mathscr{E}_x$ is *twice* the rms value of the incident wave. The expression for $\mathscr{H}_y$ in Eq. (2-81) also represents a standing wave. Note that for any fixed value d, $\mathscr{E}_x$ and $\mathscr{H}_y$ are 90° out-of-phase, which means that at any point in space, the average power in the electromagnetic field is zero. This is analogous to the 90° phase relation between voltage and current in a pure reactance wherein average power is also zero. Also note that nulls of $\mathscr{H}_y$ occur at $d = \lambda/4, 3\lambda/4, 5\lambda/4$, etc., while maximums occur at $d = 0, \lambda/2, \lambda$, etc. Thus the standing wave pattern for $\mathscr{H}_y$ is displaced one quarter wavelength relative to the $\mathscr{E}_x$ standing wave. A plot of E_x and H_y (rms values) as a function of distance d is shown in Fig. 2-24. Unlike the case of a pure traveling wave, the ratio of E_x to H_y for the standing wave is a function of position along the propagation axis and does *not* equal η, the intrinsic impedance of the medium. A plot of $\mathscr{E}_x$ and $\mathscr{H}_y$ versus position at several instants of time is shown in Fig. 2-25. The time intervals are one-eighth of a period apart and cover one period of the ac cycle. The pattern of the $\mathscr{E}_x$ curve as a function of time is similar to that of a vibrating string pinned at $d = 0$. The condition that the string's displacement is always zero at $d = 0$ is

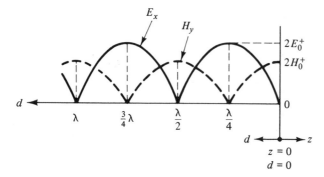

Figure 2–24 The rms values of $\mathscr{E}_x$ and $\mathscr{H}_y$ as a function of position for the case of a conducting surface located at $d = 0$.

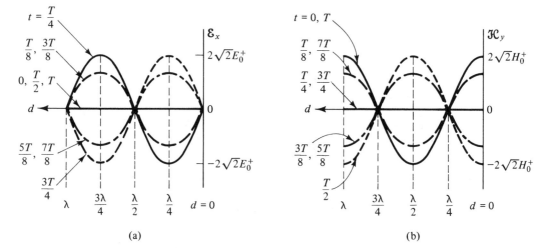

(a)   (b)

Figure 2–25 A plot of the instantaneous values of $\mathscr{E}_x$ and $\mathscr{H}_y$ as a function of position and time. The conducting surface is located at $d = 0$.

analogous to the boundary condition $\mathscr{E}_x = 0$ at the conducting surface. For any fixed value of d, both $\mathscr{E}_x$ and $\mathscr{H}_y$ are sinusoidal time functions having a period T. Observe that between a successive pair of nulls, all the sinusoidal time functions of the standing wave are in phase. That is, they all reach their maximum (and minimum) values at the same time. This contrasts with a pure traveling wave ($\alpha = 0$) wherein phase is a function of position but amplitude is not.

The analysis described here could have been carried out using phasor notation. The case of reflections at a dielectric boundary is analyzed using this approach.

Oblique incidence. Let us briefly consider the case in which the wave impinges upon the conducting surface at some angle θ_i. This angle, known as the *angle of incidence*, represents the angular distance between the propagation axis of the incident wave and a line normal to the reflecting surface. As in the case of optics, the angle of reflection θ_r equals the angle of incidence θ_i. This is true for any type polarization. The power density of the reflected wave equals that of the incident wave since a perfect conductor cannot absorb power. The proof of these results is given in Ref. 2–8.

2–7b Reflections and Refractions at a Dielectric Boundary

Consider now the case of a uniform plane wave propagating from one perfect insula-tor into another. The situation for the case of normal incidence is shown in Fig. 2–26. When the incident wave in region 1 arrives at the boundary between the two dielectrics (the $z = 0$ plane), some of it is reflected and some is transmitted into re-gion 2. The expressions for E and H in region 1 are given by Eqs. (2–48) and (2–49). In region 2, only a forward traveling wave exists since the material extends to infinity in the positve z direction. The continuity condition for the tangential components of $\mathbf{E}$ and $\mathbf{H}$ at the boundary require that $\mathbf{E}_i + \mathbf{E}_r = \mathbf{E}_t$ and $\mathbf{H}_i - \mathbf{H}_r = \mathbf{H}_t$, where the incident and reflected waves in region 1 are indicated by the subscripts i and r, respectively. The transmitted wave in region 2 is indicated by the subscript t. With $\mathbf{H}_i = \mathbf{E}_i/\eta_1$, $\mathbf{H}_r = \mathbf{E}_r/\eta_1$, and $\mathbf{H}_t = \mathbf{E}_t/\eta_2$, solving for $\mathbf{E}_t$ and $\mathbf{H}_t$ results in

$$\frac{\mathbf{E}_t}{\mathbf{E}_i} = \frac{2\eta_2}{\eta_1 + \eta_2} \quad \text{and} \quad \frac{\mathbf{H}_t}{\mathbf{H}_i} = \frac{2\eta_1}{\eta_1 + \eta_2} \tag{2–82}$$

These ratios indicate the fraction of the incident fields that are transmitted into region 2. Since $\mathbf{E}_r = \mathbf{E}_t - \mathbf{E}_i$,

$$\Gamma \equiv \frac{\mathbf{E}_r}{\mathbf{E}_i} = \frac{\mathbf{H}_r}{\mathbf{H}_i} = \frac{\eta_2 - \eta_1}{\eta_2 + \eta_1} \tag{2–83}$$

The fraction of the incident wave that is reflected at the boundary is called the *reflection coefficient* (Γ). Because it is the ratio of phasor quantities, the reflection coefficient may be complex. The magnitude of the reflection coefficient $|\Gamma|$ is the amplitude ratio of reflected to incident wave, while the angle of the reflection coefficient indicates the phase shift of the reflected wave relative to the incident wave. When η_2 and η_1 are real, Γ is real and its value is restricted to $-1 \leq \Gamma \leq +1$.

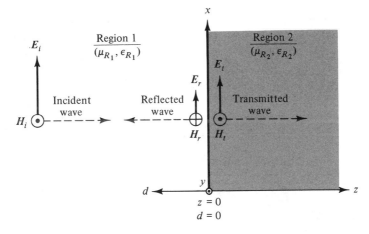

Figure 2–26 Transmission and reflection at the boundary between two ideal dielectrics (normal incidence).

At the boundary between the two dielectrics, part of the incident wave is reflected whenever $\eta_2 \neq \eta_1$. With power density proportional to the square of E, the reflected power density (p_r) is related to the incident power density (p_i) by

$$p_r = |\Gamma|^2 p_i \tag{2-84}$$

Utilizing Eqs. (2–82) and (2–83), the following relation is obtained for the transmitted power density.

$$p_t = \{1 - |\Gamma|^2\} p_i \tag{2-85}$$

This equation states that the power delivered to region 2 is equal to the incident power minus the reflected power. On the basis of a conservation of energy argument, this conclusion is plausible.

The reflected wave that is created when $\eta_2 \neq \eta_1$ produces a standing wave pattern in region 1. At a distance d to the left of the boundary plane (Fig. 2–26), the fields are given by Eqs. (2–48) and (2–49), where $z = -d$. That is,

$$\mathbf{E}_{x_1} = E_{0_1}^+ \underline{/\beta d} + E_{0_1}^- \underline{/-\beta d} \quad \text{and} \quad \mathbf{H}_{y_1} = H_{0_1}^+ \underline{/\beta d} - H_{0_1}^- \underline{/-\beta d}$$

where the subscript 1 denotes fields in region 1. Since the reflection coefficient at the boundary $\Gamma = E_{0_1}^-/E_{0_1}^+ = H_{0_1}^-/H_{0_1}^+$, the above equations may be restated as

$$\mathbf{E}_{x_1} = E_{0_1}^+ \underline{/\beta d} + \Gamma E_{0_1}^+ \underline{/-\beta d} \quad \text{and} \quad \mathbf{H}_{y_1} = H_{0_1}^+ \underline{/\beta d} - \Gamma H_{0_1}^+ \underline{/-\beta d} \tag{2-86}$$

These phasor equations represent standing waves of electric and magnetic fields. The electric field pattern (rms values) for Γ positive is shown in Fig. 2–27. If Γ were negative, the pattern would be shifted one quarter wavelength. In either case, the maximum and minimum values of E_{x_1} are given by

$$E_{\max} = \{1 + |\Gamma|\} E_{0_1}^+ \quad \text{and} \quad E_{\min} = \{1 - |\Gamma|\} E_{0_1}^+ \tag{2-87}$$

With region 1 being a perfect insulator, $\alpha = 0$ and hence all the maximums in the pattern are equal in value as are all the minimums.[11] The standing wave patterns rep-

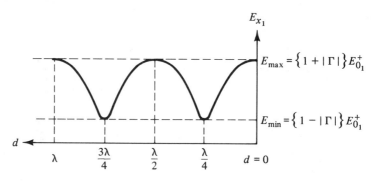

Figure 2–27 The standing wave pattern of E_{x_1} for Γ positive.

resent the amplitude variation resulting from the phasor addition of incident and reflected waves. The phasor additions for $\mathbf{E}_{x_1}$ at $d = 0$, $\lambda/8$ and $\lambda/4$ are shown in Fig. 2–28. Observe that as d increases, the incident phasor ($\mathbf{E}_i$) rotates counterclockwise while the reflected wave phasor ($\mathbf{E}_r$) rotates clockwise. The magnitude of the resultant phasor is the rms value of $\mathbf{E}_{x_1}$ at the particular position d. Note that unlike Fig. 2–24, the minimums are not zero and the maximums are not equal to $2E_0^+$. There are some similarities between the patterns in Figs. 2–24 and 2–27. Successive minimums are spaced $\lambda/2$ apart and adjacent maximums and minimums are separated by $\lambda/4$. Also, if the H_y pattern were plotted in Fig. 2–27, it would be shifted $\lambda/4$ relative to the E_x pattern.

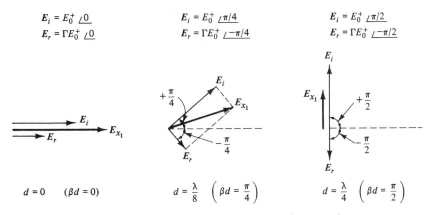

Figure 2–28 The phasor diagram of E_{x_1} at $d = 0$, $\lambda/8$, and $\lambda/4$ for the case where Γ is positive.

The magnitude of the reflection coefficient represents the fraction of the incident wave that is reflected at the boundary. Another quantity that is often used to indicate the amount of reflection is the ratio of the maximum field (rms) to the minimum field (rms) in the standing wave pattern. This quantity is called the *standing-wave ratio* (SWR) and is defined as $E_{\text{max}}/E_{\text{min}}$. From Eq. (2–87),

$$\text{SWR} = \frac{1 + |\Gamma|}{1 - |\Gamma|} \tag{2–88}$$

By virtue of its definition, SWR is always a real number that is greater than or equal to unity. It is a commonly measured quantity in high-frequency work. Since values of $|\Gamma|$ range from 0 to 1, the values of SWR range from unity to infinity. The above equation can be solved for $|\Gamma|$, which yields

$$|\Gamma| = \frac{\text{SWR} - 1}{\text{SWR} + 1} \tag{2–89}$$

[11] If $\alpha \neq 0$, the maximum and minimum values are functions of z. This case is discussed in Sec. 11.7 of Ref. 2–4.

An SWR of unity indicates no reflections and hence a pure traveling wave, while infinite SWR indicates full reflection and hence a pure standing wave. In general, the wave pattern consists of some combination of a traveling wave and a standing wave. For $|\Gamma| < 1$, there will be a net power flow in the propagation direction and therefore $\mathbf{E}_x$ and $\mathbf{H}_y$ cannot be 90° out-of-phase. The following example verifies this fact and illustrates the use of phasors in calculating the electric and magnetic standing wave patterns.

Example 2–7:

A 1250 MHz uniform plane wave propagating in free space (region 1) with an electric field equal to 10 V/m rms impinges on a semi-infinite volume of quartz (region 2) with normal incidence (Fig. 2–26).
(a) Calculate the reflection coefficient and SWR in region 1.
(b) What fraction of the incident power is transmitted into the quartz region?
(c) Calculate $\mathbf{E}_x$ and $\mathbf{H}_y$ at $d = 3$ and 6 cm.

Solution:
(a) From Table B–2 in Appendix B, $\epsilon_R = 3.8$ and $\mu_R = 1$ for quartz. Neglecting losses (that is, $\tan \delta \approx 0$), the intrinsic impedance $\eta_2 = 377/\sqrt{3.8} = 193$ ohms. Since region 1 is free space, $\eta_1 = 377$ ohms and therefore

$$\Gamma = \frac{193 - 377}{193 + 377} = 0.32\underline{/\pi} \quad \text{and} \quad \text{SWR} = \frac{1 + 0.32}{1 - 0.32} = 1.94.$$

(b) Since $|\Gamma|^2 = (0.32)^2 = 0.10$, 10 percent of the incident power is reflected. Therefore, $1 - |\Gamma|^2 = 0.90$, which means that 90 percent of the incident power is transmitted to region 2.
(c) For free space, $\eta_1 = 377$ ohms and $\lambda = \lambda_0 = 24$ cm at 1250 MHz. With $\beta = 2\pi/\lambda$, $E_{0_1}^+ = 10$ V/m, $H_{0_1}^+ = 10/377$ A/m and $\Gamma = -0.32$, Eq. (2–86) reduces to
$\mathbf{E}_{x_1} = 10\underline{/\pi d/12} - 3.2\underline{/-\pi d/12}$ and $\mathbf{H}_{y_1} = 0.027\underline{/\pi d/12} + 0.009\underline{/-\pi d/12}$.
At $d = 3$ cm,
$\mathbf{E}_{x_1} = 10\underline{/\pi/4} - 3.2\underline{/-\pi/4} = 10.5\underline{/1.09 \text{ rad}}$ or $10.5\underline{/62.7°}$ V/m
$\mathbf{H}_{y_1} = 0.027\underline{/\pi/4} + 0.009\underline{/-\pi/4} = 0.028\underline{/27.5°}$ A/m
At $d = 6$ cm,
$\mathbf{E}_{x_1} = 10\underline{/\pi/2} - 3.2\underline{/-\pi/2} = 13.2\underline{/90°}$ V/m
$\mathbf{H}_{y_1} = 0.027\underline{/\pi/2} + 0.009\underline{/-\pi/2} = 0.018\underline{/90°}$ A/m
Note that at this position, $\mathbf{E}_{x_1}$ is a maximum and $\mathbf{H}_{y_1}$ is a minimum.

As discussed earlier, $\mathbf{E}_x$ and $\mathbf{H}_y$ are in phase for a pure traveling wave and 90° out-of-phase for a pure standing wave. In the more general situation, the phase difference is a function of position along the propagation axis. In the above example, $\mathbf{E}_x$ and $\mathbf{H}_y$ are in phase at $d = 6$ cm but 35.2° out-of-phase at $d = 3$ cm. However, the net power flow at both points must be the same since the region is lossless. In both cases, the net power density must equal $\{1 - |\Gamma|^2\}p_i$, where $p_i = (10)(0.027) = 0.27$ W/m² is the power density of the incident wave. Since $|\Gamma| = 0.32$, the net power density at all points must be 0.24 W/m². Let us verify this at $d = 3$ cm. The net power density is given by $E_x H_y \cos \theta_{pf}$, where θ_{pf} is the power factor angle between $\mathbf{E}_x$ and $\mathbf{H}_y$. Its value at $d = 3$ cm is $(10.5)(0.028) \cos 35.2° = 0.24$ W/m². A similar calculation at any other value of d yields the same result.

Oblique incidence.   The case of oblique incidence at a dielectric interface is shown in Fig. 2–29. There are two possible cases, one with the E field perpendicular to the plane of incidence and the other with the E field parallel to it.[12] The first situation is described in the figure. As in the case for a conducting surface, the angle of incidence is equal to the angle of reflection. The relationship between the angle of incidence θ_i and the angle of refraction θ_t is given by Snell's law. Namely,

$$\frac{\sin \theta_i}{\sin \theta_t} = \frac{v_1}{v_2} = \frac{\lambda_1}{\lambda_2} = \sqrt{\frac{\mu_{R_2} \epsilon_{R_2}}{\mu_{R_1} \epsilon_{R_1}}} \qquad (2\text{–}90)$$

If region 1 is free space, Snell's law reduces to

$$\frac{\sin \theta_i}{\sin \theta_t} = \frac{c}{v_2} = \frac{\lambda_0}{\lambda_2} = \sqrt{\mu_{R_2} \epsilon_{R_2}} = n_i \qquad (2\text{–}91)$$

where n_i is known as the *index of refraction* of the material. Since $\mu_R \approx 1$ for practically all dielectrics, $n_i \approx \sqrt{\epsilon_R}$. This law indicates that a wave bends toward the normal to the boundary surface when propagating into a region with a higher index of refraction. For example, if region 1 is free space and region 2 has an index of refraction $n_2 = 3$, then for $\theta_i = 60°$, $\theta_t = \arcsin 0.289 = 16.8°$. On the other hand, if region 1 has the higher index of refraction, the transmitted wave bends away from the normal. In fact, under certain circumstances, it is possible that no real solution exists

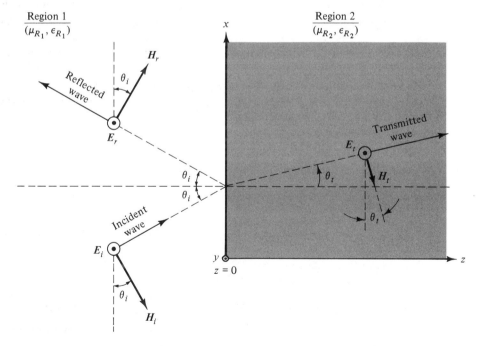

Figure 2–29   Transmission and reflection at the boundary between two ideal dielectrics (oblique incidence).

[12] Any other polarization may be considered as a linear combination of these two cases.

for θ_t. For instance, if $n_1 = 3$ and $n_2 = 1$, then for $\theta_i = 60°$, Eq. (2–90) yields $\theta_t = \arcsin 2.6$. Since this is not a real solution for θ_t, there can be no transmitted wave and hence the dielectric-air surface acts like a perfect reflector. This situation, wherein the wave does not escape the region of higher index of refraction, provides a qualitative explanation for the dielectric waveguides and resonators discussed in Chapters 5 and 9.

In general, there are both transmitted and reflected waves at the dielectric boundary. The relationships governing their amplitudes are given by Fresnel's equations. Assuming lossless, nonmagnetic dielectrics in both regions, the equations are

For $\vec{E}_i$ perpendicular to the plane of incidence,

$$\frac{E_t}{E_i} = \frac{2\,n_1 \cos\theta_i}{n_1 \cos\theta_i + n_2 \cos\theta_t} \quad \text{and} \quad \frac{E_r}{E_i} = \frac{n_1 \cos\theta_i - n_2 \cos\theta_t}{n_1 \cos\theta_i + n_2 \cos\theta_t} \quad (2\text{–}92)$$

For $\vec{E}_i$ parallel to the plane of incidence,

$$\frac{E_t}{E_i} = \frac{2\,n_1 \cos\theta_i}{n_2 \cos\theta_i + n_1 \cos\theta_t} \quad \text{and} \quad \frac{E_r}{E_i} = \frac{n_2 \cos\theta_i - n_1 \cos\theta_t}{n_2 \cos\theta_i + n_1 \cos\theta_t} \quad (2\text{–}93)$$

A particular case of interest for $\vec{E}_i$ parallel to the plane of incidence is when $n_2 \cos\theta_i = n_1 \cos\theta_t$. In this situation, $E_r = 0$ which means there is no reflected wave. The incident angle at which this occurs is called *Brewster's angle*. For an incident wave in region 1, it is equal to $\arctan(n_2/n_1)$. No comparable situation exists for $\vec{E}_i$ perpendicular to the plane of incidence since Snell's law prevents the numerator of E_r/E_i from becoming zero. This effect can be utilized to convert part of an incident circularly polarized wave to linear polarization since only one component of the incident wave will be reflected.

The above equations, as well as Snell's law, are a direct result of the conditions imposed on the electric and magnetic fields at the dielectric interface. The derivations are given in Ref. 2–8.

REFERENCES

Books

2–1. Kip, A. F., *Fundamentals of Electricity and Magnetism*, 2nd ed., McGraw-Hill Book Co., New York, 1969.

2–2. International Telephone and Telegraph Co., *Reference Data for Radio Engineers*, 5th ed., H.W. Sams & Co., Indianapolis, IN, 1968.

2–3. von Hippel, A. R., *Dielectric Materials and Applications*, The Technology Press of MIT, Cambridge, MA and John Wiley & Sons, Inc., New York, 1954.

2–4. Hayt, W. H., Jr., *Engineering Electromagnetics*, 4th ed., McGraw-Hill Book Co., New York, 1981.

2–5. Berkowitz, B., *Basic Microwaves*, Hayden Book Co., New York, 1966.

2–6. Rao, N. N., *Basic Electromagnetics with Applications*, Prentice-Hall, Inc., Englewood Cliffs, NJ, 1972.

2–7. Skilling, H. H., *Electric Transmission Lines*, McGraw-Hill Book Co., New York, 1951.

2–8. Zahn, M., *Electromagnetic Field Theory*, John Wiley & Sons, Inc., New York, 1979.

PROBLEMS

2–1. A teflon-filled, parallel-plate capacitor is fully charged by a 12 V battery. The plate separation is 0.10 cm and the dimensions of each plate is 0.8 cm by 1.2 cm. Neglecting fringe effects, calculate the energy stored in the electric field. Compare the result with that obtained from the formula for energy stored in a capacitor ($\frac{1}{2}CV^2$).

2–2. A quartz-filled, parallel-plate capacitor consists of two identical circular metal discs spaced 0.05 cm apart.
(a) Calculate the radius of the discs for $C = 6.0$ pF.
(b) What is the conductance of the structure at 3.0 GHz?

2–3. N turns of #24 copper wire are wound as a single layer on a 0.25 cm diameter glass rod. If the turns are uniformly distributed over a length of 2.0 cm, how many turns are needed for an inductance of 177 nH? Calculate the dc resistance of the coil.

2–4. The wavelength of a 600 MHz wave propagating through a nonmagnetic dielectric is 20 cm. What is the dielectric constant of the material?

2–5. An electromagnetic wave propagates through a lossless insulator with a velocity 1.8×10^{10} cm/s. Calculate the electric and magnetic properties of the insulator if its intrinsic impedance is 260 ohms.

2–6. An electromagnetic wave propagates through free space with a power density of 60 W/m^2. Another wave with the same power density propagates through a nonmagnetic dielectric. Show that the electric field in the dielectric is less than that in free space.

2–7. A linear polarized wave with a power density of 27 W/m^2 impinges on the metal grating shown in Fig. 2–14. What is the plane of polarization of the input wave (with respect to the x axis), if the power density of the output wave is 3 W/m^2?

2–8. A 3000 MHz uniform plane wave propagates through rexolite in the positive z direction. The **E** field at $z = 0$ is $100\underline{/0}$ V/m.
(a) Calculate the rms value and phase of **E** at $z = 4$ cm.
(b) Determine the total wave attenuation (in dB) over a distance of 6 wavelengths.

2–9. The current density at the surface of a thick metal plate is 100 A/m^2. What is the skin depth if the current density at a depth of 0.001 cm is 0.28 A/m^2?

2–10. It is proposed to silver-plate a 10 foot length of stainless steel wire so as to reduce its resistance at 1000 MHz. The wire diameter is 0.20 cm.
(a) Approximate the minimum plating thickness required to insure that the 1000 MHz current in the stainless steel is negligible.
(b) Assuming sufficient plating, calculate the 1000 MHz resistance of the wire. Compare the result to the resistance of the wire before plating.

2–11. An electromagnetic wave propagating in free space is reflected by a metal plate. The distance between successive nulls in the standing wave pattern is 12 cm. Calculate the frequency of the wave.

2–12. A 2000 MHz standing wave pattern exists in a nonmagnetic dielectric. What is ϵ_R if the distance between a maximum and an adjacent minimum is 1.5 cm?

2–13. Referring to Fig. 2–26, region 2 is free space and the properties of region 1 are $\mu_{R_1} = 1$ and $\epsilon_{R_1} = 9$. The net electric field at the boundary ($d = 0$) is $30\underline{/0}$ V/m. Use the phasor method to calculate $\mathbf{E}_{x_1}$ and $\mathbf{H}_{y_1}$ at $d = 1$ cm and 2 cm if the wave frequency is 1250 MHz. How far from the boundary is the first magnetic field maximum?

2–14. Regions 1 and 2 of Fig. 2–26 contain nonmagnetic dielectrics with $\epsilon_{R_1} = 6$ and $\epsilon_{R_2} = 3$. Calculate the reflection coefficient and SWR for a wave propagating from region 1 toward the dielectric interface.

2–15. A 500 MHz incident wave propagates as shown in Fig. 2–26. Region 1 contains a lossless insulator and region 2 is free space. Determine μ_{R_1} and ϵ_{R_1} if the SWR in region 1 is 1.60 and the wavelength is 35 cm. Assume $\epsilon_{R_1} > \mu_{R_1}$.

2–16. An electromagnetic wave with a power density of 5.0 W/m² impinges on a dielectric boundary causing a SWR of 1.90. Calculate the power density of the wave transmitted into the dielectric.

2–17. Calculate the SWR for the cases of 25 percent and 50 percent power reflected at a dielectric boundary.

2–18. Referring to Fig. 2–29, $\theta_i = 35°$ and region 2 is free space. What is the minimum value of index of refraction for region 1 that results in no transmission into the free space region?

3

Transmission-Line Theory

The theory of electric waves along uniform transmission lines is reviewed in this chapter. A uniform line is defined as one whose dimensions and electrical properties are identical at all planes transverse to the direction of propagation. The analysis includes a study of the reflection characteristics of terminated lines. The results allow us to apply ac circuit concepts to lines whose lengths are *not* negligible compared to the operating wavelength. (The restriction regarding line lengths was discussed in Sec 1–1). An interesting consequence of this analysis is that the impedance of a circuit can be dramatically altered by the addition of a small length of transmission line. This impedance transforming property of a line is a powerful design tool at microwave frequencies. Several illustrative examples are given in this and subsequent chapters.

3–1 CIRCUIT REPRESENTATION OF TRANSMISSION LINES

Transmission lines provide one method of transmitting electrical energy between two points in space, antennas being the other (Appendix F). Figure 1–4 shows four types of lines used at microwave frequencies. The open two-wire line is the most popular at the lower frequencies, the TV twin-lead being a familiar example. UHF and cable TV systems utilize low-loss coaxial cable as a transmission line. Modern microwave practice involves considerable use of coaxial lines at frequencies up to 30 GHz and hollow waveguides from 3 to 300 GHz.

In principle, any transmission line can be analyzed by solving Maxwell's equations and applying the appropriate boundary conditions for the particular line geometry. An example of this is the analysis of hollow waveguides described in Sec. 5–5. A simpler technique that utilizes ac circuit concepts is given in this chapter. As mentioned in Sec. 1–2, this technique was introduced by Lord Kelvin and developed fully by Oliver Heaviside. Essentially, it is an extension of ac circuit theory to lines having distributed circuit elements. A disadvantage of this method is that it reveals little about the electromagnetic field pattern or other possible modes of propagation. However, it does describe the impedance and propagation characteristics of the line for the principal mode of transmission and hence is of considerable engineering value.

The following quantities may be defined for a uniform transmission line.

R' ≡ Series resistance per unit length of line (ohm/m)

G' ≡ Shunt conductance per unit length of line (mho/m)

L' ≡ Series inductance per unit length of line (H/m)

C' ≡ Shunt capacitance per unit length of line (F/m)

The quantity R' is related to the dimensions and conductivity of the metallic conductors. Because of skin effect, it is also a function of frequency. G' is related to the loss tangent of the insulating material between the conductors.[1] L' is associated with the magnetic flux linking the conductors, while C' is associated with the charge on the conductors. Expressions for the distributed elements of various transmission lines are given in Chapter 5.

With this concept of distributed elements, a uniform transmission line may be modeled by the circuit representation in Fig. 3–1. The line is pictured as a cascade of identical sections, each Δz long. Each section consists of series inductance and resistance ($L'\Delta z$ and $R'\Delta z$) as well as shunt capacitance and conductance ($C'\Delta z$ and $G'\Delta z$). Since Δz can always be chosen small compared to the operating wavelength, an individual section of line may be analyzed using ordinary ac circuit theory. In the derivation that follows, $\Delta z \rightarrow 0$ and hence the results are valid at all frequencies.

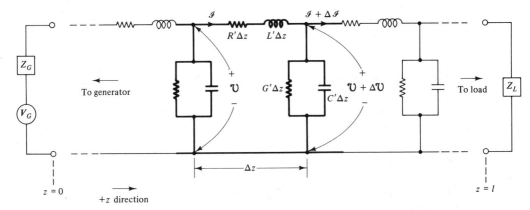

Figure 3–1 Circuit representation of a uniform transmission line.

[1] It is important to note that G' is *not* the reciprocal of R'. They are independent quantities, R' being related to the properties of the two conductors and G' to the characteristics of the insulating material between them.

In the figure, $\mathcal{V}$ and $\mathcal{I}$ represent the time-varying voltage and current at the input of a line section, while $\mathcal{V} + \Delta\mathcal{V}$ and $\mathcal{I} + \Delta\mathcal{I}$ represent the output values. The positive z direction is taken as horizontal and to the right, that is, from the generator toward the load. Also indicated are the assumed positive directions for the currents and voltages.

Applying Kirchhoff's voltage and current laws to the line section yields

$$\mathcal{V} = (R'\Delta z)\mathcal{I} + (L'\Delta z)\frac{\partial \mathcal{I}}{\partial t} + (\mathcal{V} + \Delta\mathcal{V})$$

and

$$\mathcal{I} = (G'\Delta z)(\mathcal{V} + \Delta\mathcal{V}) + (C'\Delta z)\frac{\partial}{\partial t}(\mathcal{V} + \Delta\mathcal{V}) + (\mathcal{I} + \Delta\mathcal{I})$$

Simplifying and recognizing that as $\Delta z \to 0$, $\mathcal{V} + \Delta\mathcal{V} \to \mathcal{V}$ results in the following partial differential equations.

$$-\frac{\partial \mathcal{V}}{\partial z} = R'\mathcal{I} + L'\frac{\partial \mathcal{I}}{\partial t} \quad \text{and} \quad -\frac{\partial \mathcal{I}}{\partial z} = G'\mathcal{V} + C'\frac{\partial \mathcal{V}}{\partial t} \tag{3-1}$$

By taking $\partial/\partial z$ of the first equation and $\partial/\partial t$ of the second equation and eliminating $\partial\mathcal{I}/\partial z$ and $\partial^2\mathcal{I}/\partial z\,\partial t$, a second-order differential equation for voltage is obtained.

$$\frac{\partial^2 \mathcal{V}}{\partial z^2} = L'C'\frac{\partial^2 \mathcal{V}}{\partial t^2} + (R'C' + G'L')\frac{\partial \mathcal{V}}{\partial t} + R'G'\mathcal{V} \tag{3-2}$$

Solving for current in a similar manner yields

$$\frac{\partial^2 \mathcal{I}}{\partial z^2} = L'C'\frac{\partial^2 \mathcal{I}}{\partial t^2} + (R'C' + G'L')\frac{\partial \mathcal{I}}{\partial t} + R'G'\mathcal{I} \tag{3-3}$$

The solution of either of these second-order equations and Eq. (3–1), together with the electrical properties of the generator and load, allow us to determine the instantaneous voltage and current at any time t and any place z along the uniform transmission line.

For the case of perfect conductors $(R' = 0)$ and insulators $(G' = 0)$, the above equations reduce to

$$\frac{\partial^2 \mathcal{V}}{\partial z^2} = L'C'\frac{\partial^2 \mathcal{V}}{\partial t^2} \quad \text{and} \quad \frac{\partial^2 \mathcal{I}}{\partial z^2} = L'C'\frac{\partial^2 \mathcal{I}}{\partial t^2} \tag{3-4}$$

while Eqs. (3–1) reduce to

$$-\frac{\partial \mathcal{V}}{\partial z} = L'\frac{\partial \mathcal{I}}{\partial t} \quad \text{and} \quad -\frac{\partial \mathcal{I}}{\partial z} = C'\frac{\partial \mathcal{V}}{\partial t} \tag{3-5}$$

Equations (3–4) and (3–5) represent the differential equations for a lossless line. Although real lines are never without loss, there are many in which it is sufficiently small that the lossless solution represents an excellent approximation.

Equations (3–4) are forms of the well-known wave equation of mathematical physics. We have already encountered it in Eq. (2–45). It was shown that the solution represented electromagnetic waves traveling in the plus and minus z directions with a velocity given by Eq. (2–53). The solutions of Eqs. (3–4) also represent traveling waves. In this case, they are voltage and current waves that travel with a velocity given by

$$v = \frac{1}{\sqrt{L'C'}} \tag{3–6}$$

In general, Eqs. (3–4) are satisfied by single-valued functions of the form $f(t \pm \sqrt{L'C'}\, z)$, where the plus sign indicates propagation in the *negative z* direction and the minus sign propagation in the *positive z* direction. To understand the meaning of these solutions, assume $\mathcal{V} = f(t - \sqrt{L'C'}\, z)$. At the point $z = 0$, the voltage versus time function is given by $\mathcal{V} = f(t)$. Further down the z axis at a point $z = z_1$, $\mathcal{V} = f(t - \sqrt{L'C'}\, z_1)$, which is exactly the same as $f(t)$ except that it has been time delayed by $t_d = \sqrt{L'C'}\, z_1$. Thus, it appears that the voltage versus time function at $z = 0$ has moved to $z = z_1$ with a velocity $v = z_1/t_d = 1/\sqrt{L'C'}$, which is exactly Eq. (3–6). By a similar argument, the $f(t + \sqrt{L'C'}\, z)$ solution represents a voltage function traveling in the negative z direction. In like manner, the solutions of the current equation may be interpreted as forward and reverse traveling current functions having the same velocity as the voltage. A similar conclusion was arrived at regarding the $\mathcal{E}$ and $\mathcal{H}$ waves discussed in Sec. 2–4, the explanation being that $\mathcal{E}$ generated $\mathcal{H}$ and vice versa. The voltage and current waves also travel with the same velocity since $\mathcal{V}$ and $\mathcal{I}$ generate each other. A physical explanation is presented in the next section to show the reasonableness of this conclusion.

Another result given in Sec. 2–4 is that the ratio of $\mathcal{E}$ to $\mathcal{H}$ for the traveling waves is a constant (η) which is a function of the electric and magnetic properties of the medium. Similarly, the ratio of $\mathcal{V}$ to $\mathcal{I}$ for a traveling wave on a transmission line is a constant. This constant is called the *characteristic impedance* (Z_0) of the line. For a lossless line, it is given by

$$Z_0 = \sqrt{\frac{L'}{C'}} \qquad \text{ohms} \tag{3–7}$$

To verify this expression, let $\mathcal{V} = f_1(u)$ and $\mathcal{I} = f_2(u)$, where $u \equiv t - \sqrt{L'C'}\, z$. Since

$$\frac{\partial \mathcal{V}}{\partial z} = \frac{\partial f_1}{\partial u} \cdot \frac{\partial u}{\partial z} = -\sqrt{L'C'}\,\frac{\partial f_1}{\partial u} \qquad \text{and} \qquad \frac{\partial \mathcal{I}}{\partial t} = \frac{\partial f_2}{\partial u} \cdot \frac{\partial u}{\partial t} = \frac{\partial f_2}{\partial u}$$

substitution into the first of Eqs. (3–5) yields $\sqrt{L'C'}\, \partial f_1/\partial u = L'\partial f_2/\partial u$. Integration with respect to u and simplifying results in $f_1/f_2 = \mathcal{V}/\mathcal{I} = \sqrt{L'/C'}$, which is Eq. (3–7).

It will be shown that Z_0 is a function of the cross-sectional dimensions of the line as well as the electrical properties of the insulating material between the conductors.

3–2 TRANSIENTS ON A TRANSMISSION LINE

We have seen that voltage and current waves travel along a transmission line with the same velocity. The physical argument presented here is intended to verify this fact while giving additional insight into the process of wave propagation along uniform lines.

Figure 3-2a shows a 20 V battery with an internal resistance of 100 ohms connected through a switch to an infinitely long transmission line with $Z_0 = 100$ ohms. Part *b* of the figure shows the same circuit with the transmission line replaced by its equivalent circuit representation. When the switch is closed, a voltage appears immediately at the input of the transmission line. However, it cannot appear instantaneously at other points along the line, since that would require a sudden change in voltage on all the capacitances. Furthermore, since it is current that delivers the electric charge to the capacitances, a sudden increase in current through the inductances would also be necessary. Since inductance opposes a current change and capacitance opposes a voltage change, the voltage and current require a finite time to propagate along the transmission line. The propagation process can be described in

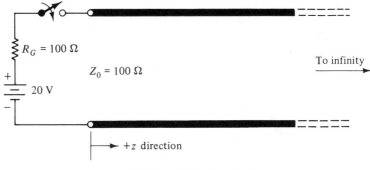

(a) Transmission-line circuit

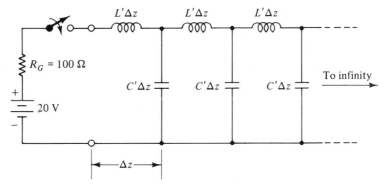

(b) Equivalent circuit representation

Figure 3–2   A dc source connected to an infinitely long, lossless line.

the following manner. When the switch is closed, the first inductance generates a back emf, in accordance with Lenz's law, to initially oppose an increase in current. Eventually, however, current flows through $L'\Delta z$ and charges the first shunt capacitance $C'\Delta z$ to a voltage V. The charged capacitor now acts like a voltage source and forces current through the next inductor. This charges the next capacitor and the process continues down the line. From this argument, it is apparent that voltage creates current and vice versa, thus requiring that they travel together along the transmission line. Since the line is infinitely long, only forward traveling waves of voltage and current exist and their ratio is given by Z_0 (100 ohms in this case). As time progresses, the battery continues to supply the current needed to charge the never-ending line of shunt capacitances. Thus, in the steady state, the infinite line presents an impedance of Z_0 to the battery. The current supplied by the battery is $20/(R_G + Z_0) = 0.10$ A. With half of the 20 V dropped across R_G, the voltage at the input to the line is 10 V. This 10 V voltage wave and its accompanying 0.10 A current wave travel in the positive z direction with a velocity given by Eq. (3–6).

Let us now look at some examples of finite length lines with various terminations.

An open-circuited line.   Consider the case of a 20 V battery with $R_G = 100$ ohms connected to an open-circuited, lossless transmission line. This situation, with specific values of l, v and Z_0, is shown at the top of Fig. 3–3. With the initially open switch closed at $t = 0$, the voltage at the input to the line immediately becomes 10 V. This occurs because at the first instant, the dc source has no indication that the line is *not* infinite in length and hence sees an input impedance $Z_0 = 100$ ohms.[2] Thus at $t = 0+$ (that is, immediately after closing the switch), the current and voltage at the input to the line are $20/(R_G + Z_0) = 0.10$ A and 10 V, respectively. These values remain constant until the battery has some indication (via a reflected wave) that the line is not infinite in length. With the velocity given as 2×10^8 m/s, it takes 10 ns for V and I to travel halfway down the 4 m line. This situation is shown in part *a* of Fig. 3–3. Part *b* shows the waves at $t = 20-$ ns (that is, slightly less than 20 ns). When the waves arrive at the open circuit, something must happen since two contradictory impedance requirements exist. First, the V/I ratio for the traveling wave must be $Z_0 = 100$ ohms. On the other hand, Ohm's law at the open-circuited end of the line requires an infinite impedance since current must be zero. The creation of reflected waves (V^-, I^-) allows both of these requirements to be satisfied. Thus at the load end $(z = 4$ m$)$,

$$V_L = V^+ + V^- \qquad \text{and} \qquad I_L = I^+ - I^- = 0$$

where the $+$ and $-$ superscripts indicate forward and reverse traveling waves.[3] V_L and I_L represent the voltage and current at the load end once the forward traveling waves have arrived, which occurs at $t = 20$ ns. The condition that $I_L = 0$ requires

---

[2] Since Z_0 represents the impedance of the line when the switch is first closed, it is sometimes referred to as the *surge impedance* of the transmission line.

[3] The reason for the minus sign in the equation for I_L is that for the forward and reverse traveling waves, I^+ and I^- are oppositely directed.

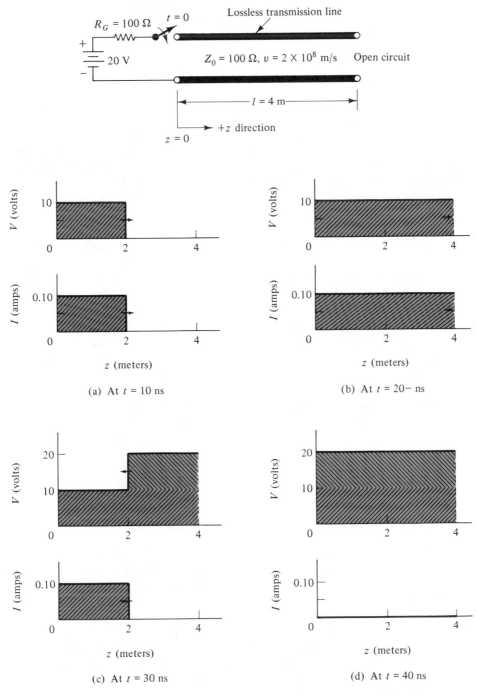

Figure 3–3 Voltage and current waves on an open-circuited transmission line. (The switch is closed at $t = 0$.)

that $I^- = I^+ = 0.10$ A. Also, with $V^+ = I^+Z_0$ and $V^- = I^-Z_0$, $V^- = V^+ = 10$ V. Therefore at $t = 20$ ns, $I_L = 0$ and $V_L = 20$ V.

If we define the reflection coefficient at the load as

$$\Gamma_L \equiv \frac{V^-}{V^+} = \frac{I^-}{I^+} \qquad (3-8)$$

then for the open-circuit case, $\Gamma_L = +1$. The open-circuit condition at the load end thus creates reflected voltage and current waves of 10 V and 0.10 A, respectively. These waves travel in the negative z direction with the same velocity as the forward waves. Parts c and d of Fig. 3–3 show the resultant voltage and current (due to the sum of the $+$ and $-$ waves) at $t = 30$ and 40 ns. As the wavefront of the 10 V, 0.10 A reflected wave moves to the left, it leaves behind a net voltage of 20 V and a net current of zero. Since $R_G = 100$ ohms, both Ohm's law and the condition that $V^-/I^- = 100$ ohms are satisfied at $t = 40$ ns, and hence no reflections are required at the generator. The process thus ends and a steady state is achieved with $V = 20$ V and $I = 0$ everywhere on the transmission line. In other words, after 40 ns, the dc source finally *sees* the open circuit and behaves accordingly. If the open circuit were replaced by a short circuit, then for $t > 40$ ns, the conditions everywhere on the line would be $V = 0$ and $I = 20/R_G = 0.20$ A (Prob. 3–3).

Resistively terminated lines.   Consider now the case of a finite length transmission line terminated with a pure resistance. This situation is shown at the top of Fig. 3–4, where R_L is the terminating or load resistance. As before, closing the switch initiates a 10 V, 0.10 A forward traveling wave. At $t = 20$ ns, the wave arrives at the load end. Since $R_L \neq Z_0$, Ohm's law can only be satisfied by assuming reflected waves. Thus at $z = 4$ m, $V_L = V^+ + V^-$ and $I_L = I^+ - I^- = (V^+ - V^-)/Z_0$. Ohm's law requires $V_L/I_L = R_L$ and hence

$$R_L = Z_0\frac{V^+ + V^-}{V^+ - V^-} = Z_0\frac{1 + \Gamma_L}{1 - \Gamma_L} \qquad (3-9)$$

Solving for the load reflection coefficient yields

$$\Gamma_L = \frac{R_L - Z_0}{R_L + Z_0} \qquad (3-10)$$

Note the similarity between this equation and Eq. (2–83) which describes the reflection coefficient for an electromagnetic wave. In the next section, Eqs. (3–9) and (3–10) are extended to include complex impedances when the source excitation is sinusoidal. For resistive terminations, Γ_L is real and can take on any value between -1 and $+1$. If $R_L = 0$ (short circuit), $\Gamma_L = -1$, while if $R_L = \infty$ (open circuit), $\Gamma_L = +1$. For the special case when $R_L = Z_0$, $\Gamma_L = 0$ and therefore no reflected waves are generated. What this means is that when the forward voltage and current waves arrive at the load, Ohm's law is automatically satisfied and reflections are not required. Referring to the circuit in Fig. 3–4 with $R_L = 100$ ohms, the steady-state condition is reached after 20 ns, namely, $V = 10$ V and $I = 0.10$ A

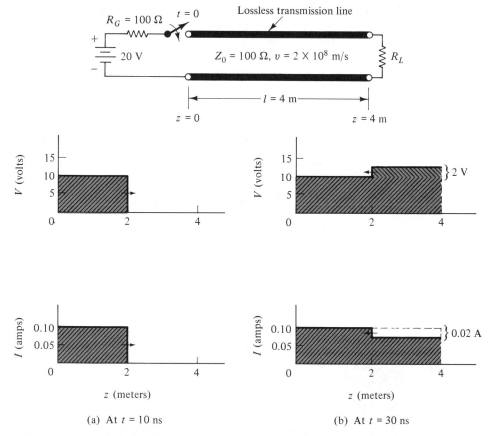

Figure 3–4   Voltage and current waves for the circuit shown when R_L = 150 ohms. (The switch is closed at t = 0.)

everywhere along the line. Note that in all these cases, the steady-state values of voltage and current are those expected from a dc analysis of the circuit.

Consider now the case where R_L = 150 ohms for the circuit shown at the top of Fig. 3–4. From Eq. (3–10), Γ_L = 0.20. With the forward wave again equal to 10 V and 0.10 A, the reflected voltage and current are 2 V and 0.02 A, respectively. Parts a and b of Fig. 3–4 shows the voltage and current along the line at t = 10 and 30 ns. At t = 10 ns, only the forward traveling waves exist, having arrived only at the halfway point of the 4 m line. At t = 30 ns, the reflected waves have been generated and have traveled halfway back toward the generator end of the line. At t = 40 ns (not shown), the reflected waves arrive at the input and the resultant voltage and current everywhere along the line become 12 V and 0.08 A. Since R_G = Z_0, no reflection is required at the generator end and the steady state is achieved after 40 ns. Again, the final values are those expected from a dc analysis of the circuit.

Multiple reflections on a transmission line. From the above cases, it is clear that when $R_G = Z_0$, the steady state is achieved after one *round trip* (40 ns, in our example). On the other hand, if $R_L = Z_0$, the steady state occurs after a *one-way trip* (20 ns, in our example). Let us now explore the situation when neither R_G nor R_L is equal to the characteristic impedance Z_0. The analysis will show that reflections occur at both ends of the line and the steady-state values are approached only as t becomes infinite.

As a specific example, consider the circuit at the top of Fig. 3–5, where $R_G = 200$ ohms, $R_L = 25$ ohms, and $Z_0 = 100$ ohms. When the switch is closed at $t = 0$, the 90 V source sees 200 ohms in series with the characteristic impedance of

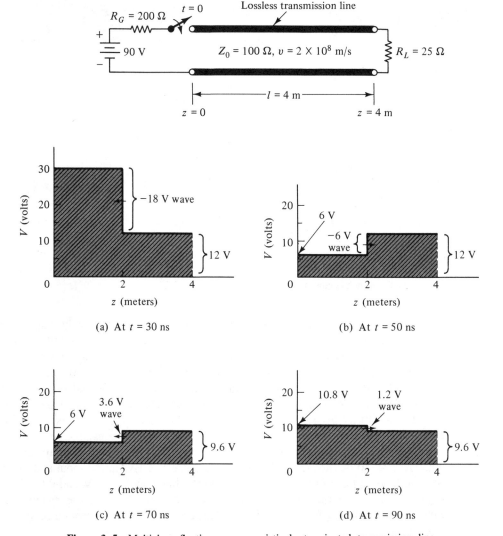

(a) At $t = 30$ ns

(b) At $t = 50$ ns

(c) At $t = 70$ ns

(d) At $t = 90$ ns

Figure 3–5 Multiple reflections on a resistively terminated transmission line. $Z_0 = 100$ ohms, $R_G = 200$ ohms, and $R_L = 25$ ohms. (The current waves are not shown.)

the line. Therefore, the current and voltage at the input end of the line ($z = 0$) are initially $I^+ = 90/300 = 0.3$ A and $V^+ = I^+ Z_0 = 30$ V. After 20 ns, the V^+ and I^+ waves arrive at the load end where the reflection coefficient $\Gamma_L = -75/125 = -0.6$ and hence $V^- = \Gamma_L V^+ = -18$ V and $I^- = \Gamma_L I^+ = -0.18$ A. At the end of 30 ns, the voltage between $z = 2$ m and $z = 4$ m is reduced to $30 - 18 = 12$ V, while the current has increased to $0.3 + 0.18 = 0.48$ A. The progress of the voltage wave along the line is shown in Fig. 3–5 for $t = 30, 50, 70,$ and 90 ns. Let us observe the voltage wave as time marches on. At the end of 40 ns, the -18 V wave arrives at the input where it sees an impedance $R_G = 200$ ohms. Since $R_G \neq Z_0$, a reflection occurs at the generator end. By analogy with Γ_L, the generator reflection coefficient Γ_G is given by

$$\Gamma_G = \frac{R_G - Z_0}{R_G + Z_0} \tag{3–11}$$

For $R_G = 200$ ohms, $\Gamma_G = 1/3$ and hence a -6 V wave is rereflected toward the load end. At $t = 50$ ns, it has progressed halfway down the line, leaving behind it a voltage of $(30 - 18 - 6) = 6$ V. This is shown in part *b* of the figure. At $t = 60$ ns, the -6 V wave arrives at the load which generates a reflected wave of value $(-6)\Gamma_L = +3.6$ V. The situations at 70 and 90 ns are also shown in the figure. Note that at $t = 90$ ns, another forward traveling wave exists having a value $(+3.6)\Gamma_G = +1.2$ V. This process continues indefinitely with the amplitude of the rereflected waves getting smaller and smaller. A plot of voltage versus time at any fixed point on the line would show that, in the limit, the voltage becomes the expected dc value (namely, $90 R_L/(R_G + R_L) = 10$ V). Such a plot at $z = 0$, the input, is shown in Fig. 3–6. Every step in voltage represents the arrival and generation of reflected waves at the input. After five *round trips* (200 ns), the voltage is within 0.10 percent of the steady-state value.

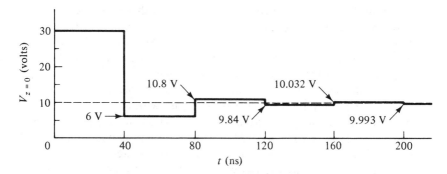

Figure 3–6   Input voltage versus time for the line shown in Fig. 3–5.

The space-time diagram developed by Bewley (Ref. 3–6) is a graphic aid in determining the voltage and current as a function of either time or position along the line. Figure 3–7 shows the diagram for the circuit conditions in Fig. 3–5. The abscissa indicates position along the line and the ordinate represents the time scale, $t = 0$ being the moment that the switch is closed. For reference, the values of Γ_G and Γ_L are given at the top of the diagram. The lines sloping downward and to the

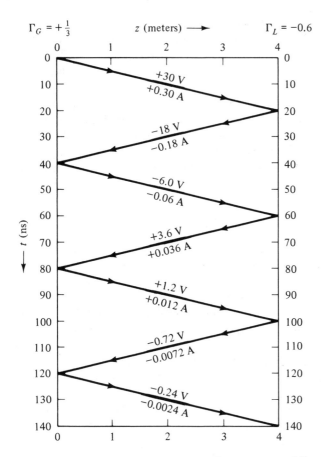

$\Gamma_G = +\frac{1}{3}$ z (meters) $\longrightarrow$ $\Gamma_L = -0.6$

+30 V
+0.30 A

−18 V
−0.18 A

−6.0 V
−0.06 A

+3.6 V
+0.036 A

+1.2 V
+0.012 A

−0.72 V
−0.0072 A

−0.24 V
−0.0024 A

Figure 3–7 Space-time diagram for the transmission-line circuit shown in Fig. 3–5.

right represent forward traveling waves, while those sloping down and to the left represent reverse waves. The voltage and current values for the particular wave are shown above and below the sloping line. As explained, the load end creates reflections equal to Γ_L of the arriving wave. Generator reflections are equal to Γ_G times the value of the wave arriving at the generator end.

To illustrate, Fig. 3–7 will be used to determine the voltage and current at $z = 2$ m. Each intersection of a sloping line with the vertical $z = 2$ m line represents the arrival of a wavefront. For $t < 10$ ns, no intersection exists and hence both V and I are zero. For $10 < t < 30$ ns, there is one intersection which means $V = 30$ V and $I = 0.30$ A. For $t > 30$ ns, the voltage is the sum of all the forward and reverse waves that have passed the $z = 2$ m location. For example, at $t = 80$ ns, $V = 30 - 18 - 6 + 3.6 = 9.6$ V. The current may be determined in a similar manner except that current values associated with reverse waves must be subtracted from those associated with the forward waves. For example, at $t = 80$ ns, $I = 0.30 - (-0.18) + (-0.06) - (+0.036) = 0.384$ A. The diagram may also be used to determine voltage and current versus z for a fixed time by drawing a horizontal line corresponding to the particular value of time. The sum of the voltages above the line correspond to the voltage at that point on the line. The same applies to the current except that, as before, reverse-traveling current waves must be sub-tracted from forward-traveling current waves.

The space-time diagram may be extended to transmission lines having disconti-nuities and branches (Chapter 3 of Ref. 3–7).

It is interesting to note that the voltage shown in Fig. 3–6 is oscillatory as it approaches its final value. The period of this *ringing* effect is 80 ns (twice the *round-trip* time) and hence its reciprocal is the natural resonant frequency of the circuit, namely, 12.5 MHz. Since $v = 2 \times 10^8$ m/s, this means that the line is $\lambda/4$ long at the resonant frequency. Thus we see that by connecting a dc source to a transmission line, high frequency oscillations are possible. Granted, the oscillation is heavily damped in this example, but the damping can be reduced by increasing the magni-tude of both reflection coefficients. In fact, if they are both unity, the oscillation will continue indefinitely (Prob. 3–6). In other words, a configuration consisting of two large reflections separated by a length of transmission line has the properties of a resonant circuit. Most of the microwave resonators described in Chapter 9 have exactly this configuration.

3–3 SINUSOIDAL EXCITATION OF TRANSMISSION LINES

Let us now turn to the important case of uniform transmission lines with sinusoidal excitation. Since our interest is in the steady-state solution, the rms-phasor method, reviewed in Sec. 1–4, will be employed.

Equations (3–1) resulted from a distributed circuit analysis of the uniform transmission line described in Fig. 3–1. Written in phasor form, they become

$$-\frac{d\mathbf{V}}{dz} = (R' + j\omega L')\mathbf{I} = Z'\mathbf{I} \quad \text{and} \quad -\frac{d\mathbf{I}}{dz} = (G' + j\omega C')\mathbf{V} = Y'\mathbf{V} \qquad (3\text{--}12)$$

where $Z' \equiv R' + j\omega L'$ is defined as the series impedance per unit length and $Y' \equiv G' + j\omega C'$ is defined as the shunt admittance per unit length.

Differentiating the first equation with respect to z and substituting $-Y'\mathbf{V}$ for $d\mathbf{I}/dz$ yields the following second-order differential equation.

$$\frac{d^2\mathbf{V}}{dz^2} = Z'Y'\mathbf{V} \qquad (3\text{--}13)$$

Its phasor solution may be written as

$$\mathbf{V} = \mathbf{V}_0^+ e^{-\gamma z} + \mathbf{V}_0^- e^{+\gamma z} = \mathbf{V}^+ + \mathbf{V}^- \qquad (3\text{--}14)$$

where γ is the propagation constant and given by

$$\gamma = \sqrt{Z'Y'} = \sqrt{(R' + j\omega L')(G' + j\omega C')} \qquad (3\text{--}15)$$

In general, γ is complex and may be written as $\gamma = \alpha + j\beta$, where as explained in Sec. 2–6 α is the attenuation constant (Np/length) and β is the phase constant (rad/length).

The phasor quantities $\mathbf{V}^+$ and $\mathbf{V}^-$ are functions of z and represent the forward and reverse voltage waves on the line. $\mathbf{V}_0^+$ and $\mathbf{V}_0^-$ represent their values at $z = 0$, the input end.[4] Substitution of Eq. (3–14) into the first of Eqs. (3–12) yields the accompanying solution for $\mathbf{I}$.

$$\mathbf{I} = \mathbf{I}_0^+ e^{-\gamma z} - \mathbf{I}_0^- e^{+\gamma z} = \mathbf{I}^+ - \mathbf{I}^- \tag{3–16}$$

where $\mathbf{I}_0^+ = \mathbf{V}_0^+ / Z_0$ and $\mathbf{I}_0^- = \mathbf{V}_0^- / Z_0$.

The quantity Z_0 is known as the *characteristic impedance* of the transmission line and is given by

$$Z_0 \equiv \sqrt{\frac{Z'}{Y'}} = \sqrt{\frac{R' + j\omega L'}{G' + j\omega C'}} \qquad \text{ohms} \tag{3–17}$$

It is the voltage-to-current ratio for the traveling waves. That is,

$$Z_0 = \frac{\mathbf{V}^+}{\mathbf{I}^+} = \frac{\mathbf{V}^-}{\mathbf{I}^-} \tag{3–18}$$

The reciprocal of Z_0 is defined as the characteristic admittance (Y_0) of the line and therefore

$$Y_0 \equiv \frac{1}{Z_0} = \sqrt{\frac{Y'}{Z'}} \qquad \text{mhos} \tag{3–19}$$

Since the voltages and currents in Eqs. (3–14) and (3–16) are phasor quantities, they are generally complex and depend upon the specific conditions at the generator and load. The determination of their values will be discussed in the next section.

From ac theory, the net average power flow at any point on the line is related to $\mathbf{V}$ and $\mathbf{I}$ by

$$P = \text{Re}(\mathbf{V} \, \mathbf{I}^*) = V \, I \cos \theta_{\text{pf}} \tag{3–20}$$

where * denotes the complex conjugate and θ_{pf} the power factor angle. V and I represent rms values.

Example 3–1:

A coaxial line has the following characteristics at 1000 MHz:
$R' = 4$ ohms/m, $L' = 450$ nH/m, $G' = 7 \times 10^{-4}$ mho/m, $C' = 50$ pF/m.
(a) Calculate Z_0, α, β, v, and λ at 1000 MHz.
(b) With $\mathbf{V}_0^+ = 10\underline{/0}$ V and $\mathbf{V}_0^- = 0$, calculate $\mathbf{V}$, $\mathbf{I}$, and P at $z = 4$ m.

Solution:
(a) At 1000 MHz,

$$Z' = 4 + j(2\pi \times 10^9)(450 \times 10^{-9}) = 4 + j2827 \quad \text{ohms/m}$$

and $Y' = 7 \times 10^{-4} + j(2\pi \times 10^9)(50 \times 10^{-12}) = 10^{-4}(7 + j3142)$ mho/m.

Therefore, $Z_0 = \sqrt{Z'/Y'} = 95\underline{/0.023°}$ ohms

---

[4] It is important to understand the terminology being used. Quantities with $+$ and $-$ superscripts are associated with the traveling waves, while those without them represent the net value of the quantity at a particular point on the line.

and $\gamma = \alpha + j\beta = \sqrt{Z'Y'} = \sqrt{-888 + j3.24} = 0.054 + j29.8$.

Thus, $\alpha = 0.054$ Np/m   or   0.47 dB/m
and $\beta = 29.8$ rad/m   or   1707°/m.
From Eq. (2-56), $\beta = 2\pi/\lambda = \omega/v$ and hence

$$v = \frac{2\pi \times 10^9}{29.8} = 2.11 \times 10^8 \text{ m/s} \qquad \text{and} \qquad \lambda = \frac{2\pi}{29.8} = 0.21 \text{ m}$$

(b) Since $V_0^- = 0$, only the forward wave exists. That is,

$$\mathbf{V} = \mathbf{V}^+ = V_0^+ e^{-\alpha z} e^{-j\beta z} = 10e^{-\alpha z}\underline{/-\beta z}$$

For $z = 4$ m, $\alpha z = 0.22$ Np   and $\beta z = 119.2$ rad.
Therefore at $z = 4$ m,

$$\mathbf{V} = 8.03\underline{/-119.2 \text{ rad}} \quad \text{V}$$

With $V_0^- = 0$, $I_0^- = V_0^-/Z_0 = 0$ and therefore at $z = 4$ m,

$$\mathbf{I} = \frac{8.03\underline{/-119.2 \text{ rad}}}{Z_0} = 0.084\underline{/-119.2 \text{ rad}} \quad \text{A}$$

Since Z_0 is practically real, the power factor angle is zero and the average power at $z = 4$ m is 0.677 W. This compares to an input power of $10(10/95) = 1.053$ W, which translates to a 36 percent power loss over 4 m. This represents a medium quality coaxial line. Lines are commercially available with $\alpha < 0.10$ dB/m at 1000 MHz.

Low-loss lines.   Certain conclusions can be drawn from the above example. First of all, for low-loss lines at high frequencies, $R' \ll \omega L'$ and $G' \ll \omega C'$. In the microwave range, these inequalities are almost always true. Applying these approximations to Eqs. (3-17) and (3-15) yields

$$Z_0 \approx \sqrt{\frac{L'}{C'}} \qquad \text{ohms} \tag{3-21}$$

$$\alpha \approx \frac{R'}{2Z_0} + \frac{G'Z_0}{2} \qquad \text{Np/length} \tag{3-22}$$

$$\beta \approx \omega\sqrt{L'C'} \qquad \text{rad/length} \tag{3-23}$$

Since $\beta = \omega/v$, $f\lambda = v$ and $f\lambda_0 = c$,

$$v = \frac{1}{\sqrt{L'C'}} \tag{3-24}$$

and

$$\lambda = \lambda_0\frac{v}{c} = \frac{\lambda_0}{c\sqrt{L'C'}} \tag{3-25}$$

The above results may be summarized as follows: For low-loss transmission lines at high frequencies,

1. The equations for Z_0, v, β, and λ are approximately the same as those for loss-less lines.

2. Since Z_0 is practically real, the average power flow in a *traveling* wave at any point on the line is simply the product of the rms voltage and current at that point. That is, $P^+ = V^+ I^+$ and $P^- = V^- I^-$.

3. For single frequency signals, the only appreciable effect of finite R' and G' is the introduction of some attenuation to the voltage and current waves as they propagate.[5] Their amplitudes are reduced by $e^{-\alpha \Delta z}$ when they travel a distance Δz, and hence the power flow is attenuated by $e^{-2\alpha \Delta z}$.

These conclusions are similar to those for electromagnetic waves in a lossy dielectric (Sec. 2–6a). The reason is that associated with the sinusoidal voltage and current waves is a sinusoidal electromagnetic wave. The only difference is that unlike those discussed in Chapter 2, this electromagnetic wave is *guided* by the transmission-line structure.

Wave propagation in coaxial lines.   Consider the coaxial line whose cross section is shown in Fig. 3–8. Its inner conductor radius is denoted by a and the inner radius of the outer conductor by b. Also shown are the electric and magnetic field lines for the TEM mode.[6] Because of skin effect, practically all of the high-frequency currents are located near the surface of the conductors. The electric and magnetic fields are given by

$$E = \frac{V}{r \ln(b/a)} \quad \text{and} \quad H = \frac{I}{2\pi r} \quad (3\text{–}26)$$

where V is the voltage between the conductors and I is the current in each conductor. These results are derived from Exs. 2–1 and 2–2 in the previous chapter and are valid for $a \leq r \leq b$.

By using Eq. (2–60) and assuming sinusoidal fields, the average power flow for a traveling electromagnetic wave may be determined. In cylindrical coordinates, an element of surface in the transverse plane is $r \, d\phi \, dr$ and therefore for the coaxial line

$$P = \int_a^b \int_0^{2\pi} \frac{VI}{2\pi r^2 \ln (b/a)} r \, d\phi \, dr = VI$$

The meaning of this result is that the power flow associated with the voltage and current waves is merely another way of viewing electromagnetic propagation in the region *between* the conductors. Both points of view are equally valid and each is used where appropriate to describe various aspects of microwave transmission.

The reader may now suspect a relationship between the velocity given by Eq. (3–24) and that given by Eq. (2–53). For the coaxial line at high frequencies, $C = C'l$ and $L = L'l$ are given by Eqs. (2–21) and (2–31). Substitution of L' and

---

[5] Another consequence of finite R' and G' is that γ and Z_0 are frequency dependent. As a result, modulated signals and pulses become distorted as they propagate along the line.

[6] The transverse electromagnetic (TEM) mode of propagation is so designated because the direction of the electric and magnetic fields are transverse to the direction of propagation.

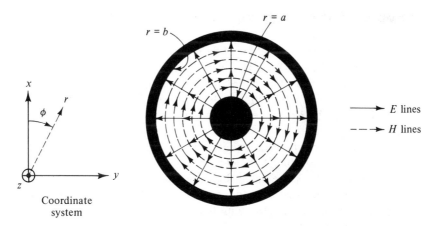

Figure 3-8   The TEM mode for a coaxial line at high frequencies. (Note: At high frequencies, the skin depth δ_s is much less than the inner conductor radius a and the thickness of the outer conductor. Therefore, the electromagnetic field is confined to the insulating region between the conductors.)

C' into Eq. (3-24) results in the following expression for the velocity of the voltage and current waves along a coaxial line.

$$v = \frac{1}{\sqrt{\mu_0 \mu_R \epsilon_0 \epsilon_R}} = \frac{c}{\sqrt{\mu_R \epsilon_R}} \qquad (3\text{-}27)$$

This is exactly the expression for velocity of an electromagnetic wave in an unbounded insulating material. In Eq. (3-27), μ_R and ϵ_R represent the relative permeability and dielectric constant of the insulating material between the two conductors. Also since $f\lambda = v$, the wavelength in the coaxial line is given by

$$\lambda = \frac{c}{f\sqrt{\mu_R \epsilon_R}} = \frac{\lambda_0}{\sqrt{\mu_R \epsilon_R}} \qquad (3\text{-}28)$$

which is the same as Eq. (2-55). With the phase constant $\beta = \omega/v$,

$$\beta = \frac{\omega}{c}\sqrt{\mu_R \epsilon_R} = \frac{2\pi}{\lambda} \qquad \text{rad/length} \qquad (3\text{-}29)$$

Let us now see if a relationship exists between the characteristic impedance of the coaxial line and the properties of the insulating material between the conductors. By using Eq. (3-21) and the above-mentioned relations for L' and C', the high-frequency characteristic impedance of a coaxial line may be written as

$$Z_0 = 60 \sqrt{\frac{\mu_R}{\epsilon_R}} \ln \frac{b}{a} = 138 \sqrt{\frac{\mu_R}{\epsilon_R}} \log \frac{b}{a} \qquad \text{ohms} \qquad (3\text{-}30)$$

where b and a are defined in Fig. 3-8. As expected, the characteristic impedance is indeed related to the properties of the insulating material.

Expressions for Z_0 of other type microwave lines are given in Chapter 5.

3-4 TERMINATED TRANSMISSION LINES

Figure 3–9a shows an infinite line driven by an ac source having an open-circuit voltage $\mathbf{V}_G$ and an internal impedance Z_G. For the infinite line, only forward traveling waves exist and therefore at $z = 0$, Eqs. (3–14) and (3–16) reduce to $\mathbf{V} = \mathbf{V}_0^+$ and $\mathbf{I} = \mathbf{I}_0^+ = \mathbf{V}_0^+/Z_0$. Thus the input impedance is equal to Z_0, the characteristic impedance of the line. The same result was obtained from the dc transient analysis in Sec. 3–2.

Suppose now that the infinite line is broken at $z = l$. Since the line to the right of $z = l$ is still infinite, its input impedance is Z_0 and therefore replacing it by a load impedance of the same value does not change any of the conditions to the left of $z = l$. This means that a finite length line *terminated in its characteristic impedance* is equivalent to an infinitely long line. Like the infinite line, a finite line terminated in Z_0 has no reflections. Also, its input impedance is equal to Z_0 and is independent of the line length l. As will be shown in the next section, this is *not* true for any other value of load impedance.

Referring to Fig. (3–9b), the input impedance is Z_0 and therefore

$$\mathbf{V}_{\text{in}} = \frac{Z_0}{Z_G + Z_0}\, \mathbf{V}_G \quad \text{and} \quad \mathbf{I}_{\text{in}} = \frac{\mathbf{V}_G}{Z_G + Z_0} \tag{3–31}$$

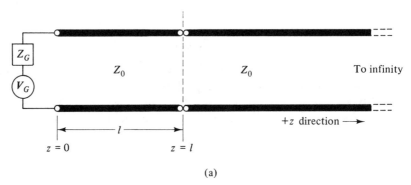

(a)

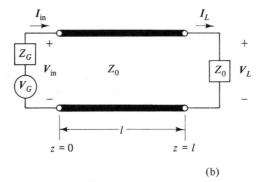

(b)

Figure 3–9   An infinite transmission line and its equivalent.

where $\mathbf{V}_{in}$ and $\mathbf{I}_{in}$ are the voltage and current phasors at $z = 0$.[7] For the reflectionless line, $\mathbf{V}_{in} = \mathbf{V}_0^+$, $\mathbf{I}_{in} = \mathbf{I}_0^+$ and therefore with the generator and line characteristics known, $\mathbf{V}_0^+$ and $\mathbf{I}_0^+$ may be determined. The voltage and current at any point on the line are readily calculated since $\mathbf{V} = \mathbf{V}_0^+ e^{-\gamma z}$ and $\mathbf{I} = \mathbf{I}_0^+ e^{-\gamma z}$. For example, at $z = l$, the load end

$$\mathbf{V}_L = \frac{Z_0}{Z_G + Z_0} \mathbf{V}_G \, e^{-\alpha l} \underline{/-\beta l} \quad \text{and} \quad \mathbf{I}_L = \frac{\mathbf{V}_G}{Z_G + Z_0} e^{-\alpha l} \underline{/-\beta l} \qquad (3\text{–}32)$$

The power absorbed by the load is given by $V_L I_L$ since with Z_0 real the power factor is unity. The product of the rms values yields

$$P_L = Z_0 \left| \frac{V_G}{Z_G + Z_0} \right|^2 e^{-2\alpha l} \qquad (3\text{–}33)$$

The magnitude sign is required since Z_G could be complex. Of course, if the line is lossless ($\alpha = 0$), P_L is exactly equal to P_{in}.

3–4a Lines Terminated in Z_L, the General Case

Figure 3–10 shows a transmission line terminated in an arbitrary load impedance Z_L. As explained in Sec. 3–2, reflections occur when $Z_L \neq Z_0$. The general expressions for voltage and current at any point on the line are given by Eqs. (3–14) and (3–16) and are repeated here.

$$\mathbf{V} = \mathbf{V}_0^+ e^{-\gamma z} + \mathbf{V}_0^- e^{+\gamma z} = \mathbf{V}^+ + \mathbf{V}^- \qquad (3\text{–}34)$$

$$\mathbf{I} = \mathbf{I}_0^+ e^{-\gamma z} - \mathbf{I}_0^- e^{+\gamma z} = \mathbf{I}^+ - \mathbf{I}^- \qquad (3\text{–}35)$$

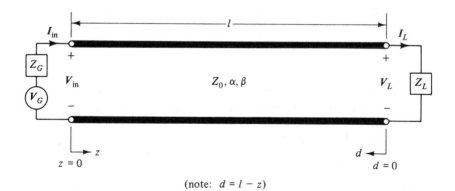

(note: $d = l - z$)

Figure 3–10 A uniform transmission line terminated in a load impedance Z_L and driven by an ac source.

---

[7] Subscripts denote electrical quantities at a specific point on the line, for example, *in* for input values and L for load values. The absence of subscripts indicate quantities at an arbitrary point on the line. Their values are generally a function of z.

The ratio of reflected to forward voltage (or current) is defined as the reflection coefficient Γ. That is,

$$\Gamma \equiv \frac{\text{Reflected Voltage (or current) at pt. } z}{\text{Forward Voltage (or current) at pt. } z} = \frac{\mathbf{V}^-}{\mathbf{V}^+} = \frac{\mathbf{I}^-}{\mathbf{I}^+} \qquad (3-36)$$

Since the voltage and current phasors are functions of position, Γ is a function of z. For example, at the load end, $z = l$ and

$$\Gamma_L = \frac{\mathbf{V}_0^- e^{+\gamma l}}{\mathbf{V}_0^+ e^{-\gamma l}} = \frac{\mathbf{V}_0^-}{\mathbf{V}_0^+} e^{2\gamma l} \qquad (3-37)$$

It is convenient to define a particular point on the line in terms of its distance from the load end. From Fig. 3–10, $d = l - z$ and therefore Eqs. (3–34) and (3–35) may be rewritten as

$$\mathbf{V} = \mathbf{V}_0^+ e^{-\gamma l}[e^{\gamma d} + \Gamma_L e^{-\gamma d}] \qquad (3-38)$$

$$\mathbf{I} = \mathbf{I}_0^+ e^{-\gamma l}[e^{\gamma d} - \Gamma_L e^{-\gamma d}] \qquad (3-39)$$

where use has been made of Eq. (3–37). Since $\mathbf{V}^+ = \mathbf{V}_0^+ e^{-\gamma l} e^{\gamma d}$ and $\mathbf{V}^- = \Gamma_L \mathbf{V}_0^+ e^{-\gamma l} e^{-\gamma d}$, the reflection coefficient at an arbitrary point d on the line is

$$\Gamma = \Gamma_L e^{-2\gamma d} \qquad (3-40)$$

which may be rewritten as

$$\Gamma = \Gamma_L e^{-2\alpha d} \underline{/-2\beta d} \qquad (3-41)$$

since $\gamma = \alpha + j\beta$. The reflection coefficient may be expressed in polar form as

$$\Gamma = |\Gamma| \underline{/\phi} = |\Gamma_L| e^{-2\alpha d} \underline{/\phi_L - 2\beta d} \qquad (3-42)$$

where $|\Gamma_L|$ and ϕ_L are the magnitude and angle of the load reflection coefficient. This equation relates the reflection coefficient at any point on the line to the load reflection coefficient. For instance, at the input terminals $d = l$ and therefore

$$\Gamma_{\text{in}} = |\Gamma_L| e^{-2\alpha l} \underline{/\phi_L - 2\beta l} \qquad (3-43)$$

This transformation of Γ_L by a length of transmission line may be interpreted in the following manner. The amplitude of the incident (forward) wave at the input is attenuated by $e^{-\alpha l}$ as it travels a distance l to the load. A fraction of this signal ($|\Gamma_L|$) is reflected and attenuated another $e^{-\alpha l}$ as it travels back to the input. Therefore the amplitude of reflected to forward voltage (or current) at the input is $|\Gamma_L| e^{-2\alpha l}$. Similarly, the phase of the incident wave is delayed βl as it travels to the load, phase shifted ϕ_L by the load and delayed another βl as it returns to the input. Thus, $\phi_{\text{in}} = \phi_L - 2\beta l$. If the line is lossless, $\alpha = 0$ and the magnitude of the reflection coefficient is the same at all points on the transmission line. The angle of the reflection coefficient ϕ, however, remains a function of position. Equation (3–42) is fundamental to the discussion of the impedance transformation and the Smith chart (Secs. 3–5 to 3–7).

When the value of Γ_L is known, Eq. (3–42) may be used to calculate the reflection coefficient at any other point on the line. The following analysis shows that Γ_L can be determined from a knowledge of Z_L and Z_0. From Eqs. (3-38) and (3-39) with $d = 0$, we have

$$\mathbf{V}_L = \mathbf{V}_0^+ e^{-\gamma l}[1 + \Gamma_L] \quad \text{and} \quad \mathbf{I}_L = \mathbf{I}_0^+ e^{-\gamma l}[1 - \Gamma_L]$$

Since $\mathbf{V}_0^+ = \mathbf{I}_0^+ Z_0$, the load impedance $(\mathbf{V}_L/\mathbf{I}_L)$ is

$$Z_L = Z_0 \frac{1 + \Gamma_L}{1 - \Gamma_L} \quad \text{or} \quad \bar{Z}_L = \frac{1 + \Gamma_L}{1 - \Gamma_L} \tag{3-44}$$

where the *bar* over the impedance denotes that it has been normalized with respect to the characteristic impedance of the line. Solving the above equation for Γ_L yields

$$\Gamma_L = \frac{Z_L - Z_0}{Z_L + Z_0} = \frac{\bar{Z}_L - 1}{\bar{Z}_L + 1} \tag{3-45}$$

Defining $Y_L \equiv 1/Z_L$, $Y_0 \equiv 1/Z_0$ and $\bar{Y}_L \equiv Y_L/Y_0$,

$$\Gamma_L = \frac{Y_0 - Y_L}{Y_0 + Y_L} = \frac{1 - \bar{Y}_L}{1 + \bar{Y}_L} \tag{3-46}$$

These equations are quite similar to Eqs. (3-9) and (3-10). In this case, they are based upon a steady-state ac analysis and hence apply to any complex impedance. This means that the reflection coefficient can be complex. The above equations show the relationship between the impedance concept of ac circuit theory and the reflection concept of wave theory.

If the impedance at any point on the line is defined as $Z \equiv \mathbf{V}/\mathbf{I}$, the above equations may be generalized in the following manner.[8]

$$Z = Z_0 \frac{1 + \Gamma}{1 - \Gamma} \quad \text{and} \quad Y = Y_0 \frac{1 - \Gamma}{1 + \Gamma} \tag{3-47}$$

where $Y \equiv 1/Z$. Solving for the reflection coefficient,

$$\Gamma = \frac{Z - Z_0}{Z + Z_0} = \frac{Y_0 - Y}{Y_0 + Y} \tag{3-48}$$

Equation (3-47) is derived by substituting Eq. (3-40) into Eqs. (3-38) and (3-39).

In Sec. 2-7, it was shown that a reflection causes standing waves. Thus if $Z_L \neq Z_0$, Γ_L is finite and standing waves of voltage and current exist along the transmission line. For a lossless line terminated in a short circuit, the patterns are the same as those in Fig. 2-24 and 2-25, where V replace E_x and I replaces H_y. As explained, successive minimums are $\lambda/2$ apart, while the distance between a maximum and an adjacent minimum is $\lambda/4$. Furthermore, the current pattern is shifted $\lambda/4$ relative to the voltage pattern. That is, wherever the rms voltage is maximum, the rms current is minimum (and vice versa).

[8] The impedance Z at any point on the line is the ratio of phasor voltage to phasor current at that point. Referring to Fig. 3-10, it represents the impedance of the circuit to the right of the point. To measure it, one would break the circuit at the desired point and connect an impedance bridge to the right hand part of the circuit. At $z = 0$, $Z = Z_{in}$, the input impedance of the complete transmission line-load impedance combination.

In general, the voltage at any point on the line is the phasor sum of the forward and reflected voltages ($\mathbf{V}^+ + \mathbf{V}^-$). In the lossless case, the change in $\mathbf{V}^+$ and $\mathbf{V}^-$ with position can be represented by two counterrotating vectors of fixed magnitudes. Figure 2–28 describes two such phasors (replace $\mathbf{E}_i$ and $\mathbf{E}_r$ by $\mathbf{V}^+$ and $\mathbf{V}^-$). At some point d, $\mathbf{V}^+$ and $\mathbf{V}^- = \Gamma \mathbf{V}^+$ are in phase, resulting in a voltage maximum. Its rms value is given by

$$V_{\max} = V^+ + |\Gamma|V^+ = \{1 + |\Gamma|\}V^+ \qquad (3\text{–}49)$$

One quarter wavelength from the maximum, the two signals are 180° out-of-phase resulting in a voltage minimum. Its rms value is

$$V_{\min} = \{1 - |\Gamma|\}V^+ \qquad (3\text{–}50)$$

For a lossless line, $|\Gamma|$ is the same for all values of d. As explained in Sec. 2–7, a standing-wave ratio (SWR) can be defined for the standing wave pattern.[9] That is,

$$\mathrm{SWR} \equiv \frac{V_{\max}}{V_{\min}} \qquad (3\text{–}51)$$

From the above expressions, the following useful equation is obtained.

$$\mathrm{SWR} = \frac{1 + |\Gamma|}{1 - |\Gamma|} \qquad (3\text{–}52)$$

By virtue of its definition, SWR is always equal to or greater than unity. If $|\Gamma| = 0$, no reflections exist and the SWR is unity, while for full reflection $|\Gamma| = 1.00$ and the SWR is infinite. Solving Eq. (3–52) for $|\Gamma|$ yields

$$|\Gamma| = \frac{\mathrm{SWR} - 1}{\mathrm{SWR} + 1} \qquad (3\text{–}53)$$

Note that in both the above equations, we are talking about the *magnitude* of Γ, which is always between zero and unity when Z_0 is real.[10] On a lossless line, all the maximums are identical and likewise all the minimums. Therefore the SWR is the same everywhere along the line. For a lossy line, this is not true and consequently the meaning and value of SWR becomes questionable (see Fig. 5.7 in Ref. 3–1).

To bring together some of the ideas discussed thus far, consider the following illustrative example.

Example 3–2:

Referring to Fig. 3–10, a 600 MHz generator with $V_G = 10$ V and $Z_G = 0$ is connected to a load impedance $Z_L = 150 + j90$ ohms via an air-insulated coaxial line. The line has a characteristic impedance of 75 ohms and is 15 cm long.
(a) Assuming $\alpha = 0$, calculate Γ_L, Γ_{in} and the SWR on the line.
(b) What is the maximum rms voltage on the line?
(c) Determine $|\Gamma_{\mathrm{in}}|$ if $\alpha = 2.0$ dB/m and the line is exactly one wavelength long.

---

[9] In some texts, SWR is written as VSWR (voltage standing-wave ratio). This can be misleading since SWR also equals $I_{\max}/I_{\min}$.

[10] An exception is when Re Z is negative. This situation occurs in negative-resistance devices such as tunnel diodes, parametric amplifiers, and masers.

Solution:

(a) $\bar{Z}_L = \dfrac{150 + j90}{75} = 2 + j1.2$ and from Eq. (3–45),

$$\Gamma_L = \frac{1 + j1.2}{3 + j1.2} = 0.48\underline{/28.4°}$$

At 600 MHz, $\lambda_0 = 50$ cm and since both μ_R and ϵ_R are unity, $\lambda = 50$ cm. Thus,

$$\beta l = (2\pi/50)15 = 0.6\pi \text{ rad or } 108°$$

and from Eq. (3–43) with $\alpha = 0$,

$$\Gamma_{in} = 0.48\underline{/28.4 - 2(108)} = 0.48\underline{/-187.6°}$$

From Eq. (3–52),

$$SWR = \frac{1 + 0.48}{1 - 0.48} = 2.85$$

(b) At the input $(d = l)$, Eq. (3–38) reduces to $\mathbf{V}_{in} = \mathbf{V}_0^+(1 + \Gamma_{in})$. With $Z_G = 0$, $\mathbf{V}_{in} = \mathbf{V}_G$, $\mathbf{V}_0^+ = \mathbf{V}_G/(1 + \Gamma_{in})$ and therefore

$$\mathbf{V}_0^+ = \frac{10\underline{/0}}{0.524 + j0.06} = 19\underline{/-6.5°} \text{ V}$$

The maximum rms voltage is given by

$$V_{max} = \{1 + |\Gamma|\}V^+$$

where for a lossless line $V^+ = V_0^+$ and $|\Gamma| = |\Gamma_{in}| = |\Gamma_L|$. Therefore,

$$V_{max} = (1.48)(19) = 28.12 \text{ V}$$

(c) For $\alpha = 2.0$ dB/m $= 0.23$ Np/m and $l = \lambda = 0.5$ m, $\alpha l = 0.115$ Np. From Eq. (3–43),

$$|\Gamma_{in}| = |\Gamma_L|e^{-2\alpha l} = 0.48e^{-0.23} = 0.381$$

In this example, $\mathbf{V}_G = \mathbf{V}_{in}$ since $Z_G = 0$. Generally, $Z_G \neq 0$ which makes solving for $\mathbf{V}_{in}$ and $\mathbf{V}_L$ a little more involved. This case is treated in part c of this section.

3–4b Some Special Cases of Terminated Lines

Let us now study four special cases of load impedances terminating a lossless transmission line, namely, $Z_L = 0$ (short circuit), $Z_L = \infty$ (open circuit), $Z_L = jX$ (pure reactance) and $Z_L = R_L$ (pure resistance).

Line terminated in a short circuit.   In this case $Z_L = 0$ and therefore $\Gamma_L = -1$ and the SWR is infinite. The standing wave patterns for rms voltage and current are the same as those in Fig. 2–24, where V replaced E_x and I replaces H_y. At $d = 0$ (the load end), $V = 0$, as it must be at a perfect short. Other voltage nulls occur at multiples of a half wavelength. Current, on the other hand, is a maximum at the load, its rms value being twice that of the incident current I_0^+. Because the voltage pattern is displaced with respect to the current pattern, the ratio of V to I is a function of d. For instance, at $d = \lambda/4$, V is a maximum and I is zero, which means that the impedance at that point is infinite. Thus, a quarter wavelength of line transforms the short circuit into an open circuit! This impedance transforming property of a transmission line is discussed in Secs. 3–5 and 3–6.

Line terminated in an open circuit.   For an open-circuited line, $Y_L = 0$ and therefore $\Gamma_L = +1$ and the SWR is infinite. The voltage and current standing wave patterns are shown in Fig. 3–11. At the open circuit, $I = 0$, by definition, while the rms voltage is a maximum and equal to $2V_0^+$. Note that a quarter wavelength from the open, $V = 0$ and therefore the impedance at that point is zero, a short circuit. As always on a lossless line, a particular rms value of voltage or current reoccurs every half wavelength.

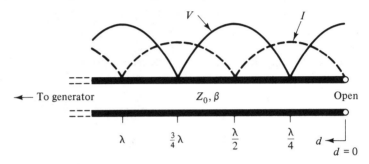

Figure 3–11   Standing-wave patterns on an open-circuited transmission line. ($\alpha = 0$.)

Reactively terminated lines.   For a transmission line terminated in a purely reactive circuit, $Z_L = jX$ and therefore

$$\Gamma_L = \frac{j\overline{X} - 1}{j\overline{X} + 1} = 1\underline{/\pi - 2 \arctan \overline{X}} \tag{3–54}$$

where $\overline{X} \equiv X/Z_0$ is the normalized load reactance. For example, if $\overline{X} = +1$, $\Gamma_L = 1\underline{/90°}$, while for $\overline{X} = -1$, $\Gamma_L = 1\underline{/-90°}$.

Note that for a pure reactance, $|\Gamma_L| = 1$ and is independent of the value of $\overline{X}$. Thus the SWR is infinite. The physical interpretation is that in the steady state, a pure reactance cannot absorb power and hence all of the incident wave must be reflected. This same argument applies to the short and open-circuited cases. Consequently, $|\Gamma_L| < 1$ only when Z_L has a resistive component.

The standing wave patterns for $Z_L = jZ_0$ ($\overline{X} = +1$) are shown in Fig. 3–12. It is left to the reader to show that the first voltage null occurs at $d = 3\lambda/8$ (Prob. 3–17). This can be verified analytically or with the aid of a phasor diagram. The use of counterrotating phasors to determine the standing wave pattern was discussed in Sec. 2–7b and Fig. 2–28.

Resistively terminated lines.   In many applications, the transmission line is terminated in a purely resistive network. That is, $Z_L = R_L$ and therefore $\Gamma_L = (R_L - Z_0)/(R_L + Z_0)$. Since R_L and Z_0 are real, Γ_L must be real. When $R_L = Z_0$, no

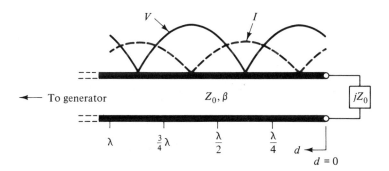

Figure 3–12 Standing-wave patterns for a reactively terminated line with $Z_L = +jZ_0$. ($\alpha = 0$.)

reflections occur. For $R_L > Z_0$, Γ_L is positive and a voltage maximum (current minimum) exists at the load. In this case,

$$\text{SWR} = \frac{R_L}{Z_0} \quad \text{since} \quad |\Gamma_L| = \frac{R_L - Z_0}{R_L + Z_0}$$

For $R_L < Z_0$, Γ_L is negative and a voltage minimum (current maximum) exists at the load. In this case,

$$\text{SWR} = \frac{Z_0}{R_L} \quad \text{since} \quad |\Gamma_L| = \frac{Z_0 - R_L}{Z_0 + R_L}$$

Thus if Z_L is *purely resistive*,

$$\text{SWR} = \frac{R_L}{Z_0} \quad \text{or} \quad \frac{Z_0}{R_L} \quad \text{whichever is greater than unity.} \qquad (3\text{–}55)$$

For any finite value of R_L, SWR is finite and $|\Gamma_L| < 1$. This means that some of the incident power is absorbed by the load.

3–4c Power Flow Along Terminated Lines

The general problem of power flow along a terminated transmission line will now be discussed. Figure 3–10 shows a line of length l driven by an ac source and terminated in a load Z_L. V_G is the open-circuit voltage of the generator and Z_G is its internal impedance. From ac theory, the average power flow into an impedance Z is given by

$$P = \text{Re} (\mathbf{VI}^*) = VI \cos \theta_{\text{pf}} \qquad (3\text{–}56)$$

where V and I are rms values, θ_{pf} is the power-factor angle and * denotes the complex conjugate. This equation also applies to the average power flow at any point along a transmission line, the direction of flow being from the generator toward the load. The voltage and current on the line are given by Eqs. (3–14) and (3–16). With $\mathbf{V}^+ = \mathbf{I}^+ Z_0$ and $\mathbf{V}^- = \Gamma\mathbf{V}^+$, they may be rewritten as

$$\mathbf{V} = \mathbf{V}^+(1 + \Gamma) \quad \text{and} \quad \mathbf{I} = \frac{\mathbf{V}^+}{Z_0}(1 - \Gamma) \qquad (3\text{–}57)$$

Substitution into Eq. (3–56) yields

$$P = \text{Re}\left[(1 - |\Gamma|^2 + \Gamma - \Gamma*)\frac{(V^+)^2}{Z_0^*}\right] \tag{3–58}$$

since $\Gamma\Gamma* = |\Gamma|^2$ and $\mathbf{V}^+\mathbf{V}^{+*} = (V^+)^2$. For low-loss, high-frequency lines, Z_0 is real and the above expression reduces to

$$P = P^+(1 - |\Gamma|^2) = P^+ - P^- \tag{3–59}$$

since $\Gamma - \Gamma*$ is imaginary. $P^+ = (V^+)^2/Z_0$ is the average power in the forward traveling wave and $P^- = |\Gamma|^2 P^+$ is the average power in the reflected wave. Equation (3–59) states that for Z_0 real, the net power flow at any point on the line is simply the difference between the power in the forward wave (P^+) and that in the reflected wave (P^-), where

$$P^- = |\Gamma|^2 P^+ \tag{3–60}$$

The same interpretation was given to the net power density of a uniform plane wave Eq. (2–85). This method of determining net power in a microwave circuit is very useful. It is the basis of the *power-wave* formulation described in Appendix D.

Equation (3–59) can be applied to any point along the line. For example, at the input and load terminals,

$$P_{\text{in}} = P_{\text{in}}^+(1 - |\Gamma_{\text{in}}|^2) \quad \text{and} \quad P_L = P_L^+(1 - |\Gamma_L|^2) \tag{3–61}$$

where Γ_{in} and Γ_L are defined in Eqs. (3–43) and (3–45). The reader is reminded that the $+$ and $-$ superscripts indicate quantities associated with the traveling waves. Thus

$$P_{\text{in}}^+ = \frac{(V_0^+)^2}{Z_0} \quad \text{and} \quad P_L^+ = \frac{(V_L^+)^2}{Z_0} = P_{\text{in}}^+ e^{-2\alpha l} \tag{3–62}$$

where P_L^+, the incident power at the load terminals, is simply P_{in}^+ attenuated by the line length l. If the line is lossless, $P_{\text{in}}^+ = P_L^+ = P^+$.

In order to calculate the *net* power flow at any point on the line, one must first determine P^+ at that point. With Z_0 real, $\mathbf{V}^+$ and $\mathbf{I}^+$ are in phase and therefore

$$P^+ = \frac{(V^+)^2}{Z_0} = \frac{(V_0^+)^2}{Z_0} e^{-2\alpha z} \tag{3–63}$$

V_0^+ (and its rms value V_0^+) is obtained by applying Kirchhoff's voltage law at the input end. From Fig. 3–10,

$$\mathbf{V}_G = \mathbf{V}_{\text{in}} + \mathbf{I}_{\text{in}} Z_G = (\mathbf{V}_0^+ + \Gamma_{\text{in}} \mathbf{V}_0^+) + \frac{Z_G}{Z_0}(\mathbf{V}_0^+ - \Gamma_{\text{in}} \mathbf{V}_0^+)$$

Solving for $\mathbf{V}_0^+$, the forward traveling voltage wave at the input,

$$\mathbf{V}_0^+ = \frac{\mathbf{V}_G Z_0}{(Z_G + Z_0)(1 - \Gamma_G \Gamma_{\text{in}})} = \frac{\mathbf{V}_G Z_0}{(Z_G + Z_0)(1 - \Gamma_G \Gamma_L e^{-2\gamma l})} \tag{3–64}$$

where Γ_G, the reflection coefficient of the generator, is defined as

$$\Gamma_G = \frac{Z_G - Z_0}{Z_G + Z_0} \tag{3–65}$$

Thus, given the generator and load conditions (V_G, Z_G, Z_L) and the line characteristics (Z_0, γ, l), V_0^+ may be determined. An alternate form for Eq. (3–64) is

$$V_0^+ = \frac{V_G(1 - \Gamma_G)}{2(1 - \Gamma_G\Gamma_{in})} = \frac{V_G(1 - \Gamma_G)}{2(1 - \Gamma_G\Gamma_L e^{-2\gamma l})} \tag{3–66}$$

Once V_0^+ is known, the power flow can be determined at any point along the line.

Some insight into the meaning of V_0^+ may be obtained by deriving Eq. (3–64) from a multiple reflection point of view. In Sec. 3–2, it was shown that the steady-state voltage and current values could be determined by considering the limiting process of multiple reflections. On the other hand, Eqs. (3–64) and (3–66) were derived from a steady-state ac analysis. We will now show that these two methods of analysis are merely alternate ways of viewing the same phenomena.

First, consider the special case when either Γ_G or Γ_L is zero (that is, no multiple reflections). Since the input impedance of the line *initially* appears to be Z_0, the voltage of the *initial* forward traveling wave is determined by the voltage-divider action between Z_G and Z_0. In the absence of multiple reflections, this is the only forward traveling wave and hence $V_0^+ = (V_G Z_0)/(Z_G + Z_0)$. This is exactly the result one obtains from Eq. (3–64) with either Γ_G or Γ_L equal to zero.

Let us now consider the situation in which neither Γ_G nor Γ_L is zero. In this case, multiple reflections occur and V_0^+ *represents the phasor sum of all the forward traveling waves* at the input terminals. To verify this, Eq. (3–64) will now be derived from a multiple reflection point of view. As before, the *initial* forward traveling wave is $(V_G Z_0)/(Z_G + Z_0)$. The next forward voltage wave is the portion of the initial wave that has been reflected at the load and rereflected at the generator. Its value is

$$(\Gamma_G\Gamma_L e^{-2\gamma l})(V_G Z_0)/(Z_G + Z_0).$$

By continuing this process, the sum of all the forward traveling voltage waves at the input becomes

$$V_0^+ = \frac{V_G Z_0}{Z_G + Z_0}[1 + \Gamma_G\Gamma_L e^{-2\gamma l} + (\Gamma_G\Gamma_L e^{-2\gamma l})^2 + \cdots]$$

For $|\Gamma_G\Gamma_L e^{-2\gamma l}| < 1$, the infinite series converges to $(1 - \Gamma_G\Gamma_L e^{-2\gamma l})^{-1}$ and therefore the above expression reduces to Eq. (3–64). Thus the multiple reflection viewpoint is consistent with that based upon a steady-state ac analysis.

It is useful and informative to have expressions for P^+ and P in terms of available generator power P_A and reflection coefficients.[11] These may be derived with the

---

[11] From circuit theory, available power (P_A) is the maximum power that a source can deliver to a passive load. This occurs when the load impedance equals the complex conjugate of Z_G.

aid of Eqs. (3–59), (3–63), and (3–66). By substituting the rms value of V_0^+ into Eq. (3–63), we obtain

$$P^+ = \frac{V_G^2}{4Z_0} \left| \frac{1 - \Gamma_G}{1 - \Gamma_G \Gamma_{\text{in}}} \right|^2 e^{-2\alpha z} \tag{3–67}$$

At the input $z = 0$,

$$P_{\text{in}}^+ = \frac{V_G^2}{4Z_0} \left| \frac{1 - \Gamma_G}{1 - \Gamma_G \Gamma_{\text{in}}} \right|^2 \tag{3–68}$$

where $\Gamma_{\text{in}} = \Gamma_L e^{-2\gamma l}$. When the line is lossless ($\alpha = 0$), $P^+ = P_{\text{in}}^+$ at any point along the line, including the load end.

The above equation may be rewritten in terms of the available power of the generator since $P_A = V_G^2/4R_G$, where R_G is the resistive portion of Z_G. Thus

$$P_{\text{in}}^+ = P_A \frac{1 - |\Gamma_G|^2}{|1 - \Gamma_G \Gamma_{\text{in}}|^2} \tag{3–69}$$

where use has been made of the following relations:

$$R_G = \frac{Z_G + Z_G^*}{2}, \quad Z_G = Z_0 \frac{1 + \Gamma_G}{1 - \Gamma_G}, \quad Z_G^* = Z_0 \frac{1 + \Gamma_G^*}{1 - \Gamma_G^*}$$

and the fact that $|1 - \Gamma_G| = |1 - \Gamma_G^*|$.

An expression for the *net* input power may be obtained by combining Eqs. (3–61) and (3–69).

$$P_{\text{in}} = P_A \frac{(1 - |\Gamma_G|^2)(1 - |\Gamma_{\text{in}}|^2)}{|1 - \Gamma_G \Gamma_{\text{in}}|^2} \tag{3–70}$$

The net power delivered to the load is

$$P_L = P_A \frac{(1 - |\Gamma_G|^2)(1 - |\Gamma_L|^2)}{|1 - \Gamma_G \Gamma_L e^{-2\gamma l}|^2} e^{-2\alpha l} \tag{3–71}$$

where use has been made of Eqs. (3–61), (3–62), and (3–69). When the line is lossless, $P_{\text{in}} = P_L$ and the above equations reduce to

$$P_{\text{in}} = P_L = P_A \frac{(1 - |\Gamma_G|^2)(1 - |\Gamma_L|^2)}{|1 - \Gamma_G \Gamma_L e^{-j2\beta l}|^2} \tag{3–72}$$

since $|\Gamma_{\text{in}}| = |\Gamma_L|$ and $\Gamma_{\text{in}} = \Gamma_L e^{-j2\beta l}$.

These equations are very important in understanding power flow in a microwave circuit. For example, if $Z_G = Z_0$, $\Gamma_G = 0$ and therefore for the lossless line case, P_L is independent of the phase constant $\beta = \omega/v$ and the length of the transmission line l. This is a distinct practical advantage since it means that load power is insensitive to changes in generator frequency and line length. Designing the circuit so that $Z_G = Z_0$ is referred to as *matching the generator to the line*. Well-designed microwave systems usually satisfy this condition. With $\Gamma_G = 0$, Eqs. (3–66) and (3–69) reduce to

$$V_0^+ = \frac{V_G}{2} \quad \text{and} \quad P_{\text{in}}^+ = P_A = \frac{V_G^2}{4Z_0} \tag{3–73}$$

Thus, with the generator matched to the line, the power associated with the incident wave at the input is simply the available generator power. As a result, Eqs. (3–70) and (3–71) reduce to

$$P_{in} = P_A(1 - |\Gamma_{in}|^2) \quad \text{and} \quad P_L = P_A e^{-2\alpha l}(1 - |\Gamma_L|^2) \tag{3–74}$$

For a lossless line,

$$P_{in} = P_L = P_A(1 - |\Gamma_L|^2) \tag{3–75}$$

Note that if the line is lossless and the generator is matched, matching the load to the line ($\Gamma_L = 0$) results in all the available generator power being delivered to the load. However if $Z_G \neq Z_0$, matching the load to the line does *not* result in $P_L = P_A$. This can be seen from Eq. (3–72). In fact, maximizing P_L requires a standing wave (that is, $\Gamma_L \neq 0$) and proper adjustment of the line length l. Examples of this are the half-wave line and the quarter-wave transformer discussed in Sec. 3–6.

The following brief example shows the usefulness of the above equations.

Example 3–3:

The transmission line circuit in Fig. 3–10 has the following parameters: $V_G = 20$ V rms, $Z_G = 100$ ohms, $f = 500$ MHz, $Z_0 = 100$ ohms, $l = 4$ m and $Z_L = 150$ ohms. Calculate the input and load power if

$$\text{(a)} \quad \alpha = 0 \qquad \text{(b)} \quad \alpha = 0.5 \text{ dB/m}.$$

Solution: The available power is $P_A = (20)^2/400 = 1.0$ W. From Eq. (3–45),
$$\Gamma_L = \frac{150 - 100}{150 + 100} = 0.20.$$
(a) With $Z_G = Z_0$ and $\alpha = 0$, P_{in} and P_L are given by Eq. (3–75). Therefore,

$$P_{in} = P_L = (1.0)[1 - (0.20)^2] = 0.96 \text{ W}.$$

(b) For $\alpha = 0.5$ dB/m, $\alpha l = 2.0$ dB or 0.23 Np. From Eq. (3–43),

$$|\Gamma_{in}| = 0.20 \, e^{-2(0.23)} = 0.126.$$

P_{in} and P_L are determined from Eq. (3–74). Thus,

$$P_{in} = 0.984 \text{ W} \quad \text{and} \quad P_L = 0.605 \text{ W}.$$

The difference, 0.379 W, is dissipated in the lossy line.

Note that with $Z_G = Z_0$, the source frequency does not enter into the power calculations. However, if $Z_G \neq Z_0$, P_L is a function of βl as seen in Eq. (3–72). Problem 3–20 gives some indication of the variation P_L with βl.

Return and reflection losses.   Two terms commonly used in conjunction with reflected signals are *return loss* and *reflection loss*. Return loss, denoted by L_R, is defined as

$$L_R \equiv 10 \log \frac{P^+}{P^-} \quad \text{dB} \tag{3–76}$$

From Eq. (3–60), $P^- = |\Gamma|^2 P^+$ and therefore

$$L_R = 10 \log \frac{1}{|\Gamma|^2} \quad \text{dB} \tag{3–77}$$

If the line is lossless, the return loss is the same everywhere along the line since $|\Gamma|$ is independent of position. For a lossy line, L_R is a function of position since from Eq. (3–42), $|\Gamma| = |\Gamma_L|e^{-2\alpha d}$, where αd is in nepers. By using Eqs. (3–43) and (3–77), a relation between the return loss at the input and its value at the load is obtained.

$$L_{R_{\text{in}}} = L_{R_{\text{load}}} + 2(8.686\alpha l) \qquad (3\text{–}78)$$

where l is the length of the lossy line. This equation states that the input return loss equals the load return loss plus the *round-trip* attenuation (in dB) due to the lossy line.

The concept of reflection loss is generally used only when $Z_G = Z_0$. It is a measure of the reduction in load power due to the load impedance having a value other than Z_0. That is,

$$\text{REF. LOSS} \equiv 10 \log \frac{P_L \text{ if } Z_L = Z_0}{P_L \text{ in } Z_L} \quad \text{dB} \qquad (3\text{–}79)$$

For $Z_G = Z_0$, the denominator is given by Eq. (3–74) while the numerator equals $P_A e^{-2\alpha l}$ and therefore

$$\text{REF. LOSS} = 10 \log \frac{1}{1 - |\Gamma_L|^2} = 10 \log \frac{(\text{SWR} + 1)^2}{4(\text{SWR})} \qquad (3\text{–}80)$$

For example, if $|\Gamma_L| = 0.707$ (an SWR of 5.83), the reflection loss is 3 dB, which for a lossless line means P_L is half the available generator power. Reflection loss is sometimes referred to as *mismatch loss*. This concept is extended in Chapter 4 to include any linear two-port network.

3–5 THE IMPEDANCE TRANSFORMATION

In discussing the short-circuited line, it was observed that the impedance became infinite at a distance one quarter wavelength from the short. The ability to change impedance by adding a length of transmission line is very important to the microwave engineer. In this section, the general impedance transformation equation is derived. Examples of its effect and use are given in this and the following section. The equation involves hyperbolic functions with complex arguments. Appendix E contains a brief review of these functions.

Derivation of the impedance transformation equation.   In the previous section, it was shown that the reflection coefficient is a function of position along the transmission line (Eq. 3–40). Since the reflection coefficient is related to impedance via Eq. (3–47), it is apparent that impedance must also be a function of position.[12]

---

[12] It is suggested that the reader review the definition of Z given in the footnote accompanying Eq. (3–47).

This relationship is called the impedance transformation equation and will now be derived. From Eqs. (3–47) and (3–40),

$$Z = Z_0 \frac{1 + \Gamma}{1 - \Gamma} = Z_0 \frac{1 + \Gamma_L e^{-2\gamma d}}{1 - \Gamma_L e^{-2\gamma d}}$$

Making use of Eq. (3–45),

$$Z = Z_0 \frac{(Z_L + Z_0)e^{\gamma d} + (Z_L - Z_0)e^{-\gamma d}}{(Z_L + Z_0)e^{\gamma d} - (Z_L - Z_0)e^{-\gamma d}}$$

Rearranging terms and using Eq. (E–3) in Appendix E yields the impedance transformation equation.

$$Z = Z_0 \frac{Z_L + Z_0 \tanh \gamma d}{Z_0 + Z_L \tanh \gamma d} \qquad \text{or} \qquad \bar{Z} = \frac{\bar{Z}_L + \tanh \gamma d}{1 + \bar{Z}_L \tanh \gamma d} \qquad (3\text{–}81)$$

where

$$\gamma = \alpha + j\beta, \qquad \bar{Z} \equiv Z/Z_0 \qquad \text{and} \qquad \bar{Z}_L \equiv Z_L/Z_0.$$

The transformation equation can also be expressed in terms of admittance. Namely,

$$Y = Y_0 \frac{Y_L + Y_0 \tanh \gamma d}{Y_0 + Y_L \tanh \gamma d} \qquad \text{or} \qquad \bar{Y} = \frac{\bar{Y}_L + \tanh \gamma d}{1 + \bar{Y}_L \tanh \gamma d} \qquad (3\text{–}82)$$

where

$$Y_0 \equiv 1/Z_0, \qquad \bar{Y} \equiv Y/Y_0 \qquad \text{and} \qquad \bar{Y}_L \equiv Y_L/Y_0.$$

Referring to Fig. 3–10, the input impedance or admittance is obtained by merely setting $d = l$.

In many practical situations, the line attenuation can be neglected and hence it is useful to write the above equations for the lossless case. For $\alpha = 0$, $\gamma d = j\beta d$ and from Eq. (E–6), $\tanh \gamma d$ becomes $j \tan \beta d$. Thus for the lossless line,

$$Z = Z_0 \frac{Z_L + jZ_0 \tan \beta d}{Z_0 + jZ_L \tan \beta d} \qquad \text{or} \qquad \bar{Z} = \frac{\bar{Z}_L + j \tan \beta d}{1 + j\bar{Z}_L \tan \beta d} \qquad (3\text{–}83)$$

and

$$Y = Y_0 \frac{Y_L + jY_0 \tan \beta d}{Y_0 + jY_L \tan \beta d} \qquad \text{or} \qquad \bar{Y} = \frac{\bar{Y}_L + j \tan \beta d}{1 + j\bar{Y}_L \tan \beta d} \qquad (3\text{–}84)$$

These transformation equations are quite remarkable. The following example shows how even a small length of low-loss line can produce a dramatic change in impedance.

Example 3–4:

A 15 cm length of air-insulated coaxial line is terminated in a short circuit. The characteristics of the line are $Z_0 = 75$ ohms and $\alpha = 0.4$ dB/m. Calculate the input impedance at 1500 MHz and 2000 MHz.

Solution: For the air-insulated line, $\mu_R = \epsilon_R = 1$ and therefore $\lambda = \lambda_0$. At 1500 MHz, $\lambda_0 = 20$ cm and therefore $\beta l = (2\pi/20)15 = 3\pi/2$ rad. Also, $\alpha l = (0.4)(0.15) = 0.06$ dB or 0.007 Np. Substituting $Z_L = 0$ and $d = l$ into Eq. (3–81) yields $Z_{in} = 75 \tanh (\alpha l + j\beta l)$. Using Eq. (E–6) and the fact that $\tanh \alpha l \approx \alpha l$,

$$Z_{in} = 75 \frac{0.007 + j \tan (3\pi/2)}{1 + j0.007 \tan (3\pi/2)}$$

Since $\tan (3\pi/2) = -\infty$, the input impedance equals $75/0.007 = 10,714$ ohms. Repeating the calculation at 2000 MHz ($\lambda_0 = 15$ cm), $\beta l = 2\pi$ rad and therefore $Z_{in} = 0.525$ ohms.

This example shows that a 15 cm section of low-loss line transforms a short circuit into 10,714 ohms at 1500 MHz! Furthermore, increasing the frequency by one-third reduces the input impedance to a mere 0.525 ohms. What is the explanation for these results? From a circuit point of view, one can say that because a transmission line contains distributed inductance and capacitance (Fig. 3–1), it is conceivable that their addition to the circuit will result in an impedance value radically different from the load impedance Z_L. In fact, impedance transformers consisting of series inductors and shunt capacitors are commonly used in radio engineering work.

An alternate way of explaining the impedance transformation effect is in terms of standing waves. For $Z_L \neq Z_0$, standing waves of voltage and current exist on the line. Because the current pattern is shifted relative to the voltage pattern, the ratio of **V** to **I** is a strong function of position d. Therefore, it *makes sense* that the impedance along the line (for example, at the input) can be significantly different from the load impedance. Conversely, without standing waves no impedance transformation can occur. To verify this, let $Z_L = Z_0$. In this case, $\Gamma_L = 0$, SWR = 1.00 and hence no standing waves exist. From Eq. (3–81), it is clear that with $Z_L = Z_0$, $Z = Z_0$, *independent* of d, which means no impedance transformation. The reason this transformation effect is not observed at low frequencies is that with $l \ll \lambda$, $\tan \beta l \approx 0$ and Eq. (3–81) reduces to $Z = Z_L$ for low-loss lines.

In the illustrative example given, an increase in frequency resulted in a large change in input impedance. The reason is that for a fixed length line, $\tan \beta l$ is a function of frequency since $\beta = \omega/v$. The fact that the input impedance is frequency dependent is a *two-edged* sword. It is an obstacle in designing broadband circuits, but is a very useful property in the design of microwave resonators and filters.

3–6 EXAMPLES OF THE IMPEDANCE TRANSFORMATION

In this section, several special cases of the impedance transformation are examined. All of the results obtained are directly applicable to a variety of microwave design problems. Situations in which the transmission line may be assumed lossless are considered first.

3–6a Impedance Transformations on a Lossless Line

For $\alpha = 0$, the impedance and admittance transformations are given by Eqs. (3–83) and (3–84). Let us now analyze some special cases of these equations.

The half-wavelength line. Figure 3–10 shows a transmission line connected to a load impedance Z_L. If the line length is a multiple of a half wavelength ($l = n\lambda/2$, n being any positive integer), the input impedance is equal to Z_L. This may be seen by substituting $\beta l = n\pi$ into Eq. (3–83). This result can be generalized to any two points that are separated by a multiple of a half wavelength. Since $\tan (\beta d + n\pi) = \tan \beta d$, impedance repeats every half wavelength along a lossless line.

An interesting case of the half-wavelength line occurs when $Z_G = Z_L$, both real. Suppose $Z_L = Z_G = 20$ ohms and $Z_0 = 100$ ohms. From Eq. (3–55), the SWR along the line is 5 to 1. With $l = n\lambda/2$, $Z_{\text{in}} = Z_L = 20$ ohms and therefore all the available generator power is delivered down the lossless line to the load impedance. Thus, despite the fact that standing waves exist on the line, maximum power is delivered to the load. The standing waves represent stored electric and magnetic energy, much like that in an L-C circuit. This situation was alluded to in Sec. 3–4c.

The quarter-wavelength line. Suppose now that the length of line shown in Fig. 3–10 is an odd multiple of a quarter wavelength. With $l = (2n - 1)\lambda/4$, n being a positive integer, $\tan \beta l \to \infty$ and Eq. (3–83) reduces to

$$Z_{\text{in}} = \frac{Z_0^2}{Z_L} \qquad \text{or} \qquad \bar{Z}_{\text{in}} = \frac{1}{\bar{Z}_L} \qquad (3\text{–}85)$$

This equation shows that a quarter-wave line transforms a large (small) value of load impedance into a small (large) value at the input. Also, an inductive (capacitive) load is transformed into a capacitive (inductive) input impedance. Furthermore, one can show that if Z_L is a series resonant circuit, Z_{in} behaves like a parallel resonant circuit, and vice versa. The quarter-wavelength line is often called an impedance inverter since it inverts the normalized load impedance. It is also referred to as a quarter-wave transformer since for Z_L real it has all the properties of an ideal transformer. Figure 3–13 shows the equivalence for specific values of Z_0 and Z_L. For the values given, the SWR along the line is five ($100/20$). In this example, $Z_L < Z_0$ and hence a voltage minimum and current maximum exist at the load. A quarter wavelength away at the input, the voltage is maximum and the current minimum. The values are given in part a of the figure. From Eq. (3–85), the input impedance is 500 ohms. Therefore, the lossless quarter-wavelength line behaves like an ideal transformer with a turns ratio $n_t = 5$, the SWR value. For $Z_L > Z_0$, the line is equivalent to a transformer with $n_t < 1$ or $N_1 < N_2$.

The quarter-wave transformer is useful in matching a resistive load to a generator, which is a necessary condition for delivering all the available generator power to the load. If the generator impedance is real, then Z_{in} must equal R_G for maximum

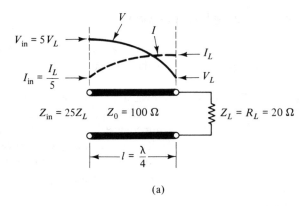

(a)

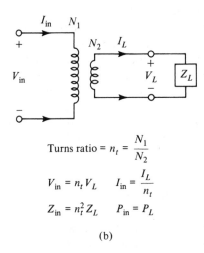

Turns ratio = $n_t = \dfrac{N_1}{N_2}$

$V_{in} = n_t V_L$ $I_{in} = \dfrac{I_L}{n_t}$

$Z_{in} = n_t^2 Z_L$ $P_{in} = P_L$

(b)

Figure 3–13 The quarter-wave transformer and its equivalent circuit when $Z_L = R_L$.

transfer of power. Solving Eq. (3–85) for Z_0 when the load is resistive ($Z_L = R_L$) yields the required value of characteristic impedance for the quarter-wave line. Namely,

$$Z_0 = \sqrt{R_G R_L} \qquad (3\text{–}86)$$

For example, a 200 ohm load can be matched to a 50 ohm generator by inserting a quarter-wavelength section of 100 ohm line between them. If the line is lossless, $P_{in} = P_L$ and therefore the available generator power (P_A) is delivered to the load. Since the line can only be a quarter-wavelength long at one frequency, the transformer is frequency sensitive. The frequency characteristics of single and multisection transformers are discussed in Sec. 4–2.

The short-circuited line.   When a lossless line is shorted at the load end, $Z_L = 0$ and from Eq. (3–83) the input impedance is

$$Z_{in} = jX_{in} = jZ_0 \tan \beta l \qquad \text{or} \qquad \bar{Z}_{in} = j\bar{X}_{in} = j \tan \beta l \qquad (3\text{–}87)$$

A plot of normalized input reactance versus βl is shown in Fig. 3–14. It is simply a graph of the tangent function. For $\beta l < \pi/2$ $(l < \lambda/4)$, the input impedance is inductive since X_{in} is positive. Thus $X_{in} = \omega L_{eq}$ and

$$L_{eq} = \frac{X_{in}}{\omega} = \frac{Z_0}{\omega} \tan \beta l \qquad \text{henries} \qquad (3\text{–}88)$$

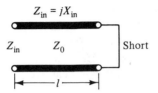

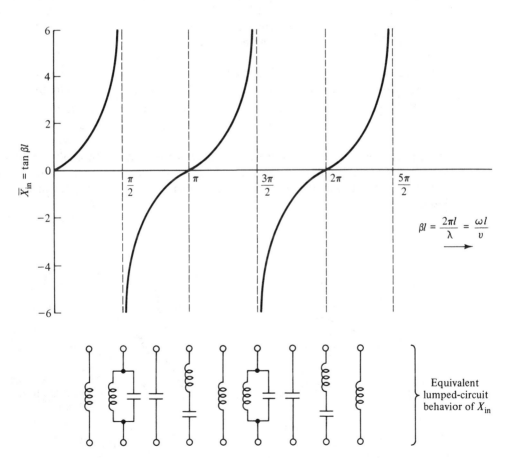

Figure 3–14 Normalized input reactance versus βl for a short-circuited, lossless transmission line.

where L_{eq} is the equivalent inductance of a shorted line with $l < \lambda/4$. For $\beta l < 0.5$ rad ($l < 0.08\lambda$), $\tan \beta l \approx \beta l$ and the above equation reduces to

$$L_{eq} \approx \frac{Z_0 l}{v} = L'l \qquad \text{henries} \qquad (3\text{–}89)$$

An inductive result for the shorted line appears logical since its configuration, shown in Fig. 3–14, is that of a one-turn loop or coil. For $\beta l \geq \pi/2$, however, our intuitive sense (or what may be called our *low-frequency common sense*) fails us. For example, when $\pi/2 < \beta l < \pi$, X_{in} is negative and hence the input impedance is capacitive. Furthermore, if $l = \lambda/4$ ($\beta l = \pi/2$), the input impedance is infinite (an open circuit), while for $l = \lambda/2$ ($\beta l = \pi$), it is zero (a short circuit). As discussed earlier, these impedance properties repeat every half wavelength.

For a fixed length l, the characteristic in Fig. 3–14 represents the variation of reactance with frequency since $\beta l = \omega l/v$. In the vicinity of $\beta l = \pi/2$ and π, the behavior of the shorted line is similar to that of resonant L-C circuits. Comparison with the reactance versus frequency characteristics of lumped-element L-C circuits (Fig. 3–15a) leads to the following equivalences:

1. At frequencies in the vicinity of $\beta l = (2n - 1)\pi/2$ (an odd number of quarter wavelengths), the shorted line behaves like a parallel resonant circuit.

2. At frequencies in the vicinity of $\beta l = n\pi$ (an integral number of half wavelengths), the shorted line behaves like a series resonant circuit.

These properties are utilized in the design of bandpass and bandstop filters at microwave frequencies. It should be noted that unlike an ideal L-C circuit, the shorted line has an infinite number of resonances, which may pose a problem to the filter designer.

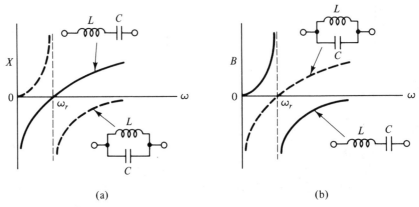

(a)   (b)

Figure 3–15   Frequency variation of reactance and susceptance for series and parallel resonant circuits (resonant frequency $\omega_r = 1\sqrt{LC}$).

The open-circuited line.   When a lossless line is open circuited at the load end, $Y_L = 0$ and from Eq. (3–84) the input admittance is

$$Y_{in} = jB_{in} = jY_0 \tan \beta l \qquad \text{or} \qquad \overline{Y}_{in} = j\overline{B}_{in} = j \tan \beta l \qquad (3\text{–}90)$$

Since the input impedance is the reciprocal of Y_{in},

$$Z_{in} = -jZ_0 \cot \beta l \qquad (3\text{–}91)$$

A plot of normalized input susceptance versus βl is the tangent function and is shown in Fig. 3–16. For $\beta l < \pi/2$ $(l < \lambda/4)$, the input is capacitive since B_{in} is positive. Thus $B_{in} = \omega C_{eq}$ and

$$C_{eq} = \frac{Y_0}{\omega} \tan \beta l \qquad \text{farads} \qquad (3\text{–}92)$$

where C_{eq} is the equivalent capacitance of an open-circuited line with $l < \lambda/4$. For $\beta l < 0.5$ rad $(l < 0.08\lambda)$, the above equation may be approximated by

$$C_{eq} \approx \frac{Y_0 l}{v} = \frac{l}{Z_0 v} = C'l \qquad \text{farads} \qquad (3\text{–}93)$$

It should be noted that C_{eq} is independent of frequency only when $\beta l < 0.5$ rad. A similar comment applies to L_{eq} of a shorted line.

The capacitive result for the open-circuited line seems plausible since its configuration, shown in Fig. 3–16, looks like a capacitor (namely, two conductors separated by an insulator). However, as before, our *low-frequency common sense* fails us for $\beta l \geq \pi/2$. In the region $\pi/2 < \beta l < \pi$, the input is inductive since B_{in} is negative. Also, $Y_{in} = 0$ (an open circuit) when l is a multiple of a half wavelength and $Y_{in} = \infty$ (a short circuit) when l is an odd multiple of $\lambda/4$. By comparing B_{in} versus βl in the vicinity of $\pi/2$ and π with the susceptance versus frequency characteristics of simple L-C circuits (Fig. 3–15b), the following equivalences can be stated.

1. At frequencies in the vicinity of $\beta l = (2n-1)\pi/2$ (an odd number of quarter wavelengths), the open-circuited line behaves like a series resonant circuit.
2. At frequencies in the vicinity of $\beta l = n\pi$ (an integral number of half wavelengths), the open-circuited line behaves like a parallel resonant circuit.

These properties are also utilized in the design of microwave filters.

Both open and shorted lines may be modeled as an infinite set of resonant L-C circuits. This point of view is described in Sec. 11.14 of Ref. 3–8.

Determining Z_0 and βl for a lossless line.   Occasionally one is interested in experimentally determining the characteristic impedance and phase constant for a low-loss transmission line. Their values at a given frequency can be obtained from two measurements on a fixed length of line (l). With the load end shorted, measure the input impedance (denoted by Z_{sc}). Next, measure the input impedance with the

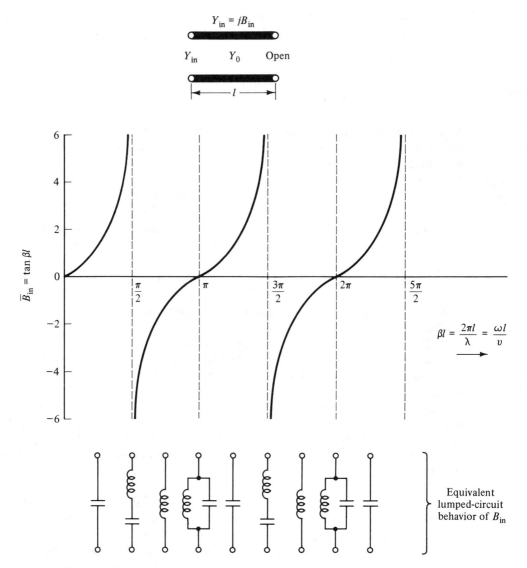

Figure 3–16   Normalized input susceptance versus βl for an open-circuited, lossless transmission line.

load end open circuited (denoted by Z_{oc}). Z_{sc} and Z_{oc} are given by Eqs. (3–87) and (3–91). Solving for Z_0 and βl yields

$$Z_0 = \sqrt{Z_{sc}\, Z_{oc}} \qquad \text{and} \qquad \beta l = \arctan \sqrt{-\frac{Z_{sc}}{Z_{oc}}} \qquad (3\text{–}94)$$

For a lossless line Z_{sc} and Z_{oc} are reactive and of opposite sign. Since impedance repeats every half wavelength there are an infinite number of solutions for βl. In the absence of additional information, the primary solution ($0 \le \beta l \le \pi$) should be used. To determine whether the correct value is in the first or second quadrant, the two possible solutions should be checked against the given values of Z_{sc} and Z_{oc} (Prob. 3–38). This procedure gives the correct value of β if one knows that the line is less than a half wavelength long. If a longer specimen of the line is subsequently measured in the same manner, a still more accurate value of β can be determined after using the previous value to estimate the number of half wavelengths contained within the line.

The application of this technique to lossy lines is described in Sec. 4.7 of Ref. 3–2.

3–6b Impedance Transformations on a Lossy Line

For $\alpha \neq 0$, the impedance and admittance transformations are given by Eqs. (3–81) and (3–82). Three specific examples will now be considered.

The short-circuited line.   For $Z_L = 0$, Eq. (3–81) reduces to $Z = Z_0 \tanh(\alpha d + j\beta d)$. By using Eq. (E–6) and letting $d = l$, the input impedance may be written as

$$Z_{in} = Z_0 \frac{\tanh \alpha l + j \tan \beta l}{1 + j \tanh \alpha l \tan \beta l} \tag{3–95}$$

If $\alpha = 0$ and $\beta l = n\pi$, then $Z_{in} = 0$, as discussed earlier. However, if the line is lossy, the input impedance is finite. For $\alpha \neq 0$ and the shorted line an integral number of half wavelengths,

$$Z_{in} = Z_0 \tanh \alpha l \approx Z_0 \alpha l \qquad \text{ohms} \tag{3–96}$$

If the line is an odd number of quarter wavelengths, $\tan \beta l \to \infty$. For a lossless line, this results in an open circuit at the input. For a lossy line, Eq. (3–95) reduces to

$$Z_{in} = \frac{Z_0}{\tanh \alpha l} \approx \frac{Z_0}{\alpha l} \qquad \text{ohms} \tag{3–97}$$

In both the above equations, the approximate expressions are valid for $\alpha l < 0.5$ Np. For the two cases considered here, the input impedance is finite and real.

The open-circuited line.   Similar relations can be developed for a lossy, open-circuited line. In this case, $Y_L = 0$ and from Eq. (3–82) with $d = l$, the input admittance becomes $Y_{in} = Y_0 \tanh(\alpha l + j\beta l)$. For $\beta l = n\pi$, the input admittance is finite and real and is given by

$$Y_{in} = Y_0 \tanh \alpha l \approx Y_0 \alpha l = \frac{\alpha l}{Z_0} \qquad \text{mhos} \tag{3–98}$$

On the other hand, if $\beta l = (2n - 1)\pi/2$, the input admittance is given by

$$Y_{in} = \frac{Y_0}{\tanh \alpha l} \approx \frac{Y_0}{\alpha l} = \frac{1}{Z_0 \alpha l} \qquad \text{mhos} \tag{3–99}$$

Since $Z_{in} = 1/Y_{in}$, these results are the same as Eqs. (3–97) and (3–96) respectively.

The shorted line as an inductor.   The input impedance of a lossless, short-circuited line is given by Eq. (3–87). For $l < \lambda/4$, the input is purely inductive. Any value of inductive reactance, no matter how large, is realizable by merely choosing βl so that $Z_0 \tan \beta l$ yields the desired value. Furthermore, the inductive reactance is achieved with no accompanying resistive component. One is tempted to say that, as long as a *low-loss line* is used, the resistive component will be negligible. The following example shows that this is not necessarily true.

Example 3–5:

The design of a microwave amplifier requires an inductor to series resonate a capacitive reactance of 15,000 ohms. Calculate the residual resistance if the inductor consists of a short-circuited line with $Z_0 = 75$ ohms, $\alpha = 0.002$ dB/cm, and $\lambda = 10$ cm.

Solution: To series resonate the $-j15,000$ ohms, the input impedance of the shorted line must be $+j15,000$ ohms. If the line were lossless, $Z_{in} = j15,000 = j75 \tan \beta l$ and therefore

$$\beta l = \arctan 200 \approx \frac{\pi}{2} - \frac{1}{200} \qquad \text{rad}$$

Since $\beta = 2\pi/\lambda$,

$$l = 2.5 \left(1 - \frac{1}{100\pi} \right) \qquad \text{cm}$$

Thus a shorted lossless line about 0.3 percent less than a quarter wavelength would provide the desired reactance without adding any resistance.

All lines however have some loss. In this case, $\alpha l = 0.005$ dB or 5.76×10^{-4} Np. Since $\tanh \alpha l \approx \alpha l$, substituting $Z_0 = 75$ ohms and $\tan \beta l = 200$ into Eq. (3–95) yields

$$Z_{in} = 75 \frac{5.76 \times 10^{-4} + j200}{1 + j0.115} = 1700 + j14,800 \qquad \text{ohms}$$

Note the effect of line loss! First of all, the reactive portion is reduced slightly. This presents no problem since a slight adjustment in line length can increase it to the required value of 15,000 ohms. Second, and much more important, a significant resistive component has been added. This means that series resonating the capacitive reactance with the shorted line leaves a residual resistance of 1700 ohms, surely not a negligible value. (With the line length adjusted for a reactance of 15,000 ohms, the residual resistance is acutally 1748 ohms.)

This example shows that despite the use of a low-loss line, the quality factor of the inductor is poor ($Q \equiv |X|/R \approx 8.7$). This is because the requirement for a large reactance necessitates that the shorted line be operated near resonance (that is, $l \approx \lambda/4$). A general expression for Q can be derived from Eq. (3–95). Assuming $\tanh \alpha l \approx \alpha l$, the quality factor of the shorted line when used as an inductor is

$$Q = \frac{\overline{X}}{\alpha l} \left[\frac{1 - (\alpha l)^2}{1 + \overline{X}^2} \right] \approx \frac{\overline{X}}{\alpha l (1 + \overline{X}^2)} \qquad (3\text{–}100)$$

where $\bar{X} \equiv X/Z_0 = \tan \beta l$, the normalized value of required reactance. For $\alpha l < 10^{-3}$ Np, the quality factor will be greater than 100 if the required reactance is less than $10\,Z_0$ (that is, $X < 10$). The maximum realizable value of inductive reactance is also limited by the attenuation. One can show, using Eq. (3–95), that the maximum possible value of $X_L \approx Z_0/2\alpha l$. A similar problem occurs when attempting to realize large values of susceptance.

In general, the short terminating the line is never perfect. That is, it has finite resistance R, where $R \ll Z_0$. The maximum Q that can be attained in this case is also limited (Prob. 3–42), as is the maximum value of X_L.

3–7 THE SMITH CHART

A graphical procedure for solving impedance transformation problems is described in this section. The technique makes use of a special impedance/admittance chart developed by P. H. Smith (Refs. 3–9 to 3–11). As in any well-structured graphical method, the advantages are a reduction in the computational effort required and an improved intuitive understanding of how the individual variables affect the desired end result. The first reason is important in analysis, while the second is vital to good synthesis and design. In practice, high-frequency circuits often contain two or more transmission lines interspersed with series and shunt elements. The Smith chart technique can significantly reduce the numerical and algebraic manipulations required to solve such problems. Of course, one can also develop computer programs to solve transmission-line circuits. In these cases, a Smith chart analysis is useful in verifying the validity of the computer solutions.

3–7a The Basis of the Smith Chart

The Smith chart is a specially constructed impedance/admittance diagram for use in solving transmission-line problems. As such, it has a pair of coordinates for plotting complex values of impedance and admittance. The chart has several useful characteristics. First, all possible values of impedance and admittance can be plotted on the chart. Second, an easy method for converting impedance to admittance (and vice versa) is available. Third, and most important, the Smith chart provides a simple graphical method for determining the impedance transformation due to a length of transmission line. An ordinary rectangular chart with resistance plotted horizontally and reactance vertically has none of these advantages.

To understand the basis of the Smith chart, consider the polar coordinate system shown in Fig. 3–17. By convention, positive angles are plotted counterclockwise. For any passive impedance (that is, Re $Z \geq 0$) and Z_0 real, the magnitude of the reflection coefficient is less than or equal to one.[13] Thus all possible values of Γ can be plotted on a polar chart having a maximum radius value of unity. Such a

[13] These two conditions are assumed in all subsequent discussions. The case wherein Z_0 is complex is discussed briefly in Sec. 9.1 of Ref. 3–3.

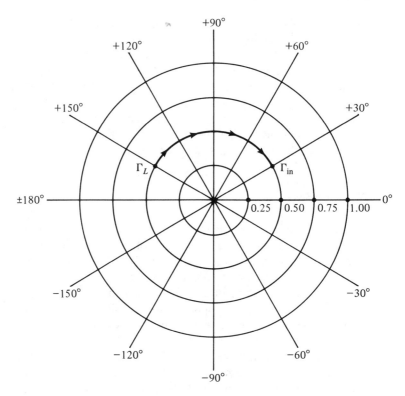

Figure 3–17 A polar chart for plotting complex reflection coefficients, $|\Gamma|\underline{/\phi}$.

chart is shown in Fig. 3–17. It can accommodate reflection coefficient values for all impedances from a short ($\Gamma = 1\underline{/180°}$) to an open ($\Gamma = 1\underline{/0}$). Furthermore, the change in reflection coefficient due to a length of transmission line (Eq. 3–43) is easily determined with the polar chart, particularly if the line is lossless.

Consider a load impedance $Z_L = 17.7 + j11.8$ ohms connected to a lossless 50 ohm line of length $l = \lambda/6$. From Eq. (3–45), $\Gamma_L = 0.5\underline{/150°}$, which is plotted in Fig. 3–17. The input reflection coefficient may be determined from Eq. (3–43). For $\alpha = 0$, $\Gamma_{in} = 0.5\underline{/150° - 120°}$, since $2\beta l = 2\pi/3$ rad or 120°. This same result can be obtained graphically using the polar chart. Since $|\Gamma_{in}| = |\Gamma_L|$ for a lossless line, merely rotate *clockwise* from the Γ_L point on the 0.5 radius circle an angular distance of 120° (namely, $2\beta l$), as indicated in Fig. 3–17. The reflection coefficient at any other point on the line is obtained by rotating clockwise $2\beta d$ where d is the distance from the load to the point.[14] Conversely, if Γ is known, Γ_L is obtained by rotating *counterclockwise* $2\beta d$.

---

[14] If the line is lossy ($\alpha \neq 0$), $|\Gamma| = |\Gamma_L|e^{-2\alpha d}$ and $|\Gamma_{in}| = |\Gamma_L|e^{-2\alpha l}$. In the example given, the Γ_{in} point would occur at the intersection of the 30° radial line and the $0.5e^{-2\alpha l}$ circle. Thus the locus of Γ points as a function of d would be a spiral of decreasing radius starting at Γ_L and ending at the Γ_{in} point. For low-loss lines, the approximation $\alpha = 0$ is usually valid.

The above example shows that the graphical solution of the reflection coefficient transformation equation is quite simple. All that is needed is a compass and a straight edge. This is very interesting, *except* that in most cases it is the *impedance* transformation that is required. In the example, Z_L and Z_0 were given and Eq. (3–45) was used to obtain Γ_L. With Γ at any other point determined graphically, its corresponding impedance can be calculated with the aid of Eq. (3–47). One might naturally ask "How can these calculations be avoided?" In other words, how can the impedance transformation be completely solved graphically? The answer is amazingly simple! Merely replace every point on the polar reflection coefficient chart by its equivalent normalized impedance. This is accomplished with Eq. (3–47) which may be rewritten as

$$\frac{Z}{Z_0} = \bar{Z} = \frac{1 + \Gamma}{1 - \Gamma} \quad \text{and} \quad \frac{Y}{Y_0} = \bar{Y} = \frac{1 - \Gamma}{1 + \Gamma} \qquad (3\text{–}101)$$

For example, $\Gamma = 1\underline{/180°}$ is equivalent to $\bar{Z} = 0$, while $\Gamma = 1\underline{/0}$ is equivalent to $\bar{Z} = \infty$, an open circuit. Also, $\Gamma = 0.5\underline{/180°}$ converts to $\bar{Z} = 1/3$, while $\Gamma = 0.5\underline{/0}$ is equivalent to $\bar{Z} = 3$. Other normalized impedance values corresponding to certain values of Γ are shown in Fig. 3–18.

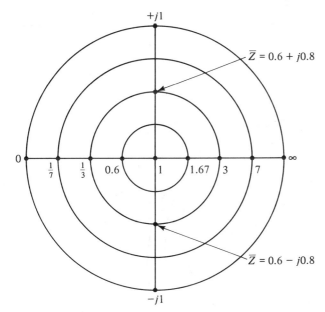

Figure 3–18 The polar chart of Fig. 3–17 with some of the reflection coefficient values replaced by their equivalent normalized impedances.

Many more values of $\bar{Z}$ may be calculated using Eq. (3–101). If all $\bar{Z}$ points having the same real parts (Re $\bar{Z}$) are joined together, the result would look like part *a* of Fig. 3–19. The circles are called *constant resistance* circles. Note that they all pass through the infinite impedance point. The outermost one is the $\bar{R} = 0$ circle and hence all impedance values on it are purely reactive (ranging from 0 to $\pm j\infty$). This agrees with the fact that for a pure reactance $|\Gamma| = 1$, which is the outermost circle

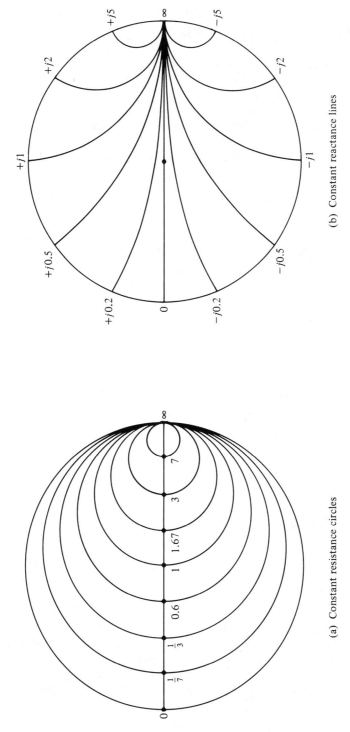

(a) Constant resistance circles

(b) Constant reactance lines

Figure 3–19 Constant resistance circles and constant reactance lines. (Note: The reactance lines are portions of circles.)

on the chart. Joining all points having the same imaginary values (Im $\bar{Z}$) results in the curved lines shown in part *b* of the figure. These are called *constant reactance lines* and in fact are portions of circles.[15] The $\bar{X} = 0$ line is the horizontal line between the zero and infinity points. It is the resistance axis of the impedance coordinates with values ranging from $\bar{R} = 0$ through $\bar{R} = 1$ (the center of the chart) to $\bar{R} = \infty$. The impedance grid formed by combining the constant resistance circles and the constant reactance lines is a Smith chart! Figure 3–20 shows a commercially available version used in many microwave laboratories. The chart was originally developed by P. H. Smith (Refs. 3–9 to 3–11). Basically, it is *a polar chart of reflection coefficient upon which a normalized impedance grid has been superimposed*. Equation (3–101) provides the conversion relationship from Γ coordinates to $\bar{Z}$ coordinates. Note that for the sake of clarity, the circles representing constant values of $|\Gamma|$ have been removed. However, the angle of reflection coefficient scale has been retained. It is the innermost of the three scales on the periphery of the chart. The two outermost scales allow us to perform the Γ transformation (and hence the Z transformation) without having to calculate $2\beta d$. The scales are given directly in fractions of a wavelength. Note that one complete rotation around the chart is a *half* wavelength since for $d = \lambda/2$, $2\beta d = 2\pi$ rad or $360°$. This makes sense since it was shown in Sec. 3–6a that impedance repeats every half wavelength. Referring to the chart in Fig. 3–20, the upper half is the positive reactance region, which means that impedance values have an inductive component. The bottom half (negative reactance) denotes impedances with a capacitive component.

At the top of the chart are the words "Impedance and Admittance Coordinates." The reason for this is that the Smith chart can also be used with normalized admittance coordinates. For any value of $\bar{Z}$, its equivalent normalized admittance ($\bar{Y} \equiv Y/Y_0$) is $180°$ away on the same $|\Gamma|$ circle, since adding $180°$ to Γ (that is, changing its sign) converts $\bar{Z}$ to $\bar{Y}$. This can be seen by comparing the expressions for $\bar{Z}$ and $\bar{Y}$ in Eq. (3–101). When using the admittance coordinates, the following comments should be kept in mind.

1. The $\bar{Y} = 0$ point corresponds to an open circuit, while the $\bar{Y} = \infty$ point corresponds to a short circuit.

2. The resistance coordinates become conductance coordinates and the reactance coordinates become susceptance coordinates. Remember that a positive susceptance value (top half of the chart) represents a capacitive component, while a negative value denotes an inductive component. This is indicated on the chart.

3. When using admittance coordinates, the angle of the reflection coefficient scale must be rotated $180°$. This can be accomplished by subtracting $180°$ from the values on the top half of the scale and adding $180°$ to the values on the bottom half.

---

[15] The proof that the locus of constant resistance and reactance values are circles and portions of circles is given in many texts. See, for example, Sec. 9.2 in Ref. 3–3 or Sec. 1.20 in Ref. 3–8.

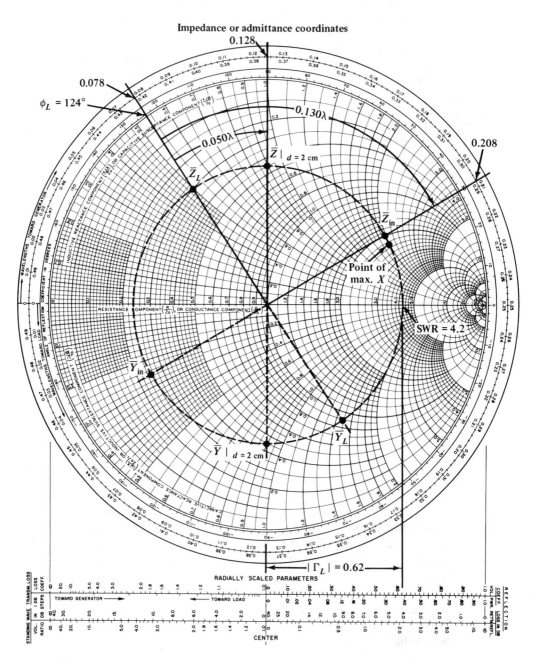

Figure 3–20 A commercially available form of the Smith chart. (Note: The data on the chart refer to Ex. 3–6.) Permission to reproduce Smith charts in this text has been granted by Phillip H. Smith, Murray Hill, N.J., under his renewal copyright issued in 1976.

Note that there are two wavelength scales on the periphery of the chart. One is labeled *Wavelengths toward Generator* and the other *Wavelengths toward Load*. The first one is used when determining the impedance at a point *nearer the input* than the known impedance. This clockwise rotation is referred to as *moving toward the generator*, the assumption being that a generator is connected to the input. The second scale is used in determining the impedance at a point *nearer the load* than the known impedance. This counterclockwise rotation is referred to as *moving toward the load*.

The following example illustrates the graphical procedure for determining the impedance and admittance transformation due to a length of transmission, line as well as other useful properties of the Smith chart.

Example 3–6:

A 5.2 cm length of lossless 100 ohm line is terminated in a load impedance $Z_L =$ 30 + j50 ohms.
(a) Calculate $|\Gamma_L|$, ϕ_L, and the SWR along the line.
(b) Determine the impedance and admittance at the input and at a point 2.0 cm from the load end. The signal frequency is 750 MHz and $\lambda = \lambda_0$.

Solution:
(a) Plot the normalized load impedance $\bar{Z} \equiv Z_L/Z_0 = 0.30 + j0.50$ on the Smith chart in Fig. 3–20. This is done by starting from the 0.30 point on the resistance axis and moving up 0.50 reactance units along the constant resistance circle. Next, draw a circle with its center at $\bar{Z} = 1$ (the center of the chart) and a radius equal to the distance between $\bar{Z} = 1$ and $\bar{Z}_L$. This is shown in the figure and will hereafter be referred to as the SWR circle. It is, in fact, the constant $|\Gamma|$ circle for the given value of load impedance. The $\bar{Z}_L$ point on the Smith chart corresponds to its Γ_L value on the polar chart. Since the angle of the reflection coefficient scale has been retained, the value of ϕ_L (124° in this case) can be obtained from the chart as shown in the figure. The value of $|\Gamma_L|$ is obtained by measuring the radius of the SWR circle on the Reflection Coefficient-Vol. scale located at the bottom of the chart. In this example problem, $|\Gamma_L| = 0.62$.

One reason that the constant $|\Gamma|$ circle is called the *SWR circle* is that its intersection with the right half of the resistance axis (that is, between 1 and ∞) yields the SWR due to $\bar{Z}_L$. For this case, the SWR is 4.2. Since SWR is so easily obtained, many engineers prefer to calculate $|\Gamma|$ from Eq. (3–53) rather than using the Reflection Coefficient-Vol. scale. The reason that the resistance axis between 1 and ∞ serves as a SWR scale is that it corresponds to positive real values of Γ. As such it represents the magnitude of Γ for all normalized impedances on the SWR circle. Equation (3–101) transforms these values of Γ into $\bar{R} = (1 + |\Gamma|)/(1 - |\Gamma|)$, which is exactly the equation for SWR (Eq. 3–52). The unity SWR circle is simply a point at the center of the chart, while the infinite SWR circle is the periphery of the chart and is equivalent to $|\Gamma| = 1.00$.

(b) A graphical method of obtaining Γ_{in} when Γ_L and βl are known has been described. For a lossless line, it consists of rotating clockwise $2\beta l$ on the constant $|\Gamma|$ circle. The procedure for obtaining $\bar{Z}_{in}$ from $\bar{Z}_L$ is *exactly the same* except that the impedance coordinates of the Smith chart are used. The steps are as follows:
1. Plot $\bar{Z}_L$ (0.30 + j0.50 in this case) and draw its SWR circle (4.20 in this case).
2. Draw a radial line from the center of the chart through $\bar{Z}_L$ to the periphery. Read the value on the *Wavelengths toward Generator* scale (0.078 in this case).

This value in itself has no physical meaning. It is merely the starting point of the clockwise rotation in the next step.

3. Since $\lambda_0 = 40$ cm at 750 MHz and the input is 5.2 cm from the load, rotate clockwise *from* 0.078 a distance $l/\lambda = 5.2/40 = 0.130$. Draw a radial line from the center of the chart through the 0.208 point on the outer scale as shown in Fig. 3–20. The intersection of the radial line with the SWR circle represents $\bar{Z}_{in}$ since it corresponds to the Γ_{in} point on the polar reflection coefficient chart. In this case, $\bar{Z}_{in} = 2 + j2$ or $Z_{in} = 200 + j200$ ohms.

To obtain the impedance at $d = 2$ cm, start from $\bar{Z}_L$ and rotate clockwise $2/40 = 0.050$ and draw a radial line through the 0.128 point as shown. Its intersection with the SWR circle yields $\bar{Z} = 0.47 + j0.93$ or $Z = 47 + j93$ ohms.

This simple graphical method of determining the impedance transformation due to a length of lossless line is the most useful characteristic of the Smith chart. The procedure for determining the admittance transformation is *exactly the same* since all admittance points are directly opposite their corresponding impedance points on the SWR circle. Thus from the figure, $\bar{Y}_L = 0.88 - j1.47$, $\bar{Y}_{in} = 0.25 - j0.25$, and at $d = 2.0$ cm, $\bar{Y} = 0.43 - j0.87$. Multiplying these values by $Y_0 = 0.01$ mho gives the unnormalized admittance values.

Part *a* of this example problem illustrates how to determine $|\Gamma_L|$, ϕ_L, and SWR when Z_L and Z_0 are known. Since the line is lossless, the SWR and $|\Gamma|$ are the same at all other points on the line. The angle of the reflection coefficient, however, is a function of position and can be read on the periphery of the chart. In this example, $\phi_{in} = 30°$ while at $d = 2$ cm, ϕ is equal to 88°.

Part *b* describes the graphical solution to the impedance/admittance transformation equation for lossless lines.[16] Given $\bar{Z}_L$, the normalized impedance at any other point on the line is obtained by rotating clockwise on a fixed SWR circle the appropriate distance d/λ. Thus for a given load impedance, the SWR circle represents the locus of all possible impedance and admittance values available on the lossless line. Stated another way, given $\bar{Z}_L$, it is the locus of all possible values of $\bar{Z}_{in}$ and $\bar{Y}_{in}$ obtainable by varying the line length l. This is an example of how a good graphical procedure can show the effect of a variable (l) on the desired result (Z_{in}). For instance, it is obvious from Fig. 3–20 that if $\bar{Z}_L = 0.30 + j0.50$, varying the line length will never result in $\bar{Z}_{in} = 0.70 + j0.40$ since it is not on the SWR circle containing $\bar{Z}_L$. If the reader is ambitious, try proving this analytically. To further emphasize the chart's usefulness, consider the ease with which the following problem is solved. Given the impedance values in Ex. 3–6, what value of line length maximizes the reactive portion of the input impedance? With $\bar{Z}_L = 0.30 + j0.50$ plotted in Fig. 3–20, the SWR circle represents all possible values of $\bar{Z}_{in}$ that can be obtained by varying l. A brief look at the chart shows that the reactive portion is maximized at the point where the SWR circle is tangent to the reactance lines. A positive reactive maximum occurs when $\bar{Z}_{in} \approx 2.3 + j2.0$. A radial line through this point intersects the *Wavelengths toward Generator* scale at 0.216. Therefore, the 100 ohm line must be $(0.216 - 0.078)$ or 0.138λ long.

---

[16] For a lossy line, a second SWR circle is required. It is obtained by multiplying $|\Gamma_L|$ by $e^{-2\alpha d}$ and converting the resulting $|\Gamma|$ to SWR. The intersection of the radial line with the circle defined by the new SWR value yields $\bar{Z}$ at d units from the load.

3–7b Typical Smith Chart Computations

In practice, high-frequency circuits often contain both lumped circuit elements and transmission-line sections. The following example problems illustrate the use of the Smith chart in these cases.

Example 3–7:

A lossless 50 ohm line terminated in $Z_L = 100 + j75$ ohms is shown at the top of Fig. 3–21. The line is 0.18λ long. Calculate the input impedance if the L-C circuit shown is inserted at a point 0.12λ from the load end.

Solution: The bottom half of Fig. 3–21 shows the Smith chart solution of the problem. The subscripts A, B, and C are used to denote the impedance and admittance at various points along the line. For example, Z_B (Y_B) represents the impedance (admittance) of the circuit to the right of plane B. As such it includes Z_L, the 0.12λ line and the 30 ohm inductor.

To obtain the impedance at plane A, first plot $\bar{Z}_L \equiv Z_L/50 = 2 + j1.5$ and draw its SWR circle (SWR = 3.3). The impedance at plane A is obtained by rotating 0.12λ toward the generator on the 3.3 SWR circle, which yields $\bar{Z}_A = 1 - j1.3$. The impedance at plane B is $\bar{Z}_A$ plus the normalized impedance of the series inductance $(+j30/50)$. Thus, $\bar{Z}_B = 1 - j1.3 + j0.6 = 1 - j0.7$. Adding inductive reactance in series is equivalent to moving upward on a constant resistance circle. As indicated on the Smith chart, $\bar{Z}_B$ is $+0.6$ units above the $\bar{Z}_A$ point on the $\bar{R} = 1$ circle. Next, the impedance at plane C is determined, which requires that the effect of the $-j200$ ohm shunt capacitance be considered. Since impedances in parallel are not directly additive, the equivalent admittance values are needed to continue the graphical procedure. The $\bar{Y}_B$ point is directly opposite the $\bar{Z}_B$ point on *its* SWR circle (SWR = 2.0). From the chart, $\bar{Y}_B = 0.67 + j0.47$. The admittance at plane C is the sum of $\bar{Y}_B$ and the normalized admittance of the shunt capacitance. Since the admittance of the capacitor is the reciprocal of $-j200$ and $Y_0 = 0.02$ mho, $\bar{Y}_C = 0.67 + j0.47 + j0.25 = 0.67 + j0.72$. On the Smith chart, this is equivalent to moving *upward* (since capacitive susceptance is positive) 0.25 susceptance units on the $\bar{G} = 0.67$ circle.

With $\bar{Y}_C$ known, the input admittance is obtained by the same graphical procedure used for impedance transformations. Draw the SWR circle for $\bar{Y}_C$ (SWR = 2.6). Move from the $\bar{Y}_C$ point a distance of 0.06λ toward the generator. The result from the chart is $\bar{Y}_{in} = 1.3 + j1.1$. The normalized input impedance is diametrically opposite the $\bar{Y}_{in}$ point on the 2.6 SWR circle. Thus $\bar{Z}_{in} = 0.45 - j0.38$. With $Z_0 = 50$ ohms, $Z_{in} = 22.5 - j19$ ohms.

The Smith chart solution reveals additional useful information about the circuit. For example, the SWR along the 0.12λ line section is 3.30, while its value on the 0.06λ section is 2.60.

Often a microwave circuit contains two or more transmission lines with different characteristic impedances. This usually requires normalizing the chart to more than one value of Z_0. The following example problem illustrates this situation.

Example 3–8:

A 50 ohm line and a 90 ohm line are connected in tandem as shown at the top of Fig. 3–22. The 50 ohm line is terminated in a 20 ohm resistance. Both lines are lossless and their lengths are indicated in the figure. Calculate the input impedance and reflection coefficient if $\lambda = 20$ cm for both lines.

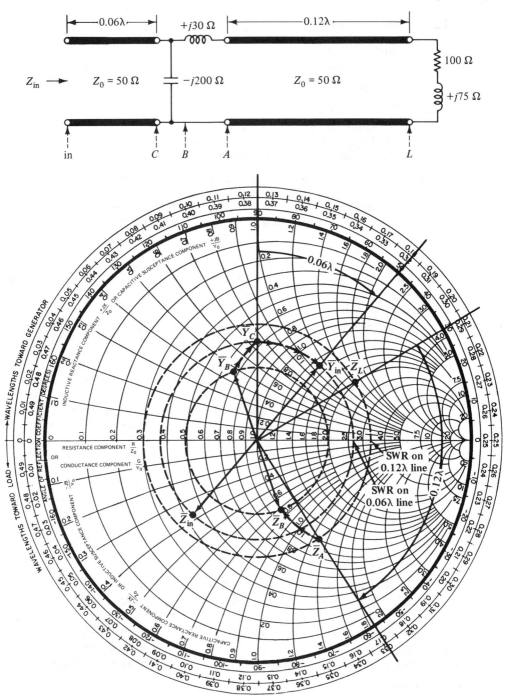

Figure 3–21   The transmission-line circuit and Smith chart solution for Ex. 3–7 in Sec. 3–7b.

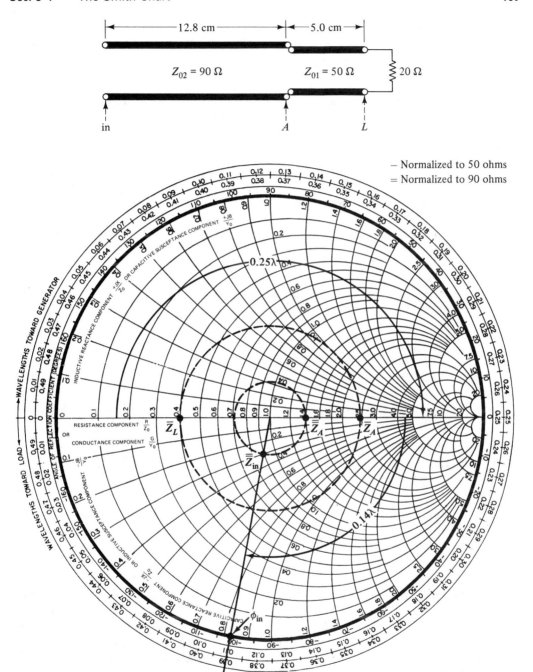

Figure 3–22 The transmission-line circuit and Smith chart solution for Ex. 3–8 in Sec. 3–7b.

Solution: Normalize the load impedance to $Z_{01} = 50$ ohms and plot on the Smith chart. This is indicated in Fig. 3–22 as $\bar{Z}_L = 20/50 = 0.4$, where the bar (–) denotes normalization with respect to Z_{01}. Next, use the impedance transformation procedure to determine Z_A. That is, draw the SWR circle for $\bar{Z}_L$ (SWR = 2.5) and rotate from $\bar{Z}_L$ 0.25 wavelengths toward the generator. The result is $\bar{Z}_A = 2.5$ or $Z_A = 125$ ohms. In order to calculate the input impedance, the transformation due to the 90 ohm line must be determined. This requires that Z_A be normalized to 90 ohms rather than 50 ohms. The determination of SWR on the 90 ohm line also requires that the data be so normalized. Denoting normalization with respect to Z_{02} by a double bar (=), $\bar{\bar{Z}}_A = 125/90 = 1.39$. Drawing the SWR circle for $\bar{\bar{Z}}_A$ (SWR = 1.39) and rotating $12.8/20 = 0.64$ wavelengths toward the generator yields the normalized input impedance $\bar{\bar{Z}}_{in}$. Note that since once around the chart is 0.5 wavelengths, a rotation of 0.14 is the same as rotating 0.64. From the Smith chart in Fig. 3–22, $\bar{\bar{Z}}_{in} = 0.9 - j0.3$ or $Z_{in} = 81 - j27$ ohms.

Since the SWR on the input line is 1.39, Eq. (3–53) yields $|\Gamma_{in}| = 0.163$. The angle of the input reflection coefficient is obtained by drawing a radial line through $\bar{\bar{Z}}_{in}$ to the periphery of the chart. Thus, $\phi_{in} = -100°$ and $\Gamma_{in} = 0.163\underline{/-100°}$.

With the completion of this and the previous two example problems, the reader should now be able to perform the following operations on a Smith chart.

1. Plot a normalized impedance or admittance.
2. Obtain the SWR and reflection coefficient angle ϕ for any value of normalized impedance or admittance. (The use of Eq. (3–53) is suggested for calculating $|\Gamma|$.)
3. Determine the impedance or admittance transformation due to a length of transmission line.
4. Convert normalized impedance to normalized admittance and vice versa.
5. Determine the effect of adding series or shunt-connected circuit elements at any point along a transmission line.

Variation of $\bar{Z}$ and $\bar{Y}$ with frequency. For purely reactive networks, the slope of the reactance or susceptance versus frequency curve is always positive. Examples of this are shown in Figs. 3–14, 3–15, and 3–16. The proof of this rule is found in most texts on network theory (see for example, Sec. 14.4 of Ref. 3–2). This condition places a restriction on the possible shape of impedance and admittance versus frequency curves. When plotted on a Smith chart, the $\bar{Z}$ and $\bar{Y}$ curves must have a *clockwise trend with increasing frequency*. This rule also holds for any combination of dissipative elements (resistors and lossy lines) and reactive elements. Some examples of $\bar{Z}$ and $\bar{Y}$ versus frequency plots for passive networks are shown in Fig. 3–23. The arrows indicate the direction of increasing frequency. Note in all cases, the clockwise tendency with increasing frequency.

An example of a frequency sensitive load impedance is shown at the top of Fig. 3–24. Normalizing to $Z_0 = 50$ ohms, $\bar{Z}_L = 0.50 + j\bar{X}$, where $\bar{X} = \{\omega L - (1/\omega C)\}/50$. For the given values of L and C, series resonance occurs at

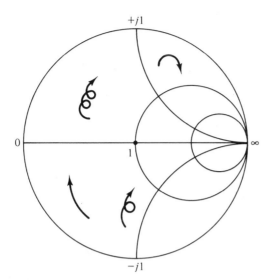

Figure 3–23 Typical Smith chart plots of $\bar{Z}$ and $\bar{Y}$ versus frequency for passive networks.

1000 MHz. The calculated values of $\bar{Z}_L$ at three frequencies are given in Table 3–1 and plotted on the Smith chart of Fig. 3–24. As is customary, the arrow on the impedance plot indicates increasing frequency. The SWR circles for the three frequencies are also shown and their values are given in the table.

Let us now determine the effect of the 4.0 cm length of lossless line on the shape of the impedance plot. The usual impedance transformation procedure is used to obtain $\bar{Z}_{in}$. With $\lambda = \lambda_0$, the table gives the amount of rotation (l/λ_0) required at each frequency. For example, a 0.133 rotation on the 2.0 SWR circle yields the $\bar{Z}_{in}$ value at 1000 MHz. This and the values at 750 and 1250 MHz are listed in the table. The resultant $\bar{Z}_{in}$ versus frequency characteristic is plotted on the Smith chart. It is interesting (and important) to note that the $\bar{Z}_{in}$ plot is spread over a greater portion of the chart than the $\bar{Z}_L$ plot. In terms of angular spread, the $\bar{Z}_{in}$ plot covers 0.151 on the wavelength scale while the $\bar{Z}_L$ plot covers 0.084. The additional 0.067 spread is due to the frequency sensitivity of the 4.0 cm length of transmission line. Although the line is of fixed length its *electrical length* increases with increasing frequency.[17] At 1250 MHz, l/λ is 0.067 larger than its value at 750 MHz. It is this frequency sensitivity that causes the $\bar{Z}_{in}$ plot to have a greater spread than the $\bar{Z}_L$ plot. The longer the line or the greater the frequency band, the larger the impedance spread and the more difficult it becomes to match an impedance over a given frequency range. Impedance matching is discussed in the next chapter.

Determination of standing wave patterns.   Voltage and current standing wave patterns can be determined with the aid of the Smith chart. The phasor voltage

---

[17] The term *electrical length* denotes either βl or l/λ. For instance, if a line section is a quarter wavelength long, $\beta l = \pi/2$ rad or 90° and $l/\lambda = 0.25$.

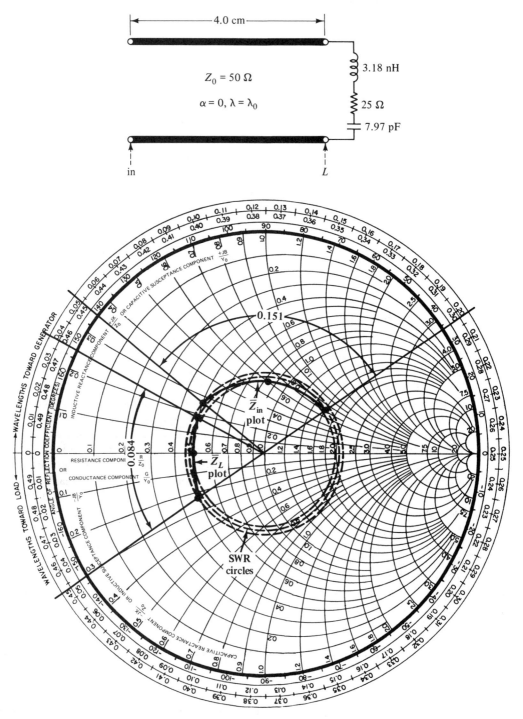

Figure 3–24 A transmission line terminated in a series resonant circuit and its corresponding Smith chart characteristics.

TABLE 3-1 Data for the circuit in Fig. 3-24.

Freq. (MHz)	λ_0 (cm)	l/λ_0	$\bar{Z}_L$	SWR	$\bar{Z}_{in}$
750	40	0.100	$0.50 - j0.23$	2.2	$0.50 + j0.27$
1000	30	0.133	0.50	2.0	$0.85 + j0.63$
1250	24	0.167	$0.50 + j0.18$	2.1	$1.60 + j0.70$

and current on a transmission line are given by Eqs. (3-34) and (3-35). Using the definition of Γ from Eq. (3-36), they may be rewritten as

$$\frac{\mathbf{V}}{\mathbf{V}^+} = 1 + \Gamma \quad \text{and} \quad \frac{\mathbf{I}}{\mathbf{I}^+} = 1 - \Gamma \quad (3-102)$$

The magnitude of these quantities are proportional to the rms voltage and current along the line. Thus the standing wave patterns are described by the following expressions.

$$\frac{V}{V^+} = |1 + \Gamma| \quad \text{and} \quad \frac{I}{I^+} = |1 - \Gamma| \quad (3-103)$$

where Γ, and hence V and I, are functions of position on the transmission line. When the line is lossless, V^+ and I^+ are independent of position.

The unity value in these expressions can be represented by the phasor $1\underline{/0}$. This is shown on the Smith chart in Fig. 3-25 as a fixed horizontal vector from the $\bar{Z} = 0$ point to $\bar{Z} = 1$, the center of the chart. Since a vector from the center of the chart to some impedance point $\bar{Z}$ is the reflection coefficient Γ, the *magnitude* of the vector sum of $1\underline{/0}$ and Γ represents the normalized rms voltage at that point on the line. The magnitude of the vector difference represents the normalized rms current at the same point. Figure 3-25 shows $\Gamma_L, -\Gamma_L, V_L/V^+$ and I_L/I^+ for the case when $\bar{Z}_L = 1 + j2$. When the line is lossless, the normalized voltage and current at any other position is obtained by rotating from the $\bar{Z}_L$ point on its SWR circle d/λ

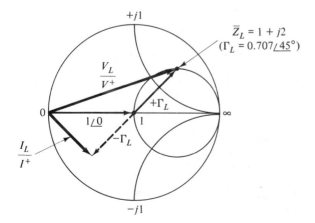

Figure 3-25 Determination of normalized load voltage and current using the Smith chart. In this example, $\bar{Z}_L = 1 + j2$.

units toward the generator and solving for $|1 + \Gamma|$ and $|1 - \Gamma|$. This is shown in Fig. 3–26 for $d = \lambda/16$, $3\lambda/16$, and $5\lambda/16$ when $\bar{Z}_L = 1 + j2$. The resultant standing wave patterns are shown at the bottom of the figure. For this case, a voltage maximum (current minimum) occurs $\lambda/16$ from the load. A quarter wavelength further at $d = 5\lambda/16$, a voltage minimum (current maximum) exists. This suggests the following procedure for determining the positions of maximum and minimum voltage and current on a transmission line. For a given load impedance, plot $\bar{Z}_L$ on the Smith chart. If $\bar{Y}_L$ is given, convert it to its equivalent impedance $\bar{Z}_L$. The clockwise angular distance from the $\bar{Z}_L$ point to the horizontal axis between $\bar{Z} = 1$ and $\bar{Z} = \infty$ is the distance to the first voltage maximum (current minimum). The distance in fractions of a wavelength is obtained by using the *Wavelength toward Generator* scale on the periphery of the chart. In like manner, the first voltage minimum (current maximum) is determined by measuring the clockwise angular distance from the $\bar{Z}_L$ point to the horizontal axis between $\bar{Z} = 0$ and $\bar{Z} = 1$. This is just another example of the usefulness of the Smith chart in solving transmission-line problems.

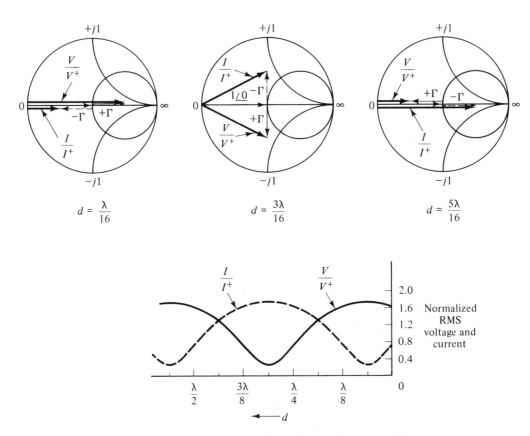

Figure 3–26 Use of the Smith chart to determine voltage and current standing-wave patterns. In this example, $\bar{Z}_L = 1 + j2$.

This chapter has reviewed the high-frequency aspects of transmission-line theory. The results are useful in the analysis and design of microwave components and systems. For a discussion of transmission-line characteristics at low frequencies, the reader is referred to any of the texts listed below.

REFERENCES

Books

3–1. Skilling, H. H., *Electric Transmission Lines,* McGraw-Hill Book Co., New York, 1951.

3–2. Johnson, W. C., *Transmission Lines and Networks,* McGraw-Hill Book Co., New York, 1950.

3–3. Chipman, R. A., *Theory and Problems of Transmission Lines,* Schaum's Outline Series, McGraw-Hill Book Co., New York, 1968.

3–4. Lance, A. L., *Introduction to Microwave Theory and Measurements,* McGraw-Hill Book Co., New York, 1964.

3–5. Karakash, J. J., *Transmission Lines and Filter Networks,* The Macmillan Co., New York, 1950.

3–6. Bewley, L. V., *Traveling Waves on Transmission Systems,* 2nd ed., John Wiley and Sons, Inc., New York, 1951.

3–7. Magnusson, P. C., *Transmission Lines and Wave Propagation,* Allyn and Bacon, Inc., Boston, MA, 1970.

3–8. Ramo, S., J. R. Whinnery, and T. Van Duzer, *Fields and Waves in Communication Electronics,* John Wiley and Sons, Inc., New York, 1965.

3–9. Smith, P. H., *Electronic Applications of the Smith Chart,* McGraw-Hill Book Co., New York, 1983.

Articles

3–10. Smith, P. H., Transmission Line Calculator. *Electronics,* 12, January 1939, pp. 29–31.

3–11. Smith, P. H., An Improved Transmission Line Calculator. *Electronics,* 17, January 1944, pp. 130–133 and 318–325.

PROBLEMS

3–1. A lossless transmission line has an inductance per unit length of 1.35 μH/m and a capacitance per unit length of 15 pF/m. Calculate the wave velocity and characteristic impedance of the line.

3–2. A lossless line is terminated in a load resistance of 50 ohms. Calculate the two possible values of Z_0 if one-third of the incident voltage wave is reflected by the load.

3–3. The open circuit in Fig. 3–3 is replaced by a short. Show that for $t > 40$ ns, $V = 0$ and $I = 0.20$ A everywhere on the line.

3–4. Referring to the circuit at the top of Fig. 3–4, let $R_L = 400$ ohms. Find the voltage and current at $z = 2$ m when $t = 5$, 15, and 35 ns.

3–5. Plot the load current versus time (from 0 to 120 ns) for the circuit shown in Fig. 3–5. Use the space-time diagram.

3–6. Replace the values of R_G, R_L, and l in Fig. 3–5 by the following: $R_G = 0$, $R_L = 0$, and $l = 60$ cm.
  (a) Plot the voltage at $z = 30$ cm versus time between 0 and 15 ns. Use the space-time diagram.
  (b) What is the frequency of oscillation? Describe two ways of reducing its value.

3–7. A uniform transmission line has the following characteristics at 200 MHz: $R' = 50$ ohm/m, $L' = 0.10$ μH/m, $G' = 0.06$ mho/m, and $C' = 200$ pF/m. What is the percent error in using Eqs. (3–22) and (3–23) rather than Eq. (3–15) to calculate the attenuation and phase constants?

3–8. (a) Find the characteristic impedance of an air-filled coaxial line having inner and outer diameters of 0.25 cm and 0.75 cm, respectively.
  (b) Determine the characteristic impedance and the wavelength at 3000 MHz when the line is teflon filled.

3–9. Referring to Fig. 3–9b, the input power is 80 W and the power absorbed by the load is 16 W. What is the power flow at a point halfway down the line?

3–10. Referring to Fig. 3–9b, a 30 V ac source is connected to a 50 ohm load via a 50 ohm lossless line. Calculate the phasor current at $z = 0.30\lambda$ when $Z_G = 25 + j25$ ohms. Determine the power absorbed by the load.

3–11. A load impedance of $30 - j75$ ohms is connected to a 75 ohm lossless line. Calculate Γ_L and Γ_{in} if the line is 0.15λ long. What is the SWR?

3–12. A 100 ohm air-insulated coaxial line is terminated by a parallel combination of an 80 ohm resistor and a 5.0 nH inductor. Calculate the input reflection coefficient at 2000 MHz if $\alpha = 1.5$ dB/m and $l = 40$ cm.

3–13. The reflection coefficient at the input of a 50 ohm, air-insulated lossless line is $0.6 + j0.8$ at 1000 MHz. What type of circuit element is connected at the load end if the line is 8.0 cm long? What is its value?

3–14. A lossless line is terminated by an impedance that reflects 16 percent of the incident power. Calculate the SWR on the line.

3–15. The SWR on a lossless 75 ohm line is 4.0. Calculate the maximum and minimum values of voltage and current on the line when the incident voltage is 30 V.

3–16. A 300 ohm line is terminated in a pure inductance. The load reflection coefficient is $1.0 \underline{/50°}$ at 1500 MHz. Calculate the inductance (in nH).

3–17. The standing wave pattern in Fig. 3–12 is for the case $Z_L = +jZ_0$. Prove that the first voltage null occurs at $d = 3\lambda/8$. Verify the result by using the counterrotating phasor diagram.

3–18. A lossless line is terminated in $\Gamma_L = 0.7\underline{/45°}$. Calculate the distance from the load to the first current minimum when $\lambda = 20$ cm. Is there a current maximum between the minimum and the load?

3–19. A 200 ohm lossless line is terminated in a pure resistance. What are the two possible values of load resistance if 25 percent of the incident power is reflected?

3–20. Referring to Fig. 3–10, $V_G = 30$ V, $Z_G = 150$ ohms, $Z_0 = 50$ ohms, $Z_L = 100$ ohms, and $\alpha = 0$. Find the change in load power when βl is increased from 1.5π to 2π rad. What would be the change in load power if Z_0 had been 150 ohms?

3–21. Referring to Fig. 3–10, $V_G = 30\underline{/0}$ V, $Z_G = 100$ ohms, $Z_L = 60 + j60$ ohms, $Z_0 = 50$ ohms, $l = 2.0$ m, $\alpha = 0$, and $\beta = 1.2\pi$ rad/m.
   (a) Calculate the phasor voltage and current at the load (V_L and $\mathbf{I}_L$).
   (b) Calculate the load and input powers.

3–22. Repeat part b of Prob. 3–21 with $\alpha = 0.40$ dB/m. Also calculate the power dissipated in the line.

3–23. Repeat Prob. 3–22 with $Z_0 = 100$ ohms.

3–24. A load impedance with a return loss of 6 dB is connected to a lossless line. Calculate the SWR on the line. Repeat for a return loss of 20 dB.

3–25. What fraction of the incident power is absorbed by the load when the load return loss is 0 dB? Repeat for an infinite return loss.

3–26. A resistive load reflects 5 percent of its incident power. Calculate the input return loss (in dB) when the line is 30 cm long and $\alpha = 0.20$ dB/cm.

3–27. Calculate the reflection loss (in dB) for the two cases in Prob. 3–24. Assume $Z_G = Z_0$.

3–28. A lossless 50 ohm line is terminated in $Z_L = 80 + j50$ ohms. Calculate the input impedance at 2000 MHz if the line is 8.4 cm long. Assume $\lambda = 0.8\lambda_0$.

3–29. A lossless, dielectric-filled coaxial line with $Z_0 = 75$ ohms is terminated by a parallel combination of a 150 ohm resistor and a 4.0 nH inductor. Calculate the input admittance at 5000 MHz if the line is 0.8 cm long. Assume a nonmagnetic dielectric with $\epsilon_R = 2.25$.

3–30. A load impedance is connected to a lossless 50 ohm line. The impedance at $d = 0.4\lambda$ from the load is $30 - j20$ ohms. Calculate the value of load impedance.

3–31. A 60 cm length of 150 ohm transmission line is terminated in a reactive load having a value $-j150$ ohms. Calculate the real part of the input impedance when $\alpha = 3$ dB/m and $\lambda = 30$ cm.

3–32. A lossless 50 ohm line is terminated in $-j80$ ohms. What is the smallest value of l/λ that results in $Z_{in} = 0$?

3–33. A 12 V, 50 ohm generator is connected to a pair of lossless 50 ohm lines each terminated with a 100 ohm resistor as shown. One line is $\lambda/4$ long and the other is $\lambda/2$ in length. Calculate the power dissipated in each load resistor.

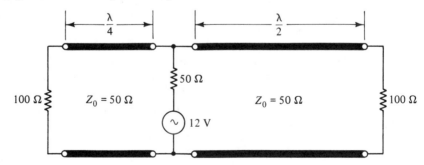

3–34. A 750 MHz, 12 V source has an internal resistance of 200 ohms.
   (a) Design a quarter-wave transformer, using an air-insulated coaxial line, to deliver all the available generator power to an 8 ohm resistive load. Calculate the length and the b/a ratio for the coaxial line.
   (b) What is the increase in load power (in dB) obtained by inserting the transformer between the generator and the load?

3–35. A one wavelength long, 90 ohm line is terminated in $Z_L = 20 - j30$ ohms. Its input terminals are connected to the load end of a 60 ohm, quarter-wavelength line. What is the impedance at the input to the 60 ohm line?

3–36. A length of 300 ohm TV twin-lead is short circuited at one end. Determine the minimum line length that results in an input impedance of $-j150$ ohms at 600 MHz. Assume $\lambda = 0.9\lambda_0$.

3–37. A length of open-circuited transmission line with $Z_0 = 90$ ohms provides the capacitive reactance for a microwave filter.

   (a) Calculate the capacitive reactance at 400 MHz when $l = 5.0$ cm and $v = 2 \times 10^8$ m/s. What is the percent error if Eq. (3–93) is used to calculate the capacitive reactance?

   (b) How much shorter would the line length be if a 30 ohm line were used to obtain the reactance value calculated in part *a?*

3–38. Determine the characteristic impedance and wave velocity of a 25 cm length of lossless transmission line from the following 300 MHz experimental data: $Z_{sc} = -j90$ ohms and $Z_{oc} = +j40$ ohms. Assume βl is less than π rad.

3–39. Given $\beta l = n\pi$ rad and $\alpha = 1.5$ dB/m, determine the minimum line length required for Z_{in} of either a shorted or open-circuited line to be within 2 percent of the characteristic impedance.

3–40. The input of a 1.4 m length of open-circuited line is connected in parallel with a 26 ohm resistor. The characteristics of the line are $Z_0 = 50$ ohms, $\alpha = 2.0$ dB/m, and $\lambda = 0.80$ m. What is the SWR when the parallel combination terminates a lossless 75 ohm line?

3–41. Derive Eq. (3–100). Calculate the Q at 2.5 GHz for a shorted, air-insulated 100 ohm line. The line is 2.90 cm long and its attenuation constant $\alpha = 0.20$ dB/m.

3–42. A lossless line of characteristic impedance Z_0 is terminated in an *imperfect* short having a resistance R, where $R \ll Z_0$. Determine the line length that results in maximum Q when the line is used as an inductor. Calculate X_L and Q if $Z_0 = 100$ ohms and $R = 1$ ohm.

3–43. A 75 ohm transmission line is terminated in a load impedance equal to $90 - j30$ ohms. Use the Smith chart to calculate the SWR and the load reflection coefficient.

3–44. A lossless 50 ohm line is terminated in an admittance Y_L. The SWR along the line is 3.0. Use the Smith chart to determine

   (a) Y_L if $\phi_L = 60°$;   (b) Y at a point $\lambda/8$ from the load.

3–45. The input impedance of a 4.8 cm, air-insulated 50 ohm line is $15 + j25$ ohms at 2.5 GHz. Use the Smith chart to determine the load impedance Z_L if

   (a) $\alpha = 0$ and   (b) $\alpha = 0.02$ Np/cm.

3–46. A lossless 40 ohm line is terminated in a 20 ohm resistor. What are the possible values of line length that result in

   (a) $Z_{in} = 40 - j28$ ohms at 1 GHz? Assume $\lambda = \lambda_0$.

   (b) $Z_{in} = 48 + j16$ ohms at 1 GHz?

   Use the Smith chart rather than the impedance transformation equation.

3–47. A shorted, air-filled, 60 ohm coaxial line is used to provide inductive reactance in a microwave filter. Use the Smith chart to determine the line length required for $Z_{in} = +j30$ ohms at 4 GHz. Check your solution with Eq. (3–87). Will the line length be smaller if a 90 ohm line is used?

3–48. Given $Z_L = 25 - j20$ ohms, $Z_0 = 50$ ohms, and $\lambda = 10$ cm, use the Smith chart to determine the smallest line length that maximizes
(a) the input conductance;   (b) the input inductive susceptance.

3–49. A 16 cm length of lossless 90 ohm line is terminated in a 135 ohm resistor. Assuming $\lambda = \lambda_0$, use the Smith chart to calculate the input impedance at 750 MHz if
(a) a 4.0 pF capacitance is connected in series at a point halfway along the line;
(b) the same capacitance is connected *across* the line at the halfway point.

3–50. A 50 ohm line is terminated in a 75 ohm resistor. Its input terminals are connected to the output of a 30 ohm line. Both lines are 0.12λ long.
(a) Find Z_{in} and Γ_{in} at the input to the 30 ohm line.
(b) What are the SWR values on the two lines?

3–51. A 2.5 cm length of teflon-filled coaxial line is terminated in a passive, frequency sensitive load. The normalized load admittance values at 4.0, 5.0, and 6.0 GHz are $0.50 + j0.60$, 0.80, and $0.50 - j0.60$, respectively. Plot $\bar{Y}_{\text{in}}$ at the three frequencies on the Smith chart. Does the admittance versus frequency plot contain a loop? Explain!

3–52. A lossless 100 ohm line is terminated in $Y_{\text{in}} = 0.03 - j0.02$ mhos. Determine the distances from the load to the first current minimum and to the first current maximum. Assume $\lambda = 15$ cm.

4

Impedance Matching and Two-Port Network Analysis

The design of impedance matching networks is an important part of microwave engineering. Some of the more widely used techniques are described in the first two sections of this chapter. The remaining sections illustrate the use of matrix methods and flow graphs in analyzing two-port networks. Attenuation, transducer loss, and insertion loss and phase are defined and expressed in terms of the matrix elements.

4–1 SOME IMPEDANCE MATCHING TECHNIQUES

Beatty (Ref. 4–14) has defined two types of impedance matching. Namely,

> **Conjugate Matching:** The matching of a load impedance to a generator for maximum transfer of power.
>
> **Z_0 Matching:** The matching of a load impedance to a transmission line to eliminate wave reflections at the load.

Conjugate matching is illustrated in Fig. 4–1. From ac circuit theory, maxi-

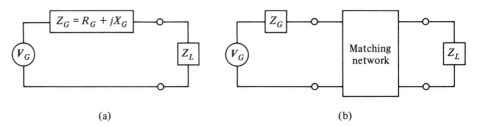

      (a)   (b)

Figure 4–1 Matching a load impedance Z_L to a generator having an internal impedance Z_G. (**Conjugate matching**)

mum power is delivered to a load when Z_L is set equal to the complex conjugate of the generator impedance. That is, $Z_L = Z_G^* = R_G - jX_G$. For this condition, the power absorbed by the load is exactly the available power of the generator (P_A). Denoting the rms value of V_G by V_G,

$$P_A = \frac{V_G^2}{4R_G} \tag{4-1}$$

In situations where the load impedance is not adjustable, a matching network may be placed between the generator and the fixed load. This is shown in part b of Fig. 4–1. The matching network is designed so that with Z_L connected at the output, its input impedance satisfies the conjugate match condition. With $Z_{in} = R_G - jX_G$, the power into the network equals P_A. If the matching network is dissipationless, all of the available generator power is delivered to the load Z_L. For a matching network with dissipation, the load power is less than P_A. When Z_G is real (that is, $Z_G = R_G$), the conjugate match condition reduces to $Z_{in} = R_G$.

The term Z_0 *matching* is used to denote matching a load impedance to the characteristic impedance of a transmission line (that is, $Z_L = Z_0$). In this case $\Gamma_L = 0$ and hence the SWR along the line is unity. If $Z_L \neq Z_0$, a matching network may be used to eliminate the standing waves on the line. This arrangement is shown in Fig. 4–2, where in most cases the network is essentially dissipationless. With Z_L connected as shown, Z_0 matching requires that the input impedance of the network equal Z_0 or $Z_{in}/Z_0 = 1 + j0$. For a Smith chart normalized to Z_0, this means that $\bar{Z}_{in} \equiv Z_{in}/Z_0$ must be at the center of the chart.

For a well-designed source, $Z_G = Z_0$, the characteristic impedance of its output line. With Z_0 real, matching the load to the line $(Z_L = Z_0)$ results in a conjugate match between the generator and the load (namely, $Z_G = Z_0 = Z_L$). If $Z_G \neq Z_0$, a matching network can be placed at the generator end to match the generator to the line. This and the load matching network are shown in Fig. 4–3. For properly designed networks, $Z_L' = Z_0$ and $Z_G' = Z_0$, where Z_L' is the impedance to the right of plane B and Z_G' is the Thevenin impedance to the left of plane A. When the matching network is dissipationless, the power available from the Thevenin equivalent must be the same as the available generator power. Hence the rms voltages are related by

$$V_G' = V_G \sqrt{\frac{Z_0}{R_G}} \tag{4-2}$$

If the generator matching network is not built into the source, it should be placed as close as possible to its output terminals. Similarly, it is good engineering practice to

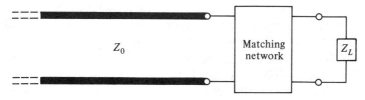

Figure 4–2 Matching a load impedance Z_L to a transmission line of characteristic impedance Z_0. (**Z_0 matching**)

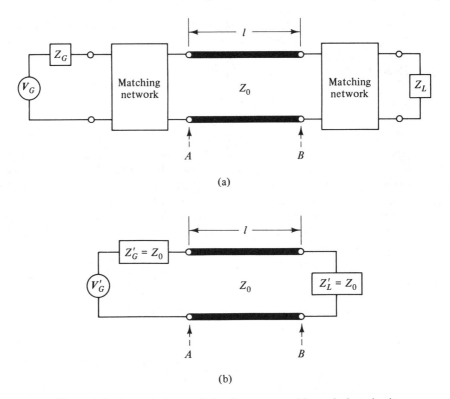

Figure 4–3   A matched transmission-line system and its equivalent circuit.

place the load matching network as close as possible to the load. This arrangement, which is shown in Fig. 4–3, has several advantages. These include

1. For dissipationless matching networks, maximum power is delivered from the generator to the load. If the transmission line is lossless, $P_L = P_A$ otherwise $P_L = P_A e^{-2\alpha l}$, where α is the attenuation constant of the line.

2. For a line with finite attenuation, power transmission is most efficient with no standing waves on the line. With SWR = 1.00, $P_L = P_{in} e^{-2\alpha l}$. For SWR $\neq$ 1.00, the load power is less. The transmission efficiency is derived in Sec. 1.37 of Ref. 4–3.

3. The load power is independent of βl, an important advantage, particularly for low-loss lines. Since $\beta l = \omega l / v$, its value is a function of frequency as well as line length. Also, temperature variations may cause a change in ϵ_R which affects v and hence βl. Some indication of the change in P_L with βl for a mismatched system is given in Prob. 3–20.

4. For a given transmitted power level, the peak voltage along the line is less when the SWR is unity. This is important in high power systems where corona loss and dielectric breakdown problems are often encountered.

5. If neither the load nor the generator is matched to the line, the transmission line may cause the output phase versus frequency characteristic to be nonlinear.

This results in modulation distortion, which reduces the information content of the signal.

6. If the SWR along the line is not unity, the impedance seen by the generator is a function of βl. For long lines and certain type sources, the generator may shift frequency intermittently or even oscillate at two frequencies simultaneously.

7. Precise measurement systems require that both the source and the load (usually a detector) be well matched to minimize errors. This is illustrated in Sec. 4–4 (Ex. 4–10).

For the reasons summarized here, many systems and components are designed to have low SWR (typically 1.25 or less) over their useful frequency range.

Recognizing the importance of impedance matching, let us consider various methods of achieving it at high frequencies. Although the following examples involve matching a load to a transmission line, the same techniques may be used to match a generator to a line or a load directly to a generator with Z_G real.

4–1a Reactive Matching Networks

One of the most widely used matching networks consists of a reactive element placed either in series or shunt with a small length of low-loss line. The design and analysis for both cases will now be explained with the aid of the Smith chart.

Series reactive matching.   Figure 4–4 illustrates the series reactance technique of matching a load Z_L to a transmission line. Specific values of Z_L and Z_0 are given in the figure. The subscripts L, A, and *in* are used to denote the impedances at the load, plane A, and the input, respectively. Note that the characteristic impedance of the line section (Z_{01}) has been chosen to equal the characteristic impedance of the input line (Z_0). This choice makes the Smith chart solution easier and usually results in a simplified configuration. With $Z_{01} = Z_0 = 50$ ohms, $\bar{Z}_L \equiv Z_L/50 = 0.50 + j0.60$. The requirement for a match condition is $\bar{Z}_{in} \equiv Z_{in}/50 = 1 + j0$. To achieve this, two steps are required. First, the line length must be chosen so that the real part of $\bar{Z}_A$ equals unity. Next a reactance X must be inserted in a series with Z_A so

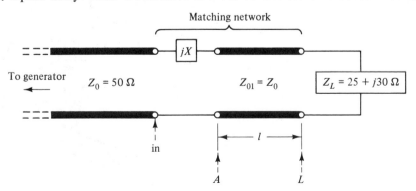

Figure 4–4   A series reactance matching network.

that the imaginary part of $\bar{Z}_{\text{in}}$ becomes zero. The Smith chart solution is outlined in Fig. 4–5. Shown is the normalized load impedance $\bar{Z}_L$ and its SWR circle. Recall that the SWR circle represents the locus of all possible impedances ($\bar{Z}_A$ in this case) that can be realized by varying the line length l. Furthermore, the $\bar{R} = 1$ circle represents all normalized impedances with a real part equal to unity. Therefore, the intersection of these two circles indicates the possible values of l that cause the real part of $\bar{Z}_A$ to equal unity. The intersection at point E yields $\bar{Z}_A = 1 + j1.1$, while the one at point F results in $\bar{Z}_A = 1 - j1.1$. The line length required to transform $\bar{Z}_L$ into $\bar{Z}_A = 1 + j1.1$ is obtained by rotating clockwise from the $\bar{Z}_L$ point to the point E. From the chart, $l = 0.065\lambda$. The line length required for $\bar{Z}_A = 1 - j1.1$ is also obtained from the chart and is $l = 0.235\lambda$. In either case, adding the appropriate amount of reactance in series with Z_A results in $\bar{Z}_{\text{in}} = 1.0$, a match condition.

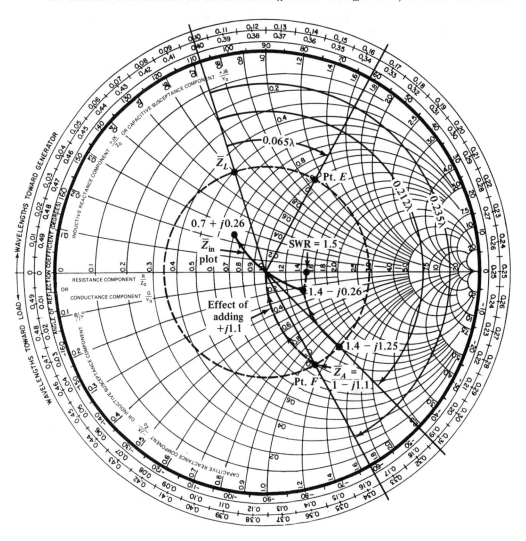

Figure 4–5 Smith chart solution for the circuit shown in Fig. 4–4.

Since the impedance at point E is inductive, a series capacitance is needed, while for point F (capacitive) a series inductance is required. Let us design the matching network using an inductor. The capacitor case is left for the reader (Prob. 4–1). Thus, $l = 0.235\lambda$ and the required value of normalized series reactance $\omega L/Z_0 = 1.1$ or $\omega L = 55$ ohms. The design procedure may be summarized as follows:

1. Rotate from $\overline{Z}_L$ on its SWR circle toward the generator to the intersection of the $\overline{R} = 1$ circle. The resultant impedance is either $\overline{Z}_A = 1 + j\overline{X}$ or $1 - j\overline{X}$.
2. Add sufficient series reactance to cancel out the imaginary portion of $\overline{Z}_A$. For $\overline{Z}_A = 1 - j\overline{X}$, the series element must be inductive. In the above example, this requires moving up 1.1 reactance units on the $\overline{R} = 1$ circle. This is indicated in Fig. 4–5. On the other hand, for $\overline{Z}_A = 1 + j\overline{X}$, the element must be capacitive, which requires moving down on the $\overline{R} = 1$ circle.
3. Calculate the line length l and the inductance or capacitance value for the specific design frequency.

To complete the example, let us assume a design frequency of 2000 MHz. Also assume an air-insulated line so that $\lambda = \lambda_0$. At 2000 MHz, $\lambda_0 = 15$ cm and therefore $l = (0.235)(15) = 3.53$ cm. Since $\omega L = 55$ ohms, $L = 55/2\pi f = 4.38$ nH. With this matching network, $\overline{Z}_{in} = 1.0$ and thus the SWR on the 50 ohm line to the left of Z_{in} is unity.[1]

One might suspect that if the matching network were used at other than the design frequency, the line would not be perfectly matched and hence the SWR would be greater than unity. This suspicion is reasonable since both l/λ_0 and $\omega L/Z_0$ are functions of frequency. Table 4–1 lists their values at 1800 MHz and 2200 MHz as well as at the design frequency. The Smith chart can be used to analyze the performance of the matching network as a function of frequency. It will be assumed that $\overline{Z}_L$ is not frequency dependent. At 1800 MHz, $\overline{Z}_A$ is obtained by rotating from $\overline{Z}_L$ 0.212λ toward the generator on its SWR circle. From the Smith chart (Fig. 4–5), $\overline{Z}_A = 1.4 - j1.25$. Adding $+j0.99$ due to the series inductance results in $\overline{Z}_{in} = 1.4 - j0.26$. Since the real part of $\overline{Z}_{in}$ is not unity and the imaginary part is finite, the line is no longer matched. The resultant SWR is obtained by rotating the $\overline{Z}_{in}$ point to the right half of the resistance axis. From the chart, the SWR $= 1.50$ at 1800 MHz. Using the same procedure at 2200 MHz, one can show that the

TABLE 4-1 Impedance and SWR Data for the Matching Network in Fig. 4–4

Freq. (MHz)	λ_0 (cm)	l/λ_0	$\dfrac{\omega L}{Z_0}$	$\overline{Z}_A$	$\overline{Z}_{in}$	SWR
1800	16.67	0.212	0.99	$1.4 - j1.25$	$1.4 - j0.26$	1.50
2000	15.00	0.235	1.10	$1.0 - j1.10$	1.00	1.00
2200	13.64	0.259	1.21	$0.7 - j0.95$	$0.7 + j0.26$	1.65

Note: $l = 3.53$ cm, $L = 4.38$ nH, and $\overline{Z}_L = 0.50 + j0.60$.

[1] The SWR along the 0.235λ section of line is not unity but 2.8 as indicated on the Smith chart. This standing wave provides the impedance transformation needed to make the real part of $\overline{Z}_{in}$ unity.

SWR = 1.65. The impedance and SWR data is listed in Table 4–1. A plot of $\bar{Z}_{in}$ versus frequency is shown on the Smith chart.

The frequency sensitivity of l/λ_0 is one of the causes for the deterioration in SWR at 1800 and 2200 MHz, the other being the frequency sensitivity of $\bar{X}_L$. This is because the fixed length of line (3.53 cm) spreads the $\bar{Z}_A$ impedance plot and therefore adding a series reactance cannot produce $\bar{Z}_{in} = 1.0$ at all frequencies. For this particular example, the choice of capacitive matching results in a shorter line length ($l = 0.065\lambda$). Thus one might expect that the resultant matching network would be *less* frequency sensitive. It is left to the reader to show that this is indeed the case (Prob. 4–1).

At microwave frequencies, lumped inductors and capacitors are difficult to construct. As a result, shorted and open-circuited lines (called *stubs*) are often used to provide the reactance needed for matching. Two examples of series matching with stubs are shown in Fig. 4–6, where l_s and Z_{0s} are the length and characteristic impedance of the stub line. By using the reactance value obtained from the matching procedure, one can determine l_s from either Eq. (3–88) or (3–92). Although the choice of Z_{0s} is arbitrary, it is best to use a *high* impedance shorted line for inductive reactance and a *low* impedance open-circuited line for capacitive reactance. This leads to the smallest value of l_s, which reduces the reactance or susceptance variation with frequency. The reactance variation of a shorted line (Fig. 3–14) approximates that of a lumped inductor when l_s is less than 0.10λ (that is, $\beta l < \pi/5$) at all frequencies of interest. Likewise, an open-circuited stub approximates a lumped capacitor when l_s is less than 0.10λ.

For the matching network just designed, a 4.38 nH inductor was required to produce unity SWR at 2000 MHz. Suppose a 100 ohm shorted line is used to create the required inductance. From Eq. (3–88),

$$(4\pi \times 10^9)(4.38 \times 10^{-9}) = 100 \tan \frac{2\pi l_s}{\lambda}$$

Hence $l_s = 0.08\lambda$, and assuming $\lambda = \lambda_0$ for the stub line, $l_s = (0.08)(15) = 1.2$ cm.[2] Note that for our choice of Z_{0s}, $l_s < 0.10\lambda$ even at the high end of the frequency band.

It is interesting to look at the matching procedure from a wave-reflection point of view. Without the matching network, the load impedance $\bar{Z}_L = 0.50 + j0.60$ causes an SWR = 2.8 and hence from Eq. (3–53), $|\Gamma_L| = 0.48$. If on the other hand, the matching network were terminated in a reflectionless load ($Z_L = Z_0$), the only reflection would be due to the series inductance. That is, $\bar{Z}_{in} = 1 + j1.1$ at 2000 MHz which also causes an SWR = 2.8. Thus the reflection due to the inductance has the same magnitude as Γ_L. Matching occurs when the two reflections as seen at the input cancel resulting in $\Gamma_{in} = 0$ and an input SWR of unity. The function of the line length l is to phase the two reflections so that they cancel at the input plane. This point of view is very useful in understanding the operation of matching networks, transformers, and microwave filters.

---

[2] The value of l_s can also be obtained from a Smith chart normalized to Z_{0s}. Determine the distance on the wavelength scale to move from $Z = 0$ (a short circuit) to the point $+j55/100 = +j0.55$. This yields $l_s = 0.08\lambda$.

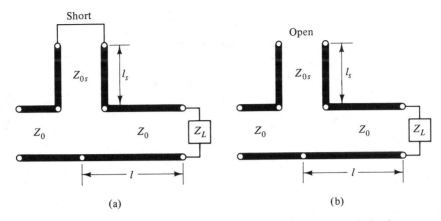

Figure 4–6   Series reactive matching with shorted and open-circuited stubs.

Shunt reactive matching.   This technique is similar to series matching except that the reactive element is placed in shunt with the line. At microwave frequencies, it is generally easier to build shunt elements into a transmission-line structure. In a waveguide system, for example, shunt reactances usually take the form of a thin metal septum (see Sec. 7–2).

A shunt reactance matching network is shown in Fig. 4–7, where again Z_{01} is set equal to Z_0. To illustrate the technique, let $Z_0 = 50$ ohms and $Z_L = 20$ ohms, resistive. The method however is applicable to any complex load impedance. First, $\bar{Z}_L = 20/50 = 0.40$ is plotted on the Smith chart (Fig. 4–8). With the matching element in shunt, the Smith chart solution must be carried out in terms of admittances. Therefore, the first step is to convert $\bar{Z}_L$ to $\bar{Y}_L (= Y_L/Y_0)$ by rotating from the $\bar{Z}_L$ point halfway around the chart on its SWR circle. From the chart, $\bar{Y}_L = 2.50$. Since matching requires $\bar{Y}_{\text{in}} = 1$, the line length l is chosen so that the real part of $\bar{Y}_A$ equals unity. As before, there are two possibilities since the SWR circle intersects the $\bar{G} = 1$ circle at two points, namely, $\bar{Y}_A = 1 + j0.95$ and $\bar{Y}_A = 1 - j0.95$. For $l = 0.09\lambda$, $\bar{Y}_A = 1 - j0.95$ and with $Y_0 = 0.02$ mho, $Y_A = 0.02 - j0.019$ mho. This means that a shunt capacitance of admittance value $+j0.019$ mho is required to cancel the inductive effect of Y_A. Thus adding a capacitive susceptance $B = \omega C = 0.019$ mho at plane A results in $Y_{\text{in}} = 0.02$ mho or $\bar{Y}_{\text{in}} = 1.0$, a matched

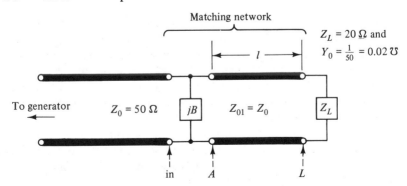

Figure 4–7   A shunt reactance matching network.

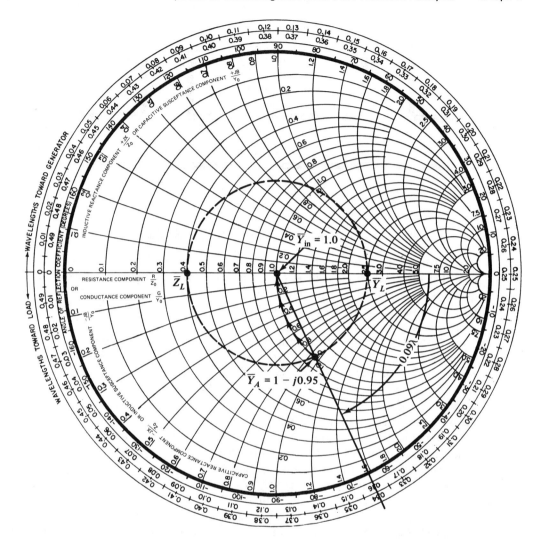

Figure 4–8 Smith chart solution for the circuit shown in Fig. 4–7.

condition. The dashed arrow on the chart from $\overline{Y}_A$ to the $\overline{Y}_{in}$ point indicates the effect of adding the capacitive susceptance. Adding $+j0.019$ mho in shunt is equivalent to moving up 0.95 susceptance units on the $\overline{G} = 1$ circle. Problem 4–3 at the end of the chapter requires the calculation of l and C at a specified design frequency. Analysis of the circuit at two other frequencies is also required. The procedure is similar to that explained for series reactive matching except that all calculations must be done in terms of admittances.

As explained earlier, shorted and open-circuited stubs may be used to obtain the required susceptance. The two shunt stub configurations are illustrated in Fig. 4–9. As before, it is best to use a high impedance shorted line for inductive susceptance and a low impedance open-circuited line for capacitive susceptance. In the above example, the required capacitance value is given by $C = 0.019/\omega$. For a

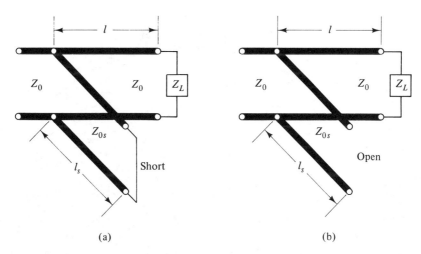

(a)   (b)

Figure 4–9   Shunt reactive matching with shorted and open-circuited stubs.

specific design frequency and a particular value of Z_{0s}, the stub length l_s can be computed from Eq. (3–92). The advantage of using a low impedance open-circuited stub is illustrated in Prob. 4–4.

The matching configurations described in Figs. 4–6 and 4–9 are known as *single-stub tuners*. In coaxial and strip transmission line systems, the shunt connection is by far the more common arrangement.

Broadband matching networks.   Series and shunt reactance matching techniques usually result in low SWR values over a narrow frequency range (typically, less than 20 percent). Improved broadband performance can often be realized by utilizing series and parallel resonant circuits. These circuits may consist of lumped inductors and capacitors or quarter-wave open and shorted stubs. In either case, their function is to reduce the reactive effects of the matching network over the frequency band of interest. The following example illustrates the design procedure.

Example 4–1:

A load impedance has been matched to a 50 ohm line at 2000 MHz using a reactance matching technique. The input impedance values (Z_D) at 1600, 2000, and 2400 MHz are $80 - j40$ ohms, 50 ohms, and $30 + j20$ ohms, respectively. Their normalized values ($\bar{Z}_D \equiv Z_D/50$) are plotted on the Smith chart in Fig. 4–10. From the chart, it appears that the SWR values range from 2.2 at the low end to 2.0 at the high end of the frequency band. It is proposed that the circuit shown in Fig. 4–11 be used to improve the input SWR. Determine the values of L, C, and l.

Solution: Since the L-C circuit is in shunt with the line, it is best to solve the problem in terms of admittance. Thus the first step is to convert the $\bar{Z}_D$ plot to a $\bar{Y}_D$ plot as explained previously. The result is shown in Fig. 4–10. Next, the line length l is chosen to produce the symmetrical admittance plot $\bar{Y}_E$. Note that Y_E is capacitive below 2000 MHz and inductive above 2000 MHz. This is desired since with the resonant circuit tuned to 2000 MHz, its reactive effect tends to cancel that due to Y_E. By trial and error, it is determined that $l = 0.90$ cm produces the desired $\bar{Y}_E$ characteristic, where $\lambda = \lambda_0$ has been assumed. For this value of l, the $\bar{Y}_E$ plot is obtained by rotating the $\bar{Y}_D$

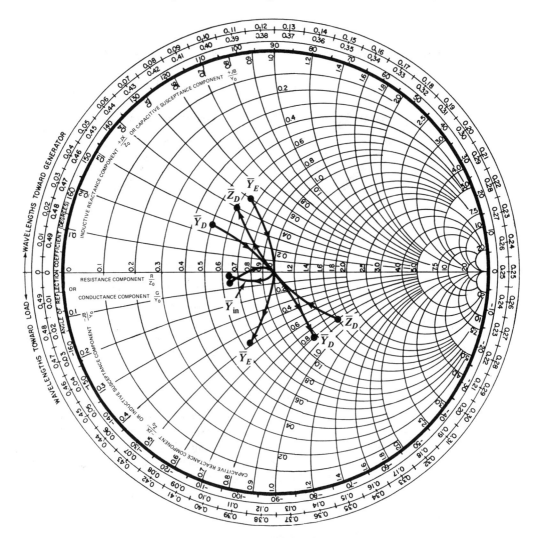

Figure 4–10 Smith chart analysis of the broadband matching network shown in Fig. 4–11. ($l = 0.90$ cm, $L = 3.25$ nH, and $C = 1.95$ pF.)

plot toward the generator $0.90/18.75 = 0.048$ at the low end and $0.90/12.50 = 0.072$ at the high end of the frequency band. At midband (2000 MHz), $\bar{Y}_E$ is unity since with $\bar{Y}_D$ at the center of the chart, $\bar{Y}_E$ is independent of l. The normalized admittance values are listed in Table 4–2.

The purpose of the resonant circuit is to offset the susceptance portion of $\bar{Y}_E$ at the band edges while leaving its midband value unchanged. Thus, the circuit must provide negative susceptance below midband, positive susceptance above midband and zero at the midband frequency. A parallel L-C circuit tuned to 2000 MHz has this characteristic. Its normalized susceptance is given by

$$\bar{B} \equiv \frac{B}{Y_0} = \frac{1}{Y_0}\left(\omega C - \frac{1}{\omega L}\right) = \frac{\omega_r C}{Y_0}\left(\frac{\omega}{\omega_r} - \frac{\omega_r}{\omega}\right) \tag{4–3}$$

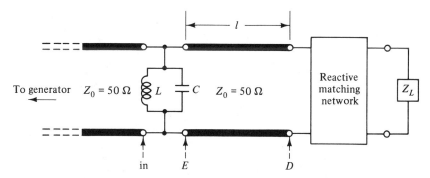

Figure 4–11 A broadband matching technique that utilizes a resonant L-C circuit.

TABLE 4–2 Admittance and SWR Data for the Broadband Matching Circuit in Fig. 4–11. The Smith Chart Solution is Shown in Fig. 4–10.

Freq. (MHz)	$\lambda = \lambda_0$ (cm)	l/λ	$\overline{Y}_D$	$\overline{Y}_E$	$\overline{Y}_{in}$	Input SWR
1600	18.75	0.048	$0.50 + j0.25$	$0.65 + j0.50$	$0.65 - j0.05$	1.55
2000	15.00	0.060	1.00	1.00	1.00	1.00
2400	12.50	0.072	$1.15 - j0.80$	$0.65 - j0.48$	$0.65 - j0.03$	1.55

where $\omega_r = 1/\sqrt{LC}$ is the resonant frequency. Its normalized susceptance slope at ω_r is given by

$$\left.\frac{d\overline{B}}{d\omega}\right|_{\omega_r} = \frac{2C}{Y_0} = 2CZ_0 \qquad (4\text{--}4)$$

The required capacitance value is obtained by equating this slope to the negative of the $\overline{Y}_E$ susceptance slope. With $Z_0 = 50$ ohms,

$$100\,C = -\frac{(-0.48 - 0.50)}{2\pi(2.40 - 1.60)10^9} \qquad \text{or} \qquad C = 1.95 \text{ pF}$$

where the $\overline{Y}_E$ susceptance slope has been approximated by $\Delta\overline{B}/\Delta\omega$ between 1600 and 2400 MHz. From the resonance condition, $L = (\omega_r^2 C)^{-1} = 3.25$ nH.

Using the above values and Eq. (4–3), $\overline{Y}_{in} = \overline{Y}_E + j\overline{B}$ may be computed at the three frequencies of interest. The results are listed in Table 4–2 and plotted in Fig. 4–10. The input SWR values are obtained from the Smith chart and listed in the table.

From this example, it is clear that the 0.90 cm length of line and the resonant circuit has improved the SWR across the 40 percent frequency band. Further improvement could have been realized if the original reactive matching network had been designed so that the $\overline{Y}_D$ plot were displaced slightly to the right on the chart. This situation and the result is sketched in Fig. 4–12. The 0.90 cm line produces the symmetrical $\overline{Y}_E$ plot shown. As before, the effect of the L-C circuit is to add inductive susceptance below midband and capacitive susceptance above midband. This

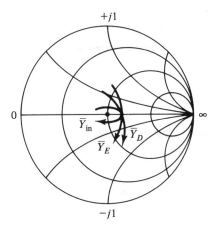

Figure 4–12 Further refinement of the broadband matching technique described in Figs. 4–10 and 4–11.

means that the low end of $\overline{Y}_E$ moves *down* and the high end moves *up* on their respective constant conductance circles. The result is shown as the $\overline{Y}_{\text{in}}$ curve. With this method an SWR of less than 1.30 can be achieved across the 40 percent band.

As indicated earlier, shorted and open-circuited stubs can be used in place of lumped L-C circuits. Their properties were discussed in Sec. 3–6. For example, the susceptance behavior of a quarter-wave shorted line is similar to that of a parallel L-C circuit. Its input admittance $Y_{\text{in}} = 1/Z_{\text{in}}$, where Z_{in} is given by Eq. (3–87). Thus the susceptance of a shorted stub of length l_s and characteristic admittance Y_{0s} is given by

$$B = -Y_{0s} \cot \frac{2\pi l_s}{\lambda} = -Y_{0s} \cot \frac{\omega l_s}{v} \qquad (4-5)$$

In order to produce the same characteristics as a parallel L-C circuit, l_s must be $\lambda/4$ long at the resonant frequency ω_r. Also, the susceptance slope of the stub must equal that of the L-C circuit at ω_r. For the shorted stub, the normalized susceptance slope at ω_r is

$$\left.\frac{d\overline{B}}{d\omega}\right|_{\omega_r} = \frac{1}{Y_0}\left.\frac{dB}{d\omega}\right|_{\omega_r} = \frac{Y_{0s}}{4 f_r Y_0} = \frac{Z_0}{4 f_r Z_{0s}} \qquad (4-6)$$

where for the broadband matching network, f_r is the midband frequency. Thus, the L-C circuit in Ex. 4–1 may be replaced by a shorted stub with $l_s = \lambda/4$ at 2000 MHz. The required value of Z_{0s} is obtained by equating Eqs. (4–4) and (4–6).

Another technique for improving the SWR characteristic of a frequency sensitive load impedance is described in Fig. 4–13. The length of the Z_{01} line section is $\lambda/2$ at the midband frequency. Impedance plots of the form shown in part *b* of the figure can be improved with this technique. If the load impedance plot ($\overline{Z}_L = Z_L/Z_0$) is not of the desired form, a length l_2 of Z_0 line can be used to rotate the plot. The choice of Z_{01} depends upon the form of the resultant $\overline{Z}_R(=Z_R/Z_0)$ plot. If $\overline{Z}_R$ is similar to plot #1, choose $Z_{01} > Z_0$, while if $\overline{Z}_R$ resembles plot #2, choose $Z_{01} < Z_0$.

The operation of the half-wave line section is best understood with the aid of the Smith chart in Fig. 4–13b. Assume that plot #1 represents $\overline{Z}_R$ and that $Z_{01} = 2Z_0$. Renormalizing Z_R to Z_{01} results in the plot labeled $\overline{\overline{Z}}_R$, where $\overline{\overline{Z}}_R \equiv Z_R/Z_{01} =$

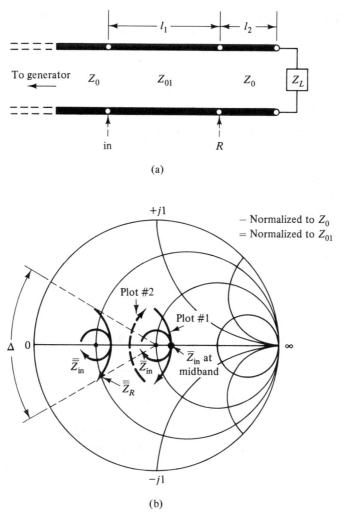

Figure 4–13 Broadband matching with a half-wavelength transmission-line section. ($l_1 = \lambda/2$ at the midband frequency.)

$0.50\ \overline{\overline{Z}}_R$. With l_1 set equal to a half wavelength at midband, it will be *less* than 0.50λ at the low frequency end and *greater* than 0.50λ at the high end. Rotating toward the generator the appropriate amounts at the three frequencies results in the $\overline{\overline{Z}}_{in}$ plot as shown. The *closing up* of the impedance plot is due to the fact that the high frequency point rotates a greater amount than the low frequency point. Normalization back to Z_0 yields the $\overline{Z}_{in}$ plot shown, where $\overline{Z}_{in} \equiv Z_{in}/Z_0 = 2\overline{\overline{Z}}_{in}$. Note that at midband $Z_{in} = Z_R$ as expected since impedance repeats every half wavelength. At the band edges, however, the $\overline{Z}_{in}$ points have much lower SWR values than the corresponding $\overline{\overline{Z}}_R$ points.

The choice $Z_{01} = 2Z_0$ is not necessarily optimum. It was selected simply to demonstrate the technique. The optimum value of Z_{01} can be determined by a trial-

and-error procedure on the Smith chart or by using the impedance transformation equation. It is chosen so that the angular spread of the resultant $\bar{\bar{Z}}_R$ plot on the fractional wavelength scale equals the difference in l_1/λ values at the band edges. That is, select Z_{01} so that

$$\frac{l_1}{\lambda_H} - \frac{l_1}{\lambda_L} = \Delta \qquad (4\text{–}7)$$

where λ_L and λ_H are respectively the wavelengths at the low and high ends of the frequency band and Δ denotes the angular spread of $\bar{\bar{Z}}_R$ on the fractional wavelength scale (see Fig. 4–13b). By choosing Z_{01} in this manner, the resultant $\bar{\bar{Z}}_{\text{in}}$ and $\bar{Z}_{\text{in}}$ plots will resemble those shown in the figure.

The performance obtained with this broadbanding technique is comparable to that produced by the resonant circuit method (Fig. 4–11). In most cases, this technique results in a simpler mechanical configuration.

4–1b Transmission-Line Sections as Matching Networks

Another widely used impedance matching technique utilizes one or more transmission-line sections in cascade with the load impedance. The design procedure for two such configurations are described here. Other designs include the quarter-wave and tapered-line transformers. These are discussed in Sec. 4–2.

The short transformer. One method of matching a complex load impedance Z_L to a Z_0 transmission line is with a short length of lossless transmission line. The matching arrangement is shown in Fig. 4–14, where Z_{01} and l are the characteristic impedance and length of the line section. The method seems plausible since, from a wave-reflection viewpoint, one can imagine Z_{01} chosen so that the reflections from the input and load planes are equal. With Z_{01} so chosen, l can then be adjusted to cancel the two reflections, resulting in an input SWR of unity. The following analysis shows that this is indeed the case, *under certain conditions*.

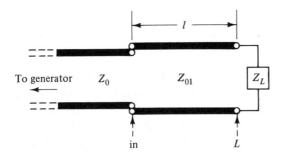

Figure 4–14 A short transformer. $(Z_L = R + jX)$

The Input impedance of the transformer is obtained from Eq. (3–83), the impedance transformation equation. Setting $Z_{\text{in}} = Z_0$ for a perfect match yields

$$Z_{\text{in}} = Z_{01} \frac{Z_L + jZ_{01} \tan \beta l}{Z_{01} + jZ_L \tan \beta l} = Z_0$$

In general, $Z_L = R + jX$ and thus

$$RZ_{01} + jZ_{01}(X + Z_{01} \tan \beta l) = Z_0 Z_{01} - Z_0 X \tan \beta l + jZ_0 R \tan \beta l$$

By separately equating real and imaginary parts, the following design equations are obtained.

$$Z_{01} = \sqrt{RZ_0 - \frac{X^2 Z_0}{Z_0 - R}} \qquad \text{and} \qquad \tan \beta l = Z_{01} \frac{Z_0 - R}{XZ_0} \qquad (4\text{–}8)$$

In order for a solution to exist, Z_{01} must be real. This occurs for either of the following set of conditions:

$$R > Z_0 \qquad \text{or} \qquad R < Z_0 \qquad \text{and} \qquad X^2 < R(Z_0 - R) \qquad (4\text{–}9)$$

For example, if $R = 20$ ohms and $Z_0 = 100$ ohms, the load cannot be matched if $|X|$ is greater than 40 ohms. Furthermore, practical values of Z_{01} are limited by other considerations (loss, power handling, and mechanical tolerances). For coaxial and strip-type transmission lines, the characteristic impedance is usually restricted to the 10 to 150 ohm range. The simplicity of this matching technique makes it attractive in most narrowband applications. It is called a *short* transformer since under certain conditions the line length l is less than a quarter wavelength.

The L-C impedance transformer.   A matching network that is used in certain low-frequency applications is the so-called *L-C transformer*. Figure 4–15 shows the connection for a resistive termination when $R_L > Z_0$. For a given value of R_L, the values of L and C required for a perfect match ($Z_{in} = Z_0$) at the design frequency are obtained from the following relations.
For $R_L > Z_0$,

$$\omega L = \sqrt{Z_0(R_L - Z_0)} \qquad \text{and} \qquad \omega C = \sqrt{\frac{R_L - Z_0}{Z_0 R_L^2}} \qquad (4\text{–}10)$$

When $R_L < Z_0$, the network must be reversed so that the series inductor is connected to R_L and the shunt capacitor is connected to the input terminals.
For $R_L < Z_0$,

$$\omega L = \sqrt{R_L(Z_0 - R_L)} \qquad \text{and} \qquad \omega C = \sqrt{\frac{Z_0 - R_L}{R_L Z_0^2}} \qquad (4\text{–}11)$$

These equations are obtained by deriving an expression for the input impedance and equating the real part to Z_0 and the imaginary part to zero. Similar expressions can be derived if the load impedance is complex.

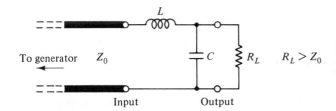

Figure 4–15 The L-C transformer. (Note: for $R_L < Z_0$, reverse the transformer so that L is connected in series with R_L and C is shunt connected to the input terminals.)

At microwave frequencies, the required inductive reactance and capacitive susceptance may be realized with shorted and open stubs. A high impedance, shorted line usually provides the inductance (Eq. 3–88), while capacitance is obtained with a low impedance, open-circuited line (Eq. 3–92). A shorted stub in series with the main transmission line often presents construction difficulties. This is particularly true in coaxial and strip-type transmission systems. These difficulties can be circumvented by cascading two transmission-line sections (Z_{01}, l_1 and Z_{02}, l_2) as shown in Fig. 4–16. The equivalence between this circuit and the L-C transformer is based upon the following approximations.

1. A *small* section of *high* impedance line is equivalent to a series inductance.
2. A *small* section of *low* impedance line is equivalent to a shunt capacitance.

The meaning of *small*, *high*, and *low* is clarified by the following derivation.

The input impedance of a lossless transmission line of length l and terminated by a load impedance Z_L is given by the impedance transformation equation (Eq. 3–83). Namely,

$$Z_{in} = Z_0 \frac{Z_L + jZ_0 \tan \beta l}{Z_0 + jZ_L \tan \beta l} \approx Z_0 \frac{Z_L + jZ_0 \beta l}{Z_0 + jZ_L \beta l}$$

where the approximate form is valid when $\beta l < \pi/6$ ($l < \lambda/12$). With this restriction, $\tan \beta l \approx \beta l$ radians to within 10 percent. When $Z_0 \gg |Z_L|$ (say, $Z_0 > 3|Z_L|$), the approximate form reduces to

$$Z_{in} \approx Z_L + jZ_0 \beta l = Z_L + j\omega \frac{Z_0 l}{v} \tag{4–12}$$

Thus a *small* section of *high* impedance line behaves like a series inductance of value $Z_0 l/v$ henries. By using the admittance transformation equation (Eq. 3–84), one can show that for $l < \lambda/12$ and $Y_0 > 3|Y_L|$, the input admittance is approximated by

$$Y_{in} \approx Y_L + jY_0 \beta l = Y_L + j\omega \frac{Y_0 l}{v} \tag{4–13}$$

Thus a *small* section of *low* impedance line behaves like a shunt capacitor of value $Y_0 l/v$ farads. Most microwave engineers find these approximations useful in conceptualizing and designing microwave circuits.

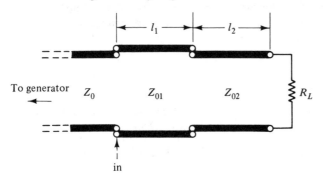

Figure 4–16   A transmission-line equivalent of the L-C transformer.

Given the above approximations, the circuit in Fig. 4–16 can be made equivalent to an L-C transformer. For example, with $Z_{02} \ll R_L$ and $Z_{01} \gg Z_{02}$, the circuit approximates the transformer in Fig. 4–15. If, on the other hand, $R_L < Z_0$, the L-C transformer must be reversed, which means that the conditions for the transmission-line equivalent are $Z_{02} \gg R_L$ and $Z_{01} \ll Z_{02}$. The procedure for designing the transmission-line equivalent of the L-C transformer is as follows:

1. Given R_L, Z_0 and the design frequency, use either Eq. (4–10) or (4–11) to calculate the values of L and C.

2. If $R_L > Z_0$, choose $Z_{02} < R_L/3$ and $Z_{01} > 9Z_{02}$. Calculate l_1 and l_2 from $L \approx Z_{01}l_1/v_1$ and $C \approx Y_{02}l_2/v_2$, where v_1 and v_2 are respectively the wave velocities in the Z_{01} and Z_{02} lines. Conversely, if $R_L < Z_0$, choose $Z_{02} > 3R_L$ and $Z_{01} < Z_{02}/9$. The line lengths are obtained from $L \approx Z_{02}l_2/v_2$ and $C \approx Y_{01}l_1/v_1$.

3. Verify that both l_1 and l_2 are less than $\lambda/12$ at the design frequency. If they are not, increase the impedance of the line that simulates the series inductor and/or decrease the impedance of the line that simulates the shunt capacitor.

In some cases, this procedure results in values of Z_{01} and Z_{02} that are impractical. As indicated earlier, TEM lines are usually restricted to the 10 to 150 ohm range. If, as a result, the above impedance inequalities cannot be satisfied, the correct values of l_1 and l_2 must be obtained from an exact analysis of the transmission-line network in Fig. 4–16. This may be done by applying the impedance transformation equation to both the Z_{01} and Z_{02} line sections and obtaining an expression for the input impedance of the network. Setting $Z_{in} = Z_0$ and separately equating real and imaginary parts yields the following design equations for l_1 and l_2 (see Prob. 4–11).

$$\tan \beta_1 l_1 = Z_{01} \left\{ \frac{(R_L - Z_0)(Z_{02}^2 - Z_0 R_L)}{(R_L Z_{01}^2 - Z_0 Z_{02}^2)(Z_0 R_L - Z_{01}^2)} \right\}^{1/2} \tag{4–14}$$

$$\tan \beta_2 l_2 = Z_{02} \left\{ \frac{(R_L - Z_0)(Z_{01}^2 - Z_0 R_L)}{(R_L Z_{01}^2 - Z_0 Z_{02}^2)(Z_0 R_L - Z_{02}^2)} \right\}^{1/2} \tag{4–15}$$

where $\beta_1 = \omega/v_1$ and $\beta_2 = \omega/v_2$. For $R_L > Z_0$, choose $Z_{01} > \sqrt{R_L Z_0}$ and as large as practical, $Z_{02} < \sqrt{R_L Z_0}$ and as small as practical, and then use the above equations to determine the exact values of l_1 and l_2. Conversely, for $R_L < Z_0$, choose $Z_{01} < \sqrt{R_L Z_0}$ and as small as practical, $Z_{02} > \sqrt{R_L Z_0}$ and as large as practical, and then solve for l_1 and l_2. This procedure minimizes the overall length of the matching network. In general, its length is smaller than that of a comparable quarter-wave transformer (Sec. 4–2). This is particularly advantageous at the lower microwave frequencies where a quarter-wavelength line can become rather large. The disadvantage of the smaller size L-C transformer is that its useful bandwidth is less than that of a quarter-wave transformer. Their bandwidth capabilities are compared in Prob. 4–13 for specific values of R_L and Z_0. Microwave engineering often involves a *design tradeoff* between the length of a device and its bandwidth capability. This will become more apparent in the discussion of multisection quarter-wave transformers (Sec. 4–2b).

The above equations are only valid when the load impedance is real (namely, $Z_L = R_L$). The matching technique in Fig. 4–16 can be extended to include complex load impedances. In this case the resultant equations are unwieldy and hence exact solutions are best obtained using a computer or a programmable calculator.

4–1c Dissipative Matching Networks

All the matching networks described thus far consist of reactive elements and lossless lines. Thus for a properly designed network (that is, unity input SWR) all the power incident on the input port of the matching network is delivered to the load. Occasionally there are microwave devices in which a low input SWR is very important, while efficient power transfer is only a secondary consideration. An example is the microwave detector used in laboratory measurements of relative power levels. To minimize measurement errors, SWR values of less than 1.25 are usually required (see Ex. 4–10 in Sec. 4–4.). Since signal generators are available with more than adequate power for microwave detection, maximum power transfer to the detector is not a primary concern. In this case a resistive-type matching network could offer some advantages. The following example illustrates the point.

Example 4–2:

The nominal impedance of a microwave detector is 200 ohms resistive. In large production lots, however, the impedance values range from 100 to 400 ohms resistive. Design a matching network that insures an input SWR of less than 1.25 no matter which detector is connected to a 50 ohm transmission line.

Solution: Without a matching network, the SWR values will range from 2.0 to 8.0. Any dissipationless network that matches the *nominal* detector impedance to the 50 ohm line will yield an input SWR of 2.0 when the 200 ohm detector is replaced by a 100 or 400 ohm unit. Furthermore, the frequency sensitivity of the matching network will result in even higher SWR values.

One solution is to place a resistor R in parallel with the detector as illustrated in Fig. 4–17. The value of R is chosen so that for the extreme values of R_L, the input SWR is less than 1.25. Referring to Eq. (3–55), this requires that

$$\frac{100R}{100 + R} > \frac{50}{1.25} \quad \text{and} \quad \frac{400R}{400 + R} < (1.25)50$$

which means that R must be greater than 66.7 ohms and less than 74.1 ohms. Table 4–3 gives Z_{in}, input SWR, and efficiency of power transfer for three values of detector impedance when $R = 70$ ohms. The second column represents the parallel combination of R_L and the 70 ohm resistor. Note that all the SWR values are less than 1.25, which satisfies the original specification.

This example shows that by using a resistive-type matching network (rather than a dissipationless type), lower SWR values as well as reduced SWR variations between

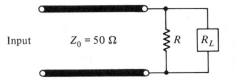

Input $Z_0 = 50\ \Omega$ R R_L **Figure 4–17**   A resistive matching network. R_L represents the detector impedance in Ex. 4–2.

TABLE 4–3 Performance of the Resistive Matching
Network in Fig. 4–17 when
$R = 70$ Ohms

R_L (ohms)	Z_{in} (ohms)	Input SWR	$\frac{P_L}{P_A} \times 100\%$
100	41.2	1.21	40.8
200	51.9	1.04	25.9
400	59.6	1.19	14.7

detectors can be achieved. Also since the matching network is resistive, it is insensitive to frequency changes, an important consideration in swept-frequency applications. This improved SWR performance is obtained at the expense of efficient power transfer. The last column in Table 4–3 shows the percentage of available power (P_A) from a 50 ohm source that is delivered to the detector. In some applications, this amount of power variation between detectors is unacceptable. Reduced power variations with comparable SWR values can be realized by using a matched attenuator as the dissipative matching network. Matched attenuators are described in Sec. 6–3a. The procedure for calculating the input SWR is explained in Ex. 6–2.

4–1d Variable Matching Networks

The need for a network capable of matching a wide range of load impedances occurs quite often in a microwave laboratory. One possibility is to use either a series or shunt matching network in which the line length and the reactance value are adjustable. This technique is the basis of the waveguide slide-screw tuner described in Refs. 4–6 and 4–7. In coaxial systems, however, it is difficult to construct a reactance that is adjustable both in value and position. As a result, the most common variable matching networks are the double-slug tuner and the double-stub tuner. Their cross-sectional views and equivalent circuits are shown in Figs. 4–18 and 4–20. They are called tuners because they are used to eliminate or *tune out* reflections due to a mismatched load.

The double-slug tuner.   This configuration (Fig. 4–18) utilizes two identical shunt capacitances of fixed values whose positions are variable. One form consists of two metal rings (or slugs) that slide along the center conductor. For purposes of explanation, their normalized susceptance values $\bar{B} = B/Y_0$ are chosen to be 0.70. The following Smith chart analysis shows that by adjusting l_1 and l_2, any load impedance with an SWR ≤ 4.0 can be matched with this tuner. To match a load having a higher SWR, one must use a larger value of $\bar{B}$. If $\bar{B}$ is chosen too large, however, the device becomes lossy and extremely sensitive to changes in position and frequency. Figure 4–19 shows the equivalent circuit of the tuner with $\bar{B} = 0.70$ and the specific load admittance to be matched. Various positions along the line are designated by L, A, B, C, and *in*. The normalized load admittance $\bar{Y}_L$ is plotted on the accompanying Smith chart. Since the $\bar{Y}_L$ point is within the 4.0 SWR circle, the tuner is capable of matching the load to the input line.

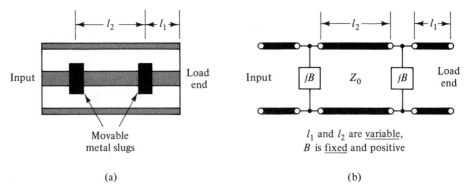

(a)   (b)

Figure 4–18 A coaxial-type double-slug tuner and its equivalent circuit. B is a fixed positive value, while l_1 and l_2 are variable.

The operation of the tuner is best understood by starting at the input and *working back* toward the load. For a perfect match, $\bar{Y}_{in} = 1.0$. Since the first shunt capacitance adds $+j0.7$, $\bar{Y}_C$ must be $1 - j0.7$, which is shown on the chart. This means that $\bar{Y}_B$ must be on an SWR circle that includes the $\bar{Y}_C$ point. In our example, it is the 2.0 SWR circle as shown. As long as $\bar{Y}_B$ is on this circle, adjustment of l_2 can produce $\bar{Y}_C = 1 - j0.7$. Now since the second capacitance also has a normalized admittance of $+j0.7$, $\bar{Y}_A = \bar{Y}_B - j0.7$. Shifting all admittance values on the 2.0 SWR circle by $-j0.7$ yields the possible values of $\bar{Y}_A$ that can be matched. These are defined by the $\bar{Y}_A$ circle shown on the Smith chart. The outer edge of this circle is tangent to the 4.0 SWR circle. This means that any normalized load admittance *on* or *inside* the 4.0 SWR circle can be transformed to a point on the $\bar{Y}_A$ circle by simply adjusting the line length l_1. Since $\bar{Y}_B = \bar{Y}_A + j0.7$, the resultant $\bar{Y}_B$ point must lie on the 2.0 SWR circle. Proper adjustment of l_2 yields $\bar{Y}_C = 1 - j0.7$ and hence $\bar{Y}_{in} = 1.0$, a matched condition.

To illustrate let $\bar{Y}_L = 3 + j1$. This value and its SWR circle (SWR = 3.4) are plotted on the Smith chart. By setting $l_1 = 0.082\lambda$, the $\bar{Y}_L$ point is transformed to a point on the $\bar{Y}_A$ circle, namely, $\bar{Y}_A = 1.30 - j1.45$. Since $\bar{Y}_B = \bar{Y}_A + j0.7$, $\bar{Y}_B = 1.3 - j0.75$ and is shown on the chart. As expected it lies on the 2.0 SWR circle. Setting $l_2 = 0.024\lambda$ transforms $\bar{Y}_B$ to $\bar{Y}_C = 1 - j0.7$, which results in $\bar{Y}_{in} = 1.0$.

Note that the 3.4 SWR circle intersects the $\bar{Y}_A$ circle twice. Thus, there exists another set of values for l_1 and l_2 that lead to a matched condition (Prob. 4–14).

The operation of the double-slug tuner can also be described from a wave-reflection point of view. Example 4–5 in Sec. 4–3a explains that when two reflections (with SWR values S_1 and S_2) are spaced on a lossless line so that their reflected waves add, the resultant input SWR = $S_1 S_2$. Conversely, when the reflected waves are out-of-phase, the input SWR = S_2/S_1, where it is assumed $S_2 \geq S_1$. Thus, depending upon the spacing between the two reflections, the input SWR can range from

$$\text{SWR}_{min} = \frac{S_2}{S_1} \qquad \text{to} \qquad \text{SWR}_{max} = S_1 S_2 \qquad \text{where} \qquad S_2 \geq S_1 \quad (4\text{--}16)$$

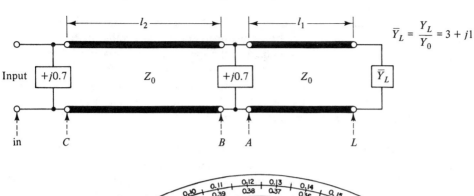

$$\overline{Y}_L = \frac{Y_L}{Y_0} = 3 + j1$$

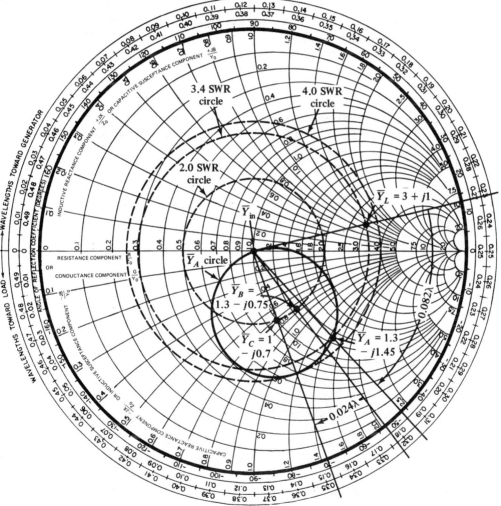

Figure 4–19 Smith chart analysis of the double-slug tuner with $\overline{B} = 0.7$. In this example, $\overline{Y}_L = 3 + j1$.

Since a normalized susceptance $+j0.7$ shunted across a matched line causes an SWR = 2.0, two of them can, by adjusting l_2, create any value of SWR between unity and 4.0. With the value set equal to the load SWR, the spacing l_1 between the load and the tuner can be adjusted so that the reflection from Y_L cancels the reflection from the double-slug tuner resulting in an input SWR of unity. In other words, l_2 is adjusted so that the reflection magnitude of the tuner equals $|\Gamma_L|$, while l_1 is adjusted so that the two reflections cancel.

The double-stub tuner.   For this tuner, shown in Fig. 4–20, the two susceptance values (B_1 and B_2) are variable but their positions on the line are fixed. The values may be positive or negative and are controlled by adjusting the lengths of the short-circuited shunt stubs (l_1 and l_2). The spacing between the stubs l is usually $\lambda/8$ or $3\lambda/8$ at the design frequency. For ease of construction, the latter spacing is used at the higher microwave frequencies.

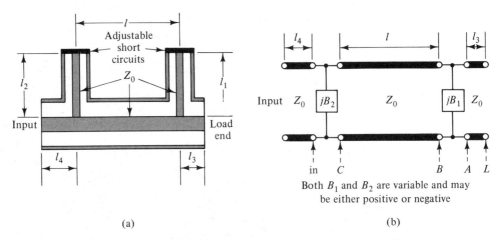

(a)   (b)

Figure 4–20   A coaxial-type double-stub tuner and its equivalent circuit. The lengths l, l_3, and l_4 are fixed, while l_1 and l_2 are variable. In some commercial units, $l_4 - l_3 = \lambda/4$.

As in the previous case, the operation of this tuner is best understood by starting at the input and *working back* toward the load. Figure 4–21 illustrates its operation when $l = \lambda/8$. The various positions along the line (L, A, B, C, and in) are indicated in Fig. 4–20. For a perfect match, $\overline{Y}_{in}$ is unity and therefore $\overline{Y}_C = 1 - j\overline{B}_2$. This means that the $\overline{Y}_C$ point is located on the $\overline{G} = 1$ circle of the Smith chart. For $l = \lambda/8$, the corresponding $\overline{Y}_B$ point must lie on the rotated $\overline{G} = 1$ circle as indicated. Since $\overline{Y}_B = \overline{Y}_A + j\overline{B}_1$, $\overline{B}_1$ can be adjusted so that for a given value of $\overline{Y}_A$, the $\overline{Y}_B$ point will indeed lie on the rotated $\overline{G} = 1$ circle. From the Smith chart, it is clear that this can be achieved only if Re $\overline{Y}_A \leq 2.0$. Thus any value of $\overline{Y}_A$ within the shaded region *cannot* be matched by this double-stub tuner. In many commercially available units, the difference in lengths of the end sections ($l_4 - l_3$) is $\lambda/4$ at the design frequency. Thus if a particular load impedance cannot be matched, one can simply reverse the tuner (that is, connect the load to the l_4 line), which shifts the $\overline{Y}_A$ point out of the forbidden region.

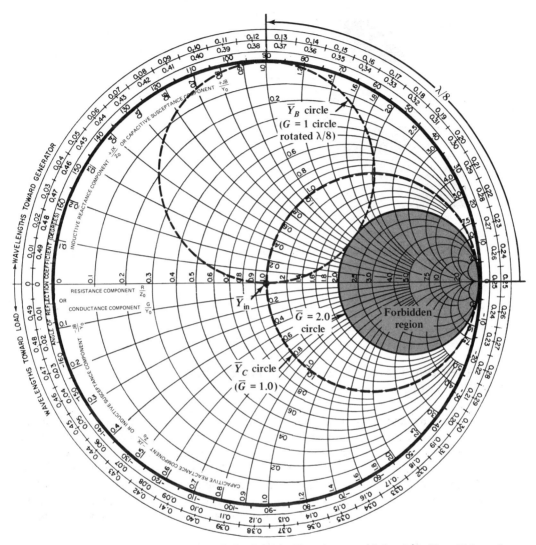

Figure 4–21 Smith chart analysis of the double-stub tuner with $l = \lambda/8$. (Note: Values of $\overline{Y}_A$ within the *forbidden* region cannot be matched when $l = \lambda/8$ or $3\lambda/8$.)

For the tuner described here, the stub spacing l was set equal to $\lambda/8$. The general case of the double-stub tuner as well as the triple-stub tuner are analyzed by Collin (Ref. 4–7).

The E-H tuner. This waveguide-type tuner is shown in Fig. 4–22. It consists of a hybrid tee (described in Sec. 8–2 and Fig. 8–9) and a pair of adjustable short circuits (described in Sec. 7–3c). Its approximate equivalent circuit is shown in part *b* of the figure. The series reactance value X_s is controlled by l_s while the shunt reactance value X_p is controlled by l_p. Practically any load impedance can be matched by the E-H tuner since X_s and X_p are independently adjustable to any

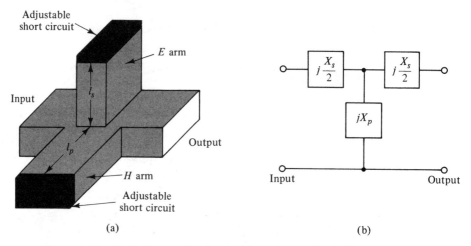

Figure 4–22　The E-H waveguide tuner and its equivalent circuit. The values of X_s and X_p are variable and may be either positive or negative.

positive or negative value. Additional information on this waveguide device may be found in Refs. 4–6 and 4–7.

4–2 QUARTER-WAVE AND TAPERED-LINE IMPEDANCE TRANSFORMERS

Quarter-wave and tapered-line transformers are used to match a resistive load to an input transmission line (Z_0 matching). When the generator is matched to the input line (that is, $Z_G = Z_0$), this results in maximum power transfer to the load. These transformers are also used as matching networks between transmission lines having different characteristic impedances. For many narrowband applications, a single-section quarter-wave transformer is adequate. Multisection and tapered-line transformers are used when a very good match is required over a broad frequency range.

4–2a The Single-Section Quarter-Wave Transformer

In Sec. 3–6a and Fig. 3–13, it was explained that a resistively terminated quarter-wave line section has the properties of an ideal transformer. In order to perfectly match a load resistance R_L to a Z_0 input line, the characteristic impedance of the quarter-wave line must be

$$Z_{01} = \sqrt{Z_0 R_L} \qquad (4-17)$$

Since the fixed length line can only be a quarter wavelength long at one frequency, the match becomes imperfect at other than the design frequency. The Smith chart is useful in analyzing this variation in input SWR. As an example, consider the situation shown in Fig. 4–23. With $R_L = 200$ ohms and $Z_0 = 50$ ohms, a perfect match is obtained by choosing $Z_{01} = \sqrt{(50)(200)} = 100$ ohms for the quarter-wave line

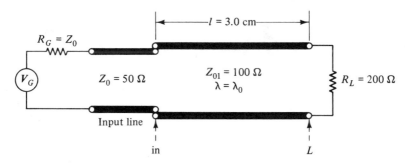

Figure 4-23   A quarter-wave transformer for $R_L = 200$ ohms, $R_G = Z_0 = 50$ ohms, and a design frequency of 2500 MHz.

section. For a design frequency of 2500 MHz and assuming an air-insulated line, $l = \lambda_0/4 = 3.0$ cm as indicated in the figure. To verify that $Z_{in} = 50$ ohms and the SWR on the input line is unity at 2500 MHz, consider the Smith chart analysis in Fig. 4-24. Plotted on the chart is $\bar{R}_L \equiv R_L/Z_{01} = 2.0$. A single bar ($-$) over the impedance symbol denotes normalization to Z_{01}, while a double bar ($=$) indicates an impedance normalized to Z_0. Since the line is a quarter wavelength long at 2500 MHz, Z_{in} is obtained by rotating 0.25λ toward the generator on the 2.0 SWR circle. This yields $\bar{Z}_{in} = 0.50$ or $Z_{in} = 50$ ohms. The SWR on the input line is obtained by normalizing Z_{in} to Z_0. With $Z_0 = 50$ ohms, $\bar{\bar{Z}}_{in} \equiv Z_{in}/Z_0 = 1.0$, which is the center of the chart, thus verifying that the input SWR is unity at the design frequency.

Table 4-4 in conjunction with the Smith chart is used to analyze the transformer at 2000 and 3000 MHz. At 2000 MHz, the 3.0 cm line is only 0.20λ long and therefore $\bar{Z}_{in}$ is given by point A on the chart. Point B represents $\bar{Z}_{in}$ at 3000 MHz since $l = 0.30\lambda$. These impedance values are listed in Table 4-4. Renormalizing the input impedance to Z_0 requires multiplying $\bar{Z}_{in}$ by Z_{01} (100 ohms) and dividing by Z_0 (50 ohms). These values ($\bar{\bar{Z}}_{in}$) are shown in the table and plotted on the Smith chart. Note that it is customary to place an arrow on the impedance (or admittance) plot to indicate increasing frequency. From the chart, the input SWR at both edges of the frequency band equals 1.60. Since *without* the transformer the SWR would have been $200/50 = 4.0$, a substantial improvement has been realized over the 40 percent frequency band.

The SWR versus frequency characteristic of the quarter-wave transformer is a function of the particular values of R_L and Z_0. In general, the greater the SWR before inserting transformer, the larger the SWR variation with frequency after the transformer is inserted. An expression can be derived for the input reflection coefficient by using the impedance transformation equation and the relationship between reflection coefficient and impedance (Prob. 4-19). For a single-section transformer with $Z_{01} = \sqrt{Z_0 R_L}$, the magnitude of the input reflection coefficient is

$$|\Gamma_{in}| = \left\{ 1 + \frac{4N}{(N-1)^2 \cos^2 \beta l} \right\}^{-1/2} \approx \frac{|N-1|}{2\sqrt{N}} |\cos \beta l| \qquad (4-18)$$

where $l = \lambda/4$ at the design frequency and $N \equiv R_L/Z_0$. The approximate form is valid for βl near $\pi/2$ rad. For the example just completed, βl equals 0.4π and 0.6π

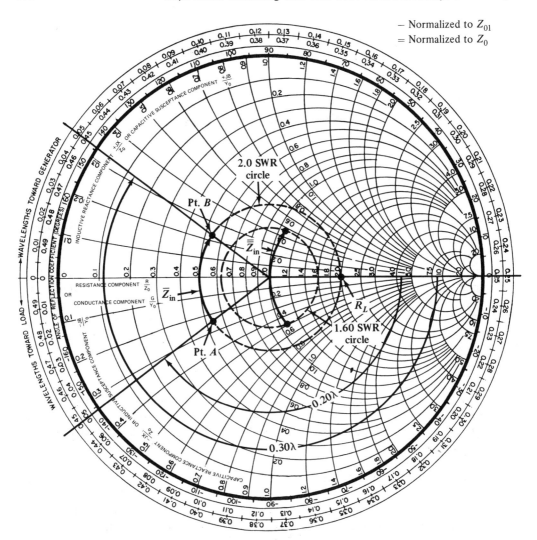

Figure 4-24 Smith chart analysis of the quarter-wave transformer in Fig. 4-23.

TABLE 4-4 Impedance and SWR Data for the Single-Section Quarter-Wave Transformer Shown in Fig. 4-23

Freq. (MHz)	$\lambda = \lambda_0$ (cm)	l/λ	$\bar{Z}_{in}$	$\bar{\bar{Z}}_{in}$	Input SWR
2000	15	0.20	$0.53 - j0.24$	$1.06 - j0.48$	1.60
2500	12	0.25	0.50	1.00	1.00
3000	10	0.30	$0.53 + j0.24$	$1.06 + j0.48$	1.60

at the band edges and $N = 4.0$. Therefore, $|\Gamma_{in}| = 0.226$ at 2000 and 3000 MHz, which corresponds to the SWR values in Table 4–4.

The reflection characteristics of the single-section transformer is shown in Fig. 4–25. For a fixed length l, it describes the variation of $|\Gamma_{in}|$ with frequency since $\beta l = \omega l / v$. Note that at the design frequency ($\beta l = \pi/2$), $\Gamma_{in} = 0$, a perfect match. Also shown in the figure are the reflection characteristics for two- and three-section transformers. These will be described next.

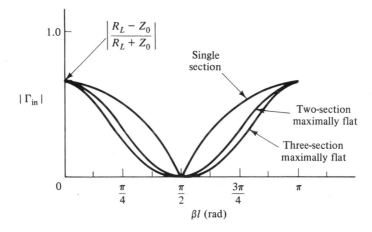

Figure 4–25   A comparison of reflection characteristics for one-, two-, and three-section maximally flat quarter-wave transformers.

4–2b Multisection Quarter-Wave Transformers

Many microwave applications require very low values of SWR over a wide frequency range. This can be achieved with a multisection quarter-wave transformer. At first glance, it may seem odd that a multisection transformer can give better results than a single section. After all, it was the frequency sensitivity of l/λ that degraded the SWR at the band edges. It would seem, for example, that two quarter-wave sections would be worse than one. This is not the case! In fact, if the frequency sensitivity of the two lines can be arranged to offset each other, the SWR performance will be significantly better than a single section. To understand this, let us first look at the quarter-wave transformer from a wave-reflection point of view. For the single section unit shown in Fig. 4–23, wave reflections occur at planes L and in. Since Z_0 and Z_{01} are real, the reflection at the input plane is equivalent to an SWR of 2.0 (namely, Z_{01}/Z_0). Similarly, the reflection at the load plane is also equivalent to an SWR of 2.0 (namely, R_L/Z_{01}). Due to the quarter wavelength spacing, the load reflection arrives at the input plane 180° out-of-phase with the input reflection resulting in complete cancellation and an input SWR of unity.[3] At other than the design frequency, the round-trip phase delay $2\beta l$ is not 180° and only partial cancellation

---

[3] Multiple reflections have been neglected in this discussion. An analysis that includes their effect is found in Sec. 5.7 of Ref. 4–7. For $l = \lambda/4$, the result is exactly as indicated here.

occurs, which results in an input SWR greater than unity. In our example, the SWR is 1.60 at the band edges.

Consider now a transformer that consists of two quarter-wave line sections having characteristic impedances of 70.7 and 141.4 ohms, respectively. This situation is shown in Fig. 4–26, where each line section is $\lambda/4$ long at the design frequency. The Z_{02} line causes the impedance at plane A to be $Z_{02}^2/R_L = 100$ ohms. The Z_{01} line transforms Z_A to $Z_{01}^2/Z_A = 50$ ohms at the input plane resulting in an input SWR of unity. At other than the design frequency, each transformer section causes a reflection. For the chosen values of Z_{01} and Z_{02}, the reflections are identical. As indicated in the figure, the center-line spacing between the two transformer sections is $\lambda/4$ at the design frequency. At other than the design frequency, the center-line spacing differs slightly from the $\lambda/4$ value, which results in partial cancellation of the two identical reflections. In other words, the two-section transformer has been arranged so that the frequency sensitivity of the individual quarter-wave lines tend to offset each other. As a result, its input SWR at the band edges is significantly lower than that of a single-section quarter-wave transformer. What has been described here is a two-section, maximally flat, quarter-wave transformer. It is one example of a large class of multisection transformers, the most common being the Butterworth (maximally flat) and Tchebyscheff (equal-ripple) designs.

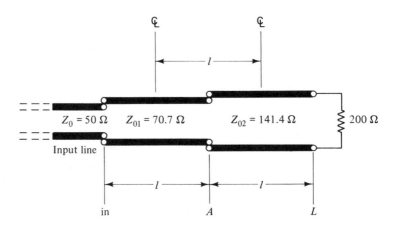

Figure 4–26 A two-section maximally flat quarter-wave transformer for $R_L = 200$ ohms and $Z_0 = 50$ ohms. ($l = \lambda/4$ at the design frequency.)

The Butterworth (or binomial) transformer. Figure 4–27 shows an n-section quarter-wave transformer for matching a resistive load R_L to an input line of characteristic impedance Z_0. The length of all line sections are identical, namely, $\lambda/4$ at the design frequency. In the Butterworth or maximally flat design, the characteristic impedances of the line sections are selected so that $|\Gamma_{in}|$ and its first $(n-1)$ derivatives with respect to frequency equal zero at the design frequency. This insures the flattest possible $|\Gamma_{in}|$ versus ω characteristic at that frequency. The greater the number of sections, the flatter the characteristic and therefore the lower the variation of input SWR with frequency. Figure 4–25 compares the reflection

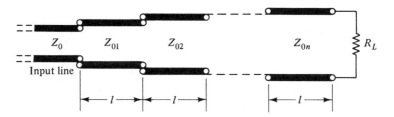

Figure 4-27   The general multisection quarter-wave transformer. ($l = \lambda/4$ at the design frequency.)

characteristics of two- and three-section maximally flat designs with that of the single-section transformer.

The theory of maximally flat transformers is given in Refs. 4-7 and 4-15. An approximate analysis shows that the impedances of the line sections can be selected so that

$$|\Gamma_{in}| = \left|\frac{R_L - Z_0}{R_L + Z_0}\right| |\cos \beta l|^n \qquad (4-19)$$

where n is the number of sections and $\beta l = \omega l/v$. At the design frequency, $\beta l = \pi/2$. This form satisfies the above stated conditions for a maximally flat design. The required impedance values for up to eight sections are given in Sec. 6.05 of Ref. 4-5. These values can be closely approximated by choosing the characteristic impedances of the individual line sections in accordance with the following relation.

$$Z_{0k} = Z_0\left(\frac{R_L}{Z_0}\right)^{M_k/2^n} \qquad k = 1, 2, \ldots n \qquad (4-20)$$

where M_k is related to the coefficients of the binomial expansion. For this reason, the maximally flat design is often referred to as the *binomial transformer*. M_k is given by

$$M_k = C_1 + C_2 + \ldots + C_k \qquad k = 1, 2, \ldots n \qquad (4-21)$$

where $C_1, C_2, \ldots C_n$ are the binomial coefficients. Pascal's triangle provides a convenient way of determining these coefficients for any given value of n. The triangle is shown below for up to $n = 3$.

$n = 0$	1	
$n = 1$	1 1	$(4-22)$
$n = 2$	1 2 1	
$n = 3$	1 3 3 1	

It can be extended to higher values of n by noting that each coefficient is equal to the sum of the two nearest numbers in the row above it. Therefore, for $n = 4$, the coefficients are 1, 4, 6, 4, 1. Note that in all cases $C_1 = 1$. The remaining values can be obtained from Pascal's triangle or the following relation.

$$C_k = \frac{n!}{(n - k + 1)!(k - 1)!} \qquad k = 2, 3, \ldots n \qquad (4-23)$$

For example, if $R_L = 10$ ohms and $Z_0 = 160$ ohms, the impedance values for a three-section transformer are obtained in the following manner. With $n = 3$, $C_1 = 1$, $C_2 = C_3 = 3$ and $2^n = 8$. From Eq. (4–21) for $k = 1$, 2, and 3,

$$M_1 = 1 \quad , \quad M_2 = 1 + 3 = 4, \quad \text{and} \quad M_3 = 1 + 3 + 3 = 7$$

The impedance values are obtained from Eq. (4–20). Thus, $Z_{01} = 160(1/16)^{1/8} = 113$ ohms, $Z_{02} = 160(1/16)^{4/8} = 40$ ohms, and $Z_{03} = 160(1/16)^{7/8} = 14$ ohms.

The following example illustrates the design of a two-section binomial transformer and compares its performance to that of a single-section design.

Example 4–3:

(a) Design a two-section, maximally flat, quarter-wave transformer to match a 200 ohm resistive load to a 50 ohm input line at 2500 MHz. Assume $\lambda = \lambda_0$.

(b) Use a Smith chart analysis to determine the input SWR at 2000 and 3000 MHz. Compare the results with the single-section design in Fig. 4–23.

Solution:

(a) At 2500 MHz, $\lambda_0 = 12.0$ cm and with $\lambda = \lambda_0$, $l = 12/4 = 3.0$ cm. For the two-section transformer, $n = 2$ and therefore $C_1 = 1$ and $C_2 = 2$. From Eq. (4–21), $M_1 = 1$ and $M_2 = 1 + 2 = 3$. For $R_L = 200$ ohms and $Z_0 = 50$ ohms, Eq. (4–20) yields

$$Z_{01} = 50(4)^{1/4} = 70.7 \text{ ohms} \quad \text{and} \quad Z_{02} = 50(4)^{3/4} = 141.4 \text{ ohms}$$

The transformer design is shown in Fig. 4–26, where in this case $l = 3.0$ cm.

(b) The performance of the transformer over the 2000 to 3000 MHz may be analyzed with the aid of the Smith chart in Fig. 4–28. For convenience, the required information and results are listed in Table 4–5. The values of $\bar{\bar{Z}}_A \equiv Z_A/Z_{02}$ are obtained from the Smith chart by rotating clockwise from $\bar{R}_L \equiv R_L/Z_{02} = 1.414$ a distance l/λ at the three frequencies. The values of $\bar{Z}_A \equiv Z_A/Z_{01}$ are obtained by renormalizing $Z_A = \bar{\bar{Z}}_A Z_{02}$ to Z_{01}. Thus, $\bar{Z}_A = \bar{\bar{Z}}_A Z_{02}/Z_{01} = 2\bar{\bar{Z}}_A$. The $\bar{Z}_A$ data is listed in the table and plotted on the Smith chart. Rotating the $\bar{Z}_A$ plot clockwise a distance l/λ at the three frequencies yields the $\bar{Z}_{in}$ plot shown. Note that the end points of the $\bar{Z}_A$ plot are on a different SWR circle than the midband point. Also notice that the angular spread of $\bar{\bar{Z}}_{in}$ has shrunk relative to that of the $\bar{Z}_A$ plot. This is because the frequency sensitive of the Z_{01} line tends to offset that of the Z_{02} line. Renormalizing $Z_{in} = \bar{\bar{Z}}_{in} Z_{01}$ to $Z_0 = 50$ ohms yields $\hat{Z}_{in} \equiv Z_{in}/Z_0 = 1.414\bar{\bar{Z}}_{in}$. The values of $\hat{Z}_{in}$ are listed in the table and plotted on the chart. The SWR at the band edges is 1.15, a substantial improvement over the 1.60 value for the single-section transformer (see Table 4–4).

It should be noted in this example that the $\bar{\bar{Z}}_{in}$ and $\hat{Z}_{in}$ plots have loops. This can be verified by calculating additional impedance values at other frequencies in the 2000 to 3000 MHz band. Actually this is not necessary since the impedance plots must have a clockwise trend with increasing frequency (see Sec. 3–7b). Any attempt to join together the three values of either $\bar{\bar{Z}}_{in}$ or $\hat{Z}_{in}$ *without* a loop would lead to a counterclockwise trend over some portion of the curve. In general, broadband matching methods usually result in one or more loops in the impedance plot.

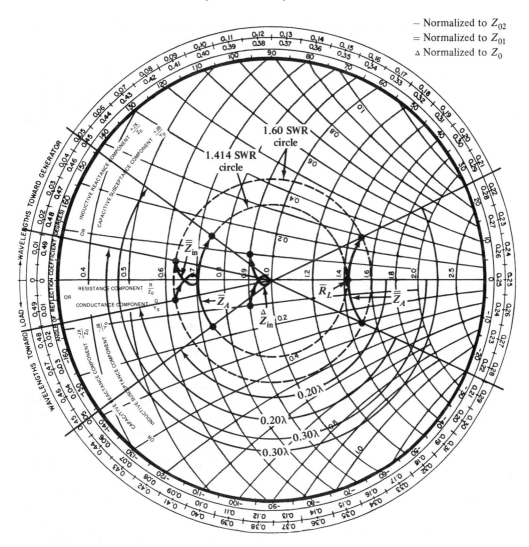

Figure 4–28 Smith chart analysis of the two-section binomial transformer described in Ex. 4–3 and Fig. 4–26. Note the use of an expanded version of the Smith chart.

TABLE 4–5 Impedance and SWR Data for the Two-Section, Maximally Flat Transformer Shown in Fig. 4–26

Freq. (MHz)	$\lambda = \lambda_0$ (cm)	l/λ	$\bar{Z}_A$	$\bar{\bar{Z}}_A$	$\bar{\bar{Z}}_{in}$	$\overset{\Delta}{Z}_{in}$	Input SWR
2000	15	0.20	$0.75 - j0.15$	$1.50 - j0.30$	$0.63 - j0.07$	$0.9 - j0.10$	1.15
2500	12	0.25	0.707	1.414	0.707	1.00	1.00
3000	10	0.30	$0.75 + j0.15$	$1.50 + j0.30$	$0.63 + j0.07$	$0.9 + j0.10$	1.15

The Tchebyscheff transformer.   Multisection quarter-wave transformers of the Tchebyscheff design are characterized by a $|\Gamma_{in}|$ versus frequency variation that has one or more ripples in its useful frequency range. The reflection characteristic for a three-section design is shown in Fig. 4–29. The useful bandwidth is defined by $\phi_0 \leq \beta l \leq \pi - \phi_0$. Note that within this band, $|\Gamma_{in}| \leq |\Gamma|_m$ and is zero at three values of βl. For an n-section transformer, Γ_{in} vanishes at n different values of βl and therefore the greater the number of sections, the more ripples in the $|\Gamma_{in}|$ characteristic.

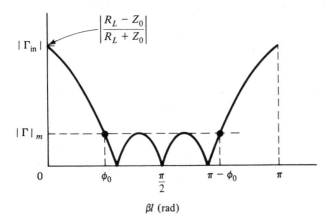

Figure 4–29   Reflection characteristic of a three-section Tchebyscheff-type quarter-wave transformer.

In order to gain some insight as to how this characteristic is achieved, consider the two-section maximally flat transformer in Fig. 4–26. As indicated in Fig. 4–28, the input impedance plot contains a single loop. An expanded view of the plot is shown in part a of Fig. 4–30 and its SWR versus frequency characteristic in part b. It appears that if somehow the impedance plot could be shifted to the right (as indicated by the dashed curve), the SWR at the band edges would be improved. Granted, the midband SWR would increase, but overall, lower SWR values would result. To achieve this requires that the values of Z_{01} and Z_{02} be adjusted slightly. Suppose the values in Fig. 4–26 are changed to $Z_{01} = 72.1$ ohms and $Z_{02} = 138.8$ ohms. The $\hat{Z}_{in}$ plot resulting from a Smith chart analysis would produce the dashed curve in Fig. 4–30a. Note that the plot lies entirely within a 1.08 SWR circle. The SWR versus frequency characteristic is shown in part b of the figure and is labeled as the *equal-ripple* design. The SWR values at midband and the band edges are all equal to 1.08, while at 2150 and 2850 MHz the SWR is unity. What we have just described is a two-section Tchebyscheff quarter-wave transformer. The equal-ripple or Tchebyscheff characteristic occurs in many other design areas such as filters, couplers, and antennas.

The difference between the maximally flat and the equal-ripple transformer is related to the choice of impedance values for the quarter-wave line sections. As indicated by the above example, the difference between the two sets of values is quite small. Design tables as well as maximum in-band SWR for up to four sections may be found in Refs. 4–5 and 4–18. In general, the greater the number of sections, the lower the value of maximum in-band SWR.

The theory of multisection Tchebyscheff transformers is given in Refs. 4–7,

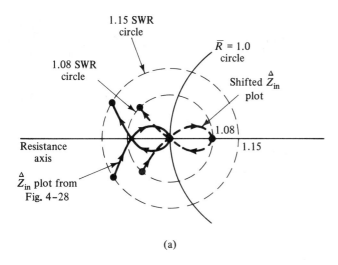

(a)

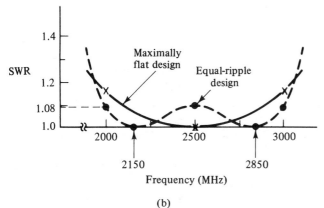

(b)

Figure 4–30 A comparison of maximally flat and equal-ripple transformer designs (two sections).

4–15, 4–16, and 4–17. An approximate analysis (Ref. 4–7) based on small reflections yields the following relationship between useful bandwidth and the maximum in-band reflection coefficient $|\Gamma|_m$.

$$|\Gamma|_m T_n\left(\frac{1}{\cos \phi_0}\right) = \left|\frac{R_L - Z_0}{R_L + Z_0}\right| \qquad (4\text{--}24)$$

where T_n denotes the Tchebyscheff polynomial of degree n and $|\Gamma|_m$ and ϕ_0 are defined in Fig. 4–29.

Polynomials for $n = 1$ to 4 and a useful recurrence relation are listed below.

$$T_1(x) = x$$
$$T_2(x) = 2x^2 - 1$$
$$T_3(x) = 4x^3 - 3x \qquad (4\text{--}25)$$
$$T_4(x) = 8x^4 - 8x^2 + 1$$
$$T_n(x) = 2xT_{n-1} - T_{n-2}$$

Equation (4–24) is useful for $|\Gamma|_m \le 0.10$ and $0.10 < R/Z_0 < 10$. Accurate values of maximum in-band SWR are given in Refs. 4–5 and 4–18. The following brief example illustrates the use of these relationships.

Example 4–4:

A three-section Tchebyscheff quarter-wave transformer has been designed for a useful frequency band of 2000 to 3000 MHz. Calculate the maximum in-band SWR when $R_L = 200$ ohms and $Z_0 = 50$ ohms.

Solution: With $\phi_0 = \beta_1 l$ at 2000 MHz and $\pi - \phi_0 = \beta_2 l$ at 3000 MHz,

$$\frac{\pi - \phi_0}{\phi_0} = \frac{\beta_2}{\beta_1} = \frac{3000}{2000} \quad \text{or} \quad \phi_0 = 0.4\pi \text{ rad}$$

where it has been assumed that wave velocity is independent of frequency. From Eq. (4–24),

$$|\Gamma|_m T_3(3.24) = 0.60 \quad \text{or} \quad |\Gamma|_m = \frac{0.60}{126} = 0.005$$

and hence the maximum in-band SWR is 1.01.

For a fixed number of sections, the Tchebyscheff design always results in the lowest SWR across the given frequency band. Stated conversely, for a maximum allowable in-band SWR, the Tchebyscheff transformer yields the widest possible bandwidth. For systems in which all frequencies in the band are equally important, the Tchebyscheff design is usually best. There are occasions, however, when performance at the band edges is not as critical as at the design frequency. In these cases, the maximally flat design may be preferred.

All quarter-wave transformers require abrupt changes in characteristic impedance. The resultant discontinuity in the transverse dimensions of the transmission line produces reactive effects. These effects and their equivalent circuits are described in Secs. 6–4 and 7–2. In most cases, the reactive behavior is equivalent to that of a shunt capacitance. To compensate for this, the length of the quarter-wave sections must be reduced a few percent. This is discussed in Sec. 6.08 of Ref. 4–5. It should also be mentioned that the quarter-wave transformer theory discussed here assumes that both characteristic impedance and wave velocity are *independent* of frequency. For coaxial and strip transmission lines at high frequencies, this is indeed the case. Although it is *not* true for waveguide transmission (see Sec. 5–4), the theory remains reasonably accurate and is therefore a useful design tool. The theory and design of waveguide transformers are discussed in Refs. 4–5 and 4–19.

4–2c Tapered-Line Transformers

Another method of changing impedance levels in a transmission system involves the use of a continuously tapered line. One such example is illustrated in Fig. 4–31a, where the impedance of the coaxial line is gradually transformed from R_L to Z_0 by tapering the diameter of the center conductor. The input SWR remains low as long as the taper length l is much greater than the operating wavelength. The higher the

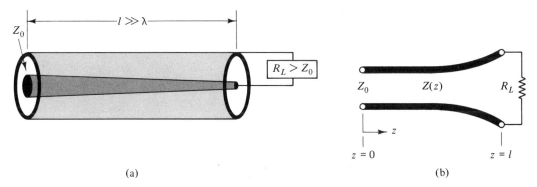

Figure 4–31 A tapered coaxial line and its equivalent circuit.

frequency, the better this condition is satisfied. Thus, the transmission characteristic of a tapered-line transformer resembles that of a highpass filter. Quarter-wave transformers, on the other hand, have the characteristics of a bandpass filter.

Part *b* of Fig. 4–31 shows the representation of a tapered-line transformer whose characteristic impedance $Z(z)$ varies from Z_0 at $z = 0$ to R_L at $z = l$. The tapered line may be viewed as an infinite number of line sections of length dz. If the change in characteristic impedance between sections is denoted as dZ, the associated reflection coefficient is given by

$$d\Gamma = \frac{\bar{Z} + d\bar{Z} - \bar{Z}}{\bar{Z} + d\bar{Z} + \bar{Z}} \approx \frac{d\bar{Z}}{2\bar{Z}} = \frac{1}{2}\frac{d(\ln \bar{Z})}{dz} dz \qquad (4\text{–}26)$$

where Γ and $\bar{Z}$ are functions of z and the bar $(-)$ denotes impedance normalization with respect to Z_0. Assuming a lossless line, the reflection coefficient referred to the input plane $(z = 0)$ becomes

$$d\Gamma_{\text{in}} = \frac{1}{2} e^{-j2\beta z} \frac{d(\ln \bar{Z})}{dz} dz \qquad (4\text{–}27)$$

The total input reflection coefficient may be approximated by summing the reflections associated with all the incremental impedance changes between $z = 0$ and $z = l$. Thus,

$$\Gamma_{\text{in}} = \frac{1}{2} \int_0^l \left\{ e^{-j2\beta z} \frac{d(\ln \bar{Z})}{dz} \right\} dz \qquad (4\text{–}28)$$

where $\bar{Z}$ is a continuous function of z between $z = 0$ and $z = l$. This expression is approximate because multiple reflections and the transmission loss associated with each incremental reflection have been neglected. (The exact analysis is outlined in Sec. 5.15 of Ref. 4–7.) The $|\Gamma_{\text{in}}|$ versus βl characteristic of the taper depends upon the way in which $\bar{Z}$ varies with z. Several designs including exponential, Gaussian, and Tchebyscheff tapers are analyzed in Refs. 4–7, 4–9, 4–20, and 4–21. The reflection characteristic for an exponential taper is shown in Fig. 4–32. Note that for $l \gg \lambda$ (that is, $\beta l \gg 2\pi$), the input reflection coefficient is quite small.

For a fixed length, multisection quarter-wave transformers yield lower SWR values over a specified frequency band than tapered-line transformers. Furthermore,

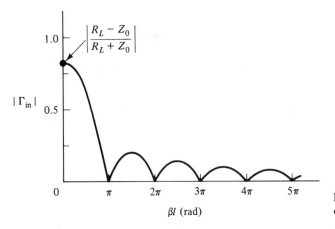

Figure 4–32   Input reflection coefficient of an exponentially tapered line.

tapered lines are usually more costly to manufacture. However, the absence of sharp discontinuities in the physical structure of tapered-line transformers does make them an attractive choice in most high power applications.

4–3 TWO-PORT NETWORK ANALYSIS WITH TRANSMISSION MATRICES

Matrix methods are invaluable in analyzing complex microwave networks, one reason being that matrix operations are readily performed on a computer or programmable calculator. At microwave frequencies, the most widely used are the ABCD matrix, the wave-transmission matrix, and the scattering matrix. The latter is discussed in Appendix D and is particularly useful in analyzing multiports. Its application to two-port networks is described in the next section. Both ABCD and wave-transmission matrices are discussed in Appendix C and should be reviewed before proceeding with the material presented here. This section illustrates how the characteristics of a two-port network are related to the matrix elements. In addition, the meaning of attenuation, insertion loss and phase, transducer loss, and mismatch loss are explained.

4–3a ABCD Matrix Analysis of Two-Port Networks

As explained in Appendix C, the terminal voltages and currents of a two-port network are related by the following equations.

$$\begin{aligned} V_1 &= A\,V_2 + B\,I_2 \\ I_1 &= C\,V_2 + D\,I_2 \end{aligned} \quad \text{or} \quad \begin{Bmatrix} V_1 \\ I_1 \end{Bmatrix} = \begin{bmatrix} A & B \\ C & D \end{bmatrix} \begin{Bmatrix} V_2 \\ I_2 \end{Bmatrix} \qquad (4\text{–}29)$$

where the phasor voltages and currents are indicated in Fig. 4–33. The ABCD matrix is a transmission-type matrix since in a cascade of two-port networks, the *output* voltage and current of one network represents the *input* voltage and current of the following one.

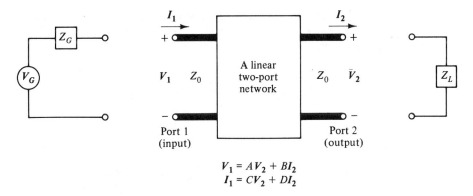

$$V_1 = AV_2 + BI_2$$
$$I_1 = CV_2 + DI_2$$

Figure 4–33   A linear two-port network and its $ABCD$ representation.

The impedance and reflection characteristics of a two-port network can be expressed in terms of its ABCD matrix elements. These expressions are derived in Appendix C and repeated here for convenience. With the output port terminated by a load impedance Z_L, the input impedance is

$$Z_{\text{in}} = \frac{AZ_L + B}{CZ_L + D} \quad \text{or} \quad \bar{Z}_{\text{in}} = \frac{\bar{A}\bar{Z}_L + \bar{B}}{\bar{C}\bar{Z}_L + \bar{D}} \tag{4-30}$$

where $\bar{Z}$ indicates an impedance normalized to Z_0 and $\bar{A}$, $\bar{B}$, $\bar{C}$, and $\bar{D}$ are the normalized forms of the matrix elements. Namely, $\bar{A} = A$, $\bar{B} = B/Z_0$, $\bar{C} = CZ_0$, and $\bar{D} = D$. When the output is match terminated ($Z_L = Z_0$), $\bar{Z}_{\text{in}}$ and the input reflection coefficient (referenced to Z_0) are given by

$$\bar{Z}_{\text{in}} = \frac{\bar{A} + \bar{B}}{\bar{C} + \bar{D}} \quad \text{and} \quad \Gamma_{\text{in}} = \frac{\bar{A} + \bar{B} - \bar{C} - \bar{D}}{\bar{A} + \bar{B} + \bar{C} + \bar{D}} \tag{4-31}$$

With a generator of impedance Z_G connected to the input, the Thevenin impedance looking into the output port is

$$Z'_G = \frac{B + DZ_G}{A + CZ_G} \quad \text{or} \quad \bar{Z}'_G = \frac{\bar{B} + \bar{D}\bar{Z}_G}{\bar{A} + \bar{C}\bar{Z}_G} \tag{4-32}$$

When the generator is matched to the input line ($Z_G = Z_0$), $\bar{Z}'_G$ and the reflection coefficient looking into the output port become

$$\bar{Z}'_G = \frac{\bar{B} + \bar{D}}{\bar{A} + \bar{C}} \quad \text{and} \quad \Gamma_{\text{out}} = \frac{\bar{B} + \bar{D} - \bar{A} - \bar{C}}{\bar{A} + \bar{B} + \bar{C} + \bar{D}} \tag{4-33}$$

Example 4–5:

Show that the overall SWR due to a pair of dissipationless mismatches having individual SWR values S_1 and S_2 ($S_2 \geq S_1$) and separated by a length of lossless line (Z_0) can range from a minimum value of S_2/S_1 up to a maximum value of $S_1 S_2$. Represent the individual mismatches by ideal transformers having turns ratios of n_{t_1} and n_{t_2}.

Solution: From Table C–1, the matrix elements for an ideal transformer are $A = \bar{A} = n_t$, $B = 0$, $C = 0$, and $D = \bar{D} = 1/n_t$. From Eq. (4–31), $\bar{Z}_{\text{in}} = n_t^2$ and assuming

$n_t \geq 1$, its input SWR $= n_t^2$ since $\bar{Z}_{in}$ is real and greater than unity (see Eq. 3–55). The normalized ABCD matrix for two such mismatches separated by a lossless Z_0 line of length l is obtained by matrix multiplication.

$$\begin{bmatrix} \bar{A} & \bar{B} \\ \bar{C} & \bar{D} \end{bmatrix} = \begin{bmatrix} n_{t_1} & 0 \\ 0 & \dfrac{1}{n_{t_1}} \end{bmatrix} \begin{bmatrix} \cos \beta l & j \sin \beta l \\ j \sin \beta l & \cos \beta l \end{bmatrix} \begin{bmatrix} n_{t_2} & 0 \\ 0 & \dfrac{1}{n_{t_2}} \end{bmatrix} = \begin{bmatrix} n_{t_1} n_{t_2} \cos \beta l & j \dfrac{n_{t_1}}{n_{t_2}} \sin \beta l \\ j \dfrac{n_{t_2}}{n_{t_1}} \sin \beta l & \dfrac{\cos \beta l}{n_{t_1} n_{t_2}} \end{bmatrix}$$

From Eq. (4–31),

$$|\Gamma_{in}| = \left| \frac{\left(n_{t_1} n_{t_2} - \dfrac{1}{n_{t_1} n_{t_2}} \right) \cos \beta l + j \left(\dfrac{n_{t_1}'}{n_{t_2}} - \dfrac{n_{t_2}}{n_{t_1}} \right) \sin \beta l}{\left(n_{t_1} n_{t_2} + \dfrac{1}{n_{t_1} n_{t_2}} \right) \cos \beta l + j \left(\dfrac{n_{t_1}}{n_{t_2}} + \dfrac{n_{t_2}}{n_{t_1}} \right) \sin \beta l} \right|$$

Since $n_{t_1}^2 = S_1$ and $n_{t_2}^2 = S_2$,

$$|\Gamma_{in}| = \left\{ \frac{\left(S_1 S_2 + \dfrac{1}{S_1 S_2} \right) \cos^2 \beta l + \left(\dfrac{S_1}{S_2} + \dfrac{S_2}{S_1} \right) \sin^2 \beta l - 2}{\left(S_1 S_2 + \dfrac{1}{S_1 S_2} \right) \cos^2 \beta l + \left(\dfrac{S_1}{S_2} + \dfrac{S_2}{S_1} \right) \sin^2 \beta l + 2} \right\}^{1/2}$$

By differentiating the bracketed expression with respect to βl, one can show that $|\Gamma_{in}|$ is minimum when $\beta l = \pi/2$ and maximum when $\beta l = 0$. Therefore,

$$|\Gamma_{in}|_{min} = \left\{ \frac{\dfrac{S_1}{S_2} + \dfrac{S_2}{S_1} - 2}{\dfrac{S_1}{S_2} + \dfrac{S_2}{S_1} + 2} \right\}^{1/2} = \frac{\dfrac{S_2}{S_1} - 1}{\dfrac{S_2}{S_1} + 1}$$

and

$$|\Gamma_{in}|_{max} = \left\{ \frac{S_1 S_2 + \dfrac{1}{S_1 S_2} - 2}{S_1 S_2 + \dfrac{1}{S_1 S_2} + 2} \right\}^{1/2} = \frac{S_1 S_2 - 1}{S_1 S_2 + 1}$$

Since $|\Gamma| = (SWR - 1)/(SWR + 1)$,

$$S_{min} = \frac{S_2}{S_1} \qquad \text{and} \qquad S_{max} = S_1 S_2 \qquad (4\text{–}34)$$

where it has been assumed that $S_2 \geq S_1$.

Conversely, if the maximum and minimum SWR values are known,

$$S_2 = \sqrt{S_{min} S_{max}} \qquad \text{and} \qquad S_1 = \sqrt{\frac{S_{max}}{S_{min}}} \qquad (4\text{–}35)$$

Although the above equations were derived for reflections due to transformers, they are valid for any type of *dissipationless* obstacle or discontinuity. When there are more than two mismatches, the maximum possible SWR is

$$S_{\max} = S_1 S_2 S_3 \ldots S_n \qquad (4\text{–}36)$$

where n is the number of mismatches.

The ABCD matrix formulation provides a convenient method for determining the insertion loss and phase characteristics of a two-port network. These quantities as well as attenuation, transducer loss, and mismatch loss are defined here.

Attenuation.   In Sec. 3–3, attenuation was discussed in the context of an infinite transmission line. Referring to Fig. 3–9a, an attenuation ratio can be defined that is the ratio of voltage (or current) amplitudes at any two points on the infinite line. If V_1 and V_2 are the rms voltages at $z = l_1$ and $z = l_2$ ($l_2 \geq l_1$), then from Eq. (3–14) with $V_0^- = 0$,

$$\frac{V_2}{V_1} = e^{-\alpha(l_2 - l_1)} = e^{-A_t} \qquad (4\text{–}37)$$

where α is the attenuation constant of the line (neper/length) and A_t is the total attenuation in nepers. Solving for A_t,

$$A_t \equiv \ln \frac{V_1}{V_2} = \ln \frac{I_1}{I_2} \qquad \text{Np} \qquad (4\text{–}38)$$

Since 1 Np = 8.686 dB and $\ln x = 2.303 \log x$,

$$A_t \equiv 20 \log \frac{V_1}{V_2} = 20 \log \frac{I_1}{I_2} \qquad \text{dB} \qquad (4\text{–}39)$$

These expressions also apply to a line terminated in its characteristic impedance (Fig. 3–9b) since it is equivalent to an infinite line. Furthermore, the equations are valid whether Z_0 is real, imaginary, or complex. When the real part of Z_0 is finite, the net power flow is proportional to V^2 (or I^2). Therefore, the total attenuation can be related to a power ratio. Namely,

$$A_t = \frac{1}{2} \ln \frac{P_1}{P_2} \qquad \text{Np} \qquad \text{or} \qquad A_t = 10 \log \frac{P_1}{P_2} \qquad \text{dB} \qquad (4\text{–}40)$$

where P_1 and P_2 represent the real power flow at l_1 and l_2, respectively.

A finite length line is merely one example of a two-port network. The attenuation concept may be extended to include any two-port network where V_1 and I_1 represent the rms voltage and current at the input port and V_2 and I_2 represent the output values *when the output port is terminated by the characteristic impedance of the network*. The characteristic impedance of a symmetrical network is defined as that value of terminating impedance that results in an input impedance of the same value.[4] For example, the characteristic impedance of the lowpass filter discussed in Sec. 6–6 (Fig. 6–36) is given by Eq. (6–30). Thus if the filter is terminated by Z_T or (Z_π), the total attenuation is defined by Eq. (4–38) or (4–39). Note that at certain frequencies, Z_T and Z_π are purely imaginary. In this situation, there is no real power flow

---

[4] In general, the requirement is that the network be terminated in its iterative impedance. For symmetrical networks, the iterative impedance equals the characteristic impedance. For a more complete discussion, see Chapter 12 of Ref. 4–1,

and therefore Eq. (4–40) is not valid. The point is that attenuation is fundamentally the ratio of rms voltages (or currents) at two locations in a circuit. For a two-port network, these are the input and output terminals. Except in special cases, the attenuation of a two-port is seldom specified or measured. At microwave frequencies, the loss quantities of interest are usually insertion loss and transducer loss.

Insertion loss.   The insertion loss of a two-port network is defined as

$$L_I \equiv 10 \log \frac{P_{L_b}}{P_{L_a}} \quad \text{dB} \tag{4–41}$$

where, referring to Fig. 4–34, P_{L_b} is the power delivered to a specified load (Z_L) *before* inserting the network and P_{L_a} is the power delivered to the *same* load after insertion of the network. Note that, unlike attenuation, insertion loss is expressed as a power ratio and both values are defined at the same terminals (namely, Z_L). Therefore, Eq. (4–41) may be written in terms of rms voltages or currents. Namely,

$$L_I = 20 \log \frac{V_{L_b}}{V_{L_a}} = 20 \log \frac{I_{L_b}}{I_{L_a}} \quad \text{dB} \tag{4–42}$$

The insertion loss may be expressed in terms of the ABCD matrix elements of the network by determining the rms load voltage *before* and *after* the network is inserted. This derivation is outlined in Sec. 2.15 of Ref. 4–3. The result is

$$L_I = 10 \log \left| \frac{AZ_L + B + CZ_G Z_L + DZ_G}{Z_G + Z_L} \right|^2 \tag{4–43}$$

At microwave frequencies, the situation where both Z_G and Z_L are matched to a transmission line is quite common. For this case, $Z_G = Z_L = Z_0$ (real) and the insertion loss of a network inserted anywhere along the transmission line becomes

$$L_I = 10 \log \left| \frac{A + \dfrac{B}{Z_0} + CZ_0 + D}{2} \right|^2 \tag{4–44}$$

A list of ABCD matrix elements for several two-port networks is given in Appendix C (Table C–1). Equation (4–44) may be written in terms of the normalized ABCD elements. Namely,

$$L_I = 10 \log \left| \frac{\bar{A} + \bar{B} + \bar{C} + \bar{D}}{2} \right|^2 \tag{4–45}$$

The reader is encouraged to use this form when $Z_G = Z_L = Z_0$. Also, the reader should note the magnitude signs in all the above loss equations.

Active networks are usually characterized in terms of their insertion gain (G_I). Since a gain ratio is simply the reciprocal of a loss ratio,

$$G_I \equiv 10 \log \frac{P_{L_a}}{P_{L_b}} \quad \text{dB} \tag{4–46}$$

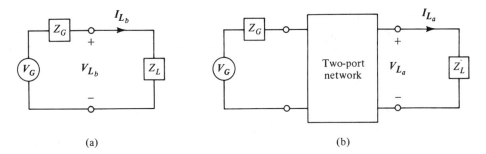

Figure 4–34   A description of the insertion loss and phase definitions given by Eqs. (4–41) and (4–54).

Transducer loss.   The insertion loss of a two-port network can be negative when Z_L is not the complex conjugate of Z_G. For example, if $Z_G = 90$ ohms and $Z_L = 10$ ohms, the insertion of a dissipationless matching network between the generator and the load *increases* the power delivered to the load. Thus $P_{L_a} > P_{L_b}$ which results in a negative insertion loss. The idea that a passive network can produce insertion gain is troublesome (at least, conceptually). This can be circumvented by characterizing the network in terms of its transducer loss. Transducer loss is defined as

$$L_T \equiv 10 \log \frac{P_A}{P_L} \qquad \text{dB} \qquad\qquad (4\text{–}47)$$

where P_A is the available power of the source [given by Eq. (4–1)] and P_L is the power that the network delivers to a specified load when the source is connected to its input port. Note that unlike insertion loss, transducer loss is always positive for a passive network since P_L is necessarily less than or equal to the available power.

The transducer loss can be expressed in terms of the ABCD matrix elements. Referring to Fig. 4–33,

$$L_T = 10 \log \frac{|AZ_L + B + CZ_GZ_L + DZ_G|^2}{4R_GR_L} \qquad\qquad (4\text{–}48)$$

where $Z_G = R_G + jX_G$ and $Z_L = R_L + jX_L$. When Z_G and Z_L are equal and real, insertion loss and transducer loss are identical. Since Eqs. (4–44) and (4–45) are based upon this assumption, they also represent the transducer loss of the network. The terms *insertion loss* and *transducer loss* are often used interchangeably by microwave workers. This is valid only when the above assumption is true. The terms *characteristic insertion loss* and *attenuation* are used by some authors (Refs. 4–10 and 4–12) to denote insertion and transducer loss when the generator and the load are matched to their respective transmission lines (that is, $Z_G = Z_0$ and $Z_L = Z_0$).

transducer gain (G_T), which is defined as

$$G_T \equiv 10 \log \frac{P_L}{P_A} \qquad \text{dB} \qquad\qquad (4\text{–}49)$$

where P_A and P_L are defined below Eq. (4–47). Therefore, transducer gain (in dB) is the negative of transducer loss (in dB) and the transducer gain ratio (P_L/P_A) is the reciprocal of the transducer loss ratio (P_A/P_L).

Mismatch loss.   The equations for insertion and transducer loss include mismatch as well as dissipative effects. This is most easily seen for the case when $Z_G = Z_0$ (real). The situation is described in Fig. 4–35. With $Z_G = Z_0$, the power associated with the incident wave (P_{in}^+) is exactly P_A, the power available from the source [Eq. (3–69) with $\Gamma_G = 0$]. Therefore the power reflected at the input port is $|\Gamma_{in}|^2 P_A$ and the net power into the network becomes

$$P_{in} = \{1 - |\Gamma_{in}|^2\}P_A \qquad (4\text{–}50)$$

In general, a portion of this power will be dissipated in the network (denoted by P_d). Thus the power delivered to the load is

$$P_L = \{1 - |\Gamma_{in}|^2\}P_A - P_d \qquad (4\text{–}51)$$

and from the transducer loss definition,

$$L_T = 10 \log \left\{ \frac{1}{1 - |\Gamma_{in}|^2 - \dfrac{P_d}{P_A}} \right\} \qquad (4\text{–}52)$$

which may be rewritten in terms of P_{in}. Thus,

$$L_T = 10 \log \left(\frac{P_{in}}{P_{in} - P_d} \right)\left(\frac{1}{1 - |\Gamma_{in}|^2} \right)$$

or $\qquad\qquad\qquad\qquad\qquad\qquad\qquad\qquad\qquad\qquad\qquad (4\text{–}53)$

$$L_T = 10 \log \left(\frac{1}{1 - \dfrac{P_d}{P_{in}}} \right) + 10 \log \left(\frac{1}{1 - |\Gamma_{in}|^2} \right)$$

The first term is the dissipative loss (in dB) and the second term is the mismatch loss (in dB). For a dissipationless network, $P_d = 0$ and the transducer loss is merely the mismatch loss of the network. Note the similarity between the expressions for mismatch loss and reflection loss [Eq. (3–80)]. Their meanings are essentially the same. The term *reflection loss* is usually applied to a load impedance wherein $Z_L \neq Z_0$. As used here, mismatch loss is the reflection loss of a two-port network terminated in Z_L *when* $Z_G = Z_0$ (real).[5] It results from Γ_{in} being non-zero, which means $Z_{in} \neq Z_0$. When Γ_{in} is zero, only the dissipative term remains in Eq. (4–53).

It should be pointed out that the reflection and loss expressions developed here are valid even if the input line section in Fig. 4–35 is eliminated. The requirement is that Z_G be real. For instance if $Z_G = 100$ ohms, simply assume a *zero* length transmission line with $Z_0 = 100$ ohms. It is interesting that the reflection concept remains useful even in the absence of transmission lines!

---

[5] In the general case (that is, $Z_G \neq Z_0$), the meaning of mismatch loss differs from that given here. Both mismatch and dissipative loss depend upon the specific values of Z_G and Z_L. This is discussed in Chapter 13 of Ref. 4–1.

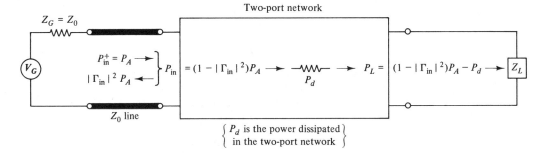

Figure 4–35   A description of transducer and mismatch loss for a passive two-port network when $Z_G = Z_0$ (real).

The following illustrative example is intended to clarify some the concepts discussed thus far.

Example 4–6:

Calculate the transducer, mismatch, and dissipative losses for the network shown in Fig. 4–36 when
(a) $l = 0$
(b) $l = \lambda/4$
(c) $l = \lambda/4$ with the generator and load interchanged.
Note that $L_T = L_l$ since $Z_G = Z_L = 50$ ohms.

Solution:
(a) With $l = 0$, the network is simply a shunt resistor of value 16.67 ohms. Normalized to 50 ohms, the matrix elements become $\bar{A} = \bar{D} = 1$, $\bar{B} = 0$, and $\bar{C} = 50/16.67 = 3$.
From Eqs. (4–45) and (4–31),

$$L_l = L_T = 10 \log \left| \frac{1 + 0 + 3 + 1}{2} \right|^2 = 7.96 \text{ dB}$$

and

$$\Gamma_{in} = \frac{1 + 0 - 3 - 1}{1 + 0 + 3 + 1} = -0.60 \qquad \text{or} \qquad |\Gamma_{in}| = 0.60$$

which is equivalent to an input SWR of 4.0.

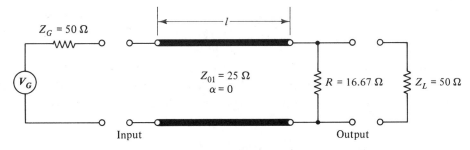

Figure 4–36   Two-port network pertaining to Ex. 4–6.

From Eq. (4–53), the mismatch loss is

$$10 \log \left(\frac{1}{1 - 0.36} \right) = 1.94 \text{ dB}$$

and hence the dissipative loss is $7.96 - 1.94 = 6.02$ dB.

(b) For $l = \lambda/4$, $\beta l = \pi/2$ and hence $\cos \beta l = 0$ and $\sin \beta l = 1.0$. The ABCD matrix of the network is the product of the matrices of the quarter-wave line and the shunt resistor. From Table C–1 and Eq. (C–4),

$$\begin{bmatrix} \bar{A} & \bar{B} \\ \bar{C} & \bar{D} \end{bmatrix} = \begin{bmatrix} 0 & j25/50 \\ j50/25 & 0 \end{bmatrix} \times \begin{bmatrix} 1 & 0 \\ 3 & 1 \end{bmatrix} = \begin{bmatrix} j1.5 & j0.5 \\ j2.0 & 0 \end{bmatrix}$$

and therefore

$$L_I = L_T = 10 \log \left| \frac{j1.5 + j0.5 + j2.0}{2} \right|^2 = 6.02 \text{ dB}$$

From Eq. (4–31), $\Gamma_{in} = 0$ and thus the mismatch loss is 0 dB. Hence the dissipative loss is 6.02 dB.

Note that the effect of the matching transformer in front of the mismatched, dissipative network (the shunt resistor) eliminates the mismatch loss, leaving only the dissipative portion. The reader is cautioned that it is possible to insert a matching network that eliminates mismatch loss, but *increases* the dissipative loss of the network (Prob. 4–34).

(c) With the load and generator interchanged, the normalized ABCD matrix becomes

$$\begin{bmatrix} \bar{A} & \bar{B} \\ \bar{C} & \bar{D} \end{bmatrix} = \begin{bmatrix} 1 & 0 \\ 3 & 1 \end{bmatrix} \times \begin{bmatrix} 0 & j0.5 \\ j2.0 & 0 \end{bmatrix} = \begin{bmatrix} 0 & j0.5 \\ j2.0 & j1.5 \end{bmatrix}$$

and therefore

$$L_I = L_T = 10 \log \left| \frac{j4.0}{2} \right|^2 = 6.02 \text{ dB}.$$

In this case, $|\Gamma_{in}| = 0.75$ and hence the input SWR is 7.0. Thus the mismatch loss is 3.59 dB and the dissipative loss is $6.02 - 3.59 = 2.43$ dB.

The results in parts *b* and *c* indicate that while the transducer (and insertion) loss value is invariant when the generator and load are interchanged, the portions attributable to mismatch and dissipative effects are not! In general, the following statements are true for linear networks consisting of reciprocal elements.[6]

1. When $Z_G = Z_L$, insertion loss, transducer loss, and insertion phase are reciprocal. In this context, reciprocal means that interchanging the generator and load leaves the particular quantity unchanged.

2. The mismatch loss and hence input SWR is not necessarily reciprocal when the network is dissipative. Also, dissipative loss is not necessarily reciprocal.

3. When the network is dissipationless, the mismatch loss and hence input SWR is reciprocal when $Z_G = Z_L = Z_0$.

[6] Reciprocity is defined in Appendix D (footnote #3). All the elements listed in Table C–1 are passive and reciprocal.

An example of the third statement is illustrated in Prob 4-26.

Insertion phase.   Referring again to Fig. 4-34, insertion phase (θ_I) is defined as the difference between the phase of the load voltage (or current) *before* the network is inserted (θ_b) and the phase of the load voltage (or current) *after* the network is inserted (θ_a). That is,

$$\theta_I \equiv \theta_b - \theta_a \qquad (4\text{-}54)$$

When Z_G and Z_L are real, $\theta_b = 0$ and $\theta_I = -\theta_a$. Thus a *positive* value of θ_I indicates phase *delay* due to the network $(\theta_a$ negative) while a *negative* value indicates phase *advance* $(\theta_a$ positive). For this case, the insertion phase is given by the phase angle of $(AZ_L + B + CZ_GZ_L + DZ_G)$. That is,

$$\theta_I = \arctan \frac{\text{Im } (AZ_L + B + CZ_GZ_L + DZ_G)}{\text{Re } (AZ_L + B + CZ_GZ_L + DZ_G)} \qquad (4\text{-}55)$$

where Im ( ) denotes the imaginary part of ( ) and Re ( ) denotes the real part of ( ). When $Z_G = Z_L = Z_0$ (real),

$$\theta_I = \arctan \frac{\text{Im } (AZ_0 + B + CZ_0^2 + DZ_0)}{\text{Re } (AZ_0 + B + CZ_0^2 + DZ_0)} \qquad (4\text{-}56)$$

or

$$\theta_I = \arctan \frac{\text{Im } (\bar{A} + \bar{B} + \bar{C} + \bar{D})}{\text{Re } (\bar{A} + \bar{B} + \bar{C} + \bar{D})} \qquad (4\text{-}57)$$

Suppose, for example, that the two-port network is simply a section of lossless transmission line having a characteristic impedance Z_{0k}, a phase constant β, and a length l. If $Z_G = Z_L = Z_0$ (real), the normalized ABCD matrix elements are $\bar{A} = \bar{D} = \cos \beta l$, $\bar{B} = j\bar{Z}_{0k} \sin \beta l$ and $\bar{C} = j(\sin \beta l)/\bar{Z}_{0k}$. For the case when $Z_{0k} = Z_0$, Eq. (4-57) yields $\theta_I = \beta l$, which means that the insertion phase is simply the phase delay associated with a wave traveling a distance l along the line. This conclusion is reasonable since with both the generator and load matched to the line section, there are no multiple reflections and the output phase is related to the one-way transit of the propagating wave. If, however, $Z_{0k} \neq Z_0$, multiple reflections exist and the insertion phase is not necessarily the same as the one-way phase delay (βl) due to the line section (see Prob. 4-35).

The following example further illustrates the relationship between multiple reflections and insertion loss and phase.

Example 4-7:
(a) Calculate the 2.0 GHz insertion loss and phase of a 3.2 nH series inductor placed in a 50 ohm system (that is, $Z_G = Z_L = Z_0 = 50$ ohms).
(b) For the circuit shown in Fig. 4-37, determine the length l of lossless line required for $L_I = 0$ dB at 2.0 GHz. Assume $\lambda = \lambda_0$.
(c) Calculate the insertion phase of the network for the length obtained in part *b*.

Solution:
(a) At 2.0 GHz, the impedance of the inductor is $j40$ ohms. With all impedances

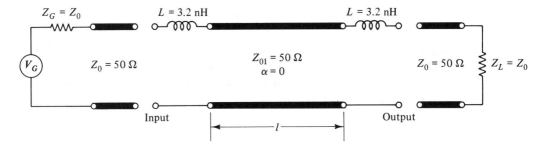

Figure 4–37   Two-port network pertaining to parts b and c of Ex. 4–7.

normalized to 50 ohms, the normalized matrix elements are $\bar{A} = \bar{D} = 1$, $\bar{B} = j\bar{X} = j0.80$, and $\bar{C} = 0$. Thus the insertion loss is

$$L_I = 10 \log \left| \frac{2 + j\bar{X}}{2} \right|^2 = 10 \log \left\{ 1 + \left(\frac{\bar{X}}{2} \right)^2 \right\} = 0.64 \text{ dB}$$

and the insertion phase is

$$\theta_I = \arctan \frac{\bar{X}}{2} = \arctan 0.40 = 0.38 \text{ rad}$$

Since the inductor is dissipationless, all of the 0.64 dB loss represents mismatch loss. From Eq. (4–53), $|\Gamma| = 0.37$.

(b) The overall normalized ABCD matrix is the product of the three individual matrices. With $Z_{01} = 50$ ohms, $\bar{Z}_{01} = 1.0$ and

$$\begin{bmatrix} \bar{A} & \bar{B} \\ \bar{C} & \bar{D} \end{bmatrix} = \begin{bmatrix} 1 & j\bar{X} \\ 0 & 1 \end{bmatrix} \begin{bmatrix} \cos \beta l & j \sin \beta l \\ j \sin \beta l & \cos \beta l \end{bmatrix} \begin{bmatrix} i & j\bar{X} \\ 0 & 1 \end{bmatrix}$$

or

$$\begin{bmatrix} \bar{A} & \bar{B} \\ \bar{C} & \bar{D} \end{bmatrix} = \begin{bmatrix} (\cos \beta l - \bar{X} \sin \beta l) & j\{2\bar{X} \cos \beta l + (1 - \bar{X}^2) \sin \beta l\} \\ j \sin \beta l & (\cos \beta l - \bar{X} \sin \beta l) \end{bmatrix}$$

For 0 dB loss, both the dissipative and mismatch losses must be zero. Therefore $\Gamma_{\text{in}} = 0$ and from Eq. (4–31), $\bar{A} + \bar{B} = \bar{C} + \bar{D}$. Since $\bar{A} = \bar{D}$, $\bar{B}$ must equal $\bar{C}$ and hence $\sin \beta l = 2\bar{X} \cos \beta l + (1 - \bar{X}^2) \sin \beta l$. Thus, $\tan \beta l = 2/\bar{X}$ or $\beta l = \arctan (2/\bar{X})$. For $\bar{X} = 0.80$, $\beta l = 2\pi l/\lambda_0 = \arctan 2.5 = 1.19$ rad. At 2.0 GHz, $\lambda_0 = 15$ cm and hence the required line length for zero loss is $l = 2.84$ cm. (This answer can be obtained directly from the Smith chart as indicated in Prob. 4– 37.)

(c) From Eq. (4–57),

$$\theta_I = \arctan \left\{ \frac{2\bar{X} \cos \beta l + (2 - \bar{X}^2) \sin \beta l}{2(\cos \beta l - \bar{X} \sin \beta l)} \right\}$$

From part b, $\tan \beta l = 2/\bar{X}$ or $\bar{X} \sin \beta l = 2 \cos \beta l$. Thus the insertion phase reduces to

$$\theta_I = \arctan (-\tan \beta l) = \pi - \beta l = 2\left(\frac{\pi}{2} - \beta l \right) + \beta l$$

which can be written as

$$\theta_I = 2\left(\frac{\pi}{2} - \arctan\frac{2}{\overline{X}}\right) + \beta l = 2\arctan\frac{\overline{X}}{2} + \beta l$$

For $\overline{X} = 0.80$ and $\beta l = 1.19$ rad,

$$\theta_I = 2(0.38) + 1.19 = 1.95 \text{ rad}$$

In part c of the above example, the overall insertion phase turned out to be the sum of the individual insertion phases (the phase of the series inductor was calculated in part a). In general, this is *not* true for two-port networks connected in tandem. Neither is the overall insertion loss the sum of the individual losses, a common misconception. For instance, in the above example, the overall loss is 0 dB when $\beta l = 1.19$ rad, while the sum of the individual losses is $0.64 + 0 + 0.64 = 1.28$ dB. The zero loss resulted from the fact that the line length between the two inductors was chosen so that the wave reflections associated with their mismatch losses canceled resulting in $\Gamma_{\text{in}} = 0$.[7]

This discussion leads us to the following conclusion. As long as two or more networks connected in cascade have mismatch losses, it is possible (in fact, probable) that the overall insertion loss (or phase) is *not* equal to the sum of the individual insertion losses (or phases). The value can be greater or less depending upon the phasing of the reflected waves or their effect on the dissipative loss of the networks. On the other hand, if the individual mismatch losses are all zero (with $Z_L = Z_G = Z_0$), the overall insertion loss is equal to the sum of the individual insertion losses and likewise the overall insertion phase is the sum of the individual insertion phases.

4–3b The Wave-Transmission Matrix

The wave-transmission matrix represents another method of characterizing two-port networks. As explained in Appendix C, the matrix relates the incident and scattered waves at the input $(\mathbf{a}_1, \mathbf{b}_1)$ to similar waves at the output $(\mathbf{a}_2, \mathbf{b}_2)$. These waves and their relationship are indicated in Fig. C–5. The equations are repeated here for convenience. Namely,

$$\mathbf{b}_1 = \mathcal{T}_{11}\mathbf{a}_2 + \mathcal{T}_{12}\mathbf{b}_2 \qquad \text{or} \qquad \begin{Bmatrix} \mathbf{b}_1 \\ \mathbf{a}_1 \end{Bmatrix} = \begin{bmatrix} \mathcal{T}_{11} & \mathcal{T}_{12} \\ \mathcal{T}_{21} & \mathcal{T}_{22} \end{bmatrix} \begin{Bmatrix} \mathbf{a}_2 \\ \mathbf{b}_2 \end{Bmatrix} \qquad (4\text{–}58)$$
$$\mathbf{a}_1 = \mathcal{T}_{21}\mathbf{a}_2 + \mathcal{T}_{22}\mathbf{b}_2$$

where the **a** and **b** waves are defined in Appendix D [Eqs. (D–1), (D–2), and (D–7)].

---

[7] In order for the wave reflections to cancel, they must be equal in amplitude and 180° out-of-phase. At first glance, one would think that cancellation occurs when the two inductors are spaced $\lambda/4$ apart. (See the *wave* explanation of the quarter-wave transformer in Sec. 4–2b.) This is not the case! The reasoning is as follows: Part of the incident wave is reflected by the first inductor (Fig. 4–37). The remainder is phase delayed $\arctan (\overline{X}/2)$ by the first inductor and βl by the line length. A portion of this wave is reflected by the second inductor and delayed an additional βl by the line length and $\arctan (\overline{X}/2)$ as it passes through the first inductor. Thus for perfect cancellation, $2\beta l + 2\arctan (\overline{X}/2) = \pi$ or $\beta l = \arctan (2/\overline{X})$, which agrees with part b of the example.

Figure 4–38 shows a pair of two-port networks connected in tandem. As indicated, the waves at the output of the first network are the same as the waves at the input of the second network. That is, $\mathbf{a}_3 = \mathbf{b}_2$ and $\mathbf{b}_3 = \mathbf{a}_2$. As a result, the relationship between the $\mathbf{a}_1$, $\mathbf{b}_1$ waves and the $\mathbf{a}_4$, $\mathbf{b}_4$ waves can be obtained by simply multiplying the individual $\mathcal{T}$ matrices. Thus,

$$\begin{Bmatrix} \mathbf{b}_1 \\ \mathbf{a}_1 \end{Bmatrix} = [\mathcal{T}]_1 \times [\mathcal{T}]_2 \begin{Bmatrix} \mathbf{a}_4 \\ \mathbf{b}_4 \end{Bmatrix} \tag{4–59}$$

Note that like the ABCD matrix, the $\mathcal{T}$ matrix is a transmission-type matrix in that for a cascade of two-port networks, the overall $\mathcal{T}$ matrix is the product of the individual $\mathcal{T}$ matrices.

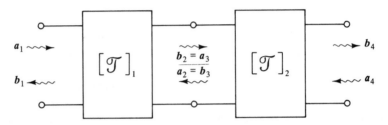

Figure 4–38 A pair of two-port networks connected in cascade. (Note: $\mathbf{a}_3 = \mathbf{b}_2$ and $\mathbf{b}_3 = \mathbf{a}_2$.) The overall transmission matrix is given by Eq. (4–59).

The various quantities (Γ_{in}, L_I, L_T and θ_I) discussed in part a of this section can also be expressed in terms of the $\mathcal{T}$ matrix elements. The relationships are summarized here.

The input reflection coefficient (referenced to Z_0) for an arbitrary load impedance is

$$\Gamma_{\text{in}} = \frac{\mathcal{T}_{11}\Gamma_L + \mathcal{T}_{12}}{\mathcal{T}_{21}\Gamma_L + \mathcal{T}_{22}} \tag{4–60}$$

Note that for $\Gamma_L = 0$, the above expression reduces to Eq. (C–19).

The insertion and transducer loss relations are given by

$$L_I = 10 \log \frac{|\mathcal{T}_{22} + \mathcal{T}_{21}\Gamma_L - \mathcal{T}_{12}\Gamma_G - \mathcal{T}_{11}\Gamma_G\Gamma_L|^2}{|1 - \Gamma_G\Gamma_L|^2} \tag{4–61}$$

$$L_T = 10 \log \frac{|\mathcal{T}_{22} + \mathcal{T}_{21}\Gamma_L - \mathcal{T}_{12}\Gamma_G - \mathcal{T}_{11}\Gamma_G\Gamma_L|^2}{(1 - |\Gamma_G|^2)(1 - |\Gamma_L|^2)} \tag{4–62}$$

When $Z_G = Z_L = Z_0$ (real), $\Gamma_G = \Gamma_L = 0$ and

$$L_I = L_T = 10 \log |\mathcal{T}_{22}|^2 \tag{4–63}$$

Also for $\Gamma_G = \Gamma_L = 0$,

$$\theta_I = \arctan \frac{\text{Im} \{\mathcal{T}_{22}\}}{\text{Re} \{\mathcal{T}_{22}\}} \tag{4–64}$$

The conversion relations between the elements of the $\mathcal{T}$ matrix and the normalized ABCD elements are given by Eqs. (C–20) and (C–21). The following example illustrates the conversion.

Example 4–8:

Figure 4–38 shows a cascade connection of two networks. Assume that the first one consists of a series inductance of reactance X_L, while the second network consists of a shunt capacitance of susceptance B_C. Also assume $Z_G = Z_L = Z_0$ (real).
(a) Obtain the individual and overall wave-transmission matrices.
(b) Obtain expressions for the input reflection coefficient (Γ_{in}) and the transmission coefficient (T)

Solution:
(a) From Table C–1, $\bar{A}_1 = \bar{D}_1 = 1$, $\bar{B}_1 = j\bar{X}_L$, $\bar{C}_1 = 0$, $\bar{A}_2 = \bar{D}_2 = 1$, $\bar{B}_2 = 0$ and $\bar{C}_2 = j\bar{B}_C$ where $\bar{X}_L \equiv X_L/Z_0$ and $\bar{B}_C \equiv B_C/Y_0$. The individual matrices can be obtained from Eq. (C–20) and the overall $\mathcal{T}$ matrix is their product. That is,

$$[\mathcal{T}] = [\mathcal{T}]_1 \times [\mathcal{T}]_2 = \begin{bmatrix} 1 - j\dfrac{\bar{X}_L}{2} & j\dfrac{\bar{X}_L}{2} \\ -j\dfrac{\bar{X}_L}{2} & 1 + j\dfrac{\bar{X}_L}{2} \end{bmatrix} \begin{bmatrix} 1 - j\dfrac{\bar{B}_C}{2} & -j\dfrac{\bar{B}_C}{2} \\ j\dfrac{\bar{B}_C}{2} & 1 + j\dfrac{\bar{B}_C}{2} \end{bmatrix}$$

and therefore

$$\begin{bmatrix} \mathcal{T}_{11} & \mathcal{T}_{12} \\ \mathcal{T}_{21} & \mathcal{T}_{22} \end{bmatrix} = \begin{bmatrix} 1 - \dfrac{\bar{X}_L\bar{B}_C}{2} - j\dfrac{\bar{X}_L + \bar{B}_C}{2} & -\dfrac{\bar{X}_L\bar{B}_C}{2} + j\dfrac{\bar{X}_L - \bar{B}_C}{2} \\ -\dfrac{\bar{X}_L\bar{B}_C}{2} - j\dfrac{\bar{X}_L - \bar{B}_C}{2} & 1 - \dfrac{\bar{X}_L\bar{B}_C}{2} + j\dfrac{\bar{X}_L + \bar{B}_C}{2} \end{bmatrix}$$

(b) From Eq. (C–19),

$$\Gamma_{\text{in}} = \frac{-\bar{X}_L\bar{B}_C + j(\bar{X}_L - \bar{B}_C)}{2 - \bar{X}_L\bar{B}_C + j(\bar{X}_L + \bar{B}_C)} \qquad \text{and} \qquad T = \frac{2}{2 - \bar{X}_L\bar{B}_C + j(\bar{X}_L + \bar{B}_C)}$$

Note that because the overall network is dissipationless, $|\Gamma_{\text{in}}|^2 + |T|^2 = 1$, which agrees with Eq. (C–16).

The situation wherein the scattering matrix is known and the wave-transmission matrix is desired (or vice versa) occurs often in microwave analysis. The conversion relations between these matrices are given in Appendix D [Eqs. (D–17) and (D–18)]. The S-parameters for various two-port networks are listed in Table D–1. S-parameters and $\mathcal{T}$ matrix elements for additional networks are given in Refs. 4–11 and 4–12. The latter reference includes the case when $Z_{01} \neq Z_{02}$. It should be noted that other references may define the wave-transmission matrix in a different manner. The reader should bear this in mind when comparing matrix elements.

4–4 S-PARAMETERS AND SIGNAL FLOW GRAPHS

Any linear multiport network may be characterized by a set of coefficients known as *S-parameters*.[8] These coefficients are the elements of the scattering matrix described in Appendix D. For a two-port network, Eq. (D–8) reduces to

$$\mathbf{b}_1 = S_{11}\mathbf{a}_1 + S_{12}\mathbf{a}_2$$
$$\mathbf{b}_2 = S_{21}\mathbf{a}_1 + S_{22}\mathbf{a}_2 \tag{4–65}$$

where the incident ($\mathbf{a}$) and scattered ($\mathbf{b}$) waves are shown in Fig. 4–39 and defined in Eqs. (D–1), (D–2), and (D–7). With $Z_{01} = Z_{02} = Z_0$,

$$\mathbf{a}_k \equiv \frac{\mathbf{V}_k^+}{\sqrt{Z_0}} = \frac{1}{2}\left(\frac{\mathbf{V}_k}{\sqrt{Z_0}} + \sqrt{Z_0}\,\mathbf{I}_k\right)$$

$$\mathbf{b}_k \equiv \frac{\mathbf{V}_k^-}{\sqrt{Z_0}} = \frac{1}{2}\left(\frac{\mathbf{V}_k}{\sqrt{Z_0}} - \sqrt{Z_0}\,\mathbf{I}_k\right) \tag{4–66}$$

where $\mathbf{V}_k$ and $\mathbf{I}_k$ represent the terminal voltage and current at port k. $\mathbf{V}_k^+$ and $\mathbf{V}_k^-$, on the other hand, are the incident and scattered voltage waves at port k. The *S*-parameters in Eq. (4–65) represent the reflection and transmission coefficients *when the network is match terminated*.[9] They are defined as

$$S_{11} \equiv \left.\frac{\mathbf{b}_1}{\mathbf{a}_1}\right|_{\mathbf{a}_2 = 0} = \text{Input Reflection Coefficient with } Z_L = Z_0. \tag{4–67}$$

$$S_{22} \equiv \left.\frac{\mathbf{b}_2}{\mathbf{a}_2}\right|_{\mathbf{a}_1 = 0} = \text{Output Reflection Coefficient with } Z_G = Z_0 \text{ and } V_G = 0 \tag{4–68}$$

$$S_{21} \equiv \left.\frac{\mathbf{b}_2}{\mathbf{a}_1}\right|_{\mathbf{a}_2 = 0} = \text{Forward Transmission Coefficient with } Z_L = Z_0 \tag{4–69}$$

$$S_{12} \equiv \left.\frac{\mathbf{b}_1}{\mathbf{a}_2}\right|_{\mathbf{a}_1 = 0} = \text{Reverse Transmission Coefficient with } Z_G = Z_0 \text{ and } V_G = 0. \tag{4–70}$$

Since $\mathbf{a}_k$ and $\mathbf{b}_k$ in Eq. (4–66) represent rms-phasor quantities, the power flow associated with the incident and scattered waves are given by

$$P_1^+ = a_1^2 = \text{Power incident on the input port.}$$

$$P_2^+ = a_2^2 = \text{Power incident on the output port.}$$

$$P_1^- = b_1^2 = \text{Power reflected from the input port.} \tag{4–71}$$

$$P_2^- = b_2^2 = \text{Power emanating from the output port.}$$

where a_1, a_2, b_1, and b_2 are respectively the rms values of $\mathbf{a}_1$, $\mathbf{a}_2$, $\mathbf{b}_1$, and $\mathbf{b}_2$. For the circuit in Fig. 4–39, a_2^2 equals the power reflected by the load Z_L, while b_2^2 is the

---

[8] Networks may also be characterized in terms of their Z and Y matrices. The conversion between these matrices and the scattering matrix are given in Refs. 4–13 and 4–26.

[9] *S*-parameters are usually referenced to the impedance of the connecting lines. In this text, all *S*-parameter values and $\mathcal{T}$ matrix elements are referenced to Z_0, the characteristics impedance of the input and output lines.

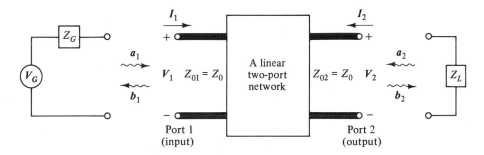

Figure 4–39   A linear two-port network and its associated input and output quantities.

power incident on the load. Thus with $\Gamma_L \equiv \mathbf{a}_2/\mathbf{b}_2$, $a_2^2 = |\Gamma_L|^2\, b_2^2$ and the net power delivered to the load is

$$P_L = b_2^2 - a_2^2 = (1 - |\Gamma_L|^2)b_2^2$$

Similarly, the net input power is

$$P_{in} = a_1^2 - b_1^2 = (1 - |\Gamma_{in}|^2)a_1^2$$

where $\Gamma_{in} \equiv \mathbf{b}_1/\mathbf{a}_1$. For a matched generator $(Z_G = Z_0)$, a_1^2 represents the available generator power P_A.

Power ratios are easily measured at microwave frequencies. The following expressions relate the S-parameter magnitudes to measurable power ratios.

$$|S_{11}|^2 = \left.\frac{\text{Power reflected from the input port}}{\text{Power incident on the input port}}\right|_{Z_L = Z_0} \tag{4–72}$$

$$|S_{22}|^2 = \left.\frac{\text{Power reflected from the output port}}{\text{Power incident on the output port}}\right|_{\substack{Z_G = Z_0 \\ V_G = 0}} \tag{4–73}$$

$$|S_{21}|^2 = \left.\frac{\text{Power delivered to matched load}}{\text{Power incident on the input port}}\right|_{Z_L = Z_0} \tag{4–74}$$

$$|S_{12}|^2 = \left.\frac{\text{Power delivered to matched load at the input port}}{\text{Power incident on the output port}}\right|_{\substack{Z_G = Z_0 \\ V_G = 0}} \tag{4–75}$$

Since the incident power equals the available generator power when $Z_G = Z_0$, $|S_{21}|^2$ represents the forward transducer power gain ratio. Its reciprocal is the transducer power loss ratio. From the loss definitions given in Eqs. (4–41) and (4–47),

$$L_T = L_I = 10 \log \frac{1}{|S_{21}|^2} \tag{4–76}$$

when $Z_G = Z_L = Z_0$. For these same conditions, $|S_{12}|^2$ represents the *reverse* transducer power gain ratio and its reciprocal the power loss ratio.

For the general case (that is, arbitrary Z_G and Z_L) the forward transducer and insertion loss are related to the S-parameters by

$$L_T = 10 \log \left\{ \frac{\left| (1 - S_{11}\Gamma_G)(1 - S_{22}\Gamma_L) - S_{12}S_{21}\Gamma_G\Gamma_L \right|^2}{|S_{21}|^2 (1 - |\Gamma_G|^2)(1 - |\Gamma_L|^2)} \right\} \qquad (4-77)$$

and

$$L_I = 10 \log \left\{ \frac{\left| (1 - S_{11}\Gamma_G)(1 - S_{22}\Gamma_L) - S_{12}S_{21}\Gamma_G\Gamma_L \right|^2}{|S_{21}|^2 |1 - \Gamma_G\Gamma_L|^2} \right\} \qquad (4-78)$$

The corresponding expressions for transducer and insertion gain are the same as above except that the bracketed terms are inverted.

The general equations for input and output reflection coefficients of the two-port network (referenced to Z_0) are as follows:

Denoting the input reflection coefficient with arbitrary Z_L as Γ_1,

$$\Gamma_1 \equiv \frac{\mathbf{b}_1}{\mathbf{a}_1} = S_{11} + \frac{S_{12}S_{21}\Gamma_L}{1 - S_{22}\Gamma_L} \qquad (4-79)$$

where

$$\Gamma_L = (Z_L - Z_0)/(Z_L + Z_0).$$

Denoting the output reflection coefficient with arbitrary Z_G and $V_G = 0$ as Γ_2,

$$\Gamma_2 \equiv \frac{\mathbf{b}_2}{\mathbf{a}_2}\bigg|_{V_G = 0} = S_{22} + \frac{S_{12}S_{21}\Gamma_G}{1 - S_{11}\Gamma_G} \qquad (4-80)$$

where

$$\Gamma_G = (Z_G - Z_0)/(Z_G + Z_0).$$

The above two equations may be deduced directly from Eq. (4–65) and the definitions of Γ_G and Γ_L (Prob. 4–39). Equations (4–77) and (4–78) may be derived using flow graph theory as illustrated in Ex. 4–9.

Flow graph analysis. Flow graph theory provides an orderly way of analyzing signal flow in a multiport network. The procedure was developed by Mason (Refs. 4–22 and 4–23) in connection with the study of feedback in linear circuits. Hunton and others (Refs. 4–24 and 4–27) have shown that S-parameters and their associated flow graphs are particularly useful in analyzing microwave networks.

The signal flow graph for a general two-port network is shown in part b of Fig. 4–40, where the scattering coefficients and the **a** and **b** waves have been previously defined. S-parameters for some basic two-ports are listed in Table D–1.

The rules for constructing a flow graph are as follows:

1. Each variable ($\mathbf{a}_1$, $\mathbf{b}_1$, $\mathbf{a}_2$, $\mathbf{b}_2$, etc.) is designated by a node.[10] Therefore, each port is represented by a node-pair (for example, the $\mathbf{a}_1$, $\mathbf{b}_1$ pair for port 1).
2. Nodes are connected to one another by directed lines that indicate the direction of power flow. A directed line is known as a *branch*. Its value is the multiplying factor that relates the two variables represented by the connected nodes.

---

[10] These nodes are *not* to be confused with the voltage nodes of the network.

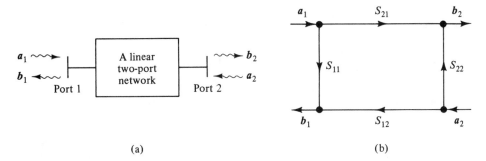

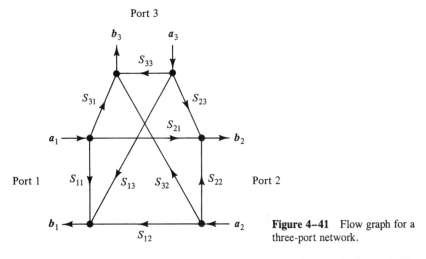

(a)   (b)

Figure 4–40   A two-port network and its signal flow graph representation.

For example, in the two-port shown, $S_{21}\mathbf{a}_1$ represents the portion of $\mathbf{b}_2$ due to the $\mathbf{a}_1$ wave.

3. The value of each node is equal to the sum of the branches *entering* it. Thus for the two-port shown, $\mathbf{b}_2 = S_{21}\mathbf{a}_1 + S_{22}\mathbf{a}_2$ and $\mathbf{b}_1 = S_{11}\mathbf{a}_1 + S_{12}\mathbf{a}_2$. These, of course, are the original signal flow equations for the two-port network [Eq. (4–65)].

The above rules apply to any multiport network. For instance, the flow graph for a three-port is indicated in Fig. 4–41. By utilizing the third rule, the signal flow equations for a three-port [(Eq. (D–8)] can be deduced directly from its flow graph.

Figure 4–41   Flow graph for a three-port network.

The flow graph for a load termination is indicated in Fig. 4–42a, where Γ_L represents the reflection coefficient of Z_L with respect to Z_0. Part *b* of the figure shows a signal generator and its flow graph representation, where Γ_G is the reflection coefficient of the generator with $V_G = 0$. The two generator representations are equivalent when

$$\mathbf{b}_G = \frac{V_G \sqrt{Z_0}}{Z_G + Z_0} \tag{4–81}$$

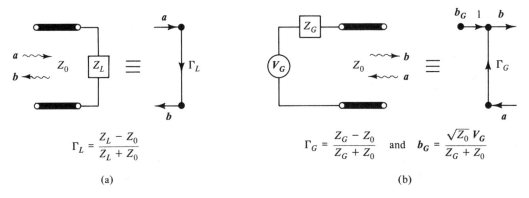

Figure 4–42   Flow graph representations of a load termination (part a) and a signal source (part b).

This can be verified by connecting the generator to a load and analyzing the resultant circuit and its flow graph representation. Referring to part a of Fig. 4–43, $\mathbf{V}_L = \mathbf{V}_G Z_L/(Z_L + Z_G)$. Also, $\mathbf{V}_L = \sqrt{Z_0}\,(\mathbf{a} + \mathbf{b}) = \mathbf{b}\sqrt{Z_0}\,(1 + \Gamma_L)$, since $\Gamma_L = \mathbf{a}/\mathbf{b}$. Equating the two expressions for $\mathbf{V}_L$ yields

$$\mathbf{b} = \frac{\mathbf{V}_G}{\sqrt{Z_0}}\frac{Z_L}{(Z_L + Z_G)(1 + \Gamma_L)} = \frac{\mathbf{V}_G}{2\sqrt{Z_0}}\frac{1 - \Gamma_G}{1 - \Gamma_G\Gamma_L} \qquad (4\text{--}82)$$

where use has been made of the relationship between impedances and reflection coefficients.[11] Referring to part b of Fig. 4–43 and applying the third rule for flow graphs, $\mathbf{b} = \mathbf{b}_G + \Gamma_G\,\mathbf{a} = \mathbf{b}_G + \Gamma_G\Gamma_L\,\mathbf{b}$. Solving for $\mathbf{b}_G$,

$$\mathbf{b}_G = (1 - \Gamma_G\Gamma_L)\mathbf{b} \qquad (4\text{--}83)$$

Substituting from Eq. (4–82) yields

$$\mathbf{b}_G = \frac{\mathbf{V}_G}{2\sqrt{Z_0}}(1 - \Gamma_G) = \frac{\mathbf{V}_G\sqrt{Z_0}}{Z_G + Z_0} \qquad (4\text{--}84)$$

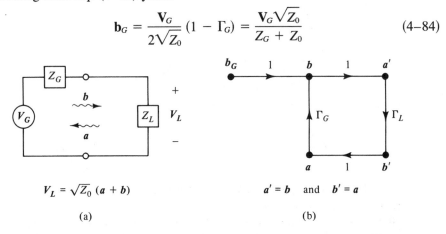

$$V_L = \sqrt{Z_0}\,(a + b)$$

(a)

$a' = b$ and $b' = a$

(b)

Figure 4–43   The connection of a signal generator to a load and its flow graph representation.

[11] Since $\mathbf{b}\sqrt{Z_0} = \mathbf{V}_L^+$, this result is the same as Eq. (3–66) where Γ_{in} is equivalent to Γ_L in Eq. (4–82).

which verifies Eq. (4–81). By setting $\Gamma_L = \Gamma_G^*$, one can show (Prob. 4–40) that the rms value of $\mathbf{b}_G$ is related to the available generator power P_A in the following manner.

$$P_A = \frac{b_G^2}{1 - |\Gamma_G|^2} \qquad \text{or} \qquad b_G = \sqrt{P_A}\sqrt{1 - |\Gamma_G|^2} \qquad (4–85)$$

The flow graph for a complex network can be constructed by joining together the flow graphs of the individual elements. For example, Fig. 4–44a shows a two-port network connected between a generator and a load. The corresponding flow graph is indicated in part b. This representation can be analyzed by either a systematic reduction of the flow graph or by use of the *nontouching loop rule*. Kuhn (Ref. 4–25) and Adam (Ref. 4–13) have described the reduction technique and its application to specific microwave problems. The second method, based on the work of Mason (Refs. 4–22 and 4–23), is summarized here. Examples of its use are given in Refs. 4–26 and 4–27.

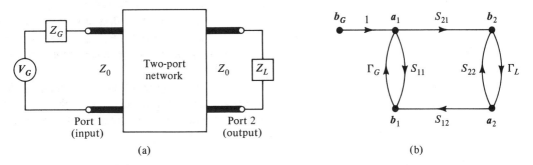

Figure 4–44 The insertion of a two-port network between a signal generator and a load. The associated flow graph is shown in part b.

Before describing the nontouching loop rule, four terms must be precisely defined.

1. **Path.** A continuous successions of branches starting at one node and following the directed arrows of the branches to another node such that no node is touched more than once. Its value is the product of all branch values along the path. Note that a path is distinct if it differs from another path by even one branch.

2. **First-order loop product.** The product of the values of all branches encountered in starting from a node and moving in the direction of the arrows back to the original node.

3. **Second-order loop product.** The product of the values of any *two* nontouching first-order loops.

4. **Third-order loop product.** The product of the values of any *three* nontouching first-order loops.

High-order loop products can be defined in a similar manner. Except for very complicated systems, they are seldom encountered in practice.

Given the above definitions, the nontouching loop rule may be used to evaluate the transfer function H between any pair of network variables. With H defined as the ratio of a dependent variable to an independent variable, the nontouching loop rule is

$$H = \frac{P_1[1 - \Sigma L(1)^{(1)} + \Sigma L(2)^{(1)} - \ldots] + P_2[1 - \Sigma L(1)^{(2)} + \Sigma L(2)^{(2)} - \ldots] + P_3[1 - \ldots]}{1 - \Sigma L(1) + \Sigma L(2) - \Sigma L(3) + \cdots} \quad (4\text{--}86)$$

where

$P_1, P_2, P_3, \ldots$	are the values of the various paths connecting the nodes representing the two variables.
$\Sigma L(1), \Sigma L(2), \ldots$	are the sum of all first-order loop products, second-order loop products, . . . , respectively.
$\Sigma L(1)^{(1)}, \Sigma L(2)^{(1)}, \ldots$	are, respectively, the sum of all first-order loop products, second-order loop products, . . . , that do *not* touch the *first* path between the nodes representing the two variables.
$\Sigma L(1)^{(2)}, \Sigma L(2)^{(2)}, \ldots$	are, respectively, the sum of all first-order loop products, second-order loop products, . . . , that do *not* touch the *second* path between the nodes representing the two variables.

The following example illustrates the use of this rather awesome-looking equation.

Example 4–9:

Use the nontouching loop rule and Fig. 4–44 to derive Eq. (4–77).

Solution: The transducer loss (in dB) of a two-port network is defined as

$$L_T \equiv 10 \log \frac{P_A}{P_L} = 10 \log \frac{(b_G/b_2)^2}{(1 - |\Gamma_G|^2)(1 - |\Gamma_L|^2)}$$

where use has been made of Eq. (4–85) and the fact that $P_L = b_2^2(1 - |\Gamma_L|^2)$. Note that b_2 and b_G are the rms values of $\mathbf{b}_2$ and $\mathbf{b}_G$, respectively. The ratio $\mathbf{b}_2/\mathbf{b}_G$ may be obtained by applying the nontouching loop rule to part b of Fig. 4–44. The first step is to identify all paths that connect $\mathbf{b}_G$ to $\mathbf{b}_2$. There is in fact only one, namely, from the $\mathbf{b}_G$ node through $\mathbf{a}_1$ to the $\mathbf{b}_2$ node. The value of the path is $1 \times S_{21} = S_{21}$. The next step is to identify the various network loops. Referring again to Fig. 4–44b, there are three first-order loops. Their values are $S_{11}\Gamma_G$, $S_{22}\Gamma_L$ and $S_{21}\Gamma_L S_{12}\Gamma_G$. Of these, only the $S_{11}\Gamma_G$ and $S_{22}\Gamma_L$ loops do not touch in any way. Therefore, only one second-order loop exists and its value is $S_{11}\Gamma_G S_{22}\Gamma_L$. No third-order loops exist since they require *three* nontouching first-order loops. With all paths and loops identified, $\mathbf{b}_2/\mathbf{b}_G$ may be obtained from Eq. (4–86). Since there are no loops that do *not* touch the path between $\mathbf{b}_G$ and $\mathbf{b}_2$, $\Sigma L(1)^{(1)} = 0$ and $\Sigma L(2)^{(1)} = 0$. Thus,

$$H = \frac{\mathbf{b}_2}{\mathbf{b}_G} = \frac{S_{21}}{1 - (S_{11}\Gamma_G + S_{22}\Gamma_L + S_{21}S_{12}\Gamma_G\Gamma_L) + S_{11}S_{22}\Gamma_G\Gamma_L} \quad (4-87)$$

Substituting $(b_G/b_2)^2 = |H|^{-2}$ into the expression for L_T leads directly to the expression for transducer loss in Eq. (4–77).

Note that when $Z_G = Z_L = Z_0$, Γ_G and Γ_L are zero and the transducer loss reduces to the expression given by Eq. (4–76).

The results of the above example can be used to derive the insertion loss expression given by Eq. (4–78). Insertion loss is defined in Eq. (4–41), where P_{L_b} and P_{L_a} represent the load power before and after the insertion of the two-port network. Figures 4–43 and 4–44 describe the two circuit configurations. In the first case, the load power can be deduced with the aid of Eq. (4–83). Namely, $b = b_G/|1 - \Gamma_G\Gamma_L|$ and therefore

$$P_{L_b} = b^2(1 - |\Gamma_L|^2) = b_G^2 \frac{1 - |\Gamma_L|^2}{|1 - \Gamma_G\Gamma_L|^2} \quad (4-88)$$

After the network is inserted, the load power can be deduced with the aid of Eq. (4–87). Namely, $b_2 = |H|b_G$ and therefore

$$P_{L_a} = b_2^2(1 - |\Gamma_L|^2) = b_G^2|H|^2(1 - |\Gamma_L|^2) \quad (4-89)$$

From the definition of insertion loss,

$$L_I = 10 \log \frac{1}{|H|^2|1 - \Gamma_G\Gamma_L|^2} \quad (4-90)$$

Substituting for H from Eq. (4–87) leads directly to the desired result, namely, Eq. (4–78).

This equation is quite useful in analyzing the uncertainty associated with S-parameter measurements when the generator and/or load are not matched. Consider the following illustrative example.

Example 4–10:

A matched attenuator has the following S-parameter values:

$$S_{11} = S_{22} = 0 \quad \text{and} \quad S_{21} = S_{12} = 0.50\underline{/90°}.$$

The $|S_{21}|$ value can be verified by measuring the insertion loss of the attenuator. This is done by noting the reduction in detected power when the attenuator is inserted between a signal generator and a microwave power detector.

(a) Calculate $|S_{21}|$ if the measured loss is 6.0 dB and both the generator and detector are perfectly matched to the input and output lines.

(b) Determine the uncertainty in the computed value of $|S_{21}|$ if the measured loss is 6.0 dB, but $|\Gamma_G| = |\Gamma_L| = 0.20$. The phase angles ($\phi_G$ and ϕ_L) of the two reflection coefficients are not known.

Solution:

(a) Since $\Gamma_G = \Gamma_L = 0$, Eq. (4–76) applies and thus

$$L_I = 6.0 = 10 \log \frac{1}{|S_{21}|^2} \qquad \text{or} \qquad |S_{21}| = 0.50$$

which agrees with the actual value given above.

(b) With $S_{11} = S_{22} = 0$, $S_{21} = S_{12} = j|S_{21}|$, and $|\Gamma_G| = |\Gamma_L| = 0.20$, Eq. (4–78) reduces to

$$L_I = 10 \log \frac{|1 + 0.04|S_{21}|^2 \underline{/\phi_G + \phi_L}|^2}{|S_{21}|^2|1 - 0.04\underline{/\phi_G + \phi_L}|^2} = 6.0$$

The upper and lower bounds on $|S_{21}|$ occur when $\phi_G + \phi_L$ equals zero (or 2π, 4π, etc.) and π (or 3π, 5π, etc.). For $\phi_G + \phi_L = 0$, the computed value is 0.527, while for $\phi_G + \phi_L = \pi$, the value is 0.476. Thus, based on the insertion loss measurement, it has been determined that the value of $|S_{21}|$ lies in the range $0.476 \le |S_{21}| \le 0.527$. In other words with $|\Gamma_G| = |\Gamma_L| = 0.20$, the deduced value has an uncertainty of about ± 5 percent.

As indicated by the above example, S-parameters and flow graphs provide a useful analytical technique for studying errors in a measurement system. They are also useful in the analysis of transistor amplifiers and oscillators. Some typical examples are given in Refs. (4–28), (4–29), and (4–30).

REFERENCES

Books

4–1. Johnson, W. C., *Transmission Lines and Networks*, McGraw-Hill Book Co., New York, 1950.

4–2. Chipman, R. A., *Theory and Problems of Transmission Lines*, Schaum's Outline Series, McGraw-Hill Book Co., New York, 1968.

4–3. Karakash, J. J., *Transmission Lines and Filter Networks*, The Macmillan Co., New York, 1950.

4–4. Wheeler, G. J., *Introduction to Microwaves*, Prentice-Hall, Inc., Englewood Cliffs, NJ, 1963.

4–5. Matthaei, G. L., L. Young, and E.M.T. Jones, *Microwave Filters Impedance-Matching Networks and Coupling Structures*, McGraw-Hill Book Co., New York, 1964.

4–6. Lance, A. L., *Introduction to Microwave Theory and Measurements*, McGraw-Hill Book Co., New York, 1964.

4–7. Collin, R. E., *Foundations for Microwave Engineering*, McGraw-Hill Book Co., New York, 1966.

4–8. Kurokawa, K., *An Introduction to the Theory of Microwave Circuits*, Academic Press, New York, 1969.

4–9. Ghose, R. N., *Microwave Circuit Theory and Analysis*, McGraw-Hill Book Co., New York, 1963.

4-10. Gray, D. A., *Handbook of Coaxial Microwave Measurements,* General Radio Co., W. Concord, MA, 1968.

4-11. Gupta, K. C., R. Garg, and R. Chadha, *Computer-Aided Design of Microwave Circuits,* Artech House, Dedham, MA, 1981.

4-12. Kerns, D. M., and R. W. Beatty, *Basic Theory of Waveguide Junctions and Introductory Microwave Network Analysis,* Pergamon Press, Ltd., Oxford, England, 1967.

4-13. Adam, S. F., *Microwave Theory and Applications,* Prentice-Hall, Inc., Englewood Cliffs, NJ, 1969.

Articles

4-14. Beatty, R. W., Insertion Loss Concepts. *Proc. IEEE,* 52, June 1964, pp. 663-671.

4-15. Riblet, H. J., General Synthesis of Quarter-Wave Impedance Transformers. *IRE Trans. Microwave Theory and Techniques,* MTT-5, January 1957, pp. 36-43.

4-16. Collin, R. E., Theory and Design of Wide-band Multisection Quarter-Wave Transformers. *Proc. IRE,* 43, February 1955, pp. 179-185.

4-17. Cohn, S. B., Optimum Design of Stepped Transmission-line Transformers. *IRE Trans. Microwave Theory and Techniques,* MTT-3, April 1955, pp. 16-21.

4-18. Young, L., Tables for Cascaded Homogeneous Quarter-Wave Transformers. *IRE Trans. Microwave Theory and Techniques,* MTT-7, April 1959, pp. 233-237. (Correction in MTT-8, March 1960, pp. 243-244.)

4-19. Young, L., Inhomogenous Quarter-Wave Transformers of Two Sections. *IRE Trans. Microwave Theory and Techniques,* MTT-8, November 1960, pp. 645-649.

4-20. Klopfenstein, R. W., A Transmission Line Taper of Improved Design. *Proc. IRE,* 44, January 1956, pp. 31-35.

4-21. Ghose, R. N., Exponential Transmission Lines as Resonators and Transformers. *IRE Trans. Microwave Theory and Techniques,* MTT-5, July 1957, pp. 213-217.

4-22. Mason, S. J., Feedback Theory—Some Properties of Signal Flow Graphs. *Proc. IRE,* 41, September 1953, pp. 1144-1156.

4-23. Mason, S. J., Feedback Theory—Further Properties of Signal Flow Graphs. *Proc. IRE,* 44, July 1956, pp. 920-926.

4-24. Hunton, J. K., Analysis of Microwave Measurement Techniques by Means of Signal Flow Graphs. *IRE Trans. Microwave Theory and Techniques,* MTT-8, March 1960, pp. 206-212.

4-25. Kuhn, N., Simplified Signal Flow Graph Analysis. *Microwave Journal,* 6, November 1963, pp. 59-66.

4-26. Anderson, R. W., *S*-Parameter Techniques for Faster, More Accurate Network Design. *Hewlett-Packard Journal,* 18, February 1967. (Also available as H-P Application Note 95-1. Another useful article is *S*-Parameter Design, H-P Application Note 154.)

4-27. Bodway, G. E., Two-Port Power Flow Analysis Using Generalized Scattering Parameters. *Microwave Journal,* 10, May 1967, pp. 61-69.

4-28. Bodway, G. E., Circuit Design and Characterization of Transistors by Means of Three-Port Scattering Parameters. *Microwave Journal,* 11, May 1968, pp. 55-63.

4-29. Weinert, F., Scattering Parameters Speed Design of High-Frequency Transistor Circuits. *Electronics,* 39, Sept. 5, 1966, pp. 78-88.

4-30. Froehner, W. H., Quick Amplifier Design with Scattering Parameters. *Electronics,* 40, Oct. 16, 1967, pp. 100-109.

PROBLEMS

4–1. Design the matching network in Fig. 4–4 using a series capacitor.
  (a) Determine l (in cm) and C (in pF) for a perfect match at 2000 MHz. Assume $\lambda = \lambda_0$.
  (b) For the above values of l and C, determine $\bar{Z}_{in}$ and the input SWR at 1800 and 2200 MHz. Assume the load impedance is not frequency sensitive.

  Compare with the results obtained by inductive matching (Table 4–1).

4–2. For the above problem, add 7.5 cm ($\lambda/2$ at 2000 MHz) to the line length l and determine the input SWR at 1800, 2000, and 2200 MHz. Compare the results with those obtained in the previous problem.

4–3. For the matching network in Fig. 4–7, assume a capacitor as the shunt element and $\lambda = \lambda_0$ for the transmission line.
  (a) Determine l (in cm) and C (in pF) for a perfect match at 3000 MHz.
  (b) For the above values of l and C, determine $\bar{Z}_{in}$ and the input SWR at 2500 and 3500 MHz.

4–4. Replace the shunt capacitor in the previous problem by an air-insulated, open-circuited coaxial stub.
  (a) Determine the stub length l_s when $Z_{0s} = 50$ ohms. Repeat for $Z_{0s} = 20$ ohms.
  (b) Which of the two cases results in a lower SWR at 3500 MHz?

4–5. Example 4–1 in the text illustrates the use of an L-C circuit as a broadband matching network (Fig. 4–11). It is proposed that the L-C circuit be replaced by an air-insulated, shorted coaxial stub.
  (a) Determine the length and characteristic impedance required to produce the same susceptance characteristics at 2000 MHz.
  (b) Calculate the input SWR at 1600 and 2400 MHz using the $\bar{Y}_E$ data in Table 4–2. Are the results comparable to those obtained with the L-C circuit?

4–6. A length of lossless 75 ohm line is terminated by a load impedance with the following values: $Z_L = 150 - j90$ ohms at 4000 MHz and $150 + j90$ ohms at 6000 MHz. Determine the line length that results in identical input impedance values at the two frequencies. Assume $\lambda = \lambda_0$. What is the impedance value?

4–7. A load impedance has the following characteristics:
  $Z_L = 40 + j30$ ohms at 5500 MHz, 60 ohms at 6000 MHz, and $40 - j30$ ohms at 6500 MHz.
  When connected directly to a 50 ohm line, the SWR at the band edges is 2.0. Insert a half-wave line section between the load and the line to improve the input SWR. Use the Smith chart with a trial and error procedure to determine the optimum value of Z_{01}. What is the input SWR at the three frequencies?

4–8. Show that when
  (a) $Z_L = R$, Eq. (4–8) reduces to the conditions for a single-section, quarter-wave transformer.
  (b) $Z_L = jX$, no real solution exists.

4–9. (a) Design a short transformer to match $Z_L = 20 + j15$ ohms to a 50 ohm line at 7500 MHz. Assume $\lambda = \lambda_0$.
  (b) Use the Smith chart to determine the input SWR at 6000 MHz.

4–10. Given $R_L = 15$ ohms and $Z_0 = 75$ ohms, design the transmission-line equivalent of an L-C transformer (Fig. 4–16) at 4.0 GHz. Assume $Z_{01} = 0.1Z_{02}$, $Z_{02} = 6R_L$, and

$\lambda = 0.8\lambda_0$. Use the approximate design procedure to determine l_1 and l_2. Compare these values with those obtained from Eqs. (4–14) and (4–15).

4–11. Utilize the impedance transformation equation to derive Eqs. (4–14) and (4–15).

4–12. Use Eqs. (4–14) and (4–15) to verify that if $\beta_1 l_1 = \beta_2 l_2$, the impedance condition for the L-C transformer reduces to $Z_{01} Z_{02} = Z_0 R_L$.

4–13. Given $R_L = 200$ ohms, $Z_0 = 50$ ohms and $l_1 = l_2$,
  (a) Design the transmission-line equivalent of an L-C transformer (Fig. 4–16) at 2500 MHz. Assume $\lambda = \lambda_0$ for both lines and $Z_{01} = 8Z_{02}$. (Hint: Use the impedance condition of Prob. 4–12.)
  (b) Use the Smith chart to determine the input SWR at 2000 and 3000 MHz. Compare the results with the quarter-wave transformer solution (Table 4–4).

4–14. A Smith chart solution to the double-slug tuner shown in Fig. 4–19 is described in the text with values of l_1 and l_2 given. Determine another set of values for l_1 and l_2 which will match the normalized load admittance $\bar{Y}_L = 3 + j1$.

4–15. Referring to Fig. 4–18, what is the minimum value of $\bar{B}$ that results in a tuner capable of matching out all load impedances with $|\Gamma_L| \leq 0.75$?

4–16. A double-slug tuner (Fig. 4–18) is used to match a load impedance to a 50 ohm line. For $B = 0.02$ mhos and $Z_0 = 50$ ohms, a perfect match is achieved with $l_1 = 0.06\lambda$ and $l_2 = 0.12\lambda$. What is the value of load impedance?

4–17. A double-stub tuner (Fig. 4–20) with $l = 3\lambda/8$ is used to match a load admittance of normalized value $\bar{Y}_L = 0.40 + j0.20$. Determine the *two* sets of values for l_1 and l_2 that produce a matched condition at the input. Assume $l_3 = 0$.

4–18. Consider the double-stub tuner in Fig. 4–20 with $l = \lambda/4$. Use the Smith chart to show that normalized admittance values $\bar{Y}_A$ *within* the $\bar{G} = 1$ circle cannot be matched by the tuner.

4–19. Derive Eq. (4–18).

4–20. Design a rexolite-filled, coaxial quarter-wave transformer to match a 10 ohm resistive load to a 80 ohm line at 3000 MHz.
  (a) Calculate the length and inner conductor diameter of the coaxial line if the inner diameter of the outer conductor is 0.60 cm.
  (b) Use the Smith chart to determine the SWR on the 80 ohm line at 2700 and 3300 MHz.

4–21. Solve Prob. 4–20 using a two-section, maximally flat transformer.

4–22. Solve Prob. 4–20 using a three-section, maximally flat transformer.

4–23. Solve Prob. 4–20 using a two-section Tchebyscheff transformer. Assume a passband of 2700 to 3300 MHz. Design values may be found in Table 6.04–1 of Ref. 4–5. Compare the results with the two-section Butterworth transformer (Prob. 4–21).

4–24. It is proposed that a multisection Tchebyscheff quarter-wave transformer be used to match a 300 ohm resistive load to a 50 ohm input line. What is the minimum number of sections required to insure an input SWR of less than 1.10 across the 4000 to 8000 MHz band?

4–25. Three shunt susceptances having normalized values $\bar{B}_1 = 0.40$, $\bar{B}_2 = 0.60$, and $\bar{B}_3 = 1.0$ are connected at different points along a lossless transmission line.
  (a) What is the input SWR if the susceptances are spaced for maximum reflection? Assume the output is match terminated.
  (b) What is the input SWR if they are spaced for minimum reflection?

4–26. Verify that the mismatch loss (and hence SWR) is reciprocal for the dissipationless network shown. Assume $Z_G = Z_L = Z_0$.

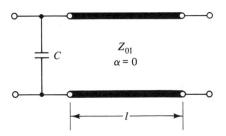

4–27. Use the ABCD matrix to prove that the insertion loss of a lossless Z_{01} transmission-line section is 0 dB if its length is a multiple of a half wavelength. Assume $Z_G = Z_L = Z_0$ and $Z_{01} \neq Z_0$.

4–28. Two shunt resistors are separated by a length of lossless 100 ohm line. Calculate the transducer and dissipative losses of the network if the line is a quarter wavelength long and both resistor values are 5 ohms. Assume $Z_G = Z_L = 50$ ohms.

4–29. Repeat Prob. 4–28 with the resistors replaced by capacitors having the same impedance magnitude as the resistors.

4–30. The element values of a certain two-port network are $A = D = -0.50$, $B = j75$ ohms, and $C = j0.01$ mhos. Assume $Z_G = Z_0 = 50$ ohms, but Z_L is unspecified. Prove that the network is dissipationless. What is the transducer loss if $Z_L = 80 + j60$ ohms?

4–31. A 2 W-50 ohm generator (i.e., $P_A = 2$ W, $Z_G = 50$ ohms) is connected to the input of the two-port network shown. The output is terminated in a 90 ohm load resistor. Determine the load power and the power dissipated in the 25 ohm resistor.

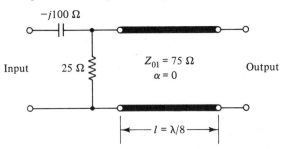

4–32. A 1 W-50 ohm generator is connected to a load impedance $Z_L = 20 - j40$ ohms through a two-port network. Calculate the transducer loss if the network consists of a series inductor $(+j80$ ohms) followed by a lossless 100 ohm line with $l = 0.40\lambda$. How much power is delivered to the load?

4–33. Calculate the insertion loss at 1.0 GHz and 7.5 GHz for the following two-port networks in a 50 ohm system (that is, $Z_G = Z_L = 50$ ohms).
(a) A 3.0 nH series inductor.
(b) A 1.0 cm length of air-insulated 20 ohm line.
(c) The above two networks connected in tandem.
Comment on the loss values in part c compared to the values obtained in part a and b.

4–34. (a) Calculate the transducer loss of the network shown for $n_t = 1.0$. What portion of the loss is mismatch loss? Assume $Z_G = Z_L = 50$ ohms.

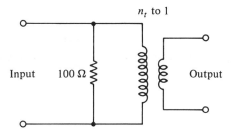

(b) Adjust the turns ratio n_t so that the mismatch loss is 0 dB. Calculate the resultant transducer loss. Comment on the *change* in dissipative loss.

4–35. Derive an expression for the insertion phase of a lossless transmission line of length l and characteristic impedance Z_{01}. Assume $Z_G = Z_L = Z_0$. Determine three conditions for which the insertion phase equals βl, the electrical length of the line.

4–36. Plot the insertion loss and phase versus frequency from 0.10 to 10.0 GHz for the R-C circuit shown. The capacitive reactance $X_C = 30$ ohms at 1.0 GHz. (Hint: B_C is directly proportional to frequency). Assume $Z_G = Z_L = 60$ ohms.

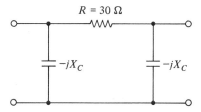

4–37. Use the Smith chart to determine the length l in Fig. 4–37 that results in an input SWR of unity at 2.0 GHz. Assume $\lambda = \lambda_0$. The answer should agree with that given in part b of Ex. 4–7 in Sec. 4–3a.

4–38. Determine the individual and overall $\mathcal{T}$ matrices for the circuit in Fig. 4–38 if the first network consists of a shunt connected 100 ohm resistor and the second network consists of a lossless $\lambda/4$ line with $Z_{01} = \sqrt{5000}$ ohms. Calculate Γ_{in} and transducer loss when $Z_G = Z_L = 50$ ohms.

4–39. Derive Eqs. (4–79) and (4–80).

4–40. Derive Eq. (4–85).

4–41. The S-parameters for a certain two-port network are
$S_{11} = 0.26 - j0.16$, $S_{12} = S_{21} = 0.42$, and $S_{22} = 0.36 - j0.57$.
(a) Determine the input reflection coefficient and transducer loss when $Z_G = Z_L = Z_0$.
(b) Repeat for $Z_G = Z_0$ and $Z_L = 3Z_0$.

4–42. Figure 4–38 shows a pair of two-port networks connected in tandem. The S-parameters for the first network are $S_{11} = 0.28$, $S_{12} = S_{21} = 0.56$, and $S_{22} = 0.28 - j0.56$. For the second network, $S_{11} = S_{22} = -j0.54$ and $S_{12} = S_{21} = 0.84$. Determine the overall $\mathcal{T}$ and scattering matrices.

4–43. Draw the flow graph representation for a pair of *identical* two-port networks connected in tandem. Label the S-parameters of the *individual* networks S_{11}, S_{12}, S_{21}, and S_{22}. Use the nontouching loop rule to derive an expression for
(a) The input reflection coefficient when the second network is match terminated. Use Eq. (4–79) to verify your result.
(b) The forward transmission coefficient of the overall network in terms of the S-parameters of the individual networks.

5

Microwave Transmission Lines

This chapter describes the characteristics of several transmission lines commonly used at microwave frequencies. For efficient transmission, coaxial lines and rectangular waveguides are usually employed.[1] Low-loss coax is used mostly at frequencies below 5 GHz, while rectangular guide is the popular choice at the higher frequencies. Circular waveguides find application where the capability of transmitting or receiving waves having more than one plane of polarization is required. Circular guides employing the TE_{01} mode are also useful for ultra low-loss transmission at high microwave frequencies.

Small sections of transmission lines are often utilized as circuit elements at microwave frequencies. Their use as inductors, capacitors, resonant circuits, and transformers has been described in Chapters 3 and 4. For these applications, strip-type transmission lines, coaxial lines, waveguide sections, and occasionally a short length of open two-wire line are employed.

5-1 THE OPEN TWO-WIRE LINE

The transmission of electrical energy with open two-wire lines is usually restricted to frequencies below 500 MHz. One of its most familiar forms is the TV twin-lead used to connect an antenna to a television set. The reason it is not used at higher frequencies is due to its tendency to radiate energy at a discontinuity or bend in the line. This radiation represents a loss in transmitted power, which generally is unacceptable. However, since small lengths are occasionally used as circuit elements, its characteristics are summarized here.

---

[1] The efficiency of transmission lines and antenna systems is compared in Appendix F.

The electric and magnetic field pattern for the open two-wire line is shown in Fig. 5–1. In most cases, the spacing between the wires s is much greater than their radius a. Typically, s is greater than $4a$. For the electric and magnetic fields shown, power flow ($\overrightarrow{E} \times \overrightarrow{H}$) is into the page. Since both electric and magnetic fields are transverse to the direction of propagation, this represents a *TEM mode*. Other modes of propagation are possible in which portions of the electric and magnetic field are longitudinally oriented. These modes can be suppressed by choosing $s \ll \lambda$.

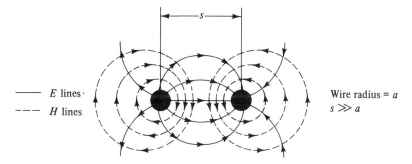

Figure 5–1   The electromagnetic field pattern for TEM transmission on an open two-wire line.

For the TEM mode, the inductance and capacitance per unit length are given by the following approximate expressions. For $s < 4a$,

$$L' \approx \frac{\mu_0 \mu_R}{\pi} \ln \frac{s}{a} \qquad \text{and} \qquad C' \approx \frac{\pi \epsilon_0 \epsilon_R}{\ln (s/a)} \qquad (5\text{–}1)$$

These represent the high-frequency values, since it has been assumed that the skin depth $\delta_s \ll a$. This means that most of the ac currents are located near the surface of the conductors. The derivation of these expressions is given in Chapter 6 of Ref. 5–1. For high-frequency, low-loss lines, the characteristic impedance Z_0 is given by Eq. (3–21). Substituting in the above expressions yields

$$Z_0 \approx 120 \sqrt{\frac{\mu_R}{\epsilon_R}} \ln \frac{s}{a} = 276 \sqrt{\frac{\mu_R}{\epsilon_R}} \log \frac{s}{a} \qquad \text{ohms} \qquad (5\text{–}2)$$

since $\mu_0 = 4\pi \times 10^{-7} \, H/m$ and $\epsilon_0 = 8.854 \times 10^{-12} \, F/m$. When rigid conductors are used, the dielectric medium surrounding the conductors is usually air and hence μ_R and ϵ_R are unity.

An expression for the velocity of propagation can be obtained by substituting Eq. (5–1) into Eq. (3–24). The result is

$$v = \frac{1}{\sqrt{\mu_0 \mu_R \epsilon_0 \epsilon_R}} = \frac{c}{\sqrt{\mu_R \epsilon_R}} \qquad (5\text{–}3)$$

which is exactly Eq. (2–53), namely, the wave velocity in an unbounded dielectric medium. This same result holds for the coaxial line in the TEM mode [Eq. (3–27)]. In fact, this conclusion can be generalized to any transmission line operating in the

TEM mode. Assuming a uniform dielectric material occupies all the insulating region between the two conductors, wave velocity, wavelength, and phase constant are given by Eqs. (2–53), (2–55), and (2–56), respectively.

Approximate expressions for the shunt conductance and series resistance per unit length of the open two-wire line are given below. With $G' = \omega C' \tan \delta$ [(see Eq. (2–23)], the shunt conductance per unit length is

$$G' \approx \omega \frac{\pi \epsilon_0 \epsilon_R}{\ln (s/a)} \tan \delta \qquad (5\text{–}4)$$

where $\tan \delta$ is the dielectric loss tangent of the insulating material. At high frequencies ($\delta_s \ll a$), the resistance of a round wire may be approximated by Eq. (2–79). Thus for the two-wire line, the series resistance per unit length is given by

$$R' \approx \frac{1}{\pi a \delta_s \sigma} \qquad (5\text{–}5)$$

where σ is the conductivity of the conducting material and δ_s is the skin depth given by Eq. (2–73). Since proximity effects have been neglected, Eq. (5–5) is only valid for $s > 4a$. Using the above expressions, the attenuation constant may be determined from Eq. (3–22). Attenuation versus frequency curves for high-frequency lines are available from the various cable manufacturers. Data sheets also provide the *velocity factor* for each line, where

$$\text{Velocity Factor} \equiv \frac{v}{c} = \frac{\lambda}{\lambda_0} \qquad (5\text{–}6)$$

and v is the wave velocity for the particular transmission line. If the conductors are completely immersed in a nonmagnetic insulator of dielectric constant ϵ_R, the velocity factor is equal to $1/\sqrt{\epsilon_R}$. In many cases, however, the insulating region contains two or more dielectric materials. Two examples are shown in Fig. 5–2. Part a shows the familiar TV twin-lead, where the two wires are separated by a thin piece of foam rubber, while part b shows a coaxial line with a loose-fitting teflon sleeve. In both examples, the velocity factor lies somewhere between $1/\sqrt{\epsilon_R}$ and unity (that is, $v = c$). For instance, the velocity factor for the TV twin-lead is practically unity (typically > 0.90) since most of the insulating region is air. On the other hand, for

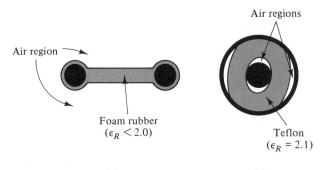

Air regions

Air region

Foam rubber
($\epsilon_R < 2.0$)

Teflon
($\epsilon_R = 2.1$)

(a)

(b)

Figure 5–2   Examples of transmission lines with a nonuniform insulating region.

the coaxial line shown in part b, a major portion of the insulating region is teflon ($\epsilon_R = 2.1$) and hence the velocity factor is slightly greater than $1/\sqrt{2.1} = 0.69$.

5–2 THE COAXIAL LINE

Unlike the open two-wire line, the coaxial line is nonradiating since the electric and magnetic fields are confined to the region between the concentric conductors. For this reason, it is widely used as a high-frequency transmission line. Before 1960, most applications were confined to frequencies below 3.0 GHz. With the advent of precision connectors and miniaturized lines, they are now used extensively in low power applications at X and K_u band and in some cases as high as K_a band.[2]

The TEM mode pattern for the coaxial line was described earlier (Fig. 3–8). The high-frequency inductance and capacitance are given by Eqs. (2–31) and (2–21). These expressions may be rewritten on a per unit length basis. Namely,

$$L' = \frac{\mu_0 \mu_R}{2\pi} \ln \frac{b}{a} \quad \text{and} \quad C' = \frac{2\pi \epsilon_0 \epsilon_R}{\ln (b/a)} \tag{5–7}$$

where the conductor radii a and b are defined in Fig. 3–8. Equations for v, λ, and β are given by Eqs. (3–27), (3–28), and (3–29), respectively. The high-frequency characteristic impedance of the coaxial line is given by Eq. (3–30) and repeated here.

$$Z_0 = 60 \sqrt{\frac{\mu_R}{\epsilon_R}} \ln \frac{b}{a} = 138 \sqrt{\frac{\mu_R}{\epsilon_R}} \log \frac{b}{a} \quad \text{ohms} \tag{5–8}$$

The shunt conductance per unit length may be obtained from Eq. (2–23).

$$G' = \omega \frac{2\pi \epsilon_0 \epsilon_R}{\ln (b/a)} \tan \delta \tag{5–9}$$

The series resistance per unit length is the sum of the resistances due to the inner and outer conductors. Since at high frequencies $\delta_s \ll a$,

$$R' \approx \frac{1}{2\pi a \delta_s \sigma} + \frac{1}{2\pi b \delta_s \sigma} = \frac{a + b}{2\pi ab \delta_s \sigma} \tag{5–10}$$

The attenuation per unit length for a low-loss line is given by Eq. (3–22). Namely,

$$\alpha \approx \frac{R'}{2Z_0} + \frac{G'Z_0}{2} = \alpha_c + \alpha_d$$

where α_c and α_d represent the portions due to the imperfect conductors and insulator, respectively. Making use of Eq. (5–8), one can derive the following expressions for α_c and α_d (Prob. 5–5).

$$\alpha_c = 13.6 \frac{\delta_s \sqrt{\epsilon_R} \{1 + (b/a)\}}{\lambda_0 b \ln (b/a)} \quad \text{dB/length} \tag{5–11}$$

[2] The more common microwave bands are listed in Table 1–1.

and

$$\alpha_d = 27.3 \frac{\sqrt{\epsilon_R}}{\lambda_0} \tan \delta \qquad \text{dB/length} \qquad (5\text{--}12)$$

where δ_s is the skin depth in the metal conductors and $\tan \delta$ is the dielectric loss tangent of the insulator. These equations are valid only if both conductors and the dielectric are nonmagnetic, which is usually the case. Values of loss tangent at 3 GHz for several dielectrics are listed in Appendix B.

An excellent source of microwave data is the Microwave Engineers' Handbook (Ref. 5–4). The section on coaxial lines contains graphs of attenuation and power capacity versus frequency for a variety of rigid and flexible cables.

Voltage breakdown in coaxial lines. The electric field at any point between the concentric conductors of a coaxial line is given by Eq. (3–26) and is repeated here.

$$E = \frac{V}{r \ln (b/a)} \qquad (5\text{--}13)$$

where r is the radius to the particular point between the conductors ($a \leq r \leq b$) and E and V represent rms values of the ac quantities. Since the largest value of electric field occurs at $r = a$, voltage breakdown of the dielectric occurs initially near the surface of the inner conductor. To avoid this condition, its peak value ($\sqrt{2}E$) at $r = a$ must be less than the dielectric strength of the insulator (E_d). Thus the maximum allowable rms voltage across the coaxial line is given by

$$V_{\text{max}} = \frac{a E_d}{\sqrt{2}} \ln \frac{b}{a} = \frac{b E_d}{\sqrt{2}} \frac{\ln (b/a)}{(b/a)} \qquad (5\text{--}14)$$

At room temperature and a pressure of one atmosphere, $E_d \approx 3 \times 10^6$ V/m for air. Therefore, for an air-insulated line with, say $b = 2$ cm and $a = 1$ cm, the applied voltage must be less than 14,700 V rms. Since standing waves are possible along the line, the maximum allowable voltage may be considerably less, namely, $V_{\text{max}}/(1 + |\Gamma_L|)$.

The maximum power that can be transmitted along a match terminated coaxial line without causing voltage breakdown in the insulator is

$$P_{\text{max}} = \frac{V_{\text{max}}^2}{Z_0} = \frac{(bE_d)^2}{120} \left(\frac{a}{b}\right)^2 \sqrt{\frac{\epsilon_R}{\mu_R}} \ln \frac{b}{a} \qquad \text{watts} \qquad (5\text{--}15)$$

where Z_0 is given by Eq. (5–8) and V_{max} by Eq. (5–14). If $Z_L \neq Z_0$, standing waves exist along the line. For large standing waves ($|\Gamma_L| \approx 1$), the actual power handling capability is one-fourth that given by this equation since the electric field at a voltage maximum is approximately twice that associated with the incident wave. In practice, an additional safety factor is usually included to account for surface roughness and other mechanical imperfections.[3]

---

[3] Since most good insulators have a higher dielectric strength (E_d) than air, it seems that higher power capability could be achieved with dielectric-filled lines. However, if the dielectric is not a perfect mechanical fit (see Fig. 5–2b), the opposite can occur! The reason is that the electric field in the air gaps can actually be greater than when the line is completely air-filled.

Since arcing is an extremely fast process, even short pulses of microwave power must be kept below this limit. Thus Eq. (5–15) represents a *peak power* limitation for the coaxial line. Its *average* power capacity is usually limited by heating effects due to the dissipation associated with α_c and α_d.

Before describing how the voltage and power handling capability of the coaxial line can be optimized by the proper choice of dimensions, a discussion of higher-order modes is necessary.

Higher-mode propagation in coaxial lines.   The principal mode of propagation in a coaxial line is the TEM mode. Its field pattern is shown in Fig. 3–8. Most coaxial components (detectors, attenuators, filters, etc.) are designed on the basis that the input signal will be in the TEM mode. There are, however, other ways in which electromagnetic energy can propagate along a coaxial line. These are called higher-order propagation modes. These higher modes can only propagate when the signal frequency exceeds a certain value. This value is known as the cutoff frequency of the mode and is denoted by f_c. At frequencies below f_c, the signal decays exponentially with distance as it attempts to propagate along the line.[4] The field patterns and cutoff frequencies of these modes can be obtained by solving Maxwell's equations and applying the boundary conditions associated with the coaxial structure. The solutions, found in Sec. 8.10 of Ref. 5–6, are beyond the scope of this discussion. However, an approximate expression for the cutoff of the TE_{11} mode in a coaxial line may be deduced by analogy with rectangular waveguide propagation. This argument is presented by Wheeler (Sec. 5.1 of Ref. 5–2). The following expressions for cutoff wavelength (λ_c) and frequency are accurate to within 5 percent for $b/a < 7$.

$$\lambda_c \approx \pi(a + b) \qquad \text{and} \qquad f_c = \frac{v}{\lambda_c} \approx \frac{c}{\pi(a + b)\sqrt{\mu_R \epsilon_R}} \qquad (5\text{–}16)$$

The TE_{11} mode is especially important since its cutoff frequency is the lowest of all the higher-order modes. A sketch of its field pattern is shown in Fig. 5–3.

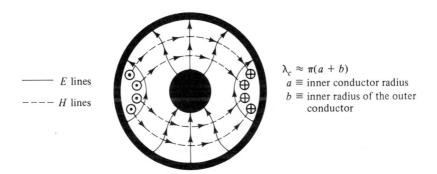

————— E lines

- - - - H lines

$\lambda_c \approx \pi(a + b)$
$a \equiv$ inner conductor radius
$b \equiv$ inner radius of the outer
conductor

Figure 5–3   The electromagnetic field pattern for the TE_{11} mode in a coaxial line.

[4] When $f < f_c$, the higher-order mode is said to be *evanescent* since its exponential decay with distance from the exciting source is associated with energy storage rather than dissipation.

In coaxial transmission, one usually desires TEM mode propagation without the presence of higher-order modes. The reason can be understood with the aid of Fig. 5–4. Part *a* of the figure shows a fixed-power, variable-frequency generator, a coaxial line, and a power meter. Part *b* shows a typical curve of output power versus frequency. As the generator frequency is increased, the power output decreases gradually due to skin effect and dielectric loss [Eqs. (5–11) and (5–12)]. Increasing the frequency further, however, leads to sudden decreases in output power over narrow frequency bands as indicated in the figure. These *resonant-like* losses are associated with higher-mode propagation in the coaxial line, particularly the TE_{11} mode. When the signal enters the coaxial line, some of it is converted from the TEM to the TE_{11} mode. At frequencies greater than the cutoff frequency of the TE_{11} mode, this power propagates to the output. Since the power meter has been designed to accept TEM signals, most of the TE_{11} mode is reflected back toward the input where it is partially re-reflected. For certain values of electrical length (βl), the resultant multiple reflections create a resonance condition with its accompanying high loss. This was explained in Sec. 3–2. To avoid this condition, use of the coaxial line should be restricted to frequencies *below* the cutoff frequency of the TE_{11} mode. Denoting the maximum operating frequency as f_{max} and using Eq. (5–16) yields

$$f_{max} < f_c = \frac{c}{\pi (a + b)\sqrt{\mu_R \epsilon_R}} \qquad (5\text{–}17)$$

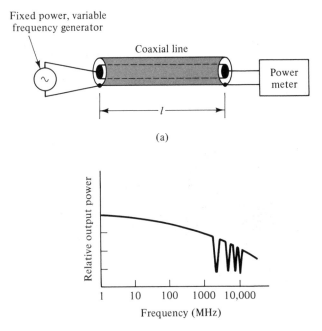

(a)

(b)

Figure 5–4   The effect of higher-mode propagation on power transmission.

In designing a coaxial line, it is customary to allow a 5 percent safety factor. As a result, f_{max} is usually defined as $0.95\,f_c$. Thus, for a given a/b ratio, the *maximum* allowable value of b that insures single-mode (TEM) propagation for all frequencies up to f_{max} is

$$b_{max} = \frac{0.95c}{\pi f_{max}\left(1 + \dfrac{a}{b}\right)\sqrt{\mu_R\,\epsilon_R}} \qquad (5-18)$$

Many coaxial components (for example, filters and transformers) utilize the radius of the inner conductor a as a design variable. To insure against TE_{11} mode propagation for any and all values of a, the worst case is assumed (namely, $a/b \approx 1$), which leads to the following conservative value for b_{max}.

$$b_{max} = \frac{0.95c}{2\pi f_{max}\sqrt{\mu_R\,\epsilon_R}} \qquad (5-19)$$

Note that in either case, the higher the required frequency range, the smaller the value of b_{max} and hence for a fixed b/a ratio, the lower the power-handling capability of the line [Eq. (5-15)]. Thus a compromise is usually required between high power and high operating frequency. This *design tradeoff* occurs quite often in electrical engineering work.

Optimizing transmission characteristics in coaxial lines.   The b/a ratio of a coaxial line and hence its characteristic impedance can be chosen to minimize attenuation, maximize voltage breakdown, or maximize power handling capability. All three cases are considered here.

1. Z_0 for minimum attenuation.   Since α_d is not a function of dimensions, the attenuation constant α is minimized when b/a is chosen to minimize α_c. For a given value of f_{max}, b_{max} is given by Eq. (5-18). By substituting b_{max} into Eq. (5-11), differentiating the result with respect to b/a and setting it equal to zero yields $b/a = 4.68$ as the optimum value. When the line is air-insulated, this is equivalent to a characteristic impedance of 93 ohms. For a teflon-filled line, the optimum Z_0 is 64 ohms.

For an air-insulated line, $\alpha_d = 0$ and hence the attenuation is given by Eq. (5-11). By substituting in the optimum value of b/a, and fixing b in accordance with Eq. (5-18), a relation between the minimum attenuation of a coaxial line and its maximum operating frequency (f_{max}) is obtained. Namely,

$$\alpha_{min} = 0.67\frac{\delta_s}{\lambda_0}f_{max} \qquad \text{dB/m} \qquad (5-20)$$

where f_{max} is in MHz and the skin depth and free space wavelength are computed at the frequency of interest. For example, suppose the required value of f_{max} is 12,000 MHz. Then for minimum attenuation, $b/a = 4.68$ and from Eq. (5-18) b is set equal to 0.623 cm. For copper conductors, δ_s is given by Eq. (2-75). Substituting into Eq. (5-20) yields an attenuation of 0.10 dB/m at 3 GHz and 0.18 dB/m at 10

GHz. In practice, surface roughness and other factors cause the actual attenuation to be slightly greater than the calculated value. It is important to note that lowering f_{max} allows the use of larger conductors [Eq. (5–18)] which reduces the attenuation.

2. Z_0 for maximum voltage breakdown. The transmission of high voltage pulses along a coaxial line usually requires that the line be optimized for maximum voltage breakdown. For a given value of f_{max}, the optimum b/a ratio is 3.59, which is equivalent to a 77 ohm, air-insulated line. This result is obtained by substituting b_{max} into Eq. (5–14), differentiating with respect to b/a and setting the result equal to zero.

3. Z_0 for maximum power handling capacity. Assuming that the maximum power that a coaxial line can handle is limited by the dielectric strength of the insulator, one would think that the optimum b/a ratio would be the same as in the previous case. However, this is not so since $P = V^2/Z_0$ and Z_0 is also a function of the b/a ratio. For a given value of f_{max}, the optimum b/a ratio is 2.09. This means that for an air-insulated line, the optimum Z_0 for maximum power handling capacity is 44 ohms. This result is obtained by substituting b_{max} into Eq. (5–15), differentiating with respect to b/a and setting the result equal to zero.

A relationship between power capacity and maximum operating frequency can be obtained by substituting the optimum value of b/a into Eq. (5–15) and setting b in accordance with Eq. (5–18). The result for an air-insulated coaxial line is

$$P_{max} = 5.3\left(\frac{E_d}{f_{max}}\right)^2 \qquad \text{watts} \qquad (5\text{--}21)$$

where f_{max} is in MHz and E_d in V/m. For example, if $f_{max} = 14,000$ MHz and $E_d = 3 \times 10^6$ V/m, $P_{max} = 243$ kW. Allowing for the possibility of high standing waves along the line, this value should be reduced by a factor of four. Therefore, the maximum power that can be safely transmitted along an air-insulated coaxial line with a maximum usable frequency of 14 GHz is about 60 kilowatts. This value can be increased somewhat by pressurizing the air in the coaxial line. For example, at a pressure of two atmospheres, the value of E_d is 50 percent greater than its value under normal atmospheric conditions.

Although the characteristic impedances obtained in the three cases discussed here are optimum for a fixed value of f_{max}, the fact of the matter is that moderate deviations from these values have only a small effect on the parameter being optimized (see Probs. 5–9 and 5–10). Furthermore, it is important as a practical matter to standardize on a value of Z_0. This is especially true in the manufacture and use of connectors, measurement equipment, and standard components. Except for the occasional use of 75 ohm line, coaxial lines and equipment are usually designed for $Z_0 = 50$ ohms. This represents a satisfactory compromise, as explained in Sec. 5.7 of Ref. 5–2.

5–3 STRIP-TYPE TRANSMISSION LINES

Strip-type transmission lines are used in the design and construction of complex microwave systems and components. Generally, the line consists of either one or two copper-clad, dielectric sheets with the desired circuit etched on one side. Its main advantage lies in the fact that the photo-etching technique used in its manufacture is both accurate and economical. A photograph of a stripline filter is shown in Fig. 5–5. Transitions between coaxial and strip transmission are shown at either end of the filter. Strip configurations may be either symmetrical or asymmetrical. In this text, the symmetrical type is called *stripline* while the asymmetric arrangement is denoted as *microstrip*. Before discussing their characteristics, formulas will be presented for the case of parallel-plate transmission (Fig. 5–6). Although seldom used as a transmission line, its characteristics are sometimes utilized in analyzing other microwave lines. For this reason, a brief summary is presented.

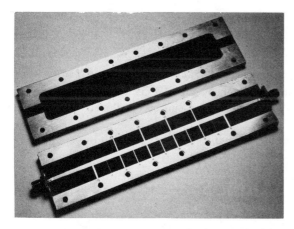

Figure 5–5 A stripline microwave filter. (Courtesy of M/A-COM, Inc., Burlington, Mass.)

Parallel-plate transmission.   A cross-sectional view of the parallel-plate transmission line is pictured in Fig. 5–6. It consists of a low-loss dielectric sandwiched between two flat metal conductors. Usually, its width is much greater than the conductor spacing ($w \gg b$) and the metal thickness is quite small ($t \ll b$). Therefore, it is assumed that the only significant fields lie in the dielectric region between the conductors and that they are uniform. A sketch of the electric and magnetic fields is shown in the figure. Since, at high frequencies, the skin depth in the conductor is small compared to its thickness, most of the current and charge reside on the inner conductor surfaces (that is, the ones in contact with the dielectric). For these conditions, the resistance and capacitance per unit length are given by

$$R' = \frac{2}{w\delta_s \sigma} \tag{5–22}$$

and

$$C' = \frac{w}{b}\epsilon_0 \epsilon_R \tag{5–23}$$

The first equation represents the resistance per length of the two conductors and follows directly from the conclusion below Eq. (2–77). Equation (5–23) derives directly from the parallel-plate capacitance formula, Eq. (2–19), where the area A is the product of the conductor length and its width. For $w \gg b$, the assumption that all the electromagnetic energy is confined to the uniform dielectric region (ϵ_R) is

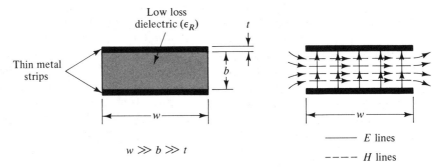

Figure 5-6 A parallel-plate transmission line and its TEM field pattern (cross-sectional view).

valid. Thus for the TEM mode shown, the velocity is given by Eq. (2–53) and the wavelength and phase constant are given by Eqs. (2–55) and (2–56), respectively. As a result, expressions for Z_0 and attenuation can be developed without having to derive equations for L' and G'.

The characteristic impedance at high frequencies is given by

$$Z_0 \approx \sqrt{\frac{L'}{C'}} = \frac{\sqrt{L'C'}}{C'} = \frac{1}{vC'}$$

With $v = c/\sqrt{\mu_R \epsilon_R}$,

$$Z_0 \approx 377 \frac{b}{w} \sqrt{\frac{\mu_R}{\epsilon_R}} \qquad \text{ohms} \qquad (5\text{–}24)$$

As before, the attenuation constant is given by $\alpha_c + \alpha_d$, where for low-loss lines, $\alpha_c = R'/2Z_0$ and $\alpha_d = G'Z_0/2$. Making use of Eqs. (5–22), (5–24), and (2–73) and the equality 1 Np = 8.686 dB results in the following expression.

$$\alpha_c = 27.3 \frac{\delta_s \sqrt{\epsilon_R}}{b\lambda_0} \qquad \text{dB/length} \qquad (5\text{–}25)$$

This equation assumes nonmagnetic conductors and dielectric. Since $\lambda_0 \propto 1/f$ and $\delta_s \propto 1/\sqrt{f}$, the attenuation due to the conductors is proportional to $\sqrt{f}$.

The attenuation per unit length due to the dielectric is given by

$$\alpha_d = \frac{G'Z_0}{2} = \frac{\omega C'}{2} Z_0 \tan \delta = \pi f \sqrt{L'C'} \tan \delta$$

where use has been made of Eq. (2–22). With $\sqrt{L'C'} = 1/v$ and $v = f\lambda$,

$$\alpha_d = \frac{\pi}{\lambda} \tan \delta = \frac{\pi}{\lambda_0} \sqrt{\epsilon_R} \tan \delta \qquad \text{Np/length}$$

or

$$\alpha_d = 27.3 \frac{\sqrt{\epsilon_R}}{\lambda_0} \tan \delta \qquad \text{dB/length} \qquad (5\text{–}26)$$

where again a nonmagnetic dielectric has been assumed. Note that this equation is identical to α_d for a coaxial line [Eq. (5–12)]. In both cases, α_d is directly proportional to frequency since λ_0 is inversely proportional to f.

Symmetrical strip transmission (stripline). For the past 25 years, most strip-transmission systems have made use of the symmetrical configuration, usually called *stripline*. Two of the best sources for information on stripline techniques are Harlan Howe's book (Ref. 5–3) and the IEEE Transactions on Microwave Theory and Techniques (Ref. 5–13). Figure 5–7 shows a cross-sectional view of the stripline structure and its TEM mode pattern. Note the similarity between its mode pattern and that of the coaxial line (Fig. 3–8). With voltage applied between the center strip and the pair of ground planes, current flows down the center strip and returns via the two ground planes. Although the structure is open at the sides, it is basically a nonradiating transmission line. In practice, however, any unbalance in the line causes energy to be radiated out the sides. To prevent this, the ground planes are shorted to each other with either screws or rivets as shown in the figure. The number and spacing of the shorting screws are adjusted to prevent higher-mode propagation in the frequency range of interest.

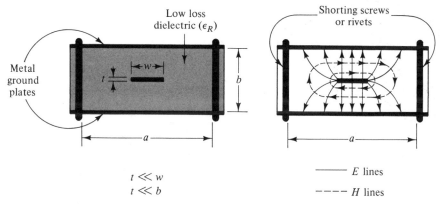

Figure 5–7   A symmetrical strip transmission line (stripline) and its TEM field pattern (cross-sectional view).

With all the fields confined to the uniform dielectric (ϵ_R), the velocity, wavelength, and phase constant for the TEM mode are given by Eqs. (2–53), (2–55), and (2–56). The characteristic impedance may be derived from the expression above Eq. (5–24). Namely,

$$Z_0 = \frac{1}{vC'} = \frac{\sqrt{\mu_R \epsilon_R}}{3 \times 10^8 C'} \qquad \text{ohms} \qquad (5\text{–}27)$$

Thus the determination of the high-frequency characteristic impedance reduces to finding the capacitance per unit length of the stripline configuration. A variety of approximate solutions are available in the literature. The one obtained by Cohn (Ref. 5–14) is probably the most widely used. The results are shown graphically in Fig. 5–8. Since the stripline is usually constructed using a pair of printed circuit boards, the thickness t is typically a few thousandths of an inch, while b ranges from 1/16 to

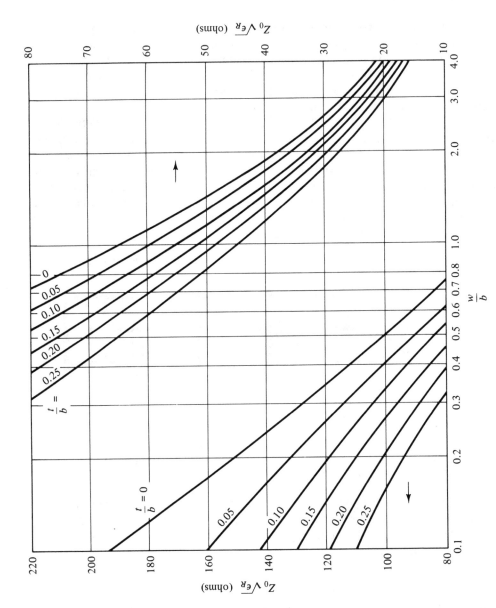

Figure 5–8 Characteristic impedance data for stripline. A nonmagnetic dielectric is assumed ($\mu_R = 1.0$). Also $a \gg w$ and $a > 2b$. (From S. Cohn, Ref. 5–14; © 1955 IRE now IEEE, with permission.)

1/4 of an inch. For this configuration, practical values of Z_0 range from about 10 to 100 ohms for most dielectrics. A discussion of the dielectric materials normally used in stripline is given in Chapter 1 of Ref. 5–3.

The stripline attenuation due to the dielectric material (α_d) is the same as for coaxial and parallel-plate transmission [Eq.(5–26)]. The attenuation associated with the conductors has been derived by Cohn (Ref. 5–14) and others. Normalized values $(\overline{\alpha}_c)$ for copper striplines are given in Fig. 5–9. Using the value of $\overline{\alpha}_c$ from the graph, α_c is obtained from the following equation,

$$\alpha_c = \overline{\alpha}_c \frac{\sqrt{f \, \epsilon_R}}{b} \qquad \text{dB/length} \qquad (5–28)$$

where the frequency f is in GHz and b is the ground-plate spacing. As before, the attenuation for low-loss lines is the sum of α_c and α_d.

The stripline configuration is basically a low-power transmission system. Typically, it is used at power levels of less than 100 W, average, and 1000 W, peak. A more complete discussion of its power capability is given in Chapter 1 of Ref. 5–3.

As with other transmission lines, stripline is capable of propagating electromagnetic energy in other modes. In order to insure only TEM mode propagation in the frequency range of interest, the transverse dimensions of the stripline must be restricted. As explained in Sec. 5–2, the dimensions must be chosen so that the cutoff frequencies for the higher-order modes are greater than the highest frequency of interest $(f_{\max})$. Two of the more troublesome higher modes are shown in Fig. 5–10. The one on the left is analogous to the TE_{11} mode in a coaxial line. For reasons that will become clear in the next section, it is sometimes called the *waveguide mode* in stripline. The shorting screws may be used to suppress this mode. Denoting the highest frequency of interest is $f_{\max}$, the spacing should be chosen so that

$$a < \frac{c}{2 f_{\max} \sqrt{\mu_R \, \epsilon_R}} \qquad (5–29)$$

These suppressors are usually placed at $\lambda/8$ intervals along the transmission direction, where λ is calculated at the highest frequency of interest. The mode shown on the right in Fig. 5–10 does not occur as often as the *waveguide* mode. It can be suppressed by adjusting the ground-plane spacing so that

$$b < \frac{c}{4 f_{\max} \sqrt{\mu_R \, \epsilon_R}} \qquad (5–30)$$

Asymmetric strip transmission (microstrip).   With the advent of low-loss, high dielectric constant materials, microstrip has become increasingly popular, particularly in the fabrication of microwave integrated circuits (see Sec. 6–7). The line consists of a thin conductor and a ground plane separated by a low-loss dielectric material. A cross-sectional view as well as the quasi-TEM mode pattern[5] is shown in Fig. 5–11. Since the top surface is easily accessible, it is very convenient to mount

---

[5] Strictly speaking, the mode is not transverse electromagnetic. The fact that the insulating region consists of two different materials (air and the dielectric) requires the presence of longitudinal field components.

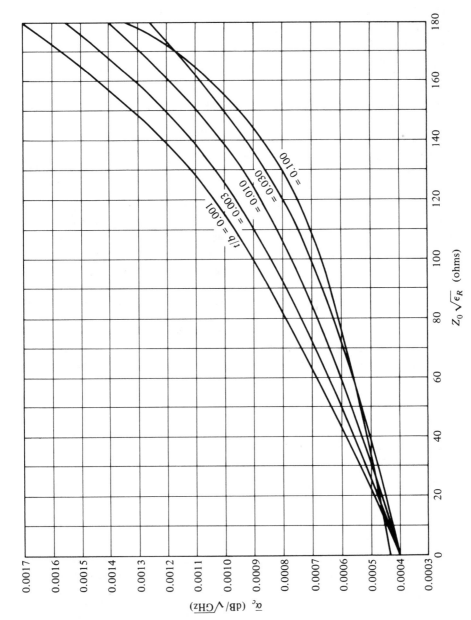

Figure 5–9 Theoretical attenuation (normalized) of copper shielded stripline. Equation (5-28) is used to calculate α_c. (From S. Cohn, Ref. 5–14; © 1955 IRE now IEEE, with permission.)

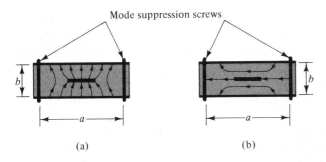

Mode suppression screws

(a)   (b)

Figure 5–10 Electric field patterns for two higher-order stripline modes.

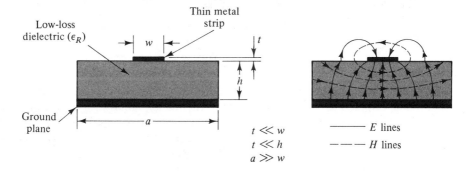

Low-loss
dielectric (ϵ_R)

Thin metal
strip

Ground
plane

$t \ll w$
$t \ll h$
$a \gg w$

———— E lines
– – – – H lines

Figure 5–11 An asymmetric strip transmission line (microstrip) and its quasi-TEM field pattern (cross-sectional view).

discrete devices and make minor adjustments in the microstrip circuit. Three commonly used dielectric materials are alumina, quartz, and Duroid®. The microwave properties of the first two are listed in Appendix B. Duroid is a registered trademark of the Rogers Corporation, Rogers, Connecticut.

Microstrip circuits and components are fabricated using printed circuit techniques. When semiconductor devices are to be integrated into the microstrip structure, silicon ($\epsilon_R = 11.8$) is often used as the dielectric. The use of high ϵ_R materials reduces the amount of fringing fields in the air region above the conductor. In most cases, the fields are negligible at a distance $2h$ above the metal conductor. To prevent radiation losses, the complete microstrip circuit is usually placed in a metal enclosure as shown in Fig. 5–12.

Since the insulating region in microstrip consists of more than one dielectric, Eqs. (2–53) and (2–55) *cannot* be used to calculate propagation velocity and wavelength. With fields in both the air and dielectric regions, one might suspect that the

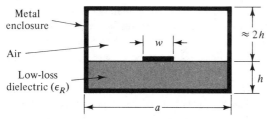

Metal
enclosure

Air

Low-loss
dielectric (ϵ_R)

$\approx 2h$

h

Figure 5–12 An enclosed microstrip configuration used to prevent radiation losses.

actual value of wavelength must lie somewhere between λ_0 and $\lambda_0/\sqrt{\epsilon_R}$. This is indeed the case. For very wide strip widths (w), the fields in the air region become negligible and the wavelength approaches the second value. The microstrip configuration has been analyzed by several workers (Refs. 5–15 to 5–22). Figure 5–13 shows the normalized wavelength (λ/λ_0) as a function of the width-to-spacing ratio (w/h) for several values of dielectric constant (ϵ_R). Curves of characteristic impedance versus w/h are given in Fig. 5–14. The results assume a nonmagnetic dielectric and a negligible strip thickness. This data is based upon the theoretical work of Bryant and Weiss (Ref. 5–18). Their analysis involves the calculation of capacitance per length for the microstrip *with* and *without* the dielectric present. The ratio of capacitances is defined as the effective dielectric constant (ϵ_{eff}) of the microstrip line.[6] The velocity of propagation is then given by $v = c/\sqrt{\epsilon_{\text{eff}}}$ and hence

$$\frac{v}{c} = \frac{\lambda}{\lambda_0} = \frac{1}{\sqrt{\epsilon_{\text{eff}}}} \tag{5–31}$$

With the characteristic impedance $Z_0 = 1/v\,C'$ [see Eq.(5–27)], its value can also be determined once ϵ_{eff} is known. For the higher values of ϵ_R, most of the energy is concentrated in the dielectric region and the value of ϵ_{eff} is nearly but slightly less than ϵ_R. The variation of ϵ_{eff} with frequency as well as dissipative losses in microstrip have been investigated by various workers (Refs. 5–19, 5–20, 5–21, and 5–22).

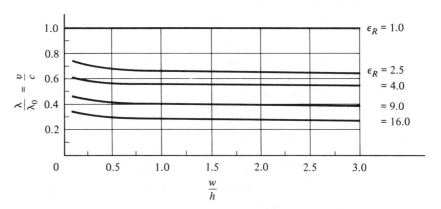

Figure 5–13   Velocity and wavelength for TEM microstrip transmission. A nonmagnetic dielectric is assumed ($\mu_R = 1.0$). (From Bryant and Weiss, Ref. 5–18; © 1968 IEEE, with permission.)

As with all transmission lines, higher-order propagation modes are possible in microstrip. Of particular concern are *TE* and *TM* surface waves between the ground plane and the dielectric material. For $\epsilon_R \gg 1$, the effect of these modes can be minimized by choosing the dielectric thickness (h) to be less than a quarter wavelength at

---

[6] The concept of an effective dielectric constant (ϵ_{eff}) for a transmission line with a nonuniform insulating region is very useful. Its value is usually obtained by solving for the static capacitance per unit length of the structure. The velocity, wavelength, and characteristic impedance of the line are then determined by assuming a *uniform* dielectric region of value ϵ_{eff} [Eqs. (5–31) and (5–27)].

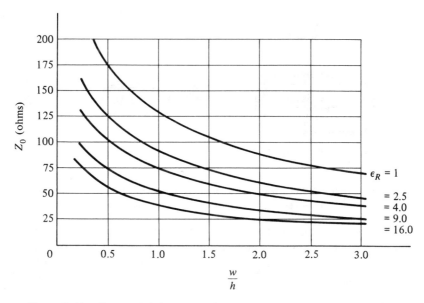

Figure 5–14 Characteristic impedance for the microstrip configuration described in Fig. 5–11. A nonmagnetic dielectric is assumed ($\mu_R = 1.0$). (From Bryant and Weiss, Ref. 5–18; © 1968 IEEE, with permission.)

the highest frequency of interest (f_{max}). Assuming a nonmagnetic dielectric, this requires

$$h < \frac{c}{4 f_{max} \sqrt{\epsilon_r}} \tag{5–32}$$

A TE mode similar to the one in stripline (Fig. 5–10a) can also be troublesome. To minimize its effect, the strip width (w) should be restricted so that

$$w < \frac{c}{2 f_{max} \sqrt{\epsilon_R}} \tag{5–33}$$

The above inequalities are approximate. More exact guidelines may be found in the literature.

Other strip transmission lines.   Figure 5–15 shows three additional transmission-line systems that utilize printed-circuit technologies. Their approximate TEM electric field patterns are also shown. The suspended-substrate stripline pictured in part a is a symmetrical configuration because the two strips are at the same potential. Consequently, the primary electric field is in the air region, while the dielectric region contains only minor fringing fields. Thus the structure is essentially an air-insulated stripline, which means that its dielectric loss (α_d) is negligible. Since ϵ_{eff} is practically unity, transverse dimensions and wavelength are larger than in a corresponding dielectric-filled stripline. This is an advantage in high-frequency applications since dimensional tolerances become less stringent. As a result, the suspended-substrate technique provides an accurate and economical method for fabricat-

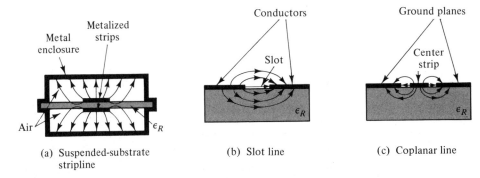

Figure 5–15 Cross-sectional views of other strip transmission-line configurations. Their approximate electric field patterns are also shown.

ing complex stripline systems at frequencies as high as 20 GHz. For additional information, the reader is referred to Refs. 5–16, 5–23, and 5–24.

The transmission line pictured in part *b* of Fig. 5–15 is known as *slot line*. It is a useful alternative to microstrip in the fabrication of microwave integrated circuits. For example, since the electric field is as shown, devices can be easily shunt connected to the line. Microstrip, on the other hand, lends itself to series-connected devices. Also, high values of Z_0 are easily realized in slot line, while low values are more easily achieved with microstrip. In slot line, the magnetic field (not shown) has a strong component in the propagation direction. Therefore, the primary transmission mode is *not* TEM but TE. This characteristic is useful when the system requires the incorporation of nonreciprocal ferrite components. The slot line has been analyzed by Cohn (Ref. 5–25) and others (Refs. 5–26 and 5–27). These articles provide detailed design information on velocity, wavelength, and impedance.

The coplanar line, shown in part *c* of the figure, consists of a thin metal strip with ground planes on either side. Also shown is the approximate electric field pattern. This structure has been analyzed by Wen (Ref. 5–28) and Davis (Ref. 5–29). It combines some of the advantages of microstrip and slot lines. For example, both series and shunt connections are easily achieved in coplanar line. Also, since a significant longitudinal magnetic field component is present, nonreciprocal ferrite components can be realized.

An excellent comparison of microstrip, slot line, and coplanar line is found in Ref. 5–7. A wealth of design data is also included.

5–4 RECTANGULAR AND CIRCULAR WAVEGUIDES

Hollow waveguides are commonly used as transmission lines at frequencies above 5 GHz. Compared to coaxial transmission, waveguides have the following advantages.

1. Higher power handling capability.
2. Lower loss per unit length.
3. A simpler, lower cost mechanical structure.

In addition, the reflections caused by the flanges used in connecting waveguide sections is usually less than that associated with coaxial connectors. The disadvantages of waveguide transmission are larger cross-sectional dimensions and a lower usable bandwidth than coaxial transmission. The fact that hollow waveguides can support electromagnetic waves is proven mathematically in Sec. 5–5. The properties of these waves are described in detail and appropriate equations are derived. In this section, the explanation is given in terms of TEM waves and the transmission-line concepts developed in Chapter 3.

5–4a Rectangular Waveguide

Figure 5–16 shows a rectangular waveguide of inner dimensions a and b. This represents the most common configuration for waveguide transmission. It is customary to label the broad dimension or guide width as a and the height as b. Typically, the conducting walls are made of brass or aluminum and the dielectric region is usually air. The following discussion shows that under certain conditions, electromagnetic waves can propagate along the inside of the waveguide. It is assumed that the wall thickness is greater than several skin depths and hence does not enter into the analysis.

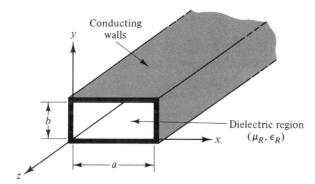

Figure 5–16 The rectangular waveguide structure.

Figure 5–17 shows a parallel-plate transmission line similar to that in Fig. 5–6. The width of the metal strips is w and their spacing is b. The E and H field patterns are also shown. Power is delivered to the load via longitudinal current flow along the two strips. Suppose now that a pair of shorted stubs of length l are connected to the parallel-plate line as shown. If the stubs are a quarter wavelength long, they will have no effect on the transmission of power since they present an infinite impedance in shunt with the line. The E and H standing wave patterns for the shorted stubs are also shown in the figure. The current flow in the stubs is reactive since no real power can be delivered to the shorts. The longitudinal current, on the other hand, can deliver power and hence is referred to as the *power* current. Note that the direction of the *reactive* and *power* currents are perpendicular to each other. One can imagine an infinite set of these quarter wavelength shorted stubs connected in shunt with the parallel-plate line. The resultant configuration is exactly the rectangular waveguide shown in Fig. 5–16 where

$$a = 2l + w$$

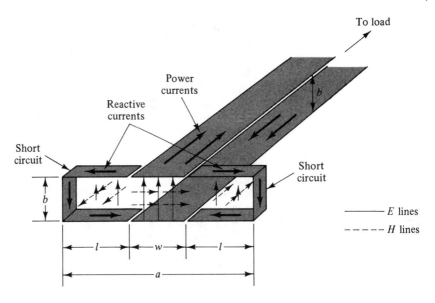

Figure 5–17   The development of rectangular waveguide from a parallel-plate transmission line.

Since $l = \lambda/4$ and the width w must be finite in order for *power* currents to flow, the condition for power transmission becomes

$$a > \frac{\lambda}{2} \tag{5–34}$$

In other words, electromagnetic wave propagation in rectangular waveguide can only occur at frequencies high enough to satisfy this inequality. As the frequency decreases, λ increases and hence the $\lambda/4$ stubs take up a greater portion of the guide width a, leaving less room w for the parallel-plate line. Decreasing the frequency to the point where $a = 2l = \lambda/2$ results in $w = 0$, which prevents the flow of power currents. The frequency at which this occurs is called the *cutoff frequency* (f_c) of the guide. Thus the waveguide has the properties of a highpass filter since power transmission is possible only when $f > f_c$ or $\lambda < \lambda_c$. Using Eq. (5–34), the cutoff wavelength is given by

$$\lambda_c = 2a \tag{5–35}$$

With $f_c \lambda_c = v$, the cutoff frequency of the guide is

$$f_c = \frac{v}{2a} = \frac{c}{2a \sqrt{\mu_R \epsilon_R}} \tag{5–36}$$

What we have described here is the TE_{10} mode of transmission in rectangular guide. Its electric and magnetic field patterns are shown in Fig. 5–18. Based upon the explanation given here, these patterns are plausible. For example, the half sine wave variation of E_y as a function of x can be attributed to the standing waves of the shorted stubs. The fact that the magnetic field is transverse to the direction of propa-

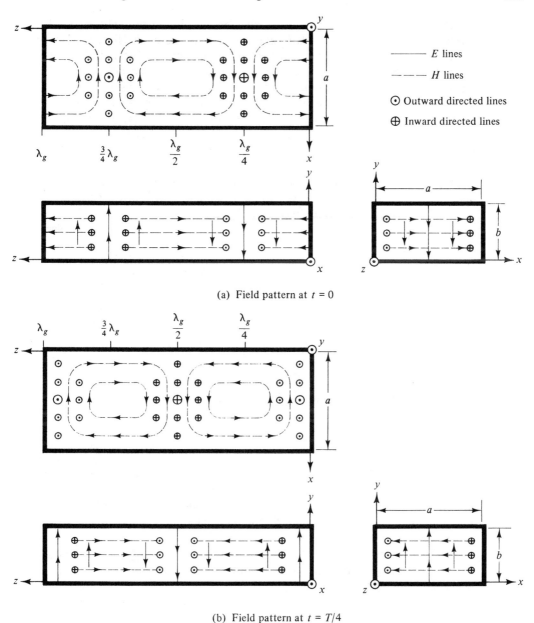

(a) Field pattern at $t = 0$

(b) Field pattern at $t = T/4$

Figure 5–18   The TE_{10} mode pattern as derived from Eq. (5–76).

gation at $x = a/2$ is due to the TEM field pattern of the parallel-plate line. Also, the z component of magnetic field near $x = 0$ and $x = a$ is a consequence of the reactive currents flowing in the shorted stubs. Note that the magnetic lines form complete loops in the x-z plane.

The TE (transverse electric) designation means that for this mode of propagation, the direction of the electric field is always and everywhere transverse to the direction of propagation. The same *cannot* be said for the magnetic field. The subscripts indicate the number of *half* sine wave variations of the field components in the x and y directions, respectively. For the mode described in Fig. 5–18, there is *one* half sine wave variation in the x direction and none in the y direction. The conduction currents in the waveguide walls for the TE_{10} mode are shown in Fig. 5–19. The width and height of the rectangular guide, denoted by a and b, represent the inner dimensions of the waveguide. Because of skin effect, practically all the conduction currents are located at the inner wall surfaces. Thus the only electrical requirement on wall thickness is that it be greater than several skin depths at the lowest frequency of interest. However, to insure mechanical rigidity, the walls are typically 40 to 120 mils thick. Note that the broad walls contain both reactive and power currents, while the narrow walls contain only reactive currents. This is consistent with the explanation based on Fig. 5–17.

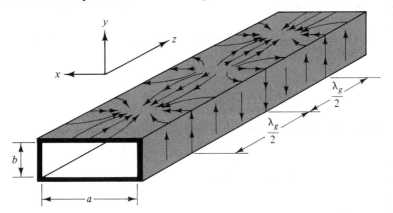

Figure 5–19 Conduction currents in rectangular guide for the TE_{10} mode.

Practically all electromagnetic transmission in rectangular guide utilizes the TE_{10} mode. It is called the *dominant* mode since it has the lowest cutoff frequency of all possible rectangular waveguide modes. Unless otherwise stated, discussions regarding waveguide transmission and components assume that only the dominant mode propagates.

The analysis in Sec. 5–5a shows that there are many other ways in which electromagnetic energy can propagate down the guide. These modes of transmission are divided into two groups, TE and TM modes. Those with the TM or *transverse magnetic* designation have magnetic field patterns that are always and everywhere transverse to the direction of propagation. A sketch of the TM_{11} mode pattern is shown in Fig. 5–20. Its cutoff wavelength is also indicated. The cutoff wavelength and frequency for any mode in rectangular waveguide (whether TE or TM) can be determined from Eqs. (5–67) and (5–68). As stated earlier, the subscripts m and n indicate the number of half sine wave variations of the field components in the x and y directions, respectively.

Sketches of some of the higher-order TE modes with their cutoff wavelengths are shown in Fig. 5–21. Additional TE and TM modes are described in Chapter 5 of Ref. 5–8. The existence of some of these modes can be verified by an argument

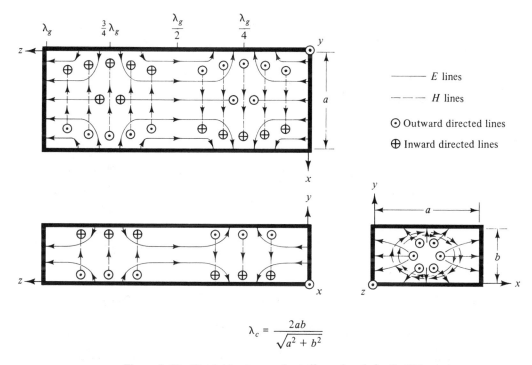

$$\lambda_c = \frac{2ab}{\sqrt{a^2 + b^2}}$$

Figure 5–20   The field pattern and cutoff wavelength for the TM_{11} mode.

similar to that used for the TE_{10} mode. For instance, the TE_{01} is merely the TE_{10} mode rotated 90°. In this case the b dimension must be wide enough to support the width w and the two quarter wavelength shorted stubs. At cutoff, $w = 0$ and therefore $\lambda_c = 2b$.

The development of the TE_{20} and the TE_{30} modes from a parallel-plate transmission line is indicated in Fig. 5–22. Only the x-y plane is shown. Part a describes the TE_{20} mode in terms of two shorted stubs, one a quarter wavelength long and the other three quarters of a wavelength long. Since both stubs present an infinite impedance, they do not interfere with transmission along the parallel-plate line. Cutoff occurs when $w = 0$ or $a = 0.25\lambda + 0.75\lambda$. Thus $\lambda_c = a$ for the TE_{20} mode. By similar reasoning, one concludes that the TE_{30} cutoff wavelength is $2a/3$ since *three* half wavelengths must be *squeezed* into the guide width a at cutoff.

Equations for velocity, wavelength, and phase constant in rectangular waveguide have been derived in Sec. 5–5a. The physical argument used thus far is helpful in showing the reasonableness of these equations. Consider the TE_{10} mode as described and developed in Fig. 5–17. It can be thought of as a TEM wave traveling down the waveguide plus two TEM waves propagating in the transverse plane. The conduction currents associated with these waves are shown in Fig. 5–23a, where I_P represents the power currents and I_R the reactive currents. The vector sum of these currents is indicated by two currents (I_T) whose direction of flow is at an angle θ

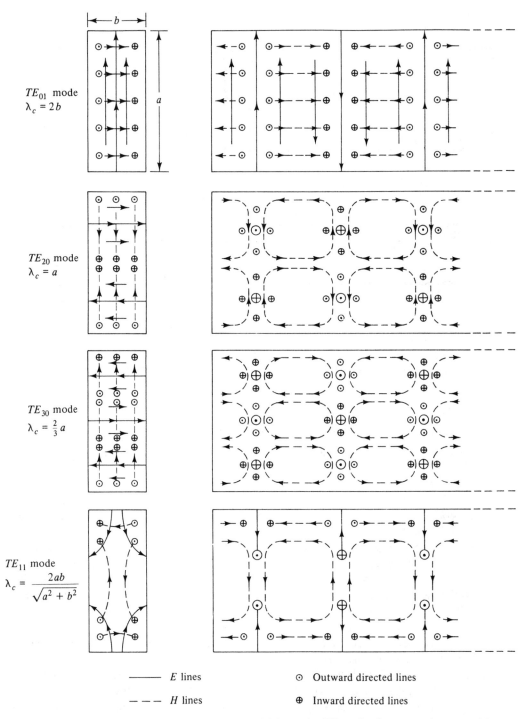

TE_{01} mode
$\lambda_c = 2b$

TE_{20} mode
$\lambda_c = a$

TE_{30} mode
$\lambda_c = \frac{2}{3}a$

TE_{11} mode
$\lambda_c = \dfrac{2ab}{\sqrt{a^2 + b^2}}$

———— E lines   ⊙ Outward directed lines

— — — H lines   ⊕ Inward directed lines

Figure 5–21   Field patterns for some higher-order TE modes in rectangular waveguide. Additional TE and TM modes are described in Chapter 5 of Ref. 5–8.

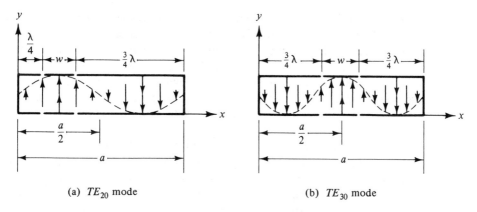

(a) TE_{20} mode   (b) TE_{30} mode

Figure 5–22   An explanation of the TE_{20} and TE_{30} modes in terms of shorted stubs and a parallel-plate line.

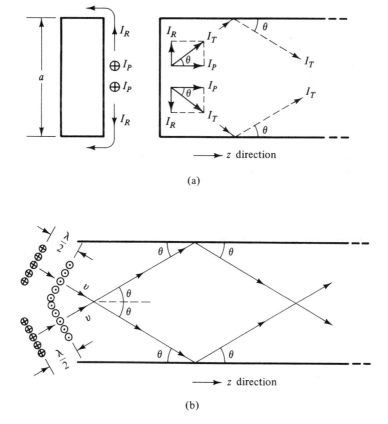

(a)

(b)

Figure 5–23   The TE_{10} mode as the interference pattern of two TEM waves.

with respect to the z axis. The electromagnetic waves associated with the two currents are shown in part b of the figure. For the sake of clarity, only the positive and negative electric field maximums are shown. Since they are TEM waves, their velocity $v = c/\sqrt{\mu_R \epsilon_R}$ and hence the wavelength $\lambda = \lambda_0/\sqrt{\mu_R \epsilon_R}$. These waves are reflected by the side walls of the metal waveguide. The continuing re-reflection of these waves causes them to propagate in the zigzag manner shown in the figure. The interference pattern of these TEM waves results in the TE_{10} mode pattern described in Fig. 5–18.

One conclusion that can be drawn from the above interpretation of a TE wave is that the velocity of energy flow down the waveguide is less than v since the effective length of travel is increased by the zigzag path of the TEM waves. This velocity (v_g) is known as the *group velocity* of the wave. An expression for v_g is given by Eq. (5–74). Since $v_g = v \cos \theta$,

$$\cos \theta = \sqrt{1 - (f_c/f)^2} \tag{5-37}$$

This equation is plausible since at cutoff ($f = f_c$), $\theta = 90°$, which means the power current I_P is zero. Thus at cutoff, the two TEM waves bounce back and forth between the side walls of the guide and hence no energy flows down the waveguide. Conversely, when $f \gg f_c$, $\theta \approx 0$ and the zigzagging effect is negligible, which means $v_g \approx v$.

One can also define a phase velocity (v_p) for the wave traveling down the waveguide. Phase velocity represents the speed with which a particular phase of the wave (for example, its maximum) travels in the propagation direction. An expression for v_p can be developed with the aid of Fig. 5–24. The E-field pattern (at $t = 0$) for one of the TEM waves is shown traveling at an angle θ with respect to the z axis. The wavefront passing point d represents the positive maximums of E traveling with a velocity v. The time required for it to move $\lambda/2$ is exactly $T/2$, where T is the ac period. Thus at $t = T/2$, the wavefront of positive maximums arrive at the plane where the negative maximums were at $t = 0$. This plane includes the point e. As a result, it appears that a positive maximum of E has moved from the point d to the point e in half a period. This distance, labeled $\lambda_g/2$, is equal to $\lambda/(2 \cos \theta)$. There-

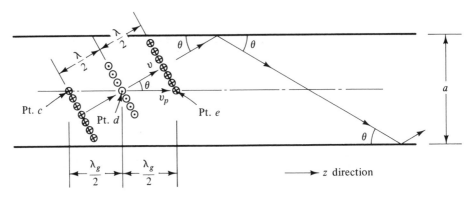

Figure 5–24 A description of phase velocity and guide wavelength in rectangular waveguide.

fore the velocity with which a positive maximum (or any other phase point) moves *along the waveguide axis* is

$$v_p = \frac{\lambda}{\frac{2\cos\theta}{T/2}} = \frac{v}{\cos\theta} = \frac{c/\sqrt{\mu_R\epsilon_R}}{\sqrt{1-(f_c/f)^2}} \qquad (5\text{--}38)$$

which is exactly Eq. (5–71). Note that v_p can be greater than the speed of light.[7] This does not contradict relativity theory because there is no energy or information transfer associated with this velocity. On the other hand, v_g must be less than the speed of light since it represents the velocity of energy flow in waveguide. For TEM lines, $f_c = 0$ and hence $v_p = v_g = v$.

Since v_p represents the phase velocity of a steady-state sinusoidal signal, it is related to guide wavelength (λ_g) by $v_p = f\lambda_g$. From Fig. 5–24,

$$\lambda_g = \frac{\lambda}{\cos\theta} = \frac{\lambda}{\sqrt{1-(f_c/f)^2}} \qquad (5\text{--}39)$$

or

$$\lambda_g = \frac{\lambda_0}{\sqrt{\mu_R\epsilon_R - (\lambda_0/\lambda_c)^2}} \qquad (5\text{--}40)$$

which is exactly Eq. (5–70). The phase constant β for the TE mode in the guide is related to v_p and λ_g by

$$\beta = \frac{\omega}{v_p} = \frac{2\pi}{\lambda_g} \qquad \text{rad/length} \qquad (5\text{--}41)$$

Although these expressions [Eqs. (5–38) to (5–41)] have been developed on the basis of TE_{10} mode propagation, they are applicable to any TE or TM mode. The equation for group velocity [Eq. (5–74)] also applies to any mode. The cutoff wavelength (λ_c) and frequency (f_c) for the particular mode can be determined from Eqs. (5–67) and (5–68). For the TE_{10} mode, $m = 1$ and $n = 0$ and hence these equations reduce to Eqs. (5–35) and (5–36), respectively. Note that as the operating frequency approaches cutoff ($f \to f_c$), $v_g \to 0$, $v_p \to \infty$, and $\lambda_g \to \infty$. A graph of normalized guide wavelength and phase velocity versus normalized frequency is shown in Fig. 5–25. Group velocity is also shown.

At frequencies below the cutoff frequency of a particular mode, the electromagnetic wave is attenuated as it attempts to propagate down the waveguide. When $f < f_c$, the phase constant β is imaginary which implies wave attenuation. When combined with Eq. (5–39), Eq. (5–41) can be rewritten as

$$\beta = j\frac{2\pi}{\lambda}\sqrt{\left(\frac{f_c}{f}\right)^2 - 1} = j\frac{2\pi f_c}{v}\sqrt{1 - \left(\frac{f}{f_c}\right)^2} = j\alpha$$

---

[7] Some authors use the analogy of an ocean wave breaking on the shoreline to describe the concept of phase velocity. See, for example, Sec. 5.4 in Ref. 5–8.

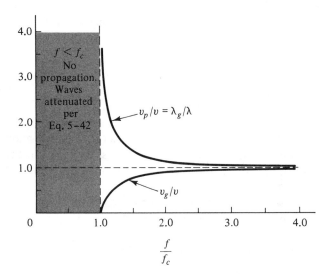

Figure 5–25 The variation of velocity and guide wavelength with frequency.

and hence the attenuation constant due to the cutoff effect is

$$\alpha = \frac{2\pi}{\lambda_c} \sqrt{1 - \left(\frac{f}{f_c}\right)^2} \quad \text{Np/length}$$

or   (5–42)

$$\alpha = \frac{54.6}{\lambda_c} \sqrt{1 - \left(\frac{f}{f_c}\right)^2} \quad \text{dB/length}$$

These expressions are identical to Eq. (5–75) and apply to any TE or TM mode. Since the derivation of these equations assumed a lossless insulator in the waveguide and perfectly conducting guide walls, this attenuation is *not* associated with dissipative losses. In general, the cutoff attenuation manifests itself in a circuit as reflection loss. Conceptually, it is identical to the stopband attenuation of a reactive filter. As stated earlier, a section of waveguide behaves like a highpass filter. When $f > f_c$, the guide exhibits very low loss, while at frequencies below f_c, the attenuation is high which results in practically full reflection (that is, $|\Gamma| \approx 1$).

Waveguide impedance.[8]   In order to use the various analytical techniques developed in Chapters 3 and 4, one usually needs to know the characteristic impedance (Z_0) and wavelength for the particular transmission line. For waveguide, the wavelength is given by Eq. (5–39) or (5–40). As explained in Sec. 5–5a, the characteristic impedance of a waveguide is *not* uniquely defined. The reason is that for a given field pattern, voltage can be defined in a variety of ways. Four possible forms of Z_0 for rectangular waveguide are given in Eqs. (5–79) and (5–81). Unless

[8] This is not to be confused with *wave impedance* which is defined as the ratio of transverse electric field to transverse magnetic field at any point in the waveguide. For TE and TM modes it is denoted as Z_{TE} and Z_{TM}, respectively.

otherwise stated, the modified power-voltage definition will be used in this text. It is repeated here for convenience.

$$Z_0 = 377 \frac{b}{a} \sqrt{\frac{\mu_R}{\epsilon_R}} \frac{\lambda_g}{\lambda} = \frac{377 \frac{b}{a} \sqrt{\frac{\mu_R}{\epsilon_R}}}{\sqrt{1 - \left(\frac{f_c}{f}\right)^2}} \quad \text{ohms} \quad (5\text{–}43)$$

Problems 5–23 and 5–25 use this form to design quarter-wave transformers in rectangular guide. If any of the other forms of Z_0 were used to solve these problems, the results would be the same! The reason is that the various forms of Z_0 differ only by a constant. Since matching involves the *ratio* of impedances, the value of the constant is immaterial.

When a waveguide must be matched to a uniquely defined impedance (for example, a coaxial line), the choice of Z_0 for the waveguide *does* become material. In this situation, the usual approach is to use whichever definition leads to the best agreement between transmission-line theory and the experimental data (see Ref. 5–30). Of course, once a choice is made, it should be used throughout the particular analysis.

It is interesting to note that below cutoff, Z_0 is imaginary. This can be understood with the aid of Fig. 5–17 since for $f < f_c$, $w = 0$, and $l < \lambda/4$. As a result, the impedance of the shorted stubs is inductive and hence the waveguide impedance is imaginary. The characteristic impedance of a reactive filter behaves similarly in its stopband. In the passband, the L/C ratio of the reactive filter is used to control its characteristic impedance. For waveguides, the dimensions (a, b) or the properties of the insulating material (μ_R, ϵ_R) may be varied to adjust Z_0.

Some of the waveguide concepts discussed thus far are highlighted in the following illustrative example.

Example 5–1:

A rectangular waveguide (Fig. 5–16) has the following characteristics:

$$b = 1.5 \text{ cm}, \quad a = 3.0 \text{ cm}, \quad \mu_R = 1, \quad \text{and} \quad \epsilon_R = 2.25.$$

(a) Calculate the cutoff wavelength and frequency for the TE_{10}, TE_{20}, and TM_{11} modes.
(b) Calculate λ_g and Z_0 at 4.0 GHz.
(c) Calculate the attenuation constant (in dB/cm) at 3.0 GHz for the dielectric-filled guide. What is the total attenuation (in dB) if the guide is 12 cm long?
(d) What is the total attenuation at frequencies much less than the TE_{10} cutoff frequency?

Solution:
(a) From Eq. (5–67), λ_c equals $2a$, a, and $2ab/\sqrt{a^2 + b^2}$ for the TE_{10}, TE_{20}, and TM_{11} modes, respectively. Thus the cutoff wavelengths are 6.0 cm for TE_{10}, 3.0 cm for TE_{20}, and 2.68 cm for TM_{11}. The cutoff frequencies are a function of the dielec-

tric properties of the insulating material as well as the guide dimensions. From Eq. (5–68),

$$f_c = 3.33 \times 10^9 \text{ Hz or } 3.33 \text{ GHz for the } TE_{10} \text{ mode.}$$

Similarly, $f_c = 6.66$ GHz for the TE_{20} mode and 7.46 GHz for the TM_{11} mode. Assuming $a > b$, the TE_{10} mode always has the lowest cutoff frequency and hence is the dominant mode.

(b) The guide wavelength is obtained from Eq. (5–40). With $\lambda_0 = 7.5$ cm at 4.0 GHz,

$$\lambda_g = \frac{7.5}{\sqrt{2.25 - \left(\dfrac{7.5}{6.0}\right)^2}} = 9.05 \text{ cm}$$

where the TE_{10} mode has been assumed.
The characteristic impedance of the guide is calculated from Eq. (5–43). Since $\lambda = 7.5/\sqrt{2.25} = 5.0$ cm,

$$Z_0 = 377 \frac{1.5}{3.0} \sqrt{\frac{1}{2.25}} \frac{9.05}{5.0} = 227 \text{ ohms}$$

Note that a reduction in the guide height b lowers the impedance of the waveguide without changing its TE_{10} cutoff frequency. This is useful when changes in impedance level are required in a waveguide system.

(c) The waveguide is below cutoff to a 3.0 GHz signal and hence the attenuation constant is given by Eq. (5–42). With $\lambda_c = 6.0$ cm for the TE_{10} mode,

$$\alpha = \frac{54.6}{6} \sqrt{1 - \left(\frac{3.0}{3.33}\right)^2} = 3.95 \text{ dB/cm}$$

and

$$A_t = \alpha l = (3.95)(12) = 47.4 \text{ dB}$$

When the guide is operating below cutoff, its characteristic impedance is imaginary. Therefore, the above figure only approximates the insertion loss of the 12 cm section in a system with Z_G and Z_L real. (The difference between L_l and A_t is explained in Sec. 4–3a.) The calculation of L_l from A_t is described in Sec. 3.07 of Ref. 5–10.

(d) When $f \ll f_c$,

$$\alpha \approx \frac{54.6}{\lambda_c} \text{ dB/length} \qquad \text{or} \qquad A_t \approx \frac{54.6}{\lambda_c} l \text{ dB.}$$

Therefore,

$$A_t = \frac{54.6}{6} (12) = 109.2 \text{ dB.}$$

Note that if the above inequality holds for the dominant mode, then α and A_t are independent of frequency for all modes.

Power handling capacity of rectangular waveguide.   The maximum power that a coaxial line can transmit before the onset of voltage breakdown was discussed in Sec. 5–2. Equation (5–21) describes how this value is related to the maximum usable frequency and the dielectric strength of the insulator. The following discussion shows that the power handling capacity of rectangular waveguide is significantly greater.

For the TE_{10} mode, the largest electric field occurs along the center line of the broad wall ($x = a/2$). This is shown in Fig. 5–18 and is given above Eq. (5–77). Since this equation gives the rms value, the peak value is $\sqrt{2} E_0 \lambda_c / \lambda_g$. To avoid voltage breakdown,

$$\sqrt{2} E_0 \frac{\lambda_c}{\lambda_g} \leq E_d$$

where E_d is the dielectric strength of the insulating material in the waveguide. From Eq. (5–80),

$$P = \frac{ab}{2Z_{TE}} \left(E_0 \frac{\lambda_c}{\lambda_g} \right)^2 \leq \frac{ab}{4Z_{TE}} E_d^2$$

where $E_0 = H_0 Z_{TE}$ and Z_{TE} is defined by Eq. (5–63). Thus for the TE_{10} mode, the maximum power capability of rectangular guide is

$$P_{\text{max}} = \frac{ab}{4Z_{TE}} E_d^2 \qquad (5\text{–}44)$$

Choosing a and b as large as possible will, of course, maximize P_{max}. However, as explained earlier (Sec. 5–2), it is desirable that only the dominant mode be allowed to propagate. To prevent higher-mode propagation, the operating frequency must be less than the cutoff frequency of the TE_{20} mode. Defining f_{max} as 0.95 of this value yields

$$f_{\text{max}} = \frac{0.95c}{a \sqrt{\mu_R \epsilon_R}} \qquad (5\text{–}45)$$

To prevent possible TE_{01} mode propagation at frequencies below f_{max}, the guide height is chosen so that $b \leq a/2$. With this restriction, the cutoff frequency of the TE_{01} mode is equal to or greater than that of the TE_{20}, and thus the above equation defines the upper frequency limit of the guide. Setting $b = a/2$ and assuming an air-filled rectangular waveguide results in the following equation for maximum power handling capacity.

$$P_{\text{max}} = 27 \left(\frac{E_d}{f_{\text{max}}} \right)^2 \sqrt{1 - \left(\frac{f_c}{f} \right)^2} \qquad \text{watts} \qquad (5\text{–}46)$$

where f_{max} is in MHz, E_d in V/m, f_c is the TE_{10} cutoff frequency, and f is the operating frequency. Since $f_c \approx 0.5 f_{\text{max}}$, f_c/f generally ranges from 0.5 to 0.8. By comparing this equation with that for an air-filled coaxial line [Eq. (5–21)], it is apparent that the power handling capacity of waveguide is significantly greater.

To get an idea of its power capacity, consider a 1 cm by 2 cm air-filled rectangular guide. From Eq. (5–45), $f_{\text{max}} = 14{,}250$ MHz. At room temperature and a pressure of one atmosphere, $E_d = 3 \times 10^6$ V/m. Assuming an operating frequency of 12,000 MHz, $P_{\text{max}} = 934$ kW since $f_c = 7500$ MHz for the TE_{10} mode. Allowing for the possibility of large standing waves, the actual power capacity is one-fourth the above value. As mentioned at the end of Sec. 5–2, P_{max} can be increased by pressurizing the air in the waveguide.

Dissipative attenuation in rectangular waveguide. Unlike the below-cutoff attenuation described earlier, this attenuation is associated with dissipative losses in the waveguide walls and the insulating material within the guide. The attenuation constant due to an imperfect, nonmagnetic dielectric in the waveguide is given by

$$\alpha_d = \frac{27.3 \sqrt{\epsilon_R} \tan \delta}{\lambda_0 \sqrt{1 - (f_c/f)^2}} \qquad \text{dB/length} \qquad (5\text{–}47)$$

where $\tan \delta$ is the dielectric loss tangent of the insulating material. Note that except for the cutoff term $\sqrt{1 - (f_c/f)^2}$, the expression for α_d is the same as for TEM lines. Since the cutoff term equals $\cos \theta$, the resultant increase in α_d can be interpreted as due to the zigzagging of the two TEM waves as they travel down the guide.

For TE_{10} mode propagation, the attenuation constant associated with imperfect conducting walls is given by

$$\alpha_c = \frac{R_s}{b\eta} \frac{1 + \frac{2b}{a}\left(\frac{f_c}{f}\right)^2}{\sqrt{1 - (f_c/f)^2}} \qquad \text{Np/length} \qquad (5\text{–}48)$$

where $\eta = 377 \sqrt{\mu_R/\epsilon_R}$, $R_s = 1/(\sigma\delta_s)$ and f_c is the TE_{10} cutoff frequency. The above equations are derived in Chapter 8 of Ref. 5–6.

In most applications, the waveguide is air-filled ($\alpha_d = 0$) and therefore the total attenuation constant is simply α_c. This attenuation is considerably less than that in a comparable coaxial line, which is another advantage of waveguide transmission.

With $\eta = 377$ ohms for air and δ_s given by Eq. (2–73), the above equation can be written as

$$\alpha_c = 27.3 \frac{\delta_s}{\lambda_0 b} \frac{1 + \frac{2b}{a}\left(\frac{f_c}{f}\right)^2}{\sqrt{1 - (f_c/f)^2}} \qquad \text{dB/length} \qquad (5\text{–}49)$$

where nonmagnetic waveguide walls have been assumed. Curves of α_c versus frequency for $a = 2.0$ cm are shown in Fig. 5–26 for three values of b/a. It is apparent from the data that increasing the guide height b reduces the attenuation. However, as explained below Eq. (5–45), it is necessary that $b \le a/2$. Letting $b = a/2$ and utilizing Eq. (5–45) reduces the above expression for α_c to

$$\alpha_c = 0.19 \frac{\delta_s}{\lambda_0} f_{max} \left\{ \frac{1 + (f_c/f)^2}{\sqrt{1 - (f_c/f)^2}} \right\} \qquad \text{dB/m} \qquad (5\text{–}50)$$

where f_{max} is in MHz and, as noted earlier, $f_c \approx 0.5 f_{max}$. Except when f approaches f_c, the value of the bracketed term is about 1.5 to 2.0. A comparison of Eqs. (5–50) and (5–20) verifies that for a given value of f_{max}, an air-filled rectangular guide has significantly lower loss than a coaxial line.

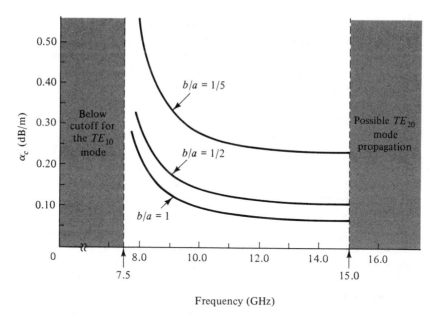

Figure 5–26 TE_{10} mode attenuation for air-filled rectangular guide with copper walls. Guide width $a = 2.0$ cm.

Standard dimensions for rectangular waveguides.

As the microwave industry matured, the need for a set of standard waveguide sizes for the various frequency bands became apparent. A complete listing of standard air-filled rectangular waveguides may be found in the Microwave Engineers' Handbook (Ref. 5–4). Table 5–1 lists some of them and their recommended frequency ranges. Note that the useful bandwidth in all cases is approximately 40 percent or half an octave. Also, the dimensional aspect ratio (a/b) is roughly two. The reason the ratios are not exactly two is that most guide sizes were chosen on the basis of extrusions that were readily available during World War II.

TABLE 5–1 Standard rectangular waveguide sizes as recommended by the Electronic Industries Association (EIA)

EIA Designation	Usable Frequency Range (GHz)	Inner Dimensions $b \times a$ (inches)	Attenuation* (dB/100 ft.)	Peak Power Capacity (Kilowatts)
WR-284	2.60–3.95	1.340 × 2.840	1.10–0.75	2200–3200
WR-187	3.95–5.85	0.872 × 1.872	2.08–1.44	1400–2000
WR-137	5.85–8.20	0.622 × 1.372	2.87–2.30	560–710
WR-90	8.20–12.40	0.400 × 0.900	6.45–4.48	200–290
WR-62	12.40–18.00	0.311 × 0.622	9.51–8.31	120–160
WR-42	18.00–26.50	0.170 × 0.420	20.7–14.8	43–58

*Theoretical values for brass waveguide.

For reasons already explained, it is desirable that only the dominant TE_{10} mode be allowed to propagate down the guide. To insure this, the usable frequency range for each guide size is restricted. That is, for a given guide width a, the operating frequency must be greater than f_c for the TE_{10} mode and less than f_c for the TE_{20} mode. Typically, the upper frequency limit is set about 5 percent below the TE_{20} cutoff frequency. The lower frequency limit is set approximately 25 percent above the TE_{10} cutoff frequency in order to avoid the high attenuation region near f_c (see Fig. 5–26). Another reason for choosing the lower limit in this manner is to avoid the rapid change of λ_g with frequency that occurs near f_c (Fig. 5–25). The choice of waveguide height b is governed by the desire for low attenuation and high power capability. However, as discussed earlier, the guide height should be no more than half the width to avoid the possibility of TE_{01} mode propagation.

Table 5–1 also lists the attenuation and power capability for each guide size. Since these quantities are frequency dependent, two values are given in each case. The first number indicates its value at the lower frequency limit and the second at the upper frequency limit. The power capacity figures include safety factors for manufacturing imperfections and the possible existence of standing waves along the guide. The attenuation values listed are for brass waveguides. Note that the higher the frequency range of interest, the smaller the guide size, which results in higher attenuation and lower power handling capacity.

5–4b Circular Waveguide

When rectangular guide is used within its recommended frequency range, the plane of polarization of the propagating wave is uniquely defined. As shown in Fig. 5–18, the electric field is directed across the narrow dimension of the waveguide. There are certain applications, however, that require dual polarization capability. For example, a waveguide connected to a circularly-polarized antenna must be able to efficiently propagate both vertically and horizontally polarized waves. A square waveguide has this capability since with $b = a$, the cutoff frequencies for the TE_{10} and TE_{01} modes are the same. To insure that only these modes can propagate, the upper frequency limit is usually set 5 percent below the cutoff frequency of the TM_{11} mode. For square guide, the TM_{11} cutoff frequency is lower than the TE_{20} and hence determines the upper frequency limit. As before, the lower limit is set about 25 percent above the cutoff frequency of the TE_{10} (and TE_{01}) mode.

Circular waveguide is the most common form of a dual-polarization transmission line. Like rectangular waveguide, it has an infinite set of TE and TM modes. The theoretical analysis detailed in Sec. 5–5b yields the field patterns and the cutoff wavelengths for the various modes. Four of the most important modes and their cutoff wavelengths are given in Fig. 5–27. Only the field patterns in the transverse plane are shown. The coefficients in the expressions for λ_c are related to the zeros of certain Bessel functions and their derivatives. Some of these coefficients are also listed in Sec. 5–5b (Table 5–3). Additional modes and their cutoff wavelengths may be found in Refs. 5–6, 5–8, and 5–9. The first mode subscript indicates the number of full sine wave variations in the circumferential direction, while the second is related to the Bessel function variations in the radial direction.

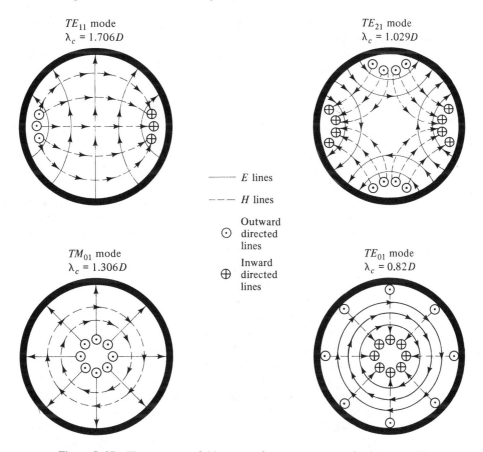

Figure 5–27 The transverse field patterns for some common circular waveguide modes. D = inner guide diameter.

Since the TE_{11} mode has the lowest cutoff frequency, it is the dominant mode in circular guide. Its electric and magnetic field pattern is similar to the TE_{10} in square guide. That is, if one imagines a gradual changing of the guide cross section from square to round, the TE_{10} mode in square guide becomes the TE_{11} mode in circular guide. In like manner, the TM_{01} mode in circular guide is analogous to the TM_{11} mode in square guide. The upper frequency limit of circular guide is restricted by the TM_{01} mode, since its cutoff frequency is the lowest of the higher-order modes.[9] It is interesting to note the similarity between the TM_{01} mode in circular guide and the TEM mode in a coaxial line. The only substantial difference is that the conduction current in the center conductor of the coaxial line is replaced by a displacement current associated with the longitudinal electric field (E_z) along the axis of the round guide. Modes with circular symmetry, such as the TM_{01} and TE_{01}, are

---

[9] In most waveguide transmission systems, energy conversion from the dominant TE_{11} mode to the TM_{01} mode is rare. Therefore, the upper frequency limit of circular guide is usually defined as 95 percent of the TE_{21} cutoff frequency.

often utilized in the design of rotary joints (Sec. 7–5). Methods for exciting the various waveguide modes are discussed in Sec. 7–1.

For waveguides containing a uniform dielectric, the cutoff wavelength and frequency of each mode are related by Eq. (5–68). Namely,

$$f_c = \frac{c}{\lambda_c \sqrt{\mu_R \epsilon_R}} \qquad (5\text{--}51)$$

where μ_R and ϵ_R refer to the properties of the dielectric region. With the cutoff parameters known, the phase velocity v_p, the guide wavelength λ_g, the phase constant β and the below-cutoff attenuation α can be determined from Eqs. (5–38), (5–40), (5–41), and (5–42), respectively.

Expressions for the TE_{11} characteristic impedance of circular waveguide can be derived using the same procedure as for rectangular guide. Four possible forms of Z_0 are given in Eqs. (5–98) and (5–99). The one based on the modified power-voltage definition is repeated here for convenience.

$$Z_0 = 382 \sqrt{\frac{\mu_R}{\epsilon_R}} \frac{\lambda_g}{\lambda} = 382 \mu_R \frac{\lambda_g}{\lambda_0} \qquad \text{ohms} \qquad (5\text{--}52)$$

For nonmagnetic dielectrics, $\mu_R = 1$.

The attenuation above cutoff is due to both dielectric and wall losses in the circular guide. The attenuation resulting from an imperfect dielectric (α_d) is given by Eq. (5–47), where f_c is the cutoff frequency of the particular waveguide mode. For air-filled circular guides, α_d is negligible and the attenuation is simply α_c, the attenuation constant due to imperfect conducting walls. Expressions for α_c are derived in Sec. 8.04 of Ref. 5–6 for both TE and TM modes. Attenuation versus frequency data for three modes is shown in Fig. 5–28 and is based upon the assumption of copper walls and a guide diameter of 3 cm. The general attenuation behavior for the TE_{11} and TM_{01} modes is about the same as for the TE_{10} mode in rectangular guide.

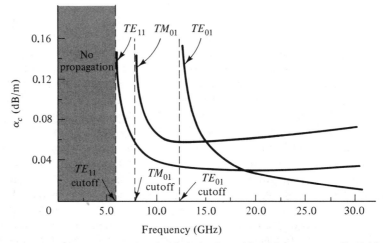

Figure 5–28 Attenuation for air-filled circular guide with a copper wall. Guide diameter $D = 3.0$ cm.

The attenuation increases rapidly as the operating frequency approaches cutoff. An explanation for this effect has already been given. When $f \gg f_c$, the attenuation increases slowly with increasing frequency. This is because the surface resistance R_s associated with skin effect is proportional to $\sqrt{f}$.

Attenuation characteristics of the TE_{01} mode.   Unlike other modes, the attenuation of the TE_{01} mode in circular guide continues to decrease as the operating frequency increases. This is due to the behavior of the conduction currents with increasing frequency. Referring to Fig. 5–27, the magnetic field at the guide walls is everywhere longitudinal for the TE_{01} mode. By virtue of Ampere's right-hand rule, this means that the wall currents are completely circumferential. Furthermore, a field analysis shows that for a fixed power level, the circumferential currents decrease with increasing frequency (Ref. 5–6). Therefore, the wall losses and hence α_c decrease indefinitely as the frequency increases (see Fig. 5–28). This characteristic of the TE_{01} mode in circular guide makes it very attractive as a low-loss transmission line at high frequencies, particularly at millimeter wavelengths. However, since it is not the dominant mode, one must take special precautions not to generate modes with a lower cutoff frequency. The generation of other modes would, of course, result in a significant loss of power in the desired TE_{01} mode. Some results and problem areas are described in Ref. 5–31.

★ 5–5 THEORY OF WAVEGUIDE TRANSMISSION

An analysis of wave transmission in rectangular and circular guides is given in this section. Maxwell's equations and the appropriate boundary conditions are used to develop the field equations for the various modes. Also obtained are expressions for cutoff, velocity, wavelength, and characteristic impedance.

In developing the field equations, it is assumed that the guide walls are perfect conductors and the region inside the waveguide is a perfect insulator. Therefore, conduction currents are confined to the inner wall surfaces and hence the wall thickness does not enter into the analysis. Further, it is assumed that the waveguide extends infinitely in both the plus and minus z direction.

For sinusoidal excitation, Maxwell's equations [Eqs. (2–38) to (2–40)] in a charge-free dielectric region may be written in phasor form. That is,

$$\nabla \cdot \vec{D} = 0, \qquad \nabla \cdot \vec{B} = 0, \qquad \nabla \times \vec{E} = -j\omega \vec{B}, \qquad \text{and} \qquad \nabla \times \vec{H} = j\omega \vec{D}$$

since ρ_v and $\vec{J}$ are zero in the dielectric. Faraday and Ampere's laws may be rewritten as

$$\nabla \times \vec{E} = -j\omega\mu_R\mu_0\vec{H} \qquad \text{and} \qquad \nabla \times \vec{H} = j\omega\epsilon_R\epsilon_0\vec{E} \qquad (5\text{–}53)$$

since $\vec{D} = \epsilon_R\epsilon_0\vec{E}$ and $\vec{B} = \mu_R\mu_0\vec{H}$. The constants μ_R and ϵ_R define the electric and magnetic properties of the dielectric region. Taking the curl of both sides of the first equation and using the vector identity, $\nabla \times \nabla \times \vec{A} \equiv \nabla(\nabla \cdot \vec{A}) - \nabla^2\vec{A}$ yields

$$\nabla(\nabla \cdot \vec{E}) - \nabla^2\vec{E} = -j\omega\mu_R\mu_0\nabla \times \vec{H}$$

where it has been assumed that the dielectric is linear, homogeneous, and isotropic.

Using Ampere's law,

$$\nabla^2 \vec{E} + \omega^2 \mu_R \mu_0 \epsilon_R \epsilon_0 \vec{E} = 0 \tag{5–54}$$

since $\nabla \cdot \vec{E} = 0$. In a similar manner,

$$\nabla^2 \vec{H} + \omega^2 \mu_R \mu_0 \epsilon_R \epsilon_0 \vec{H} = 0 \tag{5–55}$$

These are known as the *vector wave equations* for the electric and magnetic fields in a charge-free region. They will now be used to obtain the field equations in rectangular and circular waveguides.

5–5a Rectangular Waveguide Transmission

Figure 5–16 shows a rectangular waveguide of height b and width a. This is the most common configuration for waveguide transmission. Typically, the dielectric region is air. The coordinate system used in the following analysis is shown in the figure.

In rectangular coordinates, Eq. (5–54) becomes

$$\frac{\partial^2 \vec{E}}{\partial x^2} + \frac{\partial^2 \vec{E}}{\partial y^2} + \frac{\partial^2 \vec{E}}{\partial z^2} + \omega^2 \mu \epsilon \vec{E} = 0$$

where $\mu \equiv \mu_R \mu_0$ and $\epsilon \equiv \epsilon_R \epsilon_0$. For lossless propagation in the positive z direction, all three components of $\vec{E}$ must have an $e^{-j\beta z}$ functional dependence. Therefore

$$\frac{\partial^2 \vec{E}}{\partial x^2} + \frac{\partial^2 \vec{E}}{\partial y^2} + (\omega^2 \mu \epsilon - \beta^2)\vec{E} = 0 \tag{5–56}$$

Similarly,

$$\frac{\partial^2 \vec{H}}{\partial x^2} + \frac{\partial^2 \vec{H}}{\partial y^2} + (\omega^2 \mu \epsilon - \beta^2)\vec{H} = 0 \tag{5–57}$$

This pair of vector equations can be written as six scalar equations by separately equating the x, y, and z components to zero. For the z components,

$$\frac{\partial^2 E_z}{\partial x^2} + \frac{\partial^2 E_z}{\partial y^2} = -k_c^2 E_z \tag{5–58}$$

and

$$\frac{\partial^2 H_z}{\partial x^2} + \frac{\partial^2 H_z}{\partial y^2} = -k_c^2 H_z \tag{5–59}$$

where $k_c^2 \equiv \omega^2 \mu \epsilon - \beta^2$.

It can be shown that the TEM mode cannot exist in a single-conductor transmission system (see Sec. 3.6 of Ref. 5–12). Since rectangular waveguide is in this category, E_z and H_z cannot both be zero. However, if the dielectric medium is uniform, modes can exist with one of the z components equal to zero. When $E_z \equiv 0$, the modes are called TE modes since the electric field is always transverse to the propagation direction. Similarly, those with $H_z \equiv 0$ are called TM modes. Let us now consider both these cases.

Transverse electric (TE) modes.   In this case, $E_z = 0$ and $H_z \neq 0$. For propagation in the positive z direction, $H_z = \hat{H}_z e^{-j\beta z}$ where $\hat{H}_z$ is a function of x and y only. Equation (5–59) becomes

$$\frac{\partial^2 \hat{H}_z}{\partial x^2} + \frac{\partial^2 \hat{H}_z}{\partial y^2} = -k_c^2 \hat{H}_z \qquad (5\text{--}60)$$

One possible form of $\hat{H}_z$ is $\hat{H}_z = XY$, where X is a function of x only and Y is a function of y only. Substitution into the above differential equation yields

$$\frac{1}{X}\frac{d^2 X}{dx^2} + \frac{1}{Y}\frac{d^2 Y}{dy^2} = -k_c^2$$

With the first term a function of x alone and the second a function of y alone, the only way their sum can equal a constant $(-k_c^2)$ is for each term to separately equal a constant. This is the familiar *separation of variables* method. That is,

$$\frac{1}{X}\frac{d^2 X}{dx^2} = -k_x^2 \qquad \text{and} \qquad \frac{1}{Y}\frac{d^2 Y}{dy^2} = -k_y^2$$

where $k_x^2 + k_y^2 = k_c^2 = \omega^2 \mu \epsilon - \beta^2$.
Solutions to this pair of differential equations are

$$X = A \cos k_x x + B \sin k_x x \qquad \text{and} \qquad Y = C \cos k_y y + D \sin k_y y$$

where A, B, C, and D are complex constants.

These solutions will now be applied to the rectangular waveguide configuration shown in Fig. 5–16. Since perfect conductors exist at the four planes defined by $x = 0$, $x = a$, $y = 0$, and $y = b$,

$$\frac{\partial H_z}{\partial x} = 0 \quad \text{at} \quad x = 0 \quad \text{and} \quad x = a, \quad \text{and} \quad \frac{\partial H_z}{\partial y} = 0 \quad \text{at} \quad y = 0 \quad \text{and} \quad y = b$$

where use has been made of Eq. (2–43). With $H_z = \hat{H}_z e^{-j\beta z} = XY e^{-j\beta z}$, the above conditions require that $B = 0$, $k_x = m\pi/a$, $D = 0$, and $k_y = n\pi/b$, where m and n are positive integers. Setting $AC = H_0$ yields

$$H_z = H_0 e^{-j\beta z} \cos \frac{m\pi}{a} x \cos \frac{n\pi}{b} y \qquad (5\text{--}61)$$

This equation represents all possible solutions of H_z for TE waves in rectangular waveguide. The other field components can be obtained from the original curl equations [Eq. (5–53)]. For TE waves ($E_z = 0$), the phasor form of these equations reduces to

$$\beta E_y = -\omega\mu H_x \qquad , \qquad \frac{\partial H_z}{\partial y} + j\beta H_y = j\omega\epsilon E_x$$

$$\beta E_x = \omega\mu H_y \qquad , \qquad j\beta H_x + \frac{\partial H_z}{\partial x} = -j\omega\epsilon E_y$$

$$\frac{\partial E_y}{\partial x} - \frac{\partial E_x}{\partial y} = -j\omega\mu H_z \qquad , \qquad \frac{\partial H_y}{\partial x} - \frac{\partial H_x}{\partial y} = 0$$

With some algebraic manipulation, all the field components can be expressed in terms of H_z. Namely,

$$E_x = -j\frac{\omega\mu}{k_c^2}\frac{\partial H_z}{\partial y} \quad , \quad E_y = j\frac{\omega\mu}{k_c^2}\frac{\partial H_z}{\partial x}$$

$$H_x = \mp\frac{E_y}{Z_{TE}} \quad , \quad H_y = \pm\frac{E_x}{Z_{TE}}$$

(5–62)

where the upper signs are for plus z-directed waves and the lower signs for the minus z-directed waves. Z_{TE} is the *wave impedance* for TE waves. It represents the ratio of transverse electric to transverse magnetic field at any point in the waveguide and is given by

$$Z_{TE} = \frac{\omega\mu}{\beta} = \frac{\eta}{\sqrt{1 - (\lambda/\lambda_c)^2}} \quad \text{ohms}$$

(5–63)

where λ_c, the cutoff wavelength, is obtained from Eq. (5–67). The reader is cautioned not to confuse *wave* impedance (Z_{TE}) with *waveguide* impedance (Z_0), which will be described shortly. With $k_x = m\pi/a$ and $k_y = n\pi/b$, k_c^2 may be written as

$$k_c^2 \equiv k_x^2 + k_y^2 = \left(\frac{m\pi}{a}\right)^2 + \left(\frac{n\pi}{b}\right)^2$$

(5–64)

By substituting Eq. (5–61) into Eq. (5–62), the transverse field components of the TE modes for plus z-directed waves are obtained.

$$E_x = jH_0\frac{\omega\mu}{k_c^2}\frac{n\pi}{b}e^{-j\beta z}\cos\frac{m\pi}{a}x\sin\frac{n\pi}{b}y$$

$$E_y = -jH_0\frac{\omega\mu}{k_c^2}\frac{m\pi}{a}e^{-j\beta z}\sin\frac{m\pi}{a}x\cos\frac{n\pi}{b}y$$

(5–65)

$$H_x = -\frac{E_y}{Z_{TE}} \quad \text{and} \quad H_y = \frac{E_x}{Z_{TE}}$$

where H_0 is a constant that indicates the amplitude of the electromagnetic wave. The time-varying form of the field components can be obtained by multiplying the above phasors by $\sqrt{2}e^{j\omega t}$ and taking the real part thereof.

Since $\beta^2 = \omega^2\mu\epsilon - k_c^2$, the phase constant of the wave is

$$\beta = \sqrt{\omega^2\mu\epsilon - \left(\frac{2\pi}{\lambda_c}\right)^2}$$

(5–66)

where

$$\lambda_c \equiv \frac{2\pi}{k_c} = \frac{1}{\sqrt{\left(\frac{m}{2a}\right)^2 + \left(\frac{n}{2b}\right)^2}}$$

(5–67)

The following discussion explains the meaning of cutoff wavelength (λ_c) and cutoff frequency (f_c).

The integers m and n can take on all values from zero to infinity. Therefore, Eqs. (5–61) and (5–65) represent an infinite set of solutions. Each set of values for m and n describes a method or *mode* of transmitting electromagnetic energy down the rectangular waveguide. The particular TE mode is designated by the symbol TE_{mn} (or by H_{mn} in some texts) and its propagation characteristics are given by Eqs. (5–66) and (5–67). Referring to Eq. (5–65), the subscripts m and n indicate the number of *half* sine wave variations of the field components in the x and y directions, respectively. For a given mode, propagation can only occur when β is real, which requires that $\omega^2 \mu \epsilon > (2\pi/\lambda_c)^2$. This condition for propagation can be rewritten as

$$\lambda < \lambda_c \qquad \text{or} \qquad f > f_c$$

where f_c, the cutoff frequency, is given by

$$f_c = \frac{v}{\lambda_c} = \frac{c}{\lambda_c \sqrt{\mu_R \epsilon_R}} \qquad (5\text{–}68)$$

Thus for a given mode, wave propagation only occurs when the signal frequency f is greater than the cutoff frequency. Both λ_c and f_c are dependent upon the dimensions of the waveguide as well as the particular mode being considered. The cutoff frequency (but not λ_c) is also a function of the dielectric material within the guide.

For a given mode, both the phase velocity (v_p) and wavelength in the guide (λ_g) are related to the phase constant. Namely,

$$\beta = \frac{\omega}{v_p} = \frac{2\pi}{\lambda_g} \qquad \text{rad/length} \qquad (5\text{–}69)$$

Using Eq. (5–66) yields

$$\lambda_g = \frac{\lambda_0}{\sqrt{\mu_R \epsilon_R - \left(\frac{\lambda_0}{\lambda_c}\right)^2}} = \frac{\lambda}{\sqrt{1 - \left(\frac{f_c}{f}\right)^2}} \qquad (5\text{–}70)$$

where λ_0 is the free space wavelength of the signal, $\lambda = \lambda_0/\sqrt{\mu_R \epsilon_R}$ and λ_c and f_c are defined above. Note that wavelength in the guide is always greater than the wavelength in the unbounded dielectric (λ) since $f > f_c$. The phase velocity in the waveguide is given by

$$v_p = f\lambda_g = \frac{\dfrac{c}{\sqrt{\mu_R \epsilon_R}}}{\sqrt{1 - \left(\frac{f_c}{f}\right)^2}} \qquad (5\text{–}71)$$

Note that it is greater than the wave velocity in an unbounded dielectric. Phase velocity represents the speed with which a particular phase of the wave (such as, its maximum) travels in the propagation direction. With $f\lambda = v$, Eq. (5–71) can be written as

$$v_p = v\frac{\lambda_g}{\lambda} = c\frac{\lambda_g}{\lambda_0} \qquad (5\text{-}72)$$

where v is the velocity in the unbounded dielectric medium.

One can also define another velocity, known as *group velocity,* in the following manner.

$$v_g \equiv \frac{d\omega}{d\beta} = \frac{1}{d\beta/d\omega} \tag{5-73}$$

This velocity is associated with the propagation of a narrowband modulated signal. It can be shown that the velocity of energy flow in a waveguide is exactly the group velocity (Sec. 3.11 of Ref. 5–12). By substituting Eq. (5–66) into Eq. (5–73), we obtain

$$v_g = \frac{c}{\sqrt{\mu_R \epsilon_R}} \sqrt{1 - \left(\frac{f_c}{f}\right)^2} = v\frac{\lambda}{\lambda_g} \tag{5-74}$$

Note that as $f_c \rightarrow 0$, $v_g = v_p = v$. This is the situation for TEM waves because they do not exhibit the cutoff effect. From Eq. (5–74), it is clear that the velocity of energy flow is always less than the speed of light. This, of course, is consistent with the theory of relativity. Phase velocity, on the other hand, can be greater than the speed of light since it is not the velocity of energy flow or information transfer.

At frequencies below the cutoff frequency ($f < f_c$), the phase constant is imaginary and the electromagnetic wave is attenuated as it attempts to propagate in the z direction. Thus the waveguide behaves like a highpass filter. Since perfect conducting walls have been assumed, the attenuation is reflective rather than dissipative. The expression for this attenuation derives directly from Eq. (5–66). With $\beta = j\alpha$,

$$\alpha = \sqrt{\left(\frac{2\pi}{\lambda_c}\right)^2 - \omega^2 \mu\epsilon}$$

where α represents the attenuation constant of the below-cutoff waveguide. This expression can be rewritten as

$$\alpha = \frac{2\pi}{\lambda_c} \sqrt{1 - \left(\frac{f}{f_c}\right)^2} \qquad \text{Np/length}$$

or   (5-75)

$$\alpha = \frac{54.6}{\lambda_c} \sqrt{1 - \left(\frac{f}{f_c}\right)^2} \qquad \text{dB/length}$$

Note that for $f \ll f_c$, the attenuation constant is independent of frequency.

The most commonly used mode in rectangular waveguide is the TE_{10}. It is called the *dominant mode* because it has the lowest cutoff frequency. Assuming $a > b$, λ_c will have its largest value and f_c its lowest value when $m = 1$ and $n = 0$. The solution wherein both m and n are zero is a trivial case, that is, all field components are zero. The instantaneous field components for the TE_{10} mode are obtained from Eqs. (5–61) and (5–65) by multiplying by $\sqrt{2}e^{j\omega t}$ and taking the real part thereof. The results for a plus z-directed wave are

$$\mathscr{E}_z \equiv 0, \quad \mathscr{E}_x = 0, \quad \mathscr{H}_y = 0,$$

$$\mathscr{H}_z = \sqrt{2}\, H_0 \cos \frac{\pi}{a} x \cos (\omega t - \beta z)$$

$$\mathscr{E}_y = \sqrt{2}\, H_0 \frac{\pi \omega \mu}{a k_c^2} \sin \frac{\pi}{a} x \sin (\omega t - \beta z)$$

and

$$\mathscr{H}_x = - \frac{\mathscr{E}_y}{Z_{TE}}$$

where Z_{TE} is defined in Eq. (5–63).

After some algebraic manipulation, the three field components of the TE_{10} mode reduce to

$$\mathscr{E}_y = \sqrt{2}\, E_0 \frac{\lambda_c}{\lambda_g} \sin \frac{\pi}{a} x \sin (\omega t - \beta z)$$

$$\mathscr{H}_x = -\sqrt{2}\, H_0 \frac{\lambda_c}{\lambda_g} \sin \frac{\pi}{a} x \sin (\omega t - \beta z) \qquad (5\text{–}76)$$

$$\mathscr{H}_z = \sqrt{2}\, H_0 \cos \frac{\pi}{a} x \cos (\omega t - \beta z)$$

where $\lambda_c = 2a$, $E_0 = H_0 Z_{TE}$, and λ_g is defined in Eq. (5–70).

A sketch of the TE_{10} fields for a plus z-directed wave is shown in Fig. 5–18 for $t = 0$ and $t = T/4$. An end view of the electric and magnetic fields is also shown. Notice that the mode pattern moves a distance $\lambda_g/4$ in a quarter of a period. Thus the velocity is $\lambda_g/T = f\lambda_g$, which is exactly the phase velocity of the wave. Also note that E_y and H_x are maximum along the $x = a/2$ plane and their values are independent of y since $n = 0$. With $m = 1$, the field components exhibit *one* half sine wave variation in the x direction. In the propagation direction, of course, all the field components exhibit one complete sinusoidal variation per guide wavelength λ_g.

Since the skin depth is very small at microwave frequencies, practically all the conduction currents reside on the inner surfaces of the conducting walls. As explained in Sec. 2–3 [Eq. (2–42)], the surface current per unit length (K) is equal and normal to the tangential component of H at the conducting surfaces. For example, at the narrow walls, the conduction currents are in the $\pm y$ direction and $|K_y| = |H_z|$ at $x = 0$ and $x = a$. The H_x components at $y = 0$ and $y = b$ are related to conduction currents on the broad walls that are directed along the propagation direction. This longitudinal current $|K_z| = |H_x|$ is sometimes referred to as the *power current*. A sketch of the conduction currents for the TE_{10} mode is shown in Fig. 5–19.

Characteristic impedance in waveguide. In order to apply transmission-line theory to waveguides, one must be able to determine the phase constant (β) or guide wavelength (λ_g) as well as the characteristic impedance (Z_0) of the particular guide configuration. Equations (5–69) and (5–70) provide the required relationships for β and λ_g. Characteristic impedance is defined in Eq. (3–18) as V^+/I^+, where V^+ is the voltage between the conductors and I^+ is the conduction current in the propagation direction for the traveling wave. For TEM lines, Z_0 is uniquely defined since the value of $V^+ = -\int \vec{E} \cdot \vec{dl}$ is independent of the integration path. In the waveguide case, V^+ is a function of the integration path and therefore many definitions of Z_0 are possible.

For rectangular waveguide in the TE_{10} mode, the electric field is maximum at $x = a/2$. From Eq. (5–76), its rms value is $E_y = E_0 \lambda_c/\lambda_g$. Integrating directly from $y = 0$ to $y = b$ yields the voltage across the centerline of the waveguide. Namely,

$$V_{\mathbb{C}} = E_0 b \frac{\lambda_c}{\lambda_g} = H_0 Z_{TE} b \frac{\lambda_c}{\lambda_g} \tag{5–77}$$

However, this represents only *one* possible voltage value between the points $y = 0$, $x = a/2$ and $y = b$, $x = a/2$. For instance, one could integrate $\vec{E} \cdot dl$ by choosing a path along the conducting walls. In this case $V_{\mathbb{C}} = 0$. Other paths leading to different values of voltage are indicated in Fig. 5–29. Clearly, voltage and hence Z_0 does not have a unique value in waveguide. One possible definition is the ratio of $V_{\mathbb{C}}$ as defined in Eq. (5–77) to the longitudinal current I_z, where

$$I_z = \int_0^a K_z \, dx = \int_0^a H_x \, dx$$

Integration along the bottom wall ($y = 0$) yields

$$I_z = H_0 \frac{\lambda_c}{\lambda_g} \int_0^a \sin \frac{\pi}{a} x \, dx = \frac{2a}{\pi} H_0 \frac{\lambda_c}{\lambda_g} \tag{5–78}$$

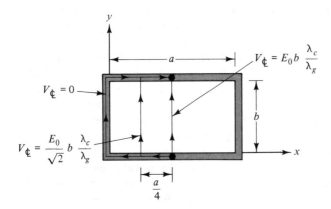

Figure 5–29 Possible integration paths for determining $V_{\mathbb{C}}$ in rectangular guide.

where H_x, the rms value of $\mathcal{H}_x$, is obtained from Eq. (5–76). Therefore,

$$Z_0 = \frac{V_{\mathfrak{C}}}{I_z} = \frac{\pi}{2}\frac{b}{a} Z_{TE} = \frac{592\dfrac{b}{a}\sqrt{\dfrac{\mu_R}{\epsilon_R}}}{\sqrt{1 - \left(\dfrac{f_c}{f}\right)^2}} \quad ohms \qquad (5\text{–}79)$$

where Z_{TE} is given by Eq. (5–63). Since $V_{\mathfrak{C}}$ and I_z are in phase, Z_0 is real for $f > f_c$. The above expression is known as the *voltage-current* definition of Z_0. Other commonly used definitions are based upon power flow along the waveguide. These are

1. **Power-Current Definition:** $\qquad\qquad Z_0 \equiv \dfrac{P}{I_z^2}$

2. **Power-Voltage Definition:** $\qquad\qquad Z \equiv \dfrac{V_{\mathfrak{C}}^2}{P}$

3. **Modified Power-Voltage Definition:** $\qquad Z_0 \equiv \dfrac{V_{\mathfrak{C}}^2}{P}$

where $V_{\mathfrak{C}}$ is given by Eq. (5–77) and I_z by Eq. (5–78). An expression for power flow in the z direction can be developed with the aid of Poynting's theorem [Eq. (2–58)]. With E_x and H_y in phase,

$$P = \int_S (\vec{E} \times \vec{H}) \cdot \vec{dS} = \int_0^a \int_0^b E_x H_y \, dy \, dx$$

and therefore

$$P = H_0^2 Z_{TE} \left(\frac{\lambda_c}{\lambda_g}\right)^2 b \int_0^a \sin^2 \frac{\pi}{a} x \, dx$$

or

$$P = \frac{1}{2} H_0^2 Z_{TE} \, ab \left(\frac{\lambda_c}{\lambda_g}\right)^2 \qquad (5\text{–}80)$$

Applying this equation to the three definitions results in the following expressions for Z_0 in rectangular guide.

Power-Current Definition: $\qquad\qquad Z_0 = 465 \dfrac{b}{a} \sqrt{\dfrac{\mu_R}{\epsilon_R}} \dfrac{\lambda_g}{\lambda} \;$ ohms

Power-Voltage Definition: $\qquad\qquad Z_0 = 754 \dfrac{b}{a} \sqrt{\dfrac{\mu_R}{\epsilon_R}} \dfrac{\lambda_g}{\lambda} \;$ ohms $\qquad (5\text{–}81)$

Modified Power-Voltage Definition: $\quad Z_0 = 377 \dfrac{b}{a} \sqrt{\dfrac{\mu_R}{\epsilon_R}} \dfrac{\lambda_g}{\lambda} \;$ ohms

where $\dfrac{\lambda_g}{\lambda} = \dfrac{1}{\sqrt{1 - (f_c/f)^2}}.$

Note that all four definitions [Eq. (5–79) and (5–81)] differ only by a constant. Since matching problems usually involve the ratio of impedances, the choice of definition is arbitrary as long as the mode type and guide configuration remain fixed. Unless otherwise stated, the modified power-voltage definition is used in this text.

In some applications, it becomes necessary to match a waveguide to a uniquely defined impedance (such as, a coaxial line). As explained in Sec. 5–4a, the usual approach in this situation is to use the definition that gives best agreement between transmission-line theory and the experimental data (Ref. 5–30).

Transverse magnetic (TM) modes. The analysis presented here is the same as for TE modes except that the field components are obtained from a knowledge of E_z rather than H_z. For propagation in the plus z direction, $E_z = \hat{E}_z e^{-j\beta z}$. Assuming $\hat{E}_z = XY$ and substituting into Eq. (5–58) yields

$$\frac{1}{X}\frac{d^2 X}{dx^2} = -k_x^2 \quad \text{and} \quad \frac{1}{Y}\frac{d^2 Y}{dy^2} = -k_y^2$$

where, as before, $k_x^2 + k_y^2 = k_c^2 = \omega^2 \mu\epsilon - \beta^2$. Thus,

$$X = A \cos k_x x + B \sin k_x x \quad \text{and} \quad Y = C \cos k_y y + D \sin k_y y$$

Referring to Fig. 5–16, the tangential component of electric field must be zero at the four conducting walls. Therefore $E_z = 0$ at $x = 0$, $x = a$, $y = 0$, and $y = b$, which requires that $A = 0$, $C = 0$, $k_x = m\pi/a$, and $k_y = n\pi/b$. Setting $BD = E_0$, the solution for E_z becomes

$$E_z = E_0 e^{-j\beta z} \sin \frac{m\pi}{a}x \sin \frac{n\pi}{b}y \tag{5–82}$$

The remaining field components can be obtained from the original curl equations [Eq. (5–53)]. For TM waves ($H_z = 0$), the phasor form of these equations reduces to

$$\frac{\partial E_z}{\partial y} + j\beta E_y = -j\omega\mu H_x \quad , \quad \beta H_y = \omega\epsilon E_x$$

$$j\beta E_x + \frac{\partial E_z}{\partial x} = j\omega\mu H_y \quad , \quad \beta H_x = -\omega\epsilon E_y \tag{5–83}$$

$$\frac{\partial E_y}{\partial x} - \frac{\partial E_x}{\partial y} = 0 \quad , \quad \frac{\partial H_y}{\partial x} - \frac{\partial H_x}{\partial y} = j\omega\epsilon E_z$$

All the field components can be expressed in terms of E_z. Namely,

$$H_x = j\frac{\omega\epsilon}{k_c^2}\frac{\partial E_z}{\partial y} \quad , \quad H_y = -j\frac{\omega\epsilon}{k_c^2}\frac{\partial E_z}{\partial x}$$

$$E_x = \pm Z_{TM} H_y \quad , \quad E_y = \mp Z_{TM} H_x \tag{5–84}$$

where the upper signs are for plus z-directed waves and the lower signs for minus z-directed waves. Z_{TM}, the wave impedance for TM waves, represents the ratio of transverse electric to transverse magnetic field at any point in the waveguide and is given by

$$Z_{TM} = \frac{\beta}{\omega\epsilon} = \eta\sqrt{1 - \left(\frac{\lambda}{\lambda_c}\right)^2} \quad \text{ohms} \tag{5–85}$$

where λ_c is obtained from Eq. (5–67). Since $k_x = m\pi/a$ and $k_y = n\pi/b$, k_c may be obtained from Eq. (5–64). By substituting Eq. (5–82) into Eq. (5–84), the transverse field components of the TM modes for plus z-directed waves are obtained.

$$\boldsymbol{H_x} = jE_0\frac{\omega\epsilon}{k_c^2}\frac{n\pi}{b}e^{-j\beta z}\sin\frac{m\pi}{a}x\cos\frac{n\pi}{b}y$$

$$\boldsymbol{H_y} = -jE_0\frac{\omega\epsilon}{k_c^2}\frac{m\pi}{a}e^{-j\beta z}\cos\frac{m\pi}{a}x\sin\frac{n\pi}{b}y \tag{5–86}$$

$$\boldsymbol{E_x} = \pm Z_{TM}\boldsymbol{H_y} \quad \text{and} \quad \boldsymbol{E_y} = \mp Z_{TM}\boldsymbol{H_x}$$

where E_0 indicates the amplitude of the wave.

With $\beta^2 = \omega^2\mu\epsilon - k_c^2$, the expressions for β, λ_c, f_c, λ_g, v_p, v_g, and α are the same as for TE waves (namely, Eqs. 5–66, 5–67, 5–68, 5–70, 5–71, 5–74, and 5–75, respectively). Note that for a given set of values for m and n, the above expressions are the same whether the mode is TE or TM. For example, the cutoff frequency for the TE_{21} mode is the same as for the TM_{21} mode.

The *TM* mode with the lowest cutoff frequency in rectangular waveguide is the TM_{11}. For *TM* modes, the case where either $m = 0$ or $n = 0$ is trivial since all the field components reduce to zero. A sketch of the TM_{11} mode is shown in Fig. 5–20. It is obtained by converting Eqs. (5–82) and (5–86) into their time-varying form and letting $t = 0$. Viewed in the transverse plane, the TM_{11} mode looks similar to the TEM mode in stripline (Fig. 5–7). The only substantial difference is that the conduction current in the center conductor of the stripline is replaced by a *displacement* current. The existence of the E_z component in the waveguide results in the cutoff effect which, of course, does not occur in TEM lines. The cutoff wavelength for the TM_{11} mode is indicated in Fig. 5–20. By proper selection of waveguide height (b), this mode can be suppressed without affecting TE_{10} propagation. For $b = a/2$, the TM_{11} cutoff frequency is about 12 percent greater than that of the TE_{20} mode, and hence does not further restrict the usable frequency range of the waveguide.

5–5b Circular Waveguide Transmission

Figure 5–30 shows a circular waveguide of inner radius a. The coordinate system used in this discussion is also shown. A perfectly conducting guide wall and a uniform, loss-free dielectric region are assumed. The analysis that follows is similar to that for rectangular waveguide transmission.

In cylindrical coordinates, the z components of Eqs. (5–54) and (5–55) may be written as

$$\frac{1}{r}\frac{\partial}{\partial r}\left(r\frac{\partial \boldsymbol{E_z}}{\partial r}\right) + \frac{1}{r^2}\frac{\partial^2 \boldsymbol{E_z}}{\partial\phi^2} = -k_c^2\boldsymbol{E_z} \tag{5–87}$$

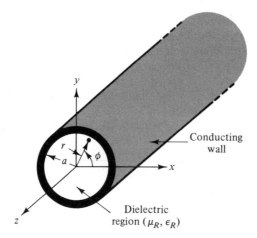

Conducting wall

Figure 5–30   The circular waveguide structure. $D = 2a =$ inner guide diameter.

Dielectric region (μ_R, ϵ_R)

and

$$\frac{1}{r}\frac{\partial}{\partial r}\left(r\frac{\partial H_z}{\partial r}\right) + \frac{1}{r^2}\frac{\partial^2 H_z}{\partial \phi^2} = -k_c^2 H_z \tag{5–88}$$

where, as before, $k_c^2 \equiv \omega^2 \mu\epsilon - \beta^2$ and an $e^{-j\beta z}$ functional dependency has been assumed.

Transverse Electric (TE) modes.   In this case, $E_z = 0$ and $H_z \neq 0$. For propagation in the positive z direction, $H_z = \hat{H}_z e^{-j\beta z}$, where $\hat{H}_z$ is a function of r and ϕ only. Therefore, Eq. (5–88) becomes

$$\frac{1}{r}\frac{\partial}{\partial r}\left(r\frac{\partial \hat{H}_z}{\partial r}\right) + \frac{1}{r^2}\frac{\partial^2 \hat{H}_z}{\partial \phi^2} = -k_c^2 \hat{H}_z \tag{5–89}$$

Using the separation of variables method as before, we assume $\hat{H}_z = R\Phi$, where R is a function of r only and Φ is a function of ϕ only. This leads to the following pair of differential equations.

$$\frac{d^2 R}{dr^2} + \frac{1}{r}\frac{dR}{dr} + \left(k_c^2 - \frac{n^2}{r^2}\right)R = 0 \qquad \text{and} \qquad \frac{d^2 \Phi}{d\phi^2} + n^2\Phi = 0$$

where n^2 is a constant resulting from the separation of variables procedure. The solutions to these equations are

$$R = AJ_n(k_c r) + BN_n(k_c r)$$
$$\Phi = C\cos n\phi + D\sin n\phi \tag{5–90}$$

where A, B, C, and D are constants and J_n and N_n represent the nth order Bessel functions of the first and second kind. These functions are plotted in Fig. 5–31 for some integer values of n. Since the field must remain finite at $r = 0$, the N_n term must be excluded and therefore $B = 0$. Also, since the field and hence Φ must be single-valued, n is restricted to integer values. Finally, one of the two Φ terms can be eliminated. Both lead to the same field pattern except that one is rotated by 90° in the ϕ direction. Setting $D = 0$ yields

$$\hat{H}_z = H_0 J_n(k_c r)\cos n\phi \tag{5–91}$$

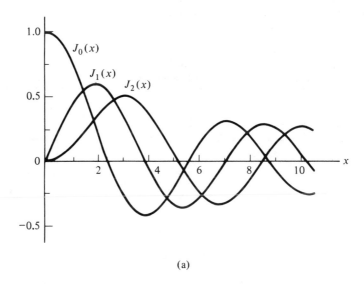

(a)

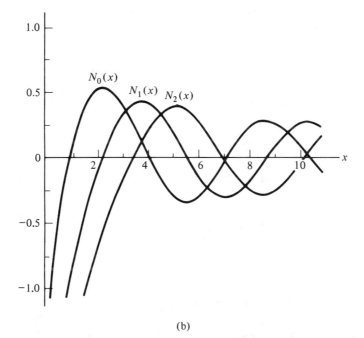

(b)

Figure 5–31   Bessel functions of the first kind (part *a*) and of the second kind (part *b*).

where $H_0 = AC$ is an arbitrary amplitude constant. At $r = a$, the boundary condition requires that $\partial H_z / \partial r = 0$, which means $J'_n(k_c a) = 0$. The prime denotes differentiation *with respect to* $k_c r$. The mth root of this equation is designated by q'_{nm} and thus $q'_{nm} = k_c a$. Several root values are listed in Table 5–2.

TABLE 5–2 Values of q'_{nm} for TE_{nm} modes and q_{nm} for TM_{nm} modes in circular guide. These represent the mth root of equations $J'_n(x) = 0$ and $J_n(x) = 0$, respectively.

q'_{nm} for TE Modes			q_{nm} for TM Modes		
n \ m	1	2	n \ m	1	2
0	3.832	7.016	0	2.405	5.520
1	1.841	5.331	1	3.832	7.016
2	3.054	6.706	2	5.135	8.417

The equation for H_z may be written as

$$H_z = \hat{H}_z e^{-j\beta z} = H_0 e^{-j\beta z} J_n(k_c r) \cos n\phi \qquad (5\text{–}92)$$

where $k_c = q'_{nm}/a$. This equation represents all possible solutions of H_z for TE_{nm} waves in circular guide. The remaining field components can be obtained from the original curl equations [Eq. (5–53)]. For TE waves ($E_z = 0$), the phasor form of these equations are

$$\beta E_\phi = -\omega\mu H_r \qquad , \qquad \frac{1}{r}\frac{\partial H_z}{\partial \phi} + j\beta H_\phi = j\omega\epsilon E_r$$

$$\beta E_r = \omega\mu H_\phi \qquad , \qquad j\beta H_r + \frac{\partial H_z}{\partial r} = -j\omega\epsilon E_\phi \qquad (5\text{–}93)$$

$$\frac{1}{r}\left\{\frac{\partial(rE_\phi)}{\partial r} - \frac{\partial E_r}{\partial \phi}\right\} = -j\omega\mu H_z \qquad , \qquad \frac{1}{r}\left\{\frac{\partial(rH_\phi)}{\partial r} - \frac{\partial H_r}{\partial \phi}\right\} = 0$$

These equations can be rewritten so that all the field components are expressed in terms of H_z. Namely,

$$E_r = -j\frac{\omega\mu}{k_c^2 r}\frac{\partial H_z}{\partial \phi} \qquad , \qquad E_\phi = j\frac{\omega\mu}{k_c^2}\frac{\partial H_z}{\partial r}$$

$$H_r = \mp\frac{E_\phi}{Z_{TE}} \qquad , \qquad H_\phi = \pm\frac{E_r}{Z_{TE}} \qquad (5\text{–}94)$$

where the upper signs are for plus z-directed waves and the lower signs for the minus z-directed waves. Z_{TE}, the wave impedance for TE waves, is given by Eq. (5–63), where for circular guides λ_c is determined from Eq. (5–96). By substituting Eq. (5–92) into Eq. (5–94), the transverse field components of the TE_{nm} modes for plus z-directed waves are obtained.[10]

___

[10] In circular guides, the mode subscripts are denoted as nm, while for rectangular guide the subscripts are mn.

$$E_r = jH_0 \frac{n\omega\mu}{k_c^2 r} e^{-j\beta z} J_n(k_c r) \sin n\phi$$

$$E_\phi = jH_0 \frac{\omega\mu}{k_c} e^{-j\beta z} J_n'(k_c r) \cos n\phi \qquad (5\text{–}95)$$

$$H_r = -\frac{E_\phi}{Z_{TE}} \qquad \text{and} \qquad H_\phi = \frac{E_r}{Z_{TE}}$$

where $k_c = q_{nm}'/a$. Since $\beta^2 = \omega^2\mu\epsilon - k_c^2$, the phase constant of the wave is given by Eq. (5–66), where in this case the cutoff wavelength is

$$\lambda_c \equiv \frac{2\pi}{k_c} = \frac{2\pi a}{q_{nm}'} = K_{nm} D \qquad (5\text{–}96)$$

$D = 2a$ is the inner diameter of the circular guide and $K_{nm} = \pi/q_{nm}'$ for TE modes. Values of K_{nm} for several modes are listed in Table 5–3. As before, the cutoff frequency can be determined from Eq. (5–68), while λ_g, v_p, v_g, and α are given by Eqs. (5–70), (5–71), (5–74), and (5–75), respectively.

TABLE 5–3 Values of K_{nm} for TE and TM modes in circular guide. Note, K_{nm} equals π/q_{nm}' for TE modes and π/q_{nm} for TM modes.

K_{nm} for TE Modes			K_{nm} for TM Modes		
n \ m	1	2	n \ m	1	2
0	0.820	0.448	0	1.306	0.569
1	1.706	0.589	1	0.820	0.448
2	1.029	0.468	2	0.612	0.373

The dominant mode in circular guide is the TE_{11}. Its transverse field pattern and cutoff wavelength are shown in Fig. 5–27. The field pattern is obtained from Eqs. (5–92) and (5–95) by multiplying each component by $\sqrt{2}\, e^{j\omega t}$ and taking the real part thereof. Thus,

$$\mathscr{E}_r = -\sqrt{2}\, E_0 \frac{\lambda_c}{\lambda_g}\left(\frac{\lambda_c}{2\pi r}\right) J_1\left(\frac{2\pi r}{\lambda_c}\right) \sin\phi \sin(\omega t - \beta z)$$

$$\mathscr{E}_\phi = -\sqrt{2}\, E_0 \frac{\lambda_c}{\lambda_g} J_1'\left(\frac{2\pi r}{\lambda_c}\right) \cos\phi \sin(\omega t - \beta z)$$

$$\mathscr{H}_z = \sqrt{2}\, H_0 J_1\left(\frac{2\pi r}{\lambda_c}\right) \cos\phi \cos(\omega t - \beta z) \qquad (5\text{–}97)$$

$$\mathscr{H}_r = -\frac{\mathscr{E}_\phi}{Z_{TE}} \qquad \text{and} \qquad \mathscr{H}_\phi = \frac{\mathscr{E}_r}{Z_{TE}}$$

where $E_0 = H_0 Z_{TE}$, λ_g is given by Eq. (5–70) and $\lambda_c = 1.706D$ for the TE_{11} mode. As indicated in the previous section, this mode is analogous to the TE_{10} mode in rectangular waveguide.

Waveguide impedance for the TE_{11} mode in circular guide can be defined in the same manner as for rectangular guide. Based on the four definitions of Z_0 in Sec. 5–5a and Eq. (5–95), expressions can be obtained for the circular guide case. For example, the voltage-current definition ($Z_0 \equiv V_{\ell}/I_z$) leads to the following relation.

$$Z_0 = \frac{2\int_0^a E_r|_{\phi=\pi/2}\, dr}{\int_0^\pi H_\phi|_{r=a}\, a d\phi} = 520 \sqrt{\frac{\mu_R}{\epsilon_R}} \frac{\lambda_g}{\lambda} \quad \text{ohms} \qquad (5\text{–}98)$$

This and the following expressions for Z_0 are derived in Ref. 5–11.

Power-Current Definition: $Z_0 = 354 \sqrt{\dfrac{\mu_R}{\epsilon_R}} \dfrac{\lambda_g}{\lambda}$ ohms

Power-Voltage Definition: $Z_0 = 764 \sqrt{\dfrac{\mu_R}{\epsilon_R}} \dfrac{\lambda_g}{\lambda}$ ohms   (5–99)

Modified Power-Voltage Definition: $Z_0 = 382 \sqrt{\dfrac{\mu_R}{\epsilon_R}} \dfrac{\lambda_g}{\lambda}$ ohms

where $\lambda_g/\lambda = 1/\sqrt{1 - (f_c/f)^2}$. As explained earlier, the modified power-voltage definition is used in this text. Note that its value is simply half that of the power-voltage definition.

The attenuation characteristics of TE_{01} mode propagation were discussed at the end of Sec. 5–4. For this mode, there are no longitudinal conduction currents since H_ϕ is zero. An expression for α_c is given in Sec. 8.04 of Ref. 5–6. Its value decreases indefinitely with increasing frequency, which makes it an attractive method for low-loss transmission.

Transverse magnetic (TM) modes. In this case, $H_z = 0$ and $E_z \neq 0$. From Eq. (5–87),

$$\frac{1}{r}\frac{\partial}{\partial r}\left(r\frac{\partial \hat{E}_z}{\partial r}\right) + \frac{1}{r^2}\frac{\partial^2 \hat{E}_z}{\partial \phi^2} = -k_c^2 \hat{E}_z \qquad (5\text{–}100)$$

where $k_c^2 = \omega^2\mu\epsilon - \beta^2$, $E_z = \hat{E}_z e^{-j\beta z}$ and $\hat{E}_z$ is a function of r and ϕ only. Using the separation of variables method and the same reasoning as for TE modes leads to

$$\hat{E}_z = E_0 J_n(k_c r) \cos n\phi \qquad (5\text{–}101)$$

The boundary condition at the conducting wall requires that $E_z = 0$ at $r = a$, which means $J_n(k_c a) = 0$. The mth root of this equation is designated by q_{nm} and thus $q_{nm} = k_c a$. Several values of q_{nm} are given in Table 5–2.

The equation for E_z becomes

$$E_z = E_0 e^{-j\beta z} J_n(k_c r) \cos n\phi \qquad (5\text{–}102)$$

where $k_c = q_{nm}/a$. The remaining TM field components can be obtained from Eq. (5–53). With $H_z = 0$, the phasor form of these equations reduces to

$$\frac{1}{r}\frac{\partial E_z}{\partial \phi} + j\beta E_\phi = -j\omega\mu H_r \quad , \quad \beta H_\phi = \omega\epsilon E_r$$

$$j\beta E_r + \frac{\partial E_z}{\partial r} = j\omega\mu H_\phi \quad , \quad \beta H_r = -\omega\epsilon E_\phi \qquad (5\text{–}103)$$

$$\frac{1}{r}\left\{\frac{\partial(rE_\phi)}{\partial r} - \frac{\partial E_r}{\partial \phi}\right\} = 0 \quad , \quad \frac{1}{r}\left\{\frac{\partial(rH_\phi)}{\partial r} - \frac{\partial H_r}{\partial \phi}\right\} = j\omega\epsilon E_z$$

These equations can be rewritten so that all the field components are expressed in terms of E_z. Namely,

$$H_r = j\frac{\omega\epsilon}{k_c^2 r}\frac{\partial E_z}{\partial \phi} \quad , \quad H_\phi = -j\frac{\omega\epsilon}{k_c^2}\frac{\partial E_z}{\partial r}$$

$$E_r = \pm Z_{TM} H_\phi \quad , \quad E_\phi = \mp Z_{TM} H_r \qquad (5\text{–}104)$$

where the upper signs are for plus z-directed waves and the lower signs for minus z-directed waves. Z_{TM}, the wave impedance for TM waves, is given by

$$Z_{TM} = \frac{\beta}{\omega\epsilon} = \eta\sqrt{1 - \left(\frac{\lambda}{\lambda_c}\right)^2} \qquad \text{ohms} \qquad (5\text{–}105)$$

where $\lambda_c = K_{nm}D$. Values of K_{nm} for several TM modes are listed in Table 5–3. As before, the cutoff frequency can be determined from Eq. (5–68), while λ_g, v_p, v_g, and α are given by Eqs. (5–70), (5–71), (5–74), and (5–75), respectively.

The transverse field components for plus z-directed TM_{nm} modes are obtained by substituting Eq. (5–102) into Eq. (5–104). With some algebraic manipulation, the expressions can be written as

$$H_r = -jH_0 n\frac{\lambda_c}{\lambda_g}\left(\frac{\lambda_c}{2\pi r}\right)e^{-j\beta z}J_n(k_c r)\sin n\phi$$

$$H_\phi = -jH_0\frac{\lambda_c}{\lambda_g}e^{-j\beta z}J_n'(k_c r)\cos n\phi \qquad (5\text{–}106)$$

$$E_r = Z_{TM}H_\phi \qquad \text{and} \qquad E_\phi = -Z_{TM}H_r$$

where $H_0 = E_0/Z_{TM}$. The lowest-order TM mode in circular guide is the TM_{01}, which has a cutoff wavelength $\lambda_c = 1.306D$. Its transverse field pattern is shown in Fig. 5–27. Note that with $n = 0$, both H_r and E_ϕ are zero. The attenuation characteristic of the TM_{01} mode is shown in Fig. 5–28. An expression for α_c is derived in Sec. 8.04 of Ref. 5–6.

In concluding this section, it should be noted that Refs. 5–6 and 5–12 provide excellent mathematical treatments of electromagnetic transmission in waveguides and are recommended for further study.

5–6 SOME SPECIAL WAVEGUIDE CONFIGURATIONS

There are many applications that require the use of special waveguide configurations such as ridge guide, dielectric guide, fin-line, and others. A comprehensive study of their properties is beyond the scope of this introductory text. Some of the more common types, however, are described briefly in this section.

Ridge waveguide.   As explained in Sec. 5–4a, the requirement of single-mode transmission limits the useful bandwidth of rectangular guide to about 40 percent. In cases where larger bandwidths are needed, ridge waveguide (Fig. 5–34) is often used. As an aid in understanding its properties, let us first consider the capacitively loaded rectangular guide shown in Fig. 5–32a, where C_p' represents the shunt capacitance per unit length. For $C_p' = 0$, the configuration reduces to the rectangular guide discussed earlier, wherein the TE_{10} and TE_{20} cutoff wavelengths are $2a$ and a, respectively. These cutoff wavelengths may be interpreted in terms of transverse TEM resonances in the waveguide. Recall that at cutoff, $\theta = 90°$ (see Fig. 5–23) and the two TEM waves propagate *transverse* to the axis of the guide. Thus at the cutoff frequency, the waveguide may be viewed as two shorted parallel-plate lines of length $a/2$ connected across the center of the guide. This is described in Fig. 5–33 for a 1 meter length of guide, where x-x denotes the center line. For the TE_{10} mode,

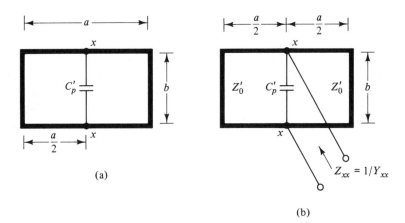

(a)

(b)

Figure 5–32   Capacitively loaded rectangular waveguide (symmetrical case).

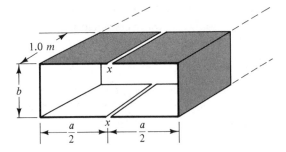

Figure 5–33   Explanation of the transverse-resonance method as applied to rectangular waveguide.

$\lambda_c = 2a$ and hence the line lengths $a/2 = \lambda_c/4$. This means that the impedance of the two quarter-wave shorted lines is infinite, as is their parallel combination (Z_{xx}), at the cutoff frequency. For the TE_{20} mode, $\lambda_c = a$ and $Z_{xx} = 0$ at cutoff since the line lengths ($a/2$) are each a half wavelength long. This result may be generalized for symmetrical structures to include all TE_{m0} modes. For m odd, $Z_{xx} = \infty$ (or $Y_{xx} = 0$) at the cutoff frequency, while for m even, $Z_{xx} = 0$. The application of these conditions to determine the cutoff frequencies of various modes is known as the *transverse-resonance method*. This technique may also be applied to other waveguide modes. For example, the cutoff condition for the TM_{01} and TE_{01} modes in circular guide can be interpreted in terms of radially propagating TEM waves.

Let us now apply this method to determine the TE_{10} and TE_{20} cutoffs for the capacitively loaded rectangular waveguide in Fig. 5–32a. Because of the capacitive loading, the field patterns for these modes are distorted versions of the TE modes in ordinary rectangular guide. The following analysis shows that the effect of C'_p is to lower the TE_{10} cutoff frequency without affecting the TE_{20} cutoff. The fact that the electric field at the center line of the guide is zero for the TE_{20} mode explains why its cutoff is unaffected by the introduction of the shunt capacitance.

For the TE_{10} mode, on the other hand, the electric field is maximum at the center and the introduction of C'_p drastically alters the cutoff frequency. By applying the transverse-resonance method to the circuit in Fig. 5–32b, the following results are obtained.

For TE_{m0} modes with m odd, $Y_{xx} = 0$ and therefore

$$-j2Y'_0 \cot \frac{2\pi}{\lambda_c}\left(\frac{a}{2}\right) + j2\pi f_c C'_p = 0$$

where Y'_0 is the characteristic admittance of a one meter wide parallel-plate line as described in Fig. 5–33. From Eq. (5–24)

$$Z'_0 = \frac{1}{Y'_0} = 377b \sqrt{\frac{\mu_R}{\epsilon_R}}$$

Since the cutoff wavelength and frequency are related by Eq. (5–68), the cutoff equation becomes

$$\tan \frac{\pi a}{\lambda_c} = \frac{\lambda_c}{C'_p} \frac{\epsilon_R}{377\pi bc} \tag{5–107}$$

Solving this transcendental equation for λ_c yields the cutoff wavelength for the TE_{10} mode as well as for the TE_{30}, TE_{50}, etc. Note that for $C'_p = 0$, the solutions are the λ_c values for ordinary rectangular guide (namely, $2a$, $2a/3$, $2a/5$, etc.). Problem 5–32 shows that for finite values of C'_p, the λ_c values increase and hence the TE_{m0} mode (m odd) cutoff frequencies decrease.

For TE_{m0} modes with m even, the transverse-resonance condition is $Z_{xx} = 0$ and therefore

$$j\frac{Z'_0}{2} \tan \frac{2\pi}{\lambda_c}\left(\frac{a}{2}\right) = 0$$

This equation states that $Z_{xx} = 0$ when the input impedance of the two shorted stubs is zero. Since the capacitance is in shunt with the stubs, this condition is independent of the value of C_p'. The above equation reduces to $\tan(\pi a/\lambda_c) = 0$, where λ_c is the cutoff wavelength for the TE_{20}, TE_{40}, . . . modes. As expected, the solutions ($\lambda_c = a$, $a/2$, etc.) are identical to the corresponding modes in ordinary rectangular guide. Thus, the introduction of capacitive loading increases the usable bandwidth of the guide since the TE_{10} cutoff frequency is decreased, while the TE_{20} cutoff is unchanged.

Ridge waveguide is essentially a form of capacitively loaded guide. A single-ridge version is pictured in Fig. 5–34. As explained in Sec. 4–1b, a small section of low impedance line behaves like a shunt capacitance. In this case, the reduced height portion of the ridge guide (d) represents the low impedance line. The cutoff wavelengths are obtained by applying the transverse-resonance method to the equivalent circuit in Fig. 5–35. An air-filled guide has been assumed. The capacitance C_f' is included to account for the fringing fields at the step discontinuity between d and b. This waveguide problem has been solved by Cohn (Ref. 5–32) and Hopfer (Ref. 5–33) for both single- and double-ridge configurations. Hopfer's results for single-ridge guide with $b/a = 0.45$ are shown in Fig. 5–36. Values of λ_c for both the TE_{10} and TE_{20} modes are given. Note that for large ridges (that is, d/b small), the TE_{10} cutoff frequency is reduced significantly, while the effect on the TE_{20} cutoff is small. As a result, the usable bandwidth of ridge guide is much greater than that of ordinary rectangular waveguide. Octave bandwidths or more are easily achieved. An additional advantage of ridge waveguide is that for a given value of TE_{10} cutoff frequency, its cross section is smaller than that of ordinary rectangular guide. Thus for a given operating frequency, the use of ridge guide reduces the required size of the waveguide structure. On the other hand, compared to ordinary rectangular guide, ridge waveguide has higher conductor losses and lower power handling capacity. Both of these problems are analyzed in Refs. 5–32 and 5–33. Also included is Z_0 data for both single- and double-ridge configurations. When using this data, note that Cohn employs the voltage-current definition while Hopfer uses the power-voltage definition (see Sec. 5–5a for these definitions).

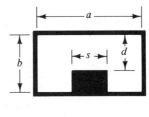

Figure 5–34 A cross-sectional view of single-ridge waveguide.

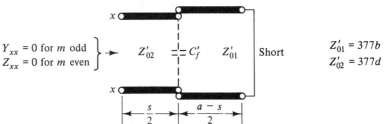

Figure 5–35 Equivalent circuit for determining the TE_{m0} cutoffs of air-filled ridge waveguide.

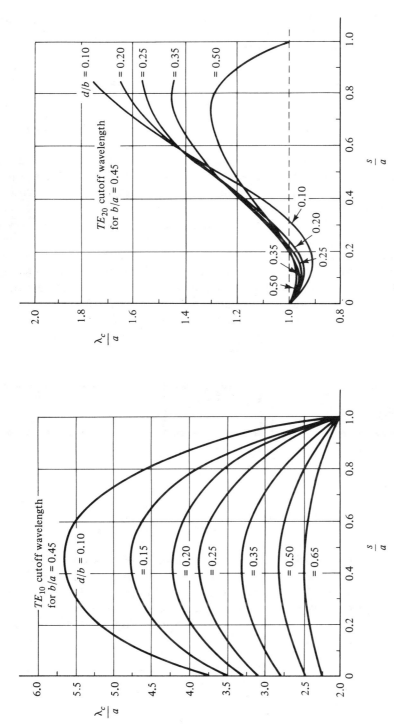

Figure 5-36 Cutoff wavelengths for the TE_{10} and TE_{20} modes in single-ridge guide (Fig. 5–34) for $b/a = 0.45$. (From S. Hopfer, Ref. 5–33; © 1955 IRE now IEEE, with permission.)

Integrated fin line.   Microstrip is widely used in the design and fabrication of microwave integrated circuits. At millimeter wavelengths, however, it is difficult to fabricate microstrip circuits with the high degree of precision required in modern systems. To overcome this and other difficulties (radiation, higher modes, etc.), integrated fin line is often utilized as a medium for constructing millimeter wave circuits.

Figure 5-37 shows cross-sectional views of three fin-line structures. In practice, the thin metallic fins are incorporated onto the dielectric slab using printed circuit techniques. As a result, complex components and circuits can be processed on a single dielectric substrate which is then inserted into the rectangular guide (see, for example, Ref. 5–34). For thin slabs with low values of ϵ_R, the fin-line structure is essentially the same as a narrow-width ridge waveguide. Thus, approximate cutoff values can be obtained from Fig. 5–36 by assuming $s/a \approx 0$. More exact values can be obtained from Refs. 5–35, 5–36, and 5–37. The usable bandwidth of fin-line is typically an octave or greater since it is basically a capacitively loaded rectangular guide. In general, its attenuation is slightly greater than that of microstrip. Due to the high field concentration at the edges of the fins, the fin-line configuration is restricted to low- and medium-power applications. The above referenced articles also include values and expressions for guide wavelength and characteristic impedance. As before, the reader is cautioned that different definitions of Z_0 are used by the various authors.

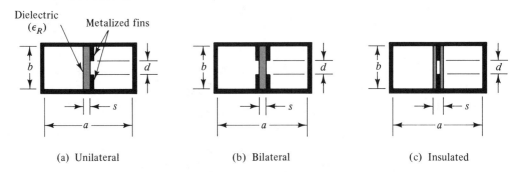

Figure 5–37   Three typical fin-line structures.

Dielectric-loaded waveguide.   Another method of capacitively loading a rectangular waveguide is to place a slab of dielectric material across the narrow dimension of the guide. The arrangement shown in Fig. 5–38 is one example of a dielectric-loaded waveguide. By applying the transverse-resonance method at the center line (Fig. 5–39), the following transcendental equations are obtained for the TE_{m0} cutoff wavelengths. For m odd,

$$\cot \pi \frac{a - s}{\lambda_c} = \sqrt{\epsilon_R} \, \tan \pi \sqrt{\epsilon_R} \, \frac{s}{\lambda_c} \qquad (5\text{–}108)$$

and for m even,

$$\tan \pi \sqrt{\epsilon_R} \, \frac{s}{\lambda_c} = -\sqrt{\epsilon_R} \, \tan \pi \frac{a - s}{\lambda_c} \qquad (5\text{–}109)$$

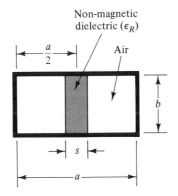

Figure 5–38   Dielectric-loaded rectangular guide (symmetrical case).

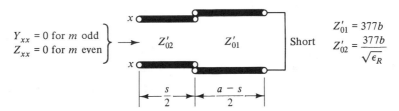

Figure 5–39   Equivalent circuit for determining the TE_{m0} cutoffs of the dielectric-loaded rectangular guide shown in Fig. 5–38.

where λ_c is related to cutoff frequency by $f_c = c/\lambda_c$. The proof of these equations is left to the reader (Prob. 5–35). By substituting specific values into the above equations, one can show that the usable bandwidth of this guide can be substantially greater than that of ordinary rectangular guide (Prob. 5–36). Thus the behavior of the dielectric-loaded guide is similar to ridge guide. However, unlike ridge guide, the dielectric loading decreases the cutoff frequency of the *cross mode* (that is, the TE_{01} mode), a possible disadvantage. If this presents a problem, the guide height b can be reduced, thus raising the cross-mode cutoff frequency. The height reduction will, however, increase conduction losses (α_c) and decrease the power capacity of the guide. Additional information on dielectric-loaded waveguides may be found in Refs. 5–38 and 5–39.

Dielectric-rod waveguide.   It is possible for an electromagnetic wave to be completely guided by a dielectric rod or slab. To understand this (at least qualitatively), recall that TE_{10} mode propagation can be described in terms of zigzagging waves being reflected by the narrow walls of a rectangular guide (Fig. 5–23). Furthermore, it was shown in Sec. 2–7b that, under certain conditions, a wave traveling from high to low index of refraction can be completely reflected at the interface. If the region surrounding the dielectric is air, complete reflection takes place when the angle of incidence (θ_i) is greater than a critical angle given by

$$\theta_c = \arcsin \sqrt{\frac{1}{\mu_R \epsilon_R}} \qquad (5\text{–}110)$$

where μ_R and ϵ_R represent the properties of the dielectric rod or slab. As in the rectangular guide case, the angle of incidence is a function of frequency and guide size. For a fixed size, θ_i will equal θ_c at some frequency. At frequencies below this critical frequency, $\theta_i < \theta_c$ which causes part of the waves to be transmitted into the air region whenever they strike the dielectric-air boundary. In this situation, the transmission loss of the dielectric guide is high. However, for frequencies greater than the critical frequency, $\theta_i > \theta_c$ and the zigzagging waves are totally contained by the dielectric resulting in low-loss transmission along the rod or slab. A sketch of this situation is shown in Fig. 5–40. These qualitative results are verified by a theoretical analysis (see, for example, Sec. 8.11 of Ref. 5–6).

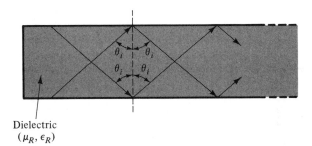

Dielectric
(μ_R, ϵ_R)

Figure 5–40 Complete reflection of wave components in a dielectric guide $(\theta_i > \theta_c)$.

The dielectric-rod waveguide is used as a low-loss transmission line at millimeter wavelengths since precise fabrication of ordinary rectangular guide for these frequencies becomes extremely difficult. In fact, the use of thin glass fibers for the efficient transmission of light has become quite common in telephone and other communication circuits. This technology is known as *fiber optics*.

In concluding this chapter, it should be mentioned that there are many other methods of guiding electromagnetic waves at microwave frequencies. These include *H* line, image line, single-wire line, radial line, and helical line. Some of these are analyzed in Refs. 5–6, 5–40, 5–41, and 5–42.

REFERENCES

Books

5–1. Skilling, H. H., *Electrical Transmission Lines,* McGraw-Hill Book Co., New York, 1951.

5–2. Wheeler, G. J., *Introduction to Microwaves,* Prentice-Hall, Inc., Englewood Cliffs, NJ, 1963.

5–3. Howe, Jr., H., *Stripline Circuit Design,* Artech House, Dedham, MA, 1974.

5–4. Saad, T. S., *Microwave Engineers' Handbook,* vol. 1, Artech House, Dedham, MA, 1971.

5–5. Southworth, G. C., *Principles and Applications of Waveguide Transmission,* D. Van Nostrand Co., Princeton, NJ, 1950.

5–6. Ramo, S., J. R. Whinnery, and T. Van Duzer, *Fields and Waves in Communication Electronics,* John Wiley and Sons, Inc., New York, 1965.

5–7. Gupta, K. C., R. Garg, and I. J. Bahl, *Microstrip Lines and Slot-lines,* Artech House, Dedham, MA, 1979.

5–8. Reich, H. J., P. F. Ordung, H. L. Krauss, and J. G. Skalnik, *Microwave Theory and Techniques,* D. Van Nostrand Co., New York, 1953.

5–9. Marcuvitz, N., *Waveguide Handbook,* Rad. Lab. Series, vol. 10, McGraw-Hill Book Co., New York, 1951.

5–10. Matthaei, G. L., L. Young, and E. M. T. Jones, *Microwave Filters, Impedance-Matching Networks and Coupling Structures,* McGraw-Hill Book Co., New York, 1964.

5–11. Schelkunoff, S. A., *Electromagnetic Waves,* D. Van Nostrand Co., Princeton, NJ, 1943.

5–12. Collin, R. E., *Foundations for Microwave Engineering,* McGraw-Hill Book Co., New York, 1966.

Articles

5–13. Special Issue on Microwave Strip Circuits. *IRE Trans. Microwave Theory and Techniques,* MTT-3, March 1955.

5–14. Cohn, S., Problems in Strip Transmission Lines. *IRE Trans. Microwave Theory and Techniques,* MTT-3, March 1955, pp. 119–126.

5–15. Silvester, P., TEM Wave Properties of Microstrip Transmission Lines. *Proc. IEE (London),* 115, January 1968, pp. 43–48.

5–16. Schneider, M. V., Microstrip Lines for Microwave Integrated Circuits. *Bell System Tech. J.,* 48, May-June 1969, pp. 1421–1444.

5–17. Stinehelfer, H. E., An Accurate Calculation of Uniform Microstrip Transmission Lines. *IEEE Trans. Microwave Theory and Techniques,* MTT-16, July 1968, pp. 439–444.

5–18. Bryant, T. G. and J. A. Weiss, Parameters of Microstrip Transmission Lines and Coupled Pairs of Microstrip Lines. *IEEE Trans. Microwave Theory and Techniques,* MTT-16, December 1968, pp. 1021–1027.

5–19. Getsinger, W. J., Microstrip Dispersion Model. *IEEE Trans. Microwave Theory and Techniques,* MTT-21, January 1973, pp. 34–39.

5–20. Mittra, R. and T. Itoh, A New Technique for the Analysis of the Dispersion Characteristics of Microstrip Lines. *IEEE Trans. Microwave Theory and Techniques,* MTT-19, January 1971, pp. 47–56.

5–21. Denlinger, E. J., A Frequency Dependent Solution for Microstrip Transmission Lines. *IEEE Trans. Microwave Theory and Techniques,* MTT-19, January 1971, pp. 30–39.

5–22. Pucel, R. A., D. J. Masse, and C. P. Hartwig, Losses in Microstrip. *IEEE Trans. Microwave Theory and Techniques,* MTT-16, June 1968, pp. 342–350.

5–23. Green, H. E., The Numerical Solution of Some Important Transmission Line Problems. *IEEE Trans. Microwave Theory and Techniques,* MTT-13, September 1965, pp. 676–692.

5–24. Gish, D. L. and O. Graham, Characteristic Impedance and Phase Velocity of a Dielectric-Supported Air Strip Transmission Line with Side Walls. *IEEE Trans. Microwave Theory and Techniques,* MTT-18, March 1970, pp. 131–148.

5–25. Cohn, S. B., Slot Line on a Dielectric Substrate. *IEEE Trans. Microwave Theory and Techniques,* MTT-17, October 1969, pp. 768–778.

5–26. Mariani, E. A., et. al., Slot Line Characteristics. *IEEE Trans. Microwave Theory and Techniques,* MTT-17, December 1969, pp. 1091–1096.

5–27. Knorr, J. B. and K. D. Kuchler, Analysis of Coupled Slots and Coplanar Strips on Dielectric Substrate. *IEEE Trans. Microwave Theory and Techniques,* MTT-23, July 1975, pp. 541–548.

5–28. Wen, C. P., Coplanar Waveguide: A Surface Strip Transmission Line Suitable for Non-reciprocal Gyromagnetic Device Applications. *IEEE Trans. Microwave Theory and Techniques,* MTT-17, December 1969, pp. 1087–1090.

5–29. Davis, M. E., E. W. Williams, and A. C. Celestini, Finite-Boundary Corrections to the Coplanar Waveguide Analysis. *IEEE Trans. Microwave Theory and Techniques,* MTT-21, September 1973, pp. 594–596.

5–30. Walker, R. M., Waveguide Impedance—Too Many Definitions. *Electronic Communicator,* 1, May-June 1966, p. 13.

5–31. Miller, S. E., Waveguide as a Communication Medium. *Bell System Tech. J.,* 33, November 1954, pp. 1209–1265.

5–32. Cohn, S. B., Properties of Ridge Waveguide. *Proc. IRE,* 35, August 1947, pp. 783–788.

5–33. Hopfer, S., The Design of Ridged Waveguide. *IRE Trans. Microwave Theory and Techniques,* MTT-3, October 1955, pp. 20–29.

5–34. Meier, P. J., Integrated Fin-Line Millimeter Components. *IEEE Trans. Microwave Theory and Techniques,* MTT-22, December 1974, pp. 1209–1216.

5–35. Saad, A. M. K. and K. Schunemann, A Single Method for Analyzing Fin-Line Structures. *IEEE Trans. Microwave Theory and Techniques,* MTT-26, December 1978, pp. 1002–1007.

5–36. Knorr, J. B. and P. M. Shayda, Millimeter-Wave Fin-Line Characteristics. *IEEE Trans. Microwave Theory and Techniques,* MTT-28, July 1980, pp. 737–743.

5–37. Shih, Y. C. and W. J. R. Hoefer, Dominant and Second-Order Mode Cutoff Frequencies in Fin Lines Calculated with a Two-Dimensional TLM Program. *IEEE Trans. Microwave Theory and Techniques,* MTT-28, December 1980, pp. 1443–1448.

5–38. Vartanian, P. H., W. P. Ayres, and A. L. Helgesson, Propagation in Dielectric Slab Loaded Rectangular Waveguide. *IRE Trans. Microwave Theory and Techniques,* MTT-6, April 1958, pp. 215–222.

5–39. Gardiol, F. E., Higher-Order Modes in Dielectrically Loaded Rectangular Waveguides. *IEEE Trans. Microwave Theory and Techniques,* MTT-16, November 1968, pp. 919–924.

5–40. Tischer, F. J., Properties of the H Guide at Microwaves and Millimeter Waves. *Proc. IEE (London),* B106 (13), January 1959, pp. 47–53.

5–41. King, D. D., Properties of Dielectric Image Lines. *IRE Trans. Microwave Theory and Techniques,* MTT-3, March 1955, pp. 75–81.

5–42. Goubau, G., Single-Conductor Surface-Wave Transmission Lines. *Proc. IRE,* 39, June 1951, pp. 619–624.

PROBLEMS

5–1. A shorted section of 300 ohm TV twin-lead is used to provide an inductive reactance of 180 ohms at 500 MHz. Calculate the line length if its velocity factor is 0.90.

5–2. An air-insulated, open two-wire line having copper conductors is used in a 900 MHz application. Calculate the total attenuation (in dB) for a 6 m section of line if the diameter of the wires is 0.20 cm and their spacing is 1.0 cm.

5–3. Calculate the b/a ratio for an air-filled 50 ohm coaxial line. If the line were dielectric-filled, would the ratio have to be increased or decreased to maintain the same impedance value?

5–4. The conductors of a teflon-filled coaxial line are silver plated. Calculate the attenuation constant (in dB/m) at 3000 MHz if $b = 0.90$ cm and $a = 0.30$ cm. Assume that the plating thickness is greater than several skin depths.

5–5. Derive Eqs. (5–11) and (5–12).

5–6. What is the maximum power handling capacity of a 25 ohm air-filled coaxial line having an inner conductor diameter of 1.0 cm? Include a safety factor for the possibility of large reflections on the line.

5–7. Calculate the TE_{11} cutoff frequency for the coaxial line described in the above problem. What would be the decrease in cutoff frequency if the line were quartz-filled and the dimensions were unchanged?

5–8. Design a 50 ohm teflon-filled coaxial line capable of single-mode propagation at K_u band. Allowing for a 5 percent safety factor at the high end of the frequency band, determine the inner and outer diameters of the conductors.

5–9. An air-insulated coaxial line capable of single-mode propagation up to 8 GHz is required in a microwave transmission system. Assuming copper conductors,
(a) Calculate the attenuation constant at 6 GHz if Z_0 is chosen optimally.
(b) What is its value if a 50 ohm line is used?

5–10. An air-insulated coaxial line is designed for single-mode propagation up to 10 GHz. What is the percent reduction in power handling capacity if Z_0 is chosen to be 50 ohms rather than 44 ohms?

5–11. A dielectric-filled parallel-plate line has a characteristic impedance of 50 ohms. How must the dimensions be changed to reduce Z_0 to 20 ohms without affecting the attenuation?

5–12. A dielectric-filled parallel-plate line has the following characteristics at 2 GHz: $\alpha_c = 0.60$ dB/m and $\alpha_d = 0.90$ dB/m. What is the attenuation constant α at 2 GHz and 8 GHz? Assume that the dielectric properties are not frequency sensitive.

5–13. Calculate the maximum ground-plane spacing for a dielectric-filled ($\epsilon_R = 2.50$) balanced stripline to be used in a C-band system.

5–14. Determine the strip width of a teflon-filled balanced stripline for $Z_0 = 50$ ohms. The ground-plane spacing is 0.25 inch and the strip thickness is 4 mils.

5–15. Calculate the attenuation constant (in dB/m) at 5 GHz for the line described in the preceding problem. Assume copper conductors.

5–16. A microstrip transmission system is constructed using 0.060 inch thick alumina. Design a single-section, quarter-wave transformer at 3 GHz to match a 20 ohm resistive load to a 75 ohm line. Assume the thickness of the center strip is negligible.

5–17. Calculate the cutoff frequencies of air-filled WR-90 guide for the TE_{10}, TE_{20}, TE_{01}, and TM_{11} modes.

5–18. Repeat the above problem if the guide is filled with a nonmagnetic dielectric of value $\epsilon_R = 4$.

5–19. Using the TM_{11} mode pattern (Fig. 5–20) as an aid, sketch the patterns for the TM_{21} and TM_{22} modes as seen in the transverse plane.

5–20. Calculate the guide wavelength (in cm) at 7 and 12 GHz for an air-filled WR-90 guide. Comment on the results.

5–21. Repeat the above calculations if the guide is quartz-filled. Comment on the change in guide wavelength at 7 GHz.

5–22. It is required that a filter be inserted into a K_u-band system so that a 13 GHz signal is attenuated 40 dB, while a 15 GHz signal is virtually unattenuated. Design the filter utilizing the cutoff effect.

5–23. Design a single-section, quarter-wave impedance transformer at 5 GHz from 2×4 cm guide to 1×4 cm guide. Assume air-filled waveguides. (Hint: Choose the width of the transformer section to be the same as that of the two guides.)

5–24. For the preceding problem, use the Smith chart to determine the SWR along the 2×4 cm guide at 4.5 GHz. Assume that the 1×4 cm guide is match terminated.

5–25. Design a single-section, quarter-wave impedance transformer at 9 GHz from 1×3 cm guide to a 1×2 cm guide. Assume air-filled waveguides. (Hint: Choose the height of the transformer section to be the same as that of the two guides.)

5–26. Verify the peak power value in Table 5–1 for an air-filled WR-137 waveguide at 8.20 GHz. Note that the value includes a safety factor that accounts for the possibility of large standing waves.

5–27. Calculate the percent increase in peak power capacity if the height of the waveguide described above is increased to equal half the guide width.

5–28. Calculate the attenuation constant (in dB/m) at 6 GHz for an air-filled, 2×4 cm rectangular guide. Compare the result with that obtained for an air-filled coaxial line having the same value of f_{max}. Assume copper conductors in both cases.

5–29. Design an air-filled rectangular waveguide with a 2 to 1 aspect ratio for use in the 40 to 60 GHz frequency band.

5–30. Calculate the cutoff frequency of the dominant mode in a 1 inch diameter, teflon-filled circular waveguide. What is its maximum operating frequency if the possibility of higher-mode propagation is to be avoided? Include a 5 percent safety factor. What would be the value of f_{max} if the possibility of TM_{01} propagation was excluded?

5–31. Calculate the guide wavelength and characteristic impedance at 6 and 8 GHz for the guide described in the previous problem. Use the modified power-voltage definition for Z_0.

5–32. An air-filled rectangular guide with an aspect ratio (a/b) of 2 has a TE_{10} cutoff frequency of 3 GHz.
   (a) What value of capacitive loading (C_p') is required to reduce the TE_{10} cutoff to 2 GHz?
   (b) For the above value of C_p', determine the TE_{20} and TE_{30} cutoff frequencies.

5–33. For $b/a = 0.45$ and $s = b$, design a single-ridge waveguide (Fig. 5–34) for use over the 3.0 to 7.0 GHz frequency band. Arrange the TE_{20} cutoff to be about 5 percent above the high-frequency end and the TE_{10} about 20 percent below the low-frequency end.

5–34. Use the transverse-resonance method to derive an equation for the TE_{10} cutoff wavelength in ridge waveguide. Assume C_f' in Fig. 5–35 is negligible and the guide is airfilled.

5–35. Derive Eqs. (5–108) and (5–109) using the transverse-resonance method.

5–36. Referring to Fig. 5–38, a 1×4 cm rectangular guide is loaded with a 0.2 cm wide dielectric having $\epsilon_R = 9$. Calculate the TE_{10} and TE_{20} cutoff frequencies.

6

Coaxial and Stripline Components

There are a large variety of coaxial and stripline components used to control and process microwave signals. Some of the more common ones are discussed in this and subsequent chapters. The material in this chapter deals specifically with one- and two-port reciprocal components. Unless otherwise indicated, TEM propagation is assumed.

6-1 TERMINATIONS

Many microwave applications require that a transmission line be terminated in a known impedance. This is particularly true in measurement systems. Among the more widely used are the matched load, the short circuit, the open circuit, and the standard mismatch.

6-1a Matched Loads

A matched load or termination is a one-port component that absorbs all the power incident upon it. This requires that its impedance equal the characteristic impedance of the line to which it is connected.

Resistor-type loads. Since Z_0 is real for low-loss lines, a matched load can be realized by terminating the line with a lumped-element resistor of value $R = Z_0$. At microwave frequencies, the resistor must be specially designed to avoid the parasitic reactances normally associated with low-frequency units. Part *a* of Fig. 6–1 shows a configuration often used in coaxial systems. Its equivalent circuit for $l \ll \lambda$ is indicated in part *b*. As explained in Sec. 1–1, the lumped-element model is only valid when the dimensions of the actual element are small compared to the operating wavelength. The resistive material is usually evaporated on a hollow dielectric rod.

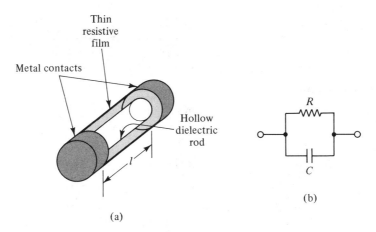

(a)

(b)

Figure 6–1   A microwave resistor and its equivalent circuit for $l \ll \lambda$.

In order to minimize capacitive effects, the wall thickness of the rod is made as small as possible, consistent with mechanical rigidity. When used as a matched load, its capacitive reactance should be at least ten times greater than R in order to insure a load SWR of less than 1.10. This means, for example, that a coaxial 50 ohm load at 3000 MHz requires a resistor with $C < 0.10\ pF$. An additional requirement is that the thickness of the resistive film be an order of magnitude less than its skin depth at the highest frequency of interest. This insures a constant resistance value over the specified frequency range. Finally, the diameter of the two metal contacts is chosen to accommodate the center conductor of the coaxial line without an abrupt change in dimensions.[1] Resistors of this form are available from many component manufacturers.

Two examples of coaxial matched loads that use this type resistor are shown in Fig. 6–2. For the one described in part *a*, the inner and outer diameters of the load

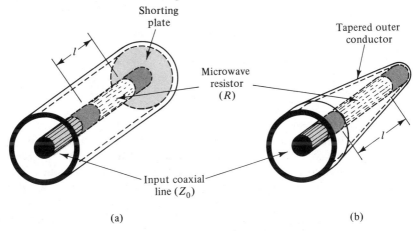

(a)

(b)

Figure 6–2   Two forms of coaxial matched loads.

[1] As explained in Sec. 6–4, an abrupt discontinuity in either conductor of a coaxial line creates reactive effects which, in turn, cause reflections.

section are the same as the coaxial line to which it is attached. With $R = Z_0$, the load SWR will be less than 1.20 as long as the resistor length l is less than 0.05λ. The following illustrative example uses transmission-line theory to show the deterioration in SWR when l exceeds this value.

Example 6–1:

A 1.0 cm long, 50 ohm film resistor is connected to an air-insulated 50 ohm coaxial line as shown in Fig. 6–2a. Treating the load section as a short-circuited lossy line, calculate its SWR at 1 and 3 GHz.

Solution: For the lossless coaxial line,

$$Z_0 = \sqrt{L'/C'} = 50 \text{ ohms} \quad \text{and} \quad v = 1/\sqrt{L'/C'} = 3 \times 10^8 \text{ m/s}.$$

Therefore, $L' = 167$ nH/m and $C' = 66.7$ pF/m. The corresponding values for the lossy-line section are the same since its dimensions are identical to those of the coaxial line. The effect of the thin dielectric tube is assumed to be negligible.

Since the only dissipative element in the lossy-line section is the series resistance,

$$G' = 0 \quad \text{and} \quad R' = R/l = 50/0.01 = 5000 \text{ ohms/m}.$$

For a short-circuited line, the impedance transformation equation [Eq. (3–81)] reduces to $Z_{in} = Z_0 \tanh \gamma l$. Using Eqs. (3–15) and (3–17) with $G' = 0$ yields

$$Z_{in} = \sqrt{L/C - jR/\omega C} \quad \tanh \sqrt{-\omega^2 LC + j\omega RC}$$

where $R = R'l$, $L = L'l$ and $C = C'l$. With $l = 0.01$ m, $R = 50$ ohms, $L = 1.67$ nH, and $C = 0.667$ pF. Therefore at 1 GHz,

$$Z_{in} = \sqrt{2500 - j11,900} \quad \tanh \sqrt{-0.044 + j0.21}$$

$$= (110\underline{/-39°})(0.47\underline{/47°}) = 51.2 + j7.2 \text{ ohms}$$

where use has been made of the third identity in Eq. (E–6) of Appendix E. Normalizing to Z_0 of the coaxial line, $\bar{Z}_{in} = 1.02 + j0.14$, which corresponds to a load SWR of 1.15. Note that with $l = 1$ cm and $\lambda_0 = 30$ cm, $l < 0.05\lambda$ at 1 GHz. Repeating the calculation for $f = 3$ GHz,

$$Z_{in} = \sqrt{2500 - j3970} \quad \tanh \sqrt{-0.396 + j0.63}$$

$$= (68.5\underline{/-28.9°})(0.95\underline{/46.3°}) = 62 + j19.5 \text{ ohms}$$

Thus, $\bar{Z}_{in} = 1.24 + j0.39$, which corresponds to a load SWR of 1.50. This deterioration in SWR is due to the fact that $l > 0.05\lambda$ at 3 GHz.

This example illustrates the effect of resistor length on the SWR versus frequency characteristic of the coaxial load. One method of improving the SWR at high frequencies is to reduce the length of the resistor element. For example, use of a 0.50 cm long resistor would reduce the SWR at 3 GHz to about 1.20. Further reduction in the resistor length, however, may not improve the SWR since the capacitive effect of the dielectric tube could become significant.

Another method of improving the SWR is to reduce the diameter of the outer conductor in the lossy-line section. It can be shown that the resultant decrease in L' reduces the load SWR (Prob. 6–2). A modified form of this technique is shown in Fig. 6–2b. In this configuration, the outer diameter is reduced gradually over the length of the lossy section.

Planar forms of the microwave resistor described in Fig. 6–1a are also commercially available. The resistive film is deposited on a thin ceramic slab typically 0.25 cm wide by 0.50 cm long. This configuration is especially suited for use in strip transmission-line systems.

Tapered loads.   The restriction on resistor length makes it difficult to achieve a low SWR above X band with resistor-type loads. The tapered load described in Fig. 6–3 avoids this difficulty. The load material chosen is quite lossy at microwave frequencies, typically, several dB per inch. For low-power applications, powdered iron or carbon in an epoxy binder is commonly used. Ceramic type materials are usually favored in high-power situations. The reason for gradually tapering the absorbing material is to minimize wave reflections. The taper length (l_1) should be at least a few wavelengths long at the lowest frequency of interest.[2] By satisfying this condition, the input SWR is essentially unity and hence the input impedance of the load equals Z_0, the characteristic impedance of the coaxial line. The purpose of the shorting plate at the end of the load section is to prevent radiation leakage out of or into the unit. The length of the fully loaded lossy section (l_2) is chosen so that the total attenuation through l_1 and l_2 is greater than 20 dB. This insures that the reflection at the shorting plate is attenuated sufficiently to minimize its effect at the input terminals. The magnitude of this effect can be calculated with the aid of the return-loss concept. For example, suppose the one-way attenuation through l_1 and l_2 is 23 dB. From Eq. (3–78), the return loss at the input is $0 + 2(23) = 46$ dB, since the return loss of a short circuit is zero. By applying Eq. (3–77) to the input, $|\Gamma_{in}| = 0.005$. This is equivalent to an input SWR of 1.01, a value that is usually negligible compared to the SWR associated with the taper itself.

A movable form of the tapered load, known as a *sliding load,* is often used in the accurate measurement of SWR and impedance.

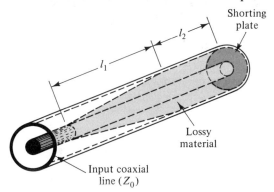

Shorting plate

l_2

l_1

Lossy material

Input coaxial line (Z_0)

Figure 6–3   A tapered coaxial load ($l_1 \gg \lambda$).

6–1b Short and Open Circuits

A coaxial short circuit may be realized by terminating the line with a metal plate as shown in Fig. 6–4. The plate creates a boundary at which the electric field associated with the TEM mode is zero. Thus $\Gamma_L = -1$, which is the reflection coefficient

[2] Note that unlike the resistor-type load, the length of a tapered load must be *long* compared to the operating wavelength. Therefore, its SWR is best at the higher frequencies while that of the resistor type is best at the lower frequencies.

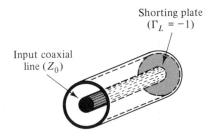

Figure 6–4 A coaxial short circuit.

of a short circuit. One might be tempted to create a short circuit by merely connecting a low-resistance wire between the inner and outer conductors. This would work at low frequencies but not at microwaves since the reactance associated with the inductance of the wire would be appreciable. Furthermore, some of the field would radiate out the end of the line, thus adding a resistive component to the terminating impedance.

Coaxial and stripline versions of an open circuit are shown in Fig. 6–5. Note that in the coaxial unit, the outer conductor extends past the end of the inner conductor. This is to prevent radiation out the end of the coaxial line. The diameter of the outer conductor D must be chosen so that the circular waveguide section is below cutoff at the highest frequency of interest. Its length l should be sufficient to attenuate the dominant mode (TE_{11}) by at least 20 dB. This insures that *all* modes will be attenuated by at least that amount, resulting in negligible radiation.

There will be some fringing of the electric field at the end of the conductor. This is indicated by the dashed lines in the figure. As a result, the open circuit appears to be slightly displaced from the end of the inner conductor. A useful rule of thumb is that the plane of the open circuit is approximately $0.6 (b - a)$ *past* the end of the inner conductor, where b and a are the radii of the outer and inner conductors, respectively. The frequency dependence of the effective open position is discussed in Ref. 6–14.

Similar comments can be made regarding the stripline open circuit in part b of Fig. 6–5. Data is available that indicates the shift in terminal plane of an open circuit in both stripline and microstrip configurations (Refs. 6–12 and 6–15).

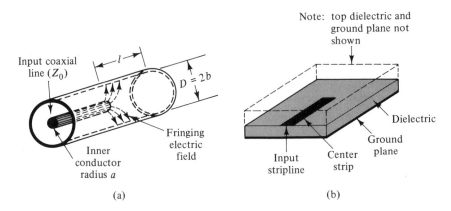

(a)   (b)

Figure 6–5 Coaxial and stripline open circuits.

6–1c Standard Mismatches

Standard mismatches are used to calibrate SWR and impedance measuring equipment. Typically, a standard mismatch consists of a known impedance whose value differs from the characteristic impedance of the connecting line. The configuration used in coaxial systems is similar to that shown for the matched load in Fig. 6–2. The resistor value can be either less than or greater than Z_0. For example, terminating a 50 ohm line with either a 25 or 100 ohm microwave resistor produces an SWR of 2.0. This SWR value will be independent of frequency as long as the resistor length is much less than 0.05λ and the capacitive effect of the dielectric tube is negligible.

6–1d Variable Reactances

At the lower end of the microwave spectrum, mechanically adjustable air capacitors are used as variable reactances. Due to the requirement that the dimensions be small compared to the operating wavelength, L-band units are typically a centimeter or less in diameter. Higher-frequency units must be even smaller, which leads to construction difficulties and severe limitations on their voltage handling capability. Two solutions that overcome these restrictions are indicated in Fig. 6–6. Both take advantage of the impedance transforming properties of a lossless line [Eq. (3–83)].

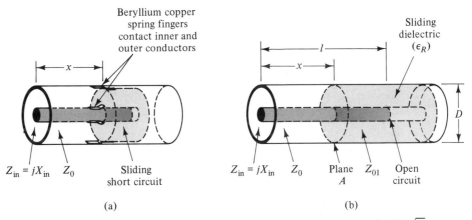

Figure 6–6   Two forms of coaxial variable reactances. (Note: $D < \lambda_0/1.706\sqrt{\epsilon_R}$ at the highest frequency of interest.)

Movable short circuit.   The sliding short circuit shown in Fig. 6–6a may be used to vary the reactance at the input terminals of a coaxial line. For an air-insulated shorted line, the input impedance is given by

$$Z_{in} = jX_{in} = jZ_0 \tan \frac{2\pi x}{\lambda_0}$$

where Z_0 is the characteristic impedance of the line. The variation of input reactance with line length is shown in Fig. 3–14. Note that all values of X_{in} from $-\infty$ to $+\infty$ are available by varying the sliding short over a half wavelength. The reader is cautioned that despite the use of low-loss lines and near perfect short circuits, large values of reactance can be accompanied by significant resistive effects. This was discussed following Ex. 3–5 in Sec. 3–6b. It should also be noted that for a fixed value of x, the frequency sensitivity of input reactance $(dX_{in}/d\omega)$ is greater than that of an equivalent lumped-element reactance (Prob. 6–4). This may be a disadvantage in broadband applications.

Another problem associated with this type of variable reactance is the ohmic contact of the sliding short. Due to the large currents at the short circuit, low-resistance contacts to the inner and outer conductors of the coaxial line are vital. This is usually accomplished with beryllium copper *spring fingers* as indicated in the figure. Unless proper care is taken, dust particles can compromise the quality of the sliding contact, which leads to erratic electrical behavior. This is usually referred to as a *noisy contact*. This problem can be avoided by using *noncontacting* sliding shorts. The principles involved are discussed in Sec. 7–3c.

Movable dielectric rod. Figure 6–6b shows another form of a coaxial variable reactance. The ohmic contact problem is avoided in this case by using a sliding dielectric tube as shown. The unit consists of a fixed length l of open-circuited line and a movable dielectric rod.[3] As the dielectric is inserted into the coaxial section, the length x of the air-filled line section decreases while that of the dielectric-filled line $(l - x)$ increases. Note that $Z_0 = \sqrt{\epsilon_R}\, Z_{01}$ since the conductor diameters are the same in both sections. The impedance at plane A of the open-circuited, dielectric-filled section is given by

$$Z_A = -jZ_{01} \cot 2\pi \frac{l - x}{\lambda} = -j\frac{Z_0}{\sqrt{\epsilon_R}} \cot 2\pi \sqrt{\epsilon_R}\, \frac{l - x}{\lambda_0}$$

where λ_0 is the free-space wavelength. The input impedance is obtained from the impedance transformation equation, where in this situation the Z_0 line is terminated by the impedance Z_A. Thus,

$$Z_{in} = Z_0 \frac{Z_A + j Z_0 \tan 2\pi \dfrac{x}{\lambda_0}}{Z_0 + jZ_A \tan 2\pi \dfrac{x}{\lambda_0}}$$

or

$$Z_{in} = jZ_0 \frac{F_0 F_1 - 1}{F_0 + F_1} \tag{6-1}$$

where $F_0 \equiv \tan 2\pi \dfrac{x}{\lambda_0}$ and $F_1 \equiv \sqrt{\epsilon_R} \tan 2\pi \sqrt{\epsilon_R}\, \dfrac{l - x}{\lambda_0}$.

With l fixed, Z_{in} is a function of x. If l is long enough, all possible values of input reactance can be realized by varying the dielectric penetration. This is verified by a numerical example at the end of the chapter (Prob. 6–5).

---

[3] As explained in Sec. 6–1b, the diameter D must be chosen so that the dominant circular waveguide mode is below cutoff at the highest frequency of interest.

Both type variable reactances described here are widely used in microwave laboratories. The double-stub tuner described in Fig. 4–20 represents one such application.

6–2 CONNECTORS AND TRANSITIONS

A variety of standard connectors are available which provide a low-SWR connection between coaxial lines. For microwave applications, the most common ones are the Type-N, the SMA, and the APC-7® precision connectors. Their electrical and mechanical features are described in this section. These connectors are also used to provide convenient input and output terminals for coaxial components and systems. Three such components are shown in Fig. 6–7.

It is important that the SWR associated with the mating of two connectors be quite low, typically 1.10 or less. Higher values of SWR can cause a significant degradation in device performance. In addition, the accuracy of microwave measurements is adversely affected by imperfect connectors.

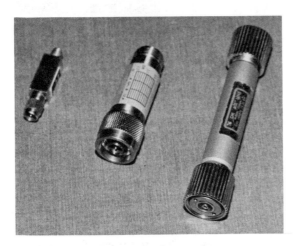

Figure 6–7 Typical microwave components with coaxial connectors. From left to right, the connector types are SMA, Type-N, and APC-7. (Courtesy of Prof. Richard Walder, S.E. Mass. Univ., N. Dartmouth, Mass.)

6–2a Dielectric Bead Supports

Most coaxial connectors use a dielectric bead to mechanically support the center conductor. Four types of low-reflection bead supports are shown in Fig. 6–8. The first one uses a half wavelength section of low-loss dielectric material. Since the inner and outer conductor diameters are the same throughout and the dielectric is nonmagnetic,

$$Z_{01} = \frac{Z_0}{\sqrt{\epsilon_R}} \tag{6–2}$$

where Z_0 and Z_{01} are the characteristic impedances of the air-insulated and the dielectric-filled lines, respectively.

Since impedance repeats every half wavelength, a matched load ($Z_L = Z_0$) connected to the air-insulated line on the right results in $Z_{in} = Z_0$ at the left edge of

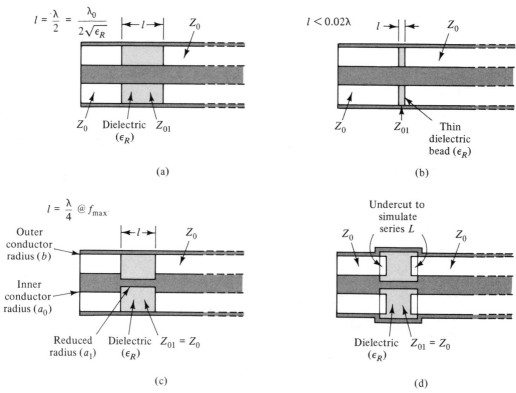

Figure 6–8   Four types of dielectric bead supports for coaxial lines.

the dielectric. Thus the SWR along the Z_0 line is unity, which means that there is no reflection loss associated with the dielectric bead.

     The disadvantage of this type bead support is that its SWR is very frequency sensitive since the bead can only be a half-wavelength long at the design frequency. Figure 6–9 shows the SWR versus frequency for two values of dielectric constant. The data was obtained by determining Z_{in}/Z_0 and the resultant input SWR. Note that for a given maximum allowable SWR (say, 1.20), the useful frequency range is larger for the lower value of dielectric constant.

     The bead support shown in Fig. 6–8b is useful at frequencies below 3 GHz. To insure low reflections, its length is typically 0.02λ or less at the highest frequency of interest. Since the bead represents a very short length of low-impedance line, it is equivalent to a shunt capacitance. Thus its SWR increases as the operating frequency is increased. Due to the electrical requirement that the bead length be less than 0.02λ, a mechanically rigid support becomes difficult to achieve at the higher microwave frequencies.

     Another type bead support is shown in Fig. 6–8c. In this case, the center conductor of dielectric section is reduced to maintain the same characteristic impedance throughout the structure. Therefore,

$$138 \log \frac{b}{a_0} = \frac{138}{\sqrt{\epsilon_R}} \log \frac{b}{a_1}$$

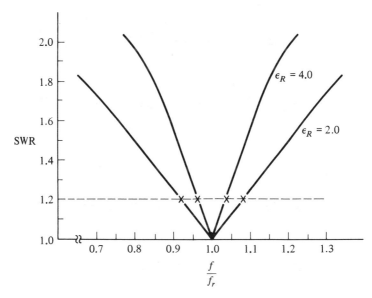

Figure 6–9 Frequency sensitivity of the half wavelength dielectric bead (Fig. 6–8a), where $l = \lambda/2$ at f_r, the design frequency.

and hence

$$a_1 = b\left(\frac{a_0}{b}\right)^{\sqrt{\epsilon_R}} \tag{6–3}$$

With the characteristic impedance the same in both the air-insulated and dielectric-filled lines, the SWR is practically unity at all frequencies. As usual, it is assumed that the radius b is chosen to avoid TE_{11} mode propagation. Some variation of SWR with frequency is caused by the capacitive effects associated with the abrupt changes in center conductor diameter. The equivalent circuit, indicated in Fig. 6–10a, is two shunt capacitances separated by the length of the dielectric-filled line. Discontinuity capacitance data is given in Fig. 6–26. Since the capacitive susceptance (ωC) increases with increasing frequency, the individual reflections of the capacitances also increase. By making the dielectric bead a quarter-wavelength long at the highest frequency of interest, the overall reflection is minimized. With the aid of the Smith chart, one can verify that near-perfect cancellation of the reflections occurs for $\bar{B}_C < 0.25$ (Prob. 6–9).

Bead supports of the type shown in Fig. 6–8d are used in both the Type-N and APC-7 connectors. Teflon and rexolite are commonly used as the dielectric material. The conductor diameters in the dielectric region are adjusted so that $Z_{01} = Z_0$. The capacitive effects of the step discontinuities are compensated for by undercutting the dielectric on both ends as shown. This creates small sections of high impedance lines which behave like series inductances [see Eq. (4–13)]. Thus the equivalent circuit becomes that shown in Fig. 6–10b. With the aid of the Smith chart, one can show that for $\bar{B}_C < 0.25$, the overall SWR of the bead is near unity when $\bar{X}_L = \bar{B}_C$. The required value of $\bar{X}_L$ can be obtained by adjusting the length of the undercut. For the Type-N connector, the teflon bead is undercut by about 0.030 cm.

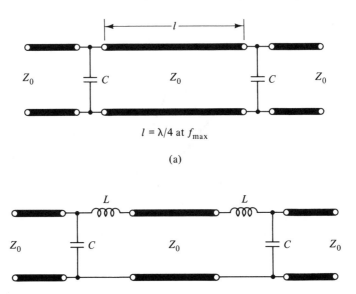

$l = \lambda/4$ at f_{max}

(a)

(b)

Figure 6–10   Equivalent circuits for the dielectric bead supports shown in parts *c* and *d* of Fig. 6–8.

The reader should note from this discussion that even the tiniest of dimensional and structural changes can produce significant electrical effects. Microwave engineers must be aware of this and account for it in their analysis and design.

6–2b Standard Coaxial Connectors

The importance of low-SWR connectors was explained at the beginning of this section. The electrical and mechanical characteristics of three such connector-pairs are described here.

The Type-N connector.   Although the Type-N connector was developed over 30 years ago, it is still one of the more popular coaxial connectors at microwave frequencies. Much work has been done over the years to improve its electrical and mechanical characteristics. One such improved version is shown in Fig. 6–11. The right-hand portion of the connector-pair is known as the *female* connector while the left-hand portion is called the *male* connector. The connector-pair shown is designed for use in 50 ohm systems. (The choice of 50 ohms was explained at the end of Sec. 5–2.) The diameters of the inner and outer conductors of the air-insulated coaxial lines are shown in the figure. Low-SWR bead supports of the type described in Fig. 6–8d are incorporated into both connectors.

The mating of the connectors is accomplished by threading them together until surface *A* of the male portion contacts firmly against surface *A'* of the female connector. For precision connectors, the tolerances on the 0.207 inch and 0.210 inch dimensions are such that when the connectors mate, surfaces *B* and *B'* are separated by a few thousandths of an inch. As a result, a small section of high impedance line

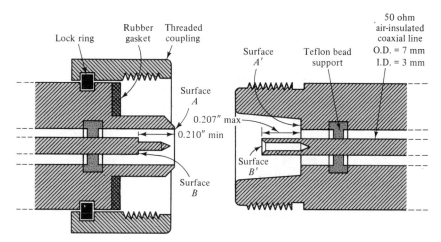

Figure 6–11   The Type-N connector-pair for 50 ohm–7mm lines.

($>$ 50 ohms) is created which, in turn, produces a small reflection. This gap, however, is necessary in order to prevent damage to the center pin of the female connector. If the tolerances were such that surfaces B and B' touched *before* surface A had seated against A', continual threading of the connectors would force surface B into the center pin of the female connector resulting in mechanical damage to the connector-pair. In some versions, the outer conductor of the male portion and the inner conductor of the female portion are slotted to improve the connection both electrically and mechanically. The slots cause a small reflection resulting in a slight degradation in electrical performance. Typical SWR versus frequency for a mated pair of precision Type-N connectors is shown in Fig. 6–12. The possibility of higher-mode propagation prevents the use of these connectors above K_u band. As a rule, Type-N connectors are seldom used much above 12 GHz.

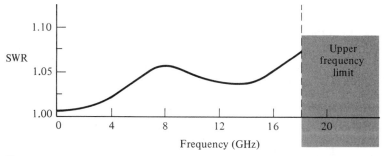

Figure 6–12   Typical SWR versus frequency for a mated pair of precision Type-N connectors.

The SMA connector.   Subminiature coaxial connectors are commonly used in low-power applications at the higher microwave frequencies, particularly C, X, and K_u band. They are designed for use with a teflon-filled 50 ohm coaxial line. The dimensions of the SMA connector-pair are given in Fig. 6–13. As in the previous case, mating of the connectors is accomplished by threading them together until sur-

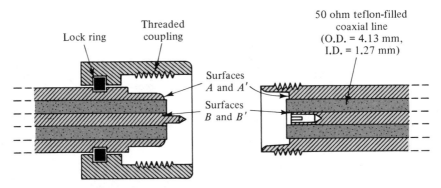

(Note: smaller diameter cables may also be used with this connector)

Figure 6–13   The SMA connector-pair for teflon-filled 50 ohm lines.

face A of the male portion contacts firmly against surface A' of the female connector. Again, the mechanical tolerances are such that with surfaces A and A' touching, surfaces B and B' are separated by a few thousandths of an inch, thus avoiding mechanical damage to the center pin of the female connector. The teflon insulators are designed to extend no more than two thousandths of an inch past surfaces A and A'. Since teflon is compressible, threading the connector-pair together so that surfaces A and A' make contact results in a firm contact between the corresponding teflon surfaces. This design precludes the possibility of an air gap between the teflon insulators which could cause significant reflections.

      The upper frequency limit of a coaxial connector is determined by the possible onset of higher-mode propagation, usually the TE_{11}. The maximum usable frequency is defined below Eq. (5–17). In practice, the SMA connector-pair may be used up to about 20 GHz. The maximum SWR of a mated connector-pair is given by the following approximate expression.

$$SWR_{max} \approx 1.05 + 0.005f \qquad (6\text{–}4)$$

where f is the operating frequency in GHz.

      A smaller version of the SMA connector-pair is also commercially available. The Omni-Spectra, Inc. version is designated as the OSSM connector and is usable up to about 30 GHz.

     The APC connector.   Two requirements for precise measurements in coaxial systems are that the reflection due to the connector-pair be small and that the connection be electrically repeatable after many connect/disconnect operations. These conditions are well-satisfied by the APC and the GR-900® connectors.[4] They are known as *sexless* connectors since both halves of the mated pair are identical. A sketch of the APC-7 type is shown in Fig. 6–14. The two halves are mated together by a threading arrangement (not shown). The dimensional tolerances are such that the B-B' surfaces touch first. The spring mechanisms behind B and B' cause them to

---

[4] The development of the APC-type connector was a result of the combined efforts of the Hewlett-Packard Co. and the Amphenol Corp. The GR-900 connector was developed by the GenRad Corp.

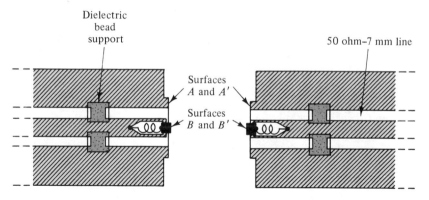

Figure 6–14   The APC-7 connector-pair for use with 50 ohm–7mm lines. (Threaded coupling arrangement is not shown.)

retract until a firm contact is made between the A and A' surfaces. Thus both inner and outer conductor surfaces are in firm contact simultaneously. This design avoids the center conductor discontinuity inherent in the Type-N and SMA connector-pairs. The low-SWR bead support shown in the figure is usually made of rexolite. Its design is similar in principle to that described in Fig. 6–8d.

The dimensions of the coaxial line associated with the APC-7 connector system is identical to that of the Type-N connector. The outer and inner conductor diameters are 0.276 inch (7 mm) and 0.120 inch (3 mm), respectively. Both types can be used up to about 18 GHz without higher-mode interference. Some workers, however, have noted resonances associated with TE_{11} propagation in the bead supports. These have generally been observed in the vicinity of 17 GHz. The SWR associated with an APC-7 connector-pair is typically less than 1.04 up to 18 GHz. The SWR value of the mated pair is repeatable to within ± 0.002 even after hundreds of connect/disconnect operations. This makes it an excellent connector for precise microwave measurement systems.

6–2c TEM to TEM Transitions

It is often necessary to affect a transition from a coaxial line to another TEM line. An important design consideration is that the reflections associated with the transition be minimized.

Coaxial to coaxial transitions.   On occasion, a low-SWR transition is required between coaxial lines having the same characteristic impedance but different dimensions. The situation is shown in Fig. 6–15a. Assuming that the dielectric material is the same throughout, the ratio of outer to inner diameters must be the same for both lines. With the characteristic impedance of the two lines equal, one might conclude that no reflection occurs at the junction. This is not the case! The discontinuity due to the change in diameters creates a shunt capacitive effect at the junction which causes a reflection. In fact, if d_1 is equal to or greater than D_2, a short circuit occurs at the junction and the SWR becomes infinite. Part b of the figure describes a method for avoiding this possibility while providing a low SWR over a wide range of

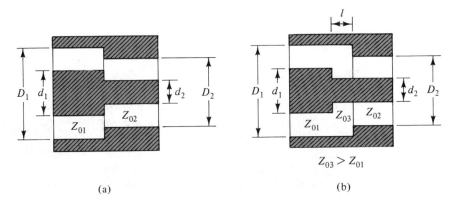

Figure 6–15 Transition between coaxial lines having different diameters but the same characteristic impedance. (With $D_1/d_1 = D_2/d_2$, $Z_{01} = Z_{02}$.)

frequencies. With the smaller center conductor extending into the larger coaxial line, a high-impedance line section is created since $D_1/d_2 > D_1/d_1$. Its behavior approximates that of a series inductance when $l \ll \lambda$. The discontinuous change in both the center and outer conductors may be represented electrically by shunt capacitances. Thus an approximate equivalent circuit for this transition is two shunt capacitors separated by a series inductor. The SWR of the transition can be minimized by choosing the inductance L so that

$$L = Z_{01}^2 (C_1 + C_2) \qquad (6\text{–}5)$$

where C_1 and C_2 are the shunt capacitances due to the step in inner and outer conductors. The required inductance value is obtained by proper selection of the line length l.

Coaxial to stripline transitions.   Coaxial equipment is often used to measure the electrical performance of stripline components and systems. This requires the use of low-SWR transitions between coaxial and strip-type transmission lines having the same characteristic impedance. Two such transitions are illustrated in Fig. 6–16. The one shown in part *a* represents a transition between a coaxial line and a symmetrical stripline. With the strip width chosen so that $Z_{01} = Z_0$, the only reflection will be due to the sudden change in conductor configuration at the junction. This discontinuity may be approximated by a shunt capacitance. As before, a series inductance may be used to compensate for this capacitive effect. A small section of the high-impedance line (Z_{02}) provides the required inductance. The high-impedance line is created by reducing the strip width near the junction. This arrangement provides a low-SWR transition over a wide frequency range.

Part *b* of Fig. 6–16 describes a typical transition between coaxial and microstrip lines. Since the electric field of the TEM mode is symmetrical in the coaxial line and asymmetrical in the microstrip, a mechanism is needed to effect the mode conversion with a minimum of reflections. The small dielectric piece performs this function. Its dimensions are adjusted experimentally for minimum SWR over the frequency band of interest. In some situations, the dielectric piece is not

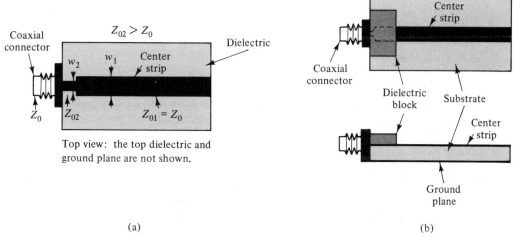

Figure 6–16   Typical transitions from a coaxial line to (*a*) stripline and (*b*) microstrip.

required. One such case is the transition between a SMA coaxial connector and an alumina microstrip line.

Many other ingenious configurations having low SWR over wide frequency ranges have been developed by practicing microwave engineers (see, for example, Chapter 2 of Ref. 6–1).

Baluns.   A *balun* is a device that provides a low-SWR transition between a *bal*anced line and an *un*balanced one. In some cases, a change in impedance level is also involved. A balanced line is defined as one in which the voltage to ground of the two conductors are equal and opposite. The two-wire line discussed in Sec. 5–1 falls into this category. In an unbalanced line, one of the two conductors is at ground potential. Coaxial, stripline, and microstrip configurations are examples of unbalanced lines.

One form of a balun is shown in Fig. 6–17a. The turns ratio of the transformer is chosen to provide a low SWR at both ports. For example, if the balanced line is

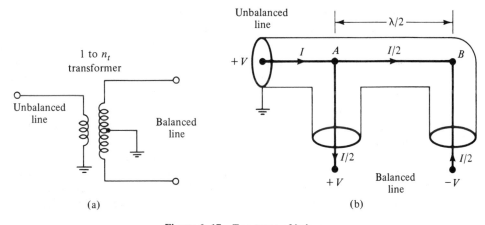

Figure 6–17   Two types of baluns.

300 ohm TV twin-lead <u>and the</u> unbalanced one is a 75 ohm coaxial line, the turns ratio required is $n_t = \sqrt{300/75} = 2$. This type of balun is fairly popular at TV and radio frequencies. Excessive dissipation and parasitic reactance effects, however, make it unsuitable for use at microwave frequencies.

There are many forms of microwave baluns. A narrow-band version is illustrated in Fig. 6–17b. Its operation depends upon the fact that both the voltage and current waves are phase delayed 180° by the half-wave line section. Since impedance repeats every half wavelength, the current I splits equally at junction A. Due to the 180° phase delay, the current direction at point B is opposite that at point A. Similarly the voltage polarity at point B is opposite that at point A. Thus the voltage and current of the unbalanced line are transformed to those required for balanced line transmission. Note that this type balun produces a four-to-one impedance transformation since the voltage on the balanced line is $2V$ and the current is $I/2$. Since the line length between A and B can only be a half wavelength at one frequency, this balun is only effective over a narrow frequency range. A more complete discussion of baluns, their use and limitations, may be found in Refs. 6–2 and 6–17.

6–3 ATTENUATORS AND PHASE SHIFTERS

The ability to control the amplitude and phase of a signal is a common requirement in microwave equipment. This section describes some of the attenuators and phase shifters used in coaxial and stripline systems.

6–3a Coaxial and Stripline Attenuators

Attenuators are used to decrease the power level of a microwave signal. Highly precise versions are utilized in the accurate measurement of power and insertion loss.

In general, attenuators may be classified as either dissipative or reflective. Both types are available with either fixed or variable values of attenuation. Some of these are described here.

The matched Tee attenuator. The Tee-type resistive network is the most widely used form of dissipative attenuator. A coaxial version is shown in Fig. 6–18a. It consists of two series resistors of the type described in Fig. 6–1a and a shunt resistor formed by depositing a thin film of resistive material on a ceramic disc. The center of the disc resistor (R_2) makes ohmic contact with the two series resistors (R_1) while its circumference contacts the outer conductor of the coaxial line. If the dimensions of the resistors are small compared to the operating wavelength and parasitic capacitive effects are neglected, the equivalent circuit reduces to that given in Fig. 6–18b.

When the attenuator is inserted between a generator having an internal impedance Z_G and a load of impedance Z_L, the power level at the load is reduced. The ABCD matrix method and Eq. (4–43) may be used to calculate the insertion loss. For the usual situation wherein $Z_G = Z_L = Z_0$, a real impedance, the insertion

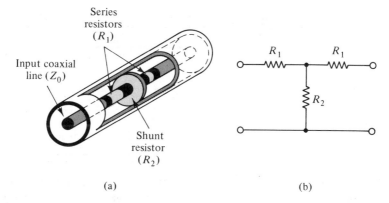

Figure 6–18   A coaxial Tee attenuator (cutaway view) and its equivalent circuit.

loss is given by Eq. (4–45). Defining $\bar{R}_1 \equiv R_1/Z_0$ and $\bar{R}_2 \equiv R_2/Z_0$, the normalized ABCD matrix of the Tee attenuator becomes

$$\begin{bmatrix} \bar{A} & \bar{B} \\ \bar{C} & \bar{D} \end{bmatrix} = \begin{bmatrix} 1 & \bar{R}_1 \\ 0 & 1 \end{bmatrix} \begin{bmatrix} 1 & 0 \\ \dfrac{1}{\bar{R}_2} & 1 \end{bmatrix} \begin{bmatrix} 1 & \bar{R}_1 \\ 0 & 1 \end{bmatrix} = \begin{bmatrix} 1 + \dfrac{\bar{R}_1}{\bar{R}_2} & 2\bar{R}_1 + \dfrac{\bar{R}_1^2}{\bar{R}_2} \\ \dfrac{1}{\bar{R}_2} & 1 + \dfrac{\bar{R}_1}{\bar{R}_2} \end{bmatrix} \tag{6–6}$$

In order for the loss to be completely dissipative, $\Gamma_{in} = 0$ or $\bar{Z}_{in} = 1$. From Eq. (4–31), this requires that $\bar{A} + \bar{B} = \bar{C} + \bar{D}$. Substituting the element values from Eq. (6–6) into this equation yields the condition required for a *matched* Tee attenuator. Namely,

$$\bar{R}_2 = \frac{1 - \bar{R}_1^2}{2\bar{R}_1} \tag{6–7}$$

With $\bar{A} + \bar{B} = \bar{C} + \bar{D}$, Eq. (4–45) reduces to

$$L_I = L_T = 10 \log |\bar{C} + \bar{D}|^2 \tag{6–8}$$

Using Eq. (6–7) and the element values from Eq. (6–6) results in the following loss equation for the matched Tee attenuator.

$$L_I = L_T = 20 \log \left[\frac{1 + \bar{R}_1}{1 - \bar{R}_1} \right] \tag{6–9}$$

Defining a voltage loss ratio $V_R \equiv$ antilog $(L_I/20)$, Eq. (6–9) may be restated as

$$\bar{R}_1 = \frac{V_R - 1}{V_R + 1} \tag{6–10}$$

Equations (6–10) and (6–7) represent the design relations for the matched Tee attenuator when $Z_G = Z_L = Z_0$, a real impedance. Table 6–1 gives the values of $\bar{R}_1$ and

TABLE 6–1 Normalized resistor values for the matched Tee attenuator

L_l (dB)	V_R	$\bar{R}_1$	$\bar{R}_2$
0	1	0	∞
3	1.41	0.17	2.86
6	2	0.33	1.33
10	3.16	0.52	0.70
20	10	0.82	0.20
30	31.6	0.94	0.06
40	100	0.98	0.02

$\bar{R}_2$ for various values of loss or attenuation.[5] Since a low-resistance value is difficult to reproduce accurately, the manufacture of large-value precision attenuators poses a problem. To circumvent this difficulty, their internal construction usually consists of two or more lower loss units (typically, 10 dB or less) connected in tandem.

The lossy-line attenuator.   At the higher microwave frequencies, the performance of the Tee attenuator deteriorates quite rapidly since the size of the resistors are no longer small compared to the operating wavelength. The distributed lossy-line attenuator complements the Tee attenuator in that it has superior performance at the higher frequencies. Standard coaxial versions are available with low SWR and flat attenuation versus frequency response from 2 to 18 GHz. A sketch of this type attenuator is given in Fig. 6–19. The center conductor of the air-insulated coaxial line consists of a thin resistive film deposited on a hollow dielectric rod. Its attenuation may be calculated with the aid of Eq. (3–15). Since $G' = 0$,

$$\gamma = \alpha + j\beta = j\omega\sqrt{L'C'}\sqrt{1 - jR'/\omega L'} \tag{6–11}$$

where R' is the resistance per unit length of the thin film. By choosing $R' \ll \omega L'$, the attenuation constant in Eq. (6–11) may be approximated by

$$\alpha \approx \frac{R'}{2}\sqrt{C'/L'} = R'/2Z_0 \tag{6–12}$$

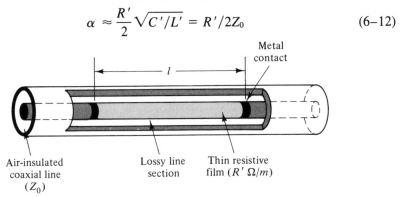

Figure 6–19   A lossy-line attenuator (cutaway view).

[5] The characteristic impedance of the matched Tee attenuator network is Z_0 since with $Z_L = Z_0$, $Z_{in} = Z_0$. Therefore, the above loss values also represent the attenuation A_l (as defined in Sec. 4–3a) of the attenuator.

where use has been made of Eq. (3–21). The total attenuation $A_t = \alpha l$, where l is the length of the resistive element. Therefore,

$$A_t \text{ (Np)} = \frac{R'l}{2Z_0} = \frac{R}{2Z_0} \qquad \text{or} \qquad A_t \text{ (dB)} = 4.34\frac{R}{Z_0} \qquad (6\text{--}13)$$

If the film thickness is made small compared to its skin depth at the highest frequency of interest, the value of R and hence A_t will be independent of frequency as long as $R' \ll \omega L'$. This inequality may be rewritten (Prob. 6–11) as

$$\lambda_0 \ll 2\pi Z_0/R' \qquad (6\text{--}14)$$

A practical lower-frequency limit for the attenuator may be defined as the frequency at which $\lambda_0 = \pi Z_0/R'$.

With $R' \ll \omega L'$ and $G' = 0$, the characteristic impedance of the lossy line is approximately $\sqrt{L'/C'}$, which is the same as the characteristic impedance of the lossless coaxial lines on either side of the attenuator section. Thus, given the condition of Eq. (6–14), the distributed lossy-line attenuator is a matched attenuator.

A lossy-line attenuator that incorporates both series and shunt distributed resistance has also been developed. It is described by S. F. Adam (Ref. 6–3).

One application of a matched attenuator is to reduce reflections associated with a mismatched load or generator. The following example describes the technique.

Example 6–2:

A resistive microwave load with $Z_L = 150$ ohms is connected to a 50 ohm coaxial line.
(a) What is the SWR along the coaxial line?
(b) Calculate the SWR if a 7 dB matched attenuator is connected between the line and the load.

Solution:
(a) For a purely resistive load, Eq. (3–55) may be used to compute the SWR. Thus,

$$\text{SWR} = 150/50 = 3.0 \qquad \text{and} \qquad |\Gamma_L| = (3 - 1)/(3 + 1) = 0.5.$$

(b) Since the attenuator is matched, the return-loss concept may be used to solve this part of the problem. Using Eqs. (3–77) and (3–78),

$$L_{R_{in}} = 10 \log \frac{1}{|0.5|^2} + 2(7) = 20 \text{ dB}$$

By applying Eq. (3–77) to the input quantities,

$$20 = 10 \log \frac{1}{|\Gamma_{in}|^2} \qquad \text{or} \qquad |\Gamma_{in}| = 0.10$$

This corresponds to an SWR of 1.22.

Thus, at the expense of a 7 dB power loss, the input SWR is reduced from 3.0 to 1.22. The reduction in SWR at the expense of decreased power is an acceptable tradeoff in most measurement applications.

The cutoff attenuator.   The cutoff attenuator is widely used as a precision variable attenuator in coaxial measurement equipment. It incorporates a below-cut-

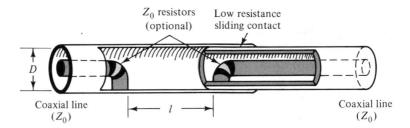

Figure 6–20   A cutoff-type variable attenuator (cutaway view). The attenuator can be matched by incorporating Z_0 resistors as shown.

off section of circular waveguide to obtain the desired attenuation. Since the waveguide is essentially dissipationless, the cutoff attenuator reflects rather than absorbs the microwave signal. For greater than 10 dB attenuation, the SWR at both ports is practically infinite (>30). In certain applications, this can be troublesome.

A cutoff-type variable attenuator is pictured in Fig. 6–20. The input coupling loop converts the incoming TEM wave into the TE_{11} mode in circular guide, while the output loop converts the attenuated TE_{11} mode back to TEM. The attenuator can be matched by incorporating Z_0 resistors as shown. The sliding cylindrical conductors allow the length l to be varied, which in turn varies the attenuation. The TE_{11} cutoff wavelength is given in Fig. 5–27 and is repeated here.

$$\lambda_c = 1.706D \qquad (6\text{–}15)$$

The attenuation of the below-cutoff waveguide section $A_t = \alpha l$, where α is the attenuation constant due to the cutoff effect [Eq. (5–75)] and l is the length of the air-filled circular guide. Thus,

$$A_t \text{ (dB)} = 54.6 \frac{l}{\lambda_c} \sqrt{1 - \left(\frac{f}{f_c}\right)^2} \qquad (6\text{–}16)$$

By choosing the guide diameter so that $\lambda_c \ll \lambda_0$ at the highest frequency of interest, $f/f_c \ll 1$ and Eq. (6–16) reduces to

$$A_t \text{ (dB)} = 54.6 \frac{l}{\lambda_c} = 32 \frac{l}{D}$$

or $\qquad (6\text{–}17)$

$$\Delta A_t \text{ (dB)} = \frac{32}{D} \Delta l$$

All three expressions for attenuation are accurate when the coupling loops are separated by at least $2D$. This is because the analysis is based on the assumption that only the TE_{11} mode generated by the input loop is coupled to the output. However, higher-order waveguide modes generated at the input are also coupled to the output. Fortunately, they decay more rapidly with distance than the TE_{11} mode, which means that Eqs. (6–16) and (6–17) are accurate when the loops are well separated. Montgomery (Ref. 6–5) discusses this point in greater detail.

Equation (6–17) indicates the major advantages of the cutoff attenuator. By choosing the guide diameter small enough, the attenuation is independent of frequency. Its value is a function of mechanical dimensions only and hence can be precisely controlled by maintaining tight tolerances on the length and diameter of the circular guide section. With ΔA_t linearly proportional to Δl, the cutoff attenuator is easily calibrated and hense particularly useful as a precision variable attenuator.

Matched variable attenuators. It is difficult to construct wideband absorptive-type variable attenuators in coaxial and stripline systems. Despite the problems, some have been successfully developed and are commercially available. Two types that make use of the lossy-line and Tee-attenuator principles discussed earlier are described in Sec. 6.1 of Ref. 6–3. A matched variable attenuator that combines a below-cutoff waveguide and a directional coupler has also been developed. It is described in Sec. 8.6 of Ref. 6–4.

A stripline version of an absorptive-type variable attenuator is shown in Fig. 6–21. It utilizes a lossy dielectric material to dissipate the microwave signal. The TEM electric field in stripline is concentrated in the vicinity of the center strip (see Fig. 5–7). When the absorbing material is situated outside this region, the attenuation is negligible. As the material is inserted into the high-field region, a portion of the TEM wave is intercepted and absorbed by the lossy dielectric, and hence the attenuation increases. Since the characteristic impedance of the stripline changes as the dielectric material is inserted, the SWR tends to increase as the attenuation increases. To minimize this effect, the two ends of the lossy material are tapered to provide a smooth impedance transformation into and out of the lossy section. SWR values of less than 1.50 are possible over a limited frequency band. In general, the SWR deteriorates at the lower frequencies. Another disadvantage of this device is that for a fixed setting the attenuation increases with increasing frequency. This makes calibration of the attenuator cumbersome. Compensation techniques are occasionally used to reduce this variation with frequency.

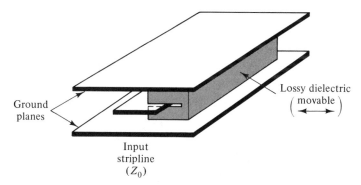

Figure 6–21   A stripline variable attenuator.

6–3b Coaxial and Stripline Phase Shifters

A phase shifter is a two-port network that is used to change the phase of the output signal. The phase delay ϕ due to a length of transmission line is equal to βl. Therefore, for coaxial and stripline transmission in the TEM mode

$$\phi = \frac{\omega}{v} l = \frac{\omega}{c} l \sqrt{\mu_R \epsilon_R} \qquad (6\text{–}18)$$

where ϕ is in radians. For a fixed frequency, ϕ can be varied by either changing the length l or the properties of the insulating space. The first method is used in the two phase shifters described in Fig. 6–22. In both cases, the lines are air-insulated. The one in part a is known as a *line stretcher*. An increase in length Δl causes an increase in phase delay at the ouput $\Delta\phi = \beta\Delta l$. This device requires low-resistance sliding contacts for both the inner and outer conductors of the coaxial line. The main disadvantage of the line stretcher is that a change in phase requires a change in the physical position of the output port. The unit shown in part b of Fig. 6–22 avoids this problem. The coaxial line is configured so that a change in line length does not alter the position of the input and output ports. Note that a change Δx causes a line length change of $2\Delta x$ and hence $\Delta\phi = 2\beta\Delta x$. This device is known as a *trombone-type* phase shifter.

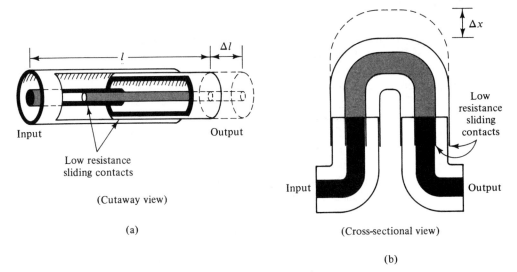

Figure 6–22 Two coaxial phase shifters. (The lines are air-insulated.)

Ideally, for both devices, the phase change is linearly proportional to a change in length. However, mismatches at either port or along the line can introduce a degree of nonlinearity. Problem 6–16 illustrates the magnitude of this effect.

A phase change can also be created by varying the properties of the insulating region between the conductors. One version uses the stripline configuration shown in Fig. 6–21 except that the lossy material is replaced by a low-loss, nonmagnetic dielectric. From Eq. (6–18), one can show that the maximum phase change that can be created by fully inserting the dielectric material into the high-field region is

$$\Delta\phi = 2\pi\left(\sqrt{\epsilon_R} - 1\right) l/\lambda_0 \tag{6–19}$$

where ϵ_R is the relative dielectric constant of the material. An L-band unit with 180° phase change capability has been built having an SWR of less than 1.15 for all phase shift settings (Ref. 6–16).

6–4 TRANSMISSION-LINE DISCONTINUITIES

A transmission-line discontinuity is defined as any interruption in the uniformity of the line. When the discontinuity extends over an appreciable portion of a wavelength in the propagation direction, it must be analyzed using transmission-line theory. Quite often, however, it occurs over a distance that is negligible compared to the operating wavelength. In these cases, its effect can be accounted for by a fairly simple equivalent circuit.

6–4a Coaxial Discontinuities

Most discontinuities involve an abrupt change in either the properties of the insulating region or the dimensions of the conductors.

Dielectric discontinuities.   Figure 6–23 illustrates a coaxial line containing two different dielectric materials. With TEM transmission assumed, the situation is the same as that described in Fig. 2–26. Thus, the boundary conditions at the dielectric discontinuity are completely satisfied by a reflected wave. By referring to the equivalent circuit (Fig. 6–23b), the reflection coefficient at the dielectric interface for a traveling wave on the Z_{01} line is given by

$$\Gamma = \frac{Z_{02} - Z_{01}}{Z_{02} + Z_{01}} \tag{6–20}$$

when the Z_{02} line is match terminated. By using the expression for Z_0 of a coaxial line [Eq. (5–8)] and assuming nonmagnetic dielectrics, the above equation reduces to

$$\Gamma = \frac{\sqrt{\epsilon_{R_1}} - \sqrt{\epsilon_{R_2}}}{\sqrt{\epsilon_{R_1}} + \sqrt{\epsilon_{R_2}}} \tag{6–21}$$

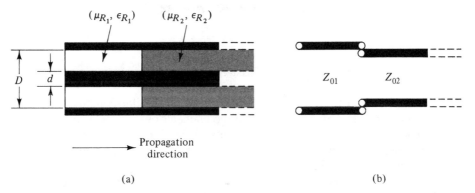

$(\mu_{R_1}, \epsilon_{R_1})$ $(\mu_{R_2}, \epsilon_{R_2})$

D d

Z_{01} Z_{02}

Propagation direction

(a)   (b)

Figure 6–23   A coaxial dielectric discontinuity (cross-sectional view) and its equivalent circuit.

and the corresponding standing-wave ratio becomes

$$\text{SWR} = \sqrt{\frac{\epsilon_{R_2}}{\epsilon_{R_1}}} \quad \text{or} \quad \sqrt{\frac{\epsilon_{R_1}}{\epsilon_{R_2}}} \qquad (6-22)$$

whichever is greater than unity.

It is important to note that if the wave enters along the Z_{02} line and a matched load terminates the Z_{01} line, the reflection coefficient is the negative of that given above, but the SWR remains the same.

A variation of the double-slug tuner discussed in Sec. 4–1d can be realized by utilizing these reflections. A sketch of the device is shown in Fig. 6–24. The two dielectric slugs are movable along the air-insulated coaxial line. Each slug is a quarter wavelength long at the design frequency. The following explanation shows that the tuner is capable of producing a reflection having any SWR value between unity and ϵ_R^2. Individually, the dielectric slugs create an SWR equal to ϵ_R. This is because each air-dielectric boundary generates a reflection having an SWR $= \sqrt{\epsilon_R}$. Since the reflection coefficient at the air-to-dielectric discontinuity is the negative of that at the dielectric-to-air discontinuity, the quarter wavelength spacing causes the two reflections to be additive. From Eq. (4–34), the overall SWR of the slug is therefore $(\sqrt{\epsilon_R})(\sqrt{\epsilon_R}) = \epsilon_R$. With two identical slugs, the spacing l_2 may be adjusted to create any SWR between unity and ϵ_R^2. Here again use has been made of Eq. (4–34). One can show that with $l_2 = 0$ or $\lambda_0/2$, the SWR is unity, while with $l_2 = \lambda_0/4$, the SWR equals ϵ_R^2 (Prob. 6–17). The tuner is thus capable of canceling any load reflection having an SWR less than or equal to ϵ_R^2. The length l_2 is set so that the tuner and load reflections are equal, while l_1 is adjusted so that the two reflections, as seen at the input, are 180° out-of-phase.

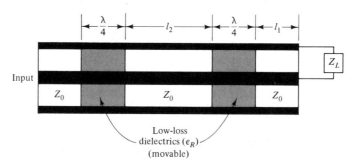

Figure 6–24   A coaxial dielectric-slug tuner (cross-sectional view).

Conductor discontinuities.   Figure 6–25 illustrates two coaxial discontinuities that involve step changes in the diameter of the conductors. The dashed lines indicate the electric field in the vicinity of the discontinuity. The distorted field pattern may be thought of as an infinite sum of higher-order modes superimposed on the coaxial TEM mode. One may say, in fact, that the discontinuity *generates* the higher-order modes. This is analogous to treating a distorted sine wave as a fundamental plus an infinite sum of harmonics. Whatever caused the distortion of the original sinusoid may be considered as the source of the harmonics. Since the distorted electric field pattern has a component in the propagation direction, the higher-

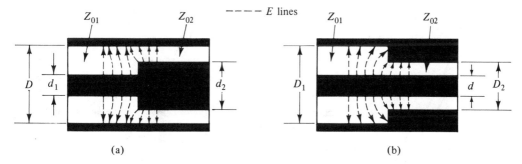

Figure 6–25 Step discontinuities in a coaxial line (cross-sectional view).

order modes are transverse magnetic (TM). Furthermore, these modes must be circularly symmetric since both the incoming TEM wave and the discontinuity itself are circularly symmetric. With the dimensions of the coaxial lines chosen so that only TEM waves can propagate in the frequency range of interest, the TM modes attenuate rapidly as a function of distance from the step discontinuity. Thus, away from the immediate vicinity of the step, the field pattern is that of an undistorted TEM wave.

The localized higher-order modes contain stored energy and hence their effect may be modeled by a reactive network. The two structures in Fig. 6–25 were originally analyzed by Whinnery, Jamieson, and Robbins (Ref. 6–18), and the equivalent circuit was shown to be simply a capacitance (C_d) shunting the coaxial lines. In the article, they stated that the net stored energy for localized TM modes is electric while that of TE modes is magnetic. Consequently, the TM modes may be modeled by an equivalent capacitance and TE modes by an equivalent inductance.[6] Curves of discontinuity capacitance for the two types of coaxial steps are given in Fig. 6–26. They are adapted from the above cited article and assume that both coaxial lines are air-insulated. Data is given for $\tau = 1$ and 5. Capacitance curves for additional τ values may be found in the original source, the *Microwave Engineers' Handbook*, vol. 1 (Ref. 6–7), and in Ref. 6–11.

The curves in Fig. 6–26 are also useful when either or both of the coaxial lines are dielectric filled. If both lines contain the same material, multiply the C_d' scales by ϵ_R to obtain the correct value of discontinuity capacitance. The same procedure applies if only the line with the larger insulating gap (the Z_{01} line, in both cases) contains the dielectric. If, on the other hand, only the line with the smaller gap (the Z_{02} line) is dielectric filled, no correction is needed and the curves may be used as given. Additional coaxial configurations are analyzed and evaluated in the above cited article.

In an earlier discussion, it was explained that a small section of low-impedance line approximates a shunt capacitance [Eq. (4–13)], while a small section of high-impedance line approximates a series inductance [Eq. (4–12)]. These equivalences find use in the design of filters and matching networks. Coaxial forms of these reactances are shown in Fig. 6–27. Assuming negligible interaction between

---

[6] For the reader interested in the mathematical details of the analysis, the original article (Ref. 6–18) as well as Sec. 11.2 of Ref. 6–6 are suggested.

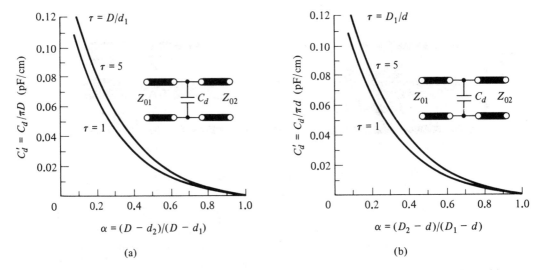

Figure 6–26 Discontinuity capacitance data for the coaxial step discontinuities pictured in Fig. 6–25. (From Whinnery, Jamieson, and Robbins, Ref. 6–18; © 1944 IRE now IEEE, with permission.)

the step discontinuities, the structure in part a may be represented by a shunt capacitance

$$C \approx \frac{l_s}{Z_{0s}v} + 2C_d \qquad (6-23)$$

where C_d is the discontinuity capacitance of each step and v is the wave velocity in the Z_{0s} line.

The equivalent circuit of the structure in part b of Fig. 6–27 is two shunt capacitances C_d separated by a series inductance L. From Eq. (4–12),

$$L \approx \frac{Z_{0s}l_s}{v} \qquad (6-24)$$

where for an air-insulated line, $v = c$. In most cases, the effect of C_d is negligible and hence the equivalent circuit reduces to a series inductance of value L.

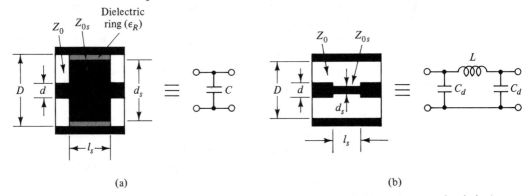

Figure 6–27 Coaxial forms of shunt capacitors and series inductors (cross-sectional view). In both cases, $l_s \ll \lambda$.

6–4b Stripline and Microstrip Discontinuities

The effect of dielectric discontinuities in stripline and microstrip is similar to that in coaxial lines. For striplines containing nonmagnetic insulators, Eqs. (6–21) and (6–22) are valid. In the microstrip case, however, they are not since the relationship between Z_0 and ϵ_R is more complex. A very good approximation for Γ is obtained by using Eq. (6–20) and Z_0 values from Fig. 5–14.

Two conductor-type discontinuities that appear fairly often in stripline circuits are illustrated in Fig. 6–28. Part *a* shows a change in width of the center conductor. In addition to causing a change in characteristic impedance, the abrupt change in strip width creates a reactive effect. Since the current pattern (dashed lines) has a transverse component, it generates a longitudinal component of magnetic field. Thus the localized higher-order modes are transverse electric (TE) which can be modeled as a series inductance. For the case of a gap in the center conductor (Fig. 6–28b), the equivalent circuit is a π network of capacitors. These and other stripline discontinuities have been analyzed by Altschuler and Oliner (Refs. 6–15 and 6–19). Equivalent circuits with element values are tabulated with experimental verification given in some cases.

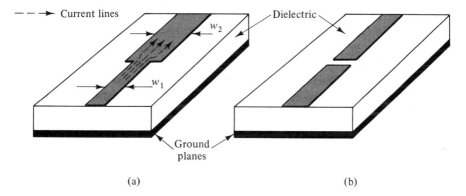

(a)   (b)

Figure 6–28 Conductor discontinuities in stripline. (Top dielectric and ground plane are not shown.)

As in the coaxial case, the equivalent circuit for small sections of low- and high-impedance lines must take into account discontinuity reactances. For example, a small length of high-impedance stripline formed by narrowing the width of the center strip over a length l_s is equivalent to a series inductance of value $L + 2L_d$, where L is given by Eq. (6–24) and L_d is the discontinuity inductance at each step in strip width.

Similar type microstrip discontinuities have been analyzed by many workers (Refs. 6–12, 6–20, 6–21, and 6–22). Equivalent circuits and calculated element values are given for several discontinuities.

One advantage of stripline and microstrip transmission is the ease with which the signal can be redirected by merely changing the direction of the center strip. The right-angle bends in Fig. 6–29 illustrate the point. The radius bend in part *a* maintains a constant strip width throughout. For low reflections, the bend radius R

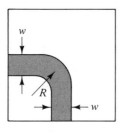

(a) Radius bend

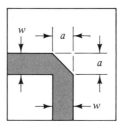

(b) Miter bend

Figure 6–29 Right-angle bends in strip-type transmission lines (top views).

should be greater than three strip widths. The miter bend in part *b* of the figure also presents a discontinuity to the propagating TEM wave. Reflections due to the bend are a function of the dimension *a*. Optimum design dimensions for stripline are given in Chapter 2 of Ref. 6–1.

Some data on right-angle bends in microstrip is found in Ref. 6–23. In addition to reflection effects, bends and other microstrip discontinuities tend to radiate some of the microwave signal. Lewin (Ref. 6–24) has calculated the amount of radiation for various discontinuities.

In summary, this section has described some typical discontinuities in TEM transmission lines. In certain cases, no higher-order modes are created and the reflection is due solely to the abrupt change in characteristic impedance. In other cases, however, higher-order modes are generated at the discontinuity. As long as these modes are nonpropagating, their effects can be represented by a simple reactive circuit. Element values for some of the more common discontinuities are available in the literature. This allows one to calculate the reflection caused by the particular discontinuity.

6–5 DC RETURNS AND BLOCKS

Certain applications require that a transmission line present either a short or open circuit to dc and low-frequency signals while simultaneously providing low-loss transmission at microwave frequencies. This requirement is quite common in semiconductor diode and transistor circuits. Some typical configurations are discussed in this section. Bias-injection circuits are also described.

6–5a DC Returns

In radio and television circuits, the problem of shorting dc voltages and currents to ground without affecting carrier transmission is solved with a shunt inductor (Fig. 6–30a). With a Z_0 source connected to the input and a Z_0 load to the output, the inductance L is selected so that $X_L \gg Z_0$ at the carrier frequencies. Because $X_L = \omega L$, the inductance presents a negligible impedance to dc and low-frequency signals. An insertion loss expression for the shunt inductor may be derived using ABCD matrix methods. The result is

$$L_I = L_T = 10 \log \left[1 + \left(\frac{\bar{B}_L}{2} \right)^2 \right] = 10 \log \left[1 + \left(\frac{Z_0}{2X_L} \right)^2 \right] \tag{6–25}$$

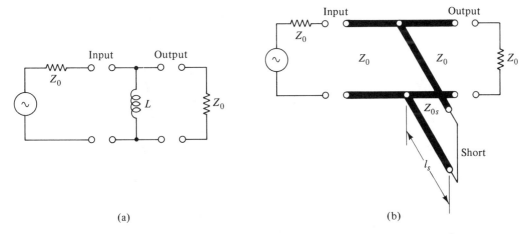

Figure 6–30 Lumped-element and transmission-line dc returns. (Note: $l_s = \lambda/4$ at f_r, the design frequency.)

where $\bar{B}_L = B_L/Y_0 = Z_0/X_L$. Since an inductance does not dissipate power, the above expression represents a mismatch loss. With a matched load at the output, $\bar{Y}_{in} = 1 - j\bar{B}_L$ and therefore the input SWR may be quickly calculated using the Smith chart. For example, if the inductor is chosen so that $\bar{B}_L = 0.4$ at the design frequency, $L_I = 0.2$ dB and the SWR $= 1.50$. At higher frequencies, the loss and SWR values are even less.

At microwave frequencies, it is difficult to realize a lumped inductor using a coil. In recent years, however, tiny coils with self-resonant frequencies above 18 GHz have become available. Typically, they consist of several turns of thin wire wound on a 1 mm diameter, nonmagnetic core and are less than 2 mm in length. Used as a dc return, they are capable of low-SWR performance (typically, less than 1.25) from 1 to 18 GHz. These coils have limited current-carrying capacity and hence are seldom used when the dc return must handle more than, say, 200 mA.

In narrow and medium bandwidth applications, a quarter-wave shorted stub may be used as a dc return. The equivalent circuit representation is given in Fig. 6–30b, where Z_{0s} is the characteristic impedance of the stub line and l_s is its length. Assuming that the source and load impedances equal Z_0, the characteristic impedance of the main line, the insertion loss is given by Eq. (6–25) except that X_L is replaced by $Z_{0s} \tan (2\pi l_s/\lambda)$, the reactance of the shorted stub. Again, the input SWR may be obtained using the Smith chart, where in this case, $\bar{Y}_{in} = 1 - j(Z_0/Z_{0s}) \cot (2\pi l_s/\lambda)$. By choosing $l_s = \lambda/4$ at the design frequency, perfect microwave transmission is achieved (namely, $L_I = 0$ dB and SWR $= 1.00$). At dc and low frequencies, the stub presents a very low impedance as required.

At other than the design frequency, microwave transmission is not quite perfect. The solid curves in Fig. 6–31 show the SWR versus normalized frequency for $Z_{0s} = Z_0$ and $3Z_0$. For larger values of Z_{0s}, the useful bandwidth is even greater. Line impedances in excess of 200 ohms, however, are impractical since the center conductor of the stub line becomes extremely thin.

A technique for further improving the bandwidth is shown in Fig. 6–32. This compensated form of the shorted stub has been analyzed by Muehe (Ref. 6–25).

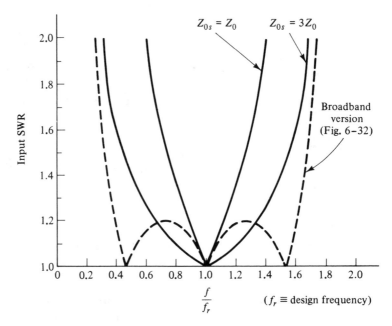

Figure 6–31 SWR versus frequency for a shorted stub shunting a transmission line. The solid curves refer to Fig. 6–30b. The dashed curve refers to Fig. 6–32 with $Z_{0s} = 2.0Z_0$ and $Z_{01} = 0.80Z_0$.

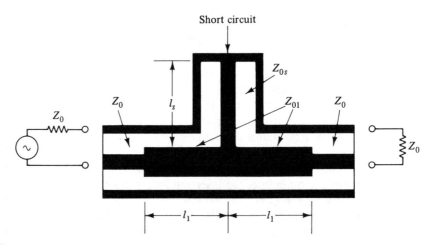

Figure 6–32 Coaxial version of a broadband quarter-wave shorted stub (cross-sectional view). (Note: $l_s = l_1 = \lambda/4$ at the design frequency.)

Conceptually, the configuration may be viewed as two quarter-wave transformers and a shorted stub. At the design frequency, the quarter-wave shorted stub presents an open circuit across the main line, and the combined length of the two Z_{01} lines is $\lambda/2$. Since impedance repeats every half wavelength, $Z_{in} = Z_0$ and hence the input SWR is unity. With $Z_{01} < Z_{0'}$ the transforming action of the Z_{01} lines reduces the generator and the load impedances as seen at the plane of the stub. This has the same effect as increasing Z_{0s} which improves the bandwidth. Furthermore, by proper choice of Z_{01} and $Z_{0s'}$ the frequency sensitivity of the two transformers will tend to cancel that of the shorted stub resulting in even broader band performance. For example, with $Z_{0s} = 2Z_0$ and $Z_{01} = 0.8Z_0$, the SWR is given by the dashed curve in Fig. 6–31. Note the significant improvement in bandwidth for, say, SWR ≤ 1.20.

Coaxial, stripline, and microstrip versions of this type dc return have been designed and incorporated into many microwave systems. Design information is available in Refs. 6–11 and 6–25 and in Chapter 2 of Ref. 6–1. Another type dc return has been described by McDermott and Levy (Ref. 6–27). Its design is based on highpass microwave filter techniques

6–5b DC Blocks

Another circuit function that oftens occurs in communication systems is that of blocking dc and low-frequency voltages while simultaneously allowing the transmission of high-frequency signals. At radio and television frequencies, this is accomplished by inserting a capacitor in series with the line. The circuit is shown in Fig. 6–33a. Its insertion loss may be determined by using the ABCD matrix method. The result is

$$L_I = L_T = 10 \log \left[1 + \left(\frac{\overline{X}_C}{2} \right)^2 \right] \tag{6–26}$$

where $\overline{X}_C \equiv X_C/Z_0 = 1/Z_0 \omega C$. The SWR associated with this mismatch loss is readily obtained using the Smith chart since $\overline{Z}_{in} = 1 - j\overline{X}_C$. A low SWR results when C is chosen large enough to insure that $X_C \ll Z_0$. For example, with $C = 18$ pF and $Z_0 = 50$ ohms, the SWR < 1.20 at frequencies above 1.0 GHz. At dc and low frequencies, the series capacitor presents a very high impedance as required. With the advent of high-Q chip capacitors whose dimensions are small compared to microwave wavelengths, this type of dc block has proven to be useful at microwave frequencies.

A transmission-line form of the dc block is shown in Fig. 6–33b. With $l_s = \lambda/4$ at the design frequency, the input impedance of the open-circuited stub is zero and hence the microwave signal passes unattenuated to the output port. Given a

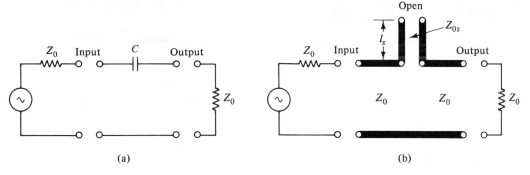

Figure 6–33 Lumped-element and transmission-line dc blocks. (Note: $l_s = \lambda/4$ at the design frequency.)

Z_0 source and Z_0 load, the insertion loss may be determined from Eq. (6–26), where $\bar{X}_C$ is replaced by $(Z_{0s}/Z_0) \cot (2\pi l_s/\lambda)$, the reactance of the stub. Since $\bar{Z}_{in}= 1 - j(Z_{0s}/Z_0) \cot (2\pi l_s/\lambda)$, the SWR is also readily determined. With $l_s = \lambda/4$ at the design frequency, $Z_{in} = 1.0$ and hence the input SWR is unity. The SWR is higher at other frequencies since the $\cot (2\pi l_s/\lambda)$ term is no longer zero. To minimize this effect and hence broadband the microwave performance, Z_{0s} should be chosen as small as possible. Note that if Z_0 and Z_{0s} were interchanged, the expression for $\bar{Z}_{in}$ of the dc block would be the same as $\bar{Y}_{in}$ of the dc return shown in Fig. 6–30b. This means that the solid curves in Fig. 6–31 also describe the frequency response of the dc block. For example, the curve labeled $Z_{0s} = 3Z_0$ represents the SWR versus frequency of a dc block with $Z_{0s} = Z_0/3$.

Coaxial and stripline forms of the transmission-line dc block are shown in Fig. 6–34. In the coaxial version, a hole of radius b_s is drilled into one half of the center conductor. The other half, with a reduced radius a_s, is inserted into the hole. The dielectric sleeve of thickness $(b_s - a_s)$ prevents ohmic contact between the conductors and contributes to the mechanical rigidity of the structure. As indicated in the figure, the open circuit is provided by the circular waveguide that is below cutoff in the frequency range of interest. The dielectric-filled section represents an open-circuited stub in series with the main line. With $Z_{0s} = (138/\sqrt{\epsilon_R}) \log (b_s/a_s)$, its value can be made small by choosing ϵ_R as large as possible (consistent with low $\tan \delta$) and making the thickness of the dielectric sleeve as small as possible (consistent with reasonable dc voltage capacity). The usual practice is to wrap a thin sheet (3 to 5 mils thick) of teflon around the reduced diameter center conductor and force it into the hole drilled in the other half of the conductor. The following illustrative example indicates its performance as a function of frequency.

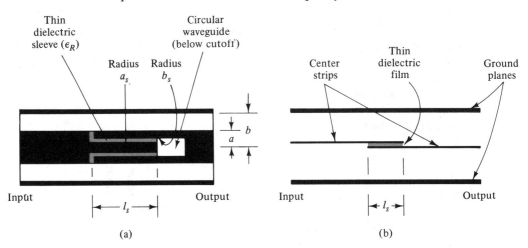

Figure 6–34 Coaxial and stripline versions of the dc block indicated in Fig. 6–33b. (Note: $l_s = \lambda/4$ at the design frequency.)

Example 6–3:

Design a dc block for use in a 7 mm, air-insulated, 50 ohm coaxial system. The design frequency is 3.0 GHz. Calculate its SWR at 2.0 GHz.

Solution: For a 50 ohm line with an outer conductor diameter of 7 mm, $50 = 138 \log(3.5/a)$. Thus the radius of the inner conductor a is 1.5 mm or 0.15 cm. Since the radius of the drilled hole must be smaller, let $b_s = 0.10$ cm. Choosing 4 mil (0.01 cm) teflon as the dielectric sleeve, $a_s = 0.10 - 0.01 = 0.09$ cm. Therefore,

$$Z_{0s} = \frac{138}{\sqrt{2.1}} \log \frac{0.10}{0.09} = 4.4 \text{ ohms}$$

At the design frequency, $\lambda_0 = 10$ cm, $\lambda = 10/\sqrt{2.1} = 6.9$ cm and hence $l_s = 6.9/4 = 1.73$ cm. With the line a quarter wavelength long, the SWR is unity at 3.0 GHz.

At 2.0 GHz, $\lambda_0 = 15$ cm, $\lambda = 10.35$ cm and therefore

$$\bar{Z}_{\text{in}} = 1 - j(4.4/50) \cot(3.46\pi/10.35) = 1 - j0.05.$$

This corresponds to an input SWR of 1.05.

A stripline version of the coaxial dc block is shown in Fig. 6–34b. The overlapping center strips are separated by a thin dielectric sheet and are a quarter wavelength long at the design frequency. Howe (Ref. 6–1) explains that although this configuration is not exactly equivalent to the coaxial version, experimental results show good agreement with calculations based on the equivalent circuit of Fig. 6–33b.

6–5c Bias-Injection Circuits

Microwave semiconductor circuits usually require the injection of a dc control voltage across a transmission line. An important design consideration is that the injection circuit not interfere with microwave transmission along the line. A lumped-element version that performs this function is shown in Fig. 6–35a. Since the inductor L_1 is a short to direct current, a dc voltage applied at point C will appear at the output terminals of the line (point B). With this arrangement, the characteristics of, say, a semiconductor device connected at point B can be controlled by a bias voltage injected at point C.[7] The series capacitor C_1 is used to block the bias voltage from the input line (point A). It also prevents dc voltages at point A from affecting a device connected at the output terminals. The capacitor C_2 is chosen so that its reactance is practically zero at microwave frequencies. Thus the microwave equivalent circuit reduces to a series capacitor (C_1) and a shunt inductor (L_1). At frequencies where X_{C_1} and B_{L_1} are negligible, the input SWR (with a Z_0 load at the output) is practically unity. The circuit is, in fact, a highpass filter and as such provides low-loss transmission at frequencies greater than some minimum value (f_{min}). The design equations for L_1 and C_1 are derived from the following relations.

$$Z_0 = \sqrt{L_1/C_1} \qquad \text{and} \qquad f_{\text{min}} = \frac{1}{\pi\sqrt{L_1 C_1}} \qquad (6\text{–}27)$$

where for $f \geq f_{\text{min}}$, the input SWR ≤ 1.25.

Microwave versions of this bias-injection circuit have been successfully built using the tiny coils and chip capacitors mentioned earlier.

---

[7] This circuit can also be used to monitor voltages. The existence of dc and low-frequency voltages at point B can be detected and measured by a meter connected at point C.

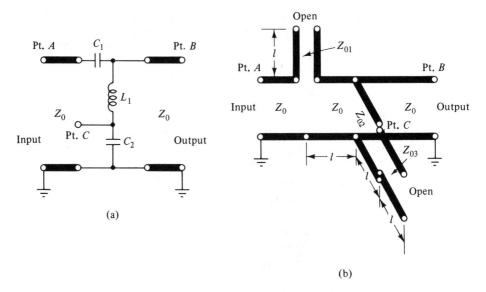

Figure 6–35 Lumped-element and transmission-line type bias-injection circuits. (Note: $l = \lambda/4$ at f_r, the design frequency.)

A circuit that utilizes transmission-line stubs to satisfy the bias-injection requirement is shown in Fig. 6–35b. As before a dc voltage at point C will appear at point B but not at point A since the quarter-wave open-circuited Z_{01} line acts as a dc block. The quarter-wave open-circuited Z_{03} line produces a short circuit at point C, thus performing the same function as the bypass capacitor (C_2) in Fig. 6–35a. By choosing $Z_{03} \ll Z_0$, point C remains a virtual short over a broad range of microwave frequencies. As a result, the Z_{02} line may be considered as a shorted stub shunting the main line. Since it also is a quarter-wavelength long, its input admittance is zero which means that microwave transmission along the main line is unaffected by its presence. Thus the Z_{02} line performs the same function as the inductor (L_1) in Fig. 6–35a. With the series and shunt stubs spaced $\lambda/4$ apart on the main line, their reflections tend to cancel over a wide frequency range. This bias-injection circuit has been analyzed by Mouw (Ref. 6–26) and shown to produce a low SWR over a ten-to-one frequency band. The design procedure is to choose Z_{02} as large as possible and Z_{01} in accordance with the relation $Z_0^2 = Z_{01}Z_{02}$. This relation causes the reflection magnitudes of the series and shunt stubs to be the same, which helps produce the broadband performance. For example, with $Z_{02} = 4Z_0$ and $Z_{01} = Z_0/4$, the SWR is less than 1.35 over a ten-to-one frequency range. Design curves and SWR characteristics are given in Ref. 6–26 and Chapter 2 of Ref. 6–1. The assumption is made that point C is a short circuit at all frequencies of interest.

6–6 LOWPASS FILTERS

The procedures for the design of lumped-element electric filters is well established. At microwave frequencies, it is often difficult to realize the high-Q inductors and capacitors needed to construct such filters. One way of circumventing this difficulty is to use small lengths of low-loss transmission lines as reactive elements. Their properties have already been discussed in Secs. 3–6a and 4–1b. This section describes how these properties can be used to obtain the transmission-line equivalent of lumped-element filters. The focus will be on the design of lowpass types. The discussion of bandpass and bandstop filters will be postponed until the subject of microwave resonators (Chapter 9) has been covered.

Lumped-element filters. T and π forms of the constant-k lowpass filter are shown in Fig. 6–36, where L_1 represents the total series inductance and C_2 the total shunt capacitance. The characteristics of the filter may be found in many texts (for example, Refs. 6–8, 6–9, and 6–10) and are summarized here.

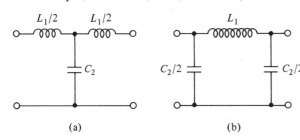

(a)   (b)

Figure 6–36 T and π versions of the constant-K lowpass filter.

At frequencies below the cutoff frequency f_c, signals pass through the filter unattenuated, while above f_c they are attenuated in accordance with the following equation.

$$A_t \text{ (Np)} = 2 \cosh^{-1}\left(\frac{f}{f_c}\right) \tag{6–28}$$

where

$$f_c = \frac{1}{\pi\sqrt{L_1 C_2}} \tag{6–29}$$

As explained in Sec. 4–3a, the attenuation A_t of a two-port is defined in terms of input and output quantities when the output is *terminated in the characteristic impedance of the network*. The characteristic impedance of the T and π lowpass filters is given by

$$Z_T = R_k \sqrt{1 - \left(\frac{f}{f_c}\right)^2} \quad \text{and} \quad Z_\pi = \frac{R_K}{\sqrt{1 - \left(\frac{f}{f_c}\right)^2}} \tag{6–30}$$

where

$$R_k \equiv \sqrt{\frac{L_1}{C_2}}$$

R_k is known as the *low frequency* characteristic impedance since both Z_T and Z_π approach it in value when $f \ll f_c$.[8]

---

[8] Most texts refer to Z_T and Z_π as image impedances. For symmetrical networks, image impedance is the same as characteristic impedance.

Designing a filter on the basis of the above equations is usually referred to as the *image-impedance* method. The reason is that Eq. (6–28) requires that the load impedance equal Z_T (or Z_π). This condition is unrealistic since the load impedance would have to be real in the passband and imaginary in the stopband. In most practical situations, the load impedance is essentially resistive over the frequency range of interest. This is particularly true for well-designed microwave systems wherein $Z_G = Z_L = Z_0$, the characteristic impedance of the connecting lines. Therefore a more useful design procedure is to define the loss characteristic of the filter in terms of insertion loss. With $Z_G = Z_L = Z_0$ (real), the insertion loss may be derived with the aid of Eq. (4–45). Note that under these conditions, insertion loss equals transducer loss. Since the networks in Fig. 6–36 are symmetrical, reciprocal, and dissipationless, the following conditions hold for the ABCD matrix elements.

1. $A = D$.   **2.** $AD - BC = 1$.
3. A and D are real.   **4.** B and C are imaginary.

Substitution into Eq. (4–45) yields the following useful equation.

$$L_I = L_T = 10 \ \log \left[1 + \left(\frac{\bar{B}}{2j} - \frac{\bar{C}}{2j} \right)^2 \right] \tag{6–31}$$

The overall ABCD matrix of the T and π filters may be obtained using the methods described in Appendix C.
For the T-type lowpass filter,

$$\bar{B} = j \frac{\omega L_1}{Z_0} \left(1 - \frac{\omega^2 L_1 C_2}{4} \right) \qquad \text{and} \qquad \bar{C} = j Z_0 \omega C_2$$

After substitution into Eq. (6–31) and considerable algebraic manipulation,

$$L_I = L_T = 10 \ \log \left[1 + \left(\frac{Z_T}{Z_0} - \frac{Z_0}{Z_T} \right)^2 \left\{ \left(\frac{f}{f_c} \right)^2 - \left(\frac{f}{f_c} \right)^4 \right\} \right] \tag{6–32}$$

where f_c and Z_T are defined in Eqs. (6–29) and (6–30).
For the π-type lowpass filter,

$$\bar{B} = j \frac{\omega L_1}{Z_0} \qquad \text{and} \qquad \bar{C} = j Z_0 \omega C_2 \left(1 - \frac{\omega^2 L_1 C_2}{4} \right)$$

A similar derivation shows that the insertion loss expression for the π filter is exactly the same as Eq. (6–32) except that Z_T is replaced by Z_π.
In some designs R_k is set equal to Z_0. For this case, the loss for both type filters reduces to

$$L_I = L_T = 10 \ \log \left[1 + \left(\frac{f}{f_c} \right)^6 \right] \tag{6–33}$$

A plot of loss versus normalized frequency is given by the solid curve in Fig. 6–37. Note that the 3 dB loss point occurs at $f = f_c$. In the stopband ($f > f_c$), the loss increases fairly rapidly with frequency. For $f/f_c > 1.5$,

$$L_I = L_T \approx 10 \ \log \left(\frac{f}{f_c} \right)^6 \tag{6–34}$$

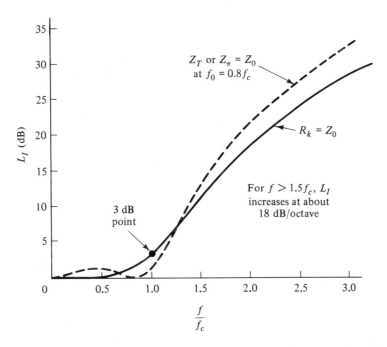

Figure 6–37 Loss characteristics of lowpass constant-K filters. For the solid curve, $R_k = Z_0$. For the dashed curve, $f_0 = 0.8f_c$, which is equivalent to $R_k = Z_0/0.6$ for a T filter and $R_k = 0.6Z_0$ for a π filter.

which means that the loss rises at a rate of 18 dB per octave. In the passband ($f < f_c$), the loss decreases gradually to zero as frequency decreases. Keep in mind that the filters are dissipationless and hence the insertion loss represents *mismatch* loss. Thus by equating the loss value to the second term of Eq. (4–53), one may calculate the corresponding value of $|\Gamma_{in}|$ and hence the input SWR (assuming $Z_L = Z_0$). For example, if $L_I = 3$ dB, $|\Gamma_{in}| = 0.707$ and SWR = 5.83. For losses in excess of 10 dB, $|\Gamma_{in}| \approx 1$, and the input SWR is practically infinite.

In microwave applications, one is usually interested in low loss in a band of frequencies slightly below f_c rather than at frequencies near zero. Equation (6–32) indicates that a zero-loss point occurs at the frequency where $Z_T = Z_0$ (or $Z_\pi = Z_0$). Denoting the zero-loss frequency as f_0 and making use of Eq. (6–30) yields

$$\sqrt{1 - \left(\frac{f_0}{f_c}\right)^2} \quad \begin{array}{ll} = Z_0/R_k & \text{for the } T \text{ filter} \\ = R_k/Z_0 & \text{for the } \pi \text{ filter} \end{array} \qquad (6\text{–}35)$$

Thus, given Z_0 and the cutoff frequency of the filter, R_k may be chosen to set f_0 as desired. The dashed curve in Fig. 6–37 indicates the loss versus frequency when $f_0 = 0.8f_c$. This is equivalent to $R_k = Z_0/0.6$ for the T filter and $R_k = 0.6Z_0$ for the π filter. Note that in this case the insertion loss at the cutoff frequency is less than 3.0 dB. Its value may be determined from

$$L_I = L_T = 10 \log\left[2 - \left(\frac{f_0}{f_c}\right)^2\right] \qquad \text{at } f = f_c \qquad (6\text{–}36)$$

In the stopband, the loss still increases at about 18 dB per octave.

Transmission-line filters. Due to the difficulty in realizing high-Q lumped reactances, microwave versions of the lowpass filter are usually constructed using sections of low-loss transmission lines. One approach utilizes shorted line sections less than a quarter wavelength long as series inductors and open-circuited line sections as shunt capacitors [Eqs. (3–88) and (3–92)]. In coaxial and stripline systems, the implementation of series-connected shorted stubs is mechanically awkward. A method that circumvents this difficulty and results in a more cost-effective design, employs small sections of low- and high-impedance lines in tandem. As explained in Sec. 4–1b, a small section of high-impedance line approximates a series inductance of value $Z_0 l / v$ [Eq. (4–12)], while a small section of low-impedance line approximates a shunt capacitance of value $l / Z_0 v$ [Eq. (4–13)]. These approximations are quite accurate for $l < \lambda/12$ and are even useful up to $l = \lambda/8$. The following example illustrates the design procedure.

Example 6–4:

Design a T-type lowpass filter with a cutoff frequency of 1.5 GHz and a zero-loss point at 1.2 GHz. The filter is to be part of a 50 ohm coaxial system in which all signal frequencies of interest are below 5.0 GHz. Calculate the insertion loss at 1.0, 1.5, 2.5, and 5.0 GHz.

Solution: A sketch of the proposed T filter is shown in Fig. 6–38. The air-insulated, high-impedance lines (Z_{01}) provide the series inductances $(L_1/2)$, while the teflon-filled, low-impedance line (Z_{02}) provides the shunt capacitance (C_2). Air-insulated 50 ohm connecting lines are also included.

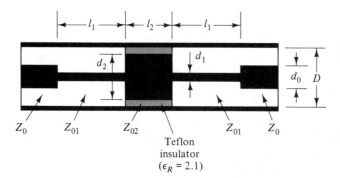

Teflon
insulator
$(\epsilon_R = 2.1)$

Figure 6–38 A coaxial version of the T-type lowpass filter (cross-sectional view).

The first step in the procedure is to design the lumped-element prototype for $Z_0 = 50$ ohms, $f_c = 1.5$ GHz, and $f_0 = 1.2$ GHz. From Eqs. (6–29) and (6–35),

$$\frac{1}{\pi \sqrt{L_1 C_2}} = 1.5 \times 10^9 \text{ Hz} \quad \text{and} \quad R_k \equiv \sqrt{L_1/C_2} = \frac{50}{\sqrt{1 - (0.8)^2}} = 83.3 \text{ ohms}$$

Therefore,

$$L_1 = 17.6 \text{ nH}, \quad L_1/2 = 8.8 \text{ nH} \quad \text{and} \quad C_2 = 2.55 \text{ pF}.$$

The next step is to design the transmission-line equivalent using coaxial sections. Since no signals exist above 5.0 GHz, the maximum frequency of interest $f_{\max}$ is set equal to

5.0 GHz. The outer conductor diameter D is chosen to insure only TEM mode propagation up to f_{max}. By applying Eq. (5–19) to the teflon-filled line,

$$D = 2b = \frac{2(0.95)(3 \times 10^{10})}{2\pi(5.0 \times 10^9)(\sqrt{2.1})} = 1.25 \text{ cm}$$

Since the teflon-filled section has the lowest TE_{11} cutoff frequency, this value of D prevents higher-mode propagation in all the lines.

Next, the line lengths are set equal to $\lambda/4$ at f_{max}, which insures that they are less than $\lambda/4$ at all frequencies of interest. At 5.0 GHz, $\lambda_0 = 6.0$ cm and therefore

$$l_1 = \frac{6.00}{4} = 1.50 \text{ cm} \qquad \text{and} \qquad l_2 = \frac{1.50}{\sqrt{2.1}} = 1.04 \text{ cm}$$

With l_1 and l_2 much less than $\lambda/4$ for $f \le f_c$, Eqs. (4–12) and (4–13) may be used to design the transmission-line equivalents of $L_1/2$ and C_2. Thus,

$$8.8 \times 10^{-9} \text{ H} = \frac{Z_{01}(1.50)}{3 \times 10^{10}} \qquad \text{and} \qquad 2.55 \times 10^{-12} \text{ F} = \frac{(1.04)\sqrt{2.1}}{Z_{02}(3 \times 10^{10})}$$

or

$$Z_{01} = 176 \text{ ohms} \qquad \text{and} \qquad Z_{02} = 19.7 \text{ ohms}$$

The inner-conductor diameters of the air- and teflon-filled lines may now be calculated using the characteristic impedance formula for coaxial lines [Eq. (5–8)]. This yields

$$d_0 = 0.54 \text{ cm}, \qquad d_1 = 0.066 \text{ cm}, \qquad \text{and} \qquad d_2 = 0.78 \text{ cm}$$

This completes the filter design since all dimensions in Fig. 6–38 have been determined. In practice, the length l_2 is shortened somewhat to compensate for the discontinuity capacitance at each end of the Z_{02} line section.

The insertion loss for the filter may be calculated using the formulas for the lumped-element prototype [Eqs. (6–32) and (6–36)]. The results are summarized in Table 6–2. Since $Z_T = 0$ at $f = f_c$, the insertion loss at 1.5 GHz is more easily calculated using Eq. (6–36). It is equal to 1.3 dB, which is equivalent to an input SWR of 3.07. Note that although Z_T is imaginary in the stopband, no difficulty is encountered in using Eq. (6–32).

TABLE 6–2 Loss characteristics of the lowpass filter with $f_0 = 1.2$ GHz, $f_c = 1.5$ GHz, and $Z_0 = 50$ ohms

Freq. (GHz)	f/f_c	Z_T (ohms)	L_I (dB) for Lumped-Element Prototype	L_I (dB) for Transmission-Line Equivalent
1.0	0.67	62	0.20	0.09
1.2	0.80	50	0	0.04
1.5	1.00	0	1.3	1.7
2.5	1.67	$j111$	15.6	13.8
5.0	3.33	$j265$	35.3	23.9

At this point, one might ask whether the results based on the lumped-element prototype accurately reflect the actual loss of the equivalent transmission-line filter.

In the passband, the results are indeed valid since both l_1 and l_2 are much less than $\lambda/4$. However, their validity is questionable at the high end of the stopband since the lines are nearly $\lambda/4$ long. An exact loss calculation can be made by deriving the ABCD matrix for the three transmission lines in tandem and using Eq. (6–31). The procedure is indicated below and the results are tabulated in the last column of Table 6–2.

The normalized ABCD matrix for the circuit in Fig. 6–38 is

$$
\begin{bmatrix} \bar{A} & \bar{B} \\ \bar{C} & \bar{D} \end{bmatrix} = \begin{bmatrix} \cos \beta_1 l_1 & j\bar{Z}_{01} \sin \beta_1 l_1 \\ j\dfrac{\sin \beta_1 l_1}{\bar{Z}_{01}} & \cos \beta_1 l_1 \end{bmatrix} \begin{bmatrix} \cos \beta_2 l_2 & j\bar{Z}_{02} \sin \beta_2 l_2 \\ j\dfrac{\sin \beta_2 l_2}{\bar{Z}_{02}} & \cos \beta_2 l_2 \end{bmatrix} \begin{bmatrix} \cos \beta_1 l_1 & j\bar{Z}_{01} \sin \beta_1 l_1 \\ j\dfrac{\sin \beta_1 l_1}{\bar{Z}_{01}} & \cos \beta_1 l_1 \end{bmatrix}
$$

Performing the matrix multiplication and noting that $\beta_1 l_1 = \beta_2 l_2 = \pi f / 2 f_{max}$ results in the following expressions for $\bar{B}$ and $\bar{C}$.

$$
\bar{B} = j\left[(2\bar{Z}_{01} + \bar{Z}_{02}) \sin\left(\frac{\pi}{2}\frac{f}{f_{max}}\right) - \left(2\bar{Z}_{01} + \bar{Z}_{02} + \frac{\bar{Z}_{01}^2}{\bar{Z}_{02}}\right) \sin^3\left(\frac{\pi}{2}\frac{f}{f_{max}}\right) \right]
$$

$$
\bar{C} = j\left[\left(\frac{2}{\bar{Z}_{01}} + \frac{1}{\bar{Z}_{02}}\right) \sin\left(\frac{\pi}{2}\frac{f}{f_{max}}\right) - \left(\frac{2}{\bar{Z}_{01}} + \frac{1}{\bar{Z}_{02}} + \frac{\bar{Z}_{02}}{\bar{Z}_{01}^2}\right) \sin^3\left(\frac{\pi}{2}\frac{f}{f_{max}}\right) \right]
$$

With $\bar{Z}_{01} = 176/50 = 3.52$ and $\bar{Z}_{02} = 19.7/50 = 0.394$, the loss values in the last column of Table 6–2 are obtained by utilizing Eq. (6–31). Note that except for the 5.0 GHz point, the results agree quite well with the prototype loss values. The discrepancy at the high end of the stopband is expected since the lines are nearly $\lambda/4$ long. The 23.9 dB value can be increased somewhat if l_2 is made $\lambda/8$ rather than $\lambda/4$ long at f_{max}. In our example, this would mean $l_2 = 0.52$ cm, $Z_{02} = 9.9$ ohms and therefore $d_2 = 0.98$ cm. Unfortunately, l_1 cannot be similarly reduced since d_1 would be less than 0.004 cm, and impractical value for the center conductor.

There is an even more important difference between the lumped-element and transmission-line versions of the filter. In the lumped-element design, the loss in the stopband increases indefinitely with increasing frequency, while in the transmission-line version, the loss starts to decrease above $f = f_{max}$. In fact, at $f = 2f_{max}$, the insertion loss is zero since both l_1 and l_2 are $\lambda/2$ long. This low-loss region in the vicinity of $2f_{max}$ limits the actual stopband range of the filter. The fact that impedance repeats every half wavelength invariable leads to spurious responses in transmission-line networks. The microwave engineer must keep this in mind when designing components and systems.

Multisection filters.   For the filters described thus far, the stopband loss increases at about 18 dB per octave. In applications where a higher loss slope is required, one must resort to more elaborate circuit configurations. One approach is to cascade a number of constant-k filter sections. This arrangement is described in Sec. 9–5a (see Fig. 9–51) with some element values listed in Table 9–2. Additional design information for both maximally-flat (Butterworth) and equal-ripple (Tchebyscheff) passband responses may be found in Chapter 4 of Ref. 6–11. In the stopband, the loss slope is approximately $6n$ dB per octave, where n is the number of filter elements.

Another standard technique employs *m*-derived sections at either end of a constant-*k* filter. A lumped-element version is shown in Fig. 6–39a. The theory of *m*-derived filters may be found in many texts (for example, Refs. 6–8, 6–9, and 6–10). The shunt portion of each end section provides a pole of attenuation ($L_1 = \infty$ dB) at a frequency given by

$$f_\infty = \frac{f_c}{\sqrt{1 - m^2}} \qquad \text{where} \qquad 0 \le m \le 1 \qquad (6\text{--}37)$$

The smaller the value of *m*, the larger the loss slope since f_∞ approaches f_c. However, if *m* is too small, the variation in passband impedance and loss is excessive. Values of *m* between 0.4 and 0.7 represent a good compromise. The following example illustrates the design procedure.

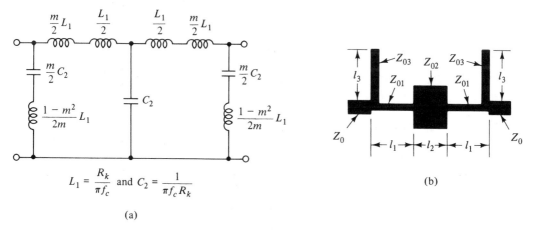

$$L_1 = \frac{R_k}{\pi f_c} \quad \text{and} \quad C_2 = \frac{1}{\pi f_c R_k}$$

(a)

(b)

Figure 6–39   A constant-*K* lowpass filter with *m*-derived end sections. The conductor pattern for a strip transmission-line version is indicated in part *b*.

Example 6–5:

Design a *T*-type constant-*k* filter with *m*-derived end sections. The design parameters are $m = 0.6$, $R_k = Z_0 = 50$ ohms, and $f_c = 1.5$ GHz. Note that for $m = 0.6$, $f_\infty = 1.875$ GHz.

Solution: From Eqs. (6–29) and (6–30),

$$1.5 \times 10^9 \text{ Hz} = \frac{1}{\pi \sqrt{L_1 C_2}} \qquad \text{and} \qquad \sqrt{\frac{L_1}{C_2}} = 50 \text{ ohms}$$

Therefore,

$$L_1 = 10.6 \text{ nH}, \qquad C_2 = 4.24 \text{ pF}, \qquad \frac{m}{2} C_2 = 1.27 \text{ pF}$$

$$\frac{1 + m}{2} L_1 = 8.48 \text{ nH} \qquad \text{and} \qquad \frac{1 - m^2}{2m} L_1 = 5.65 \text{ nH}$$

As in the previous example, the transmission-line version uses high- and low-impedance line sections to realize the series inductance and the shunt capacitance.

Figure 6–39b shows the center conductor pattern for a stripline version of the filter. The narrow width lines (Z_{01}, l_1) provide the 8.48 nH inductances, while the wide line (Z_{02}, l_2) provides the 4.24 pF capacitance. Quarter-wave, open-circuited stubs (Z_{03}, l_3) are used to simulate the shunt-connected series L-C circuits. The reactance of an open-circuited line is given by Eq. (3–91). At f_∞, the stubs must provide zero impedance and have the same reactance slope as the 1.27 pF/5.65 nH series combination. The first condition is satisfied by setting $l_3 = \lambda/4$ at f_∞. The reactance slopes at f_∞ for a series L-C circuit and a $\lambda/4$ open-circuited stub are given by

$$\left.\frac{dX}{d\omega}\right|_{\omega_\infty} = \left.\frac{d}{d\omega}\left(\omega L - \frac{1}{\omega C}\right)\right|_{\omega_\infty} = L + \frac{1}{\omega_\infty^2 C} = 2L \qquad (6\text{–}38)$$

and

$$\left.\frac{dX}{d\omega}\right|_{\omega_\infty} = \left.\frac{d}{d\omega}\left(-Z_{03}\cot\frac{\omega l}{v}\right)\right|_{\omega_\infty} = Z_{03}\frac{l}{v}\csc^2\frac{\omega_\infty l}{v} = \frac{Z_{03}}{4f_\infty} \qquad (6\text{–}39)$$

where $\omega_\infty = 2\pi f_\infty$ and $\omega_\infty l/v = \pi/2$. In our example, $L = 5.65$ nH and $f_\infty = 1.875$ GHz. Equating the two slopes yields $Z_{03} = 84.8$ ohms.

Given the properties of the particular strip transmission line being used, the length and width of all the line sections may be determined, thus completing the design (Prob. 6–29).

The loss-versus-frequency characteristics of both lumped-element and transmission-line versions may be obtained by using the ABCD matrix method. Standard programs are available for use with either computers or programmable calculators to perform the laborious calculations. The results for the filter in our example are shown in Fig. 6–40. Note that the frequency scale is normalized to f_c. The dashed curve indicates the results for a transmission-line version in which l_1 and l_2 were

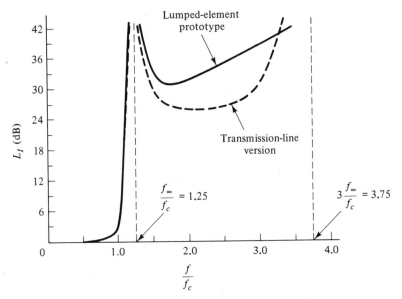

Figure 6–40   Loss characteristics for the lowpass filter described in illustrative Ex. 6–5.

made $\lambda/4$ long at 5.5 GHz and $l_3 = \lambda/4$ at $f_\infty = 1.875$ GHz. In both versions, the loss slope below f_∞ is very high. Above f_∞, the loss of the transmission-line version decreases more rapidly than the lumped-element prototype. This is because the impedance of the stubs increases more rapidly than that of the series L-C circuits. In fact at $f = 2f_\infty$, the stub impedances are infinite and hence contribute nothing to the loss of the filter. This type of lowpass filter is simple to construct in stripline and microstrip configurations.

Another lowpass design technique that is widely used at microwave frequencies makes use of Richards' transformation and Kuroda's identities. Since this method is also used to realize bandstop filters, it is discussed in Sec. 9–5a. Highpass filters can also be realized using similar design concepts. In fact, the variety of filter configurations and methods is practically endless. The subject is discussed in greater detail in Secs. 9–4 and 9–5. Extensive bibliographies on microwave filters may be found in Refs. 6–7 and 6–28. The objective here has been merely to show how transmission-line techniques can be used to extend lumped-element filter concepts into the microwave range.

6–7 MICROWAVE INTEGRATED CIRCUITS

Before 1955, microwave systems typically consisted of an interconnection of individual waveguide and/or coaxial components. Type-N connectors were used to interconnect coaxial devices, while waveguide units were joined together with specially designed flanges. This general approach is still used, particularly in medium- and high-power systems. Since considerable machining of metal parts is required, manufacturing costs are invariably high. In some cases, investment-casting techniques are utilized to reduce the cost. Difficulty in reproducing overall system performance is another problem associated with the *individual components* approach. This is due, in part, to the fact that each connector-pair introduces a mismatch which affects system performance in an unpredictable manner. In low-power systems, these problems can be minimized by using stripline and integrated-circuit techniques.

The advent of printed-circuit technology provided the impetus for the development of stripline components and systems. Initial results of this effort are described in the March 1955 issue of *IRE Transactions on Microwave Theory and Techniques*. An example of a stripline microwave system is shown in Fig. 6–41. Note that the various components are all on a single circuit board. Thus the use of connectors to join individual components is eliminated. Furthermore, the use of printed-circuit techniques (photolithography, chemical etching, etc.) results in a low-cost, highly reproducible microwave system. A variation of this approach is to use the suspended-stripline arrangement described in Fig. 5–15a. In this version, the microwave circuit is printed on a thin dielectric sheet which is suspended between two ground planes. The only function of the dielectric is to provide a mechanical support for the center conductor pattern. Essentially, it is an air-insulated stripline structure and therefore has much less dielectric loss than ordinary stripline.

Both these techniques have the disadvantage that discrete elements—such as resistors, diodes, and capacitors—must be attached separately. Quite often the para-

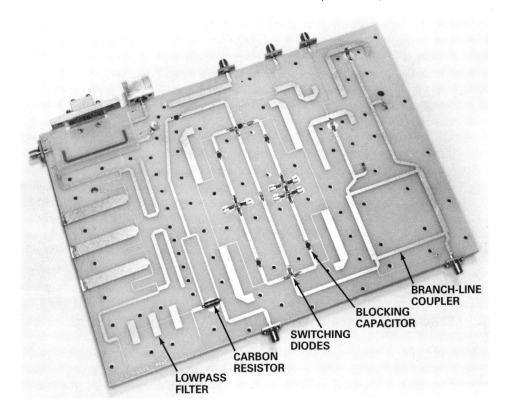

Figure 6–41 A typical microwave stripline system. The top dielectric and ground plane are not shown. (Courtesy of M/A-COM, Inc., Burlington, Mass.)

sitic inductance associated with the connecting leads degrades the overall electrical performance. Despite this fact, most low-power microwave systems are manufactured using stripline techniques.

The early 1960s saw the successful introduction of integrated circuit (IC) technology into the manufacture of multifunction electronic and computer systems. These ICs feature a high packing density of passive and active elements. Typically, one small semiconductor chip contains hundreds of diodes, transistors, and associated circuitry. In the mid-1960s, development efforts were undertaken by many companies and laboratories to apply IC techniques to microwave components and systems. The expected advantages were low production cost, small size, high reliability, and reproducible electrical performance. The technology for microwave integrated circuits (MICs) is now fairly well developed. They may be classified into two groups, hybrid and monolithic. A brief description of both types is given here. For a more detailed discussion, the reader is referred to Refs. 6–29, 6–30, and 6–31. Also, Frey (Ref. 6–13) has compiled some of the significant articles in this exciting field.

Hybrid-type integrated circuits. In general, microwave systems contain both lumped- and distributed-type circuit elements. In a hybrid integrated circuit, the distributed elements, as well as certain types of inductors, resistors, and capacitors, are formed by depositing a metal conductor pattern on a dielectric substrate. Then semiconductor devices and discrete-type resistors and capacitors are mounted on the substrate and individually bonded to the conductor pattern.[9] The semiconductor devices are usually in beam-lead packages, while the individual resistors and capacitors are tiny chips. Microstrip (Fig. 5–12) is widely used as the transmission line in MICs. Occasionally, slot line or coplanar line (Fig. 5–15) is employed. A common feature of all these lines is an open and accessible surface upon which discrete devices can be mounted. A brief description of the materials used to form these lines is given below.

Materials.   The most common substrate materials are Duroid, alumina, sapphire, and quartz. Because of its low ϵ_R value, Duroid is a popular choice for millimeter-wave circuits. Ferrite is sometimes used when the MIC includes nonreciprocal components. The material requirements include low dielectric loss, high dielectric constant, and minimal variation in ϵ_R and tan δ with frequency and temperature. One reason for using high dielectric-constant material is that wavelength and hence circuit size is reduced. Another is that radiation losses are minimized. This is an important consideration since microstrip is susceptible to radiation at discontinuities (Ref. 6–24).

The requirements for a conductor material are low resistance, minimum resistance variation with temperature, good adhesion to the substrate material, and compatibility with the photoetching method used to define the conductor pattern. In order to achieve these characteristics, a combination of metals are used to form both the conductor pattern and the ground plane. The technique is described in Fig. 6–42. First, a conductor material that adheres well to the substrate is deposited on the substrate. Typically chromium and tantalum, both poor conductors, are used. Next, a layer of low-resistance material, such as silver, copper, or gold, is deposited on the chromium or tantalum. In order to minimize conductor losses, the thickness of the poor conductor should be small compared to its skin depth, while that of the good conductor should be large compared to its skin depth. Sobol (Ref. 6–30)

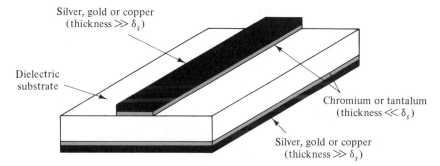

Figure 6–42   A multimaterial conductor system for microstrip transmission lines and circuits.

[9] The stripline system described earlier fits this description and hence may be classified as a microwave integrated circuit.

shows that this method works quite well up to X band for chromium thicknesses of less than one micron.

Dielectric films are sometimes used in fabricating series capacitors or to provide coupling between adjacent transmission lines. Typically, oxides of silicon, alumium, and tantalum are used. There are various techniques of depositing the film on the conductors. These include evaporation, sputtering, and anodization.

There are two ways in which resistors can be incorporated into a hybrid-type MIC. One approach is to solder or weld tiny discrete resistors directly to the conductor pattern. The other is to deposit a resistive film directly on the substrate. Most applications require surface resistivities of from 10 to 1000 ohms per square. As is the case with low-frequency resistors, they should exhibit good stability over a wide range of environmental conditions. Typical resistive materials include nichrome, chromium, and tantalum. Both the resistive material and the conductor patterns can be deposited on the substrate by either thin or thick film techniques. A comparison of the two techniques is given in Ref. 6–31.

A photograph of a hybrid-type MIC is shown in Fig. 6–43. The transmission line is microstrip with 25 mil alumina as the dielectric substrate. Discrete elements which are bonded to the conductor pattern may be seen in the photograph. Note, for instance, the small resistors that are part of the power divider network. Most of the circuit, however, is formed by the conductor pattern itself.

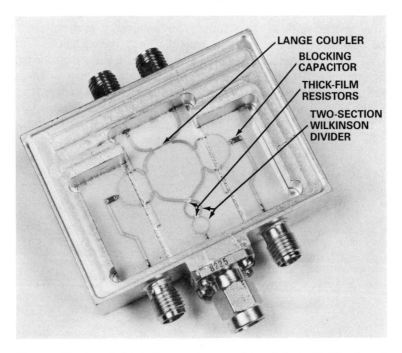

Figure 6–43 A typical hybrid-type microwave integrated circuit. (Courtesy of M/A-COM, Inc., Burlington, Mass.)

Monolithic integrated circuits.   In a monolithic MIC, the active devices are grown on or diffused into a semiconductor substrate and the passive circuitry is either deposited or grown on the substrate. This technique has been used for years in digital and computer applications to obtain very high packing densities. Since microwave circuit applications seldom require a high density of devices, most MICs are of the hybrid type. However, the monolithic approach does have the advantage that parasitic effects due to wire leads are practically eliminated since the active devices are an integral part of the semiconductor substrate. With parasitic effects being more pronounced at the higher frequencies, this advantage makes the monolithic approach quite attractive in millimeter-wave applications.

REFERENCES

Books

6–1. Howe, Jr., H., *Stripline Circuit Design,* Artech House, Dedham, MA, 1974.

6–2. Reich, H. J., P. F. Ordung, H. L. Krauss, and J. G. Skalnik, *Microwave Theory and Techniques,* D. Van Nostrand Co., New York, 1953.

6–3. Adam, S. F., *Microwave Theory and Applications,* Prentice Hall, Inc., Englewood Cliffs, NJ, 1969.

6–4. Lance, A. L., *Introduction to Microwave Theory and Measurements,* McGraw-Hill Book Co., New York, 1964.

6–5. Montgomery, C. G., *Technique of Microwave Measurements,* Rad. Lab. Series, vol. 11, McGraw-Hill Book Co., New York, 1951.

6–6. Ghose, R. N., *Microwave Circuit Theory and Analysis,* McGraw-Hill Book Co., New York, 1963.

6–7. Saad, T. S., *Microwave Engineers' Handbook,* vol. 1, Artech House, Dedham, MA, 1971.

6–8. Potter, J. L. and S. J. Fich, *Theory of Networks and Lines,* Prentice-Hall, Inc., Englewood Cliffs, NJ, 1963.

6–9. Johnson, W. C., *Transmission Lines and Networks,* McGraw-Hill Book Co., New York, 1950.

6–10. Karakash, J. J., *Transmission Lines and Filter Networks,* The Macmillan Co., New York, 1950.

6–11. Matthaei, G. L., L. Young, and E.M.T. Jones, *Microwave Filters, Impedance-Matching Networks and Coupling Structures,* McGraw-Hill Book Co., New York, 1964.

6–12. Gupta, K. C., R. Garg, and I. J. Bahl, *Microstrip Lines and Slotlines,* Artech House, Dedham, MA, 1979.

6–13. Frey, J., *Microwave Integrated Circuits,* Artech House, Dedham, MA, 1975.

Articles

6–14. DiBeneditto, J. and A. Uhlir, Jr., Frequency Dependence of 50 Ohm Coaxial Open-Circuit Reflection Standard. *IEEE Trans. Instrumentation and Measurement,* IM-30, September 1981, pp. 228–229.

6-15. Altschuler, H. M. and A. A. Oliner, Discontinuities in the Center Conductor of Symmetric Strip Transmission Line. *IRE Trans. Microwave Theory and Techniques,* MTT-8, May 1960, pp. 328–339.

6-16. Joines, W. T., A Continuously Variable Dielectric Phase Shifter. *IEEE Trans. Microwave Theory and Techniques,* MTT-19, August 1971, pp. 729–732.

6-17. Oltman, G., The Compensated Balun. *IEEE Trans. Microwave Theory and Techniques,* MTT-14, March 1966, pp.112–119.

6-18. Whinnery, J. R., H. W. Jamieson, and T. E. Robbins, Coaxial Line Discontinuities. *Proc. I.R.E.,* 32, November 1944, pp. 695–709.

6-19. Oliner, A. A., Equivalent Circuits for Discontinuities in Balanced Strip Transmission Line. *IRE Trans. Microwave Theory and Techniques,* MTT-3, March 1955, pp. 134–143.

6-20. Benedek, P. and P. Silvester, Equivalent Capacitances for Microstrip Gaps and Steps. *IEEE Trans. Microwave Theory and Techniques,* MTT-20, November 1972, pp. 729–733.

6-21. Maeda, M., An Analysis of Gap in Microstrip Transmission Lines. *IEEE Trans. Microwave Theory and Techniques,* MTT-20, June 1972, pp. 390–396.

6-22. Horton, R., Equivalent Representation of an Abrupt Impedance Step in Microstrip Line. *IEEE Trans. Microwave Theory and Techniques,* MTT-21, August 1973, pp. 562–564.

6-23. Silvester, P. and P. Benedek, Microstrip Discontinuity Capacitance for Right-Angle Bends, T-Junctions and Crossings. *IEEE Trans. Microwave Theory and Techniques,* MTT-21, May 1973, pp. 341–346.

6-24. Lewin, L., Radiation from Discontinuities in Stripline. *IEEE Monograph No. 358E,* February 1960.

6-25. Muehe, C. E., Quarter-Wave Compensation of Resonant Discontinuities. *IRE Trans. Microwave Theory and Techniques,* MTT-7, April 1959, pp. 296–297.

6-26. Mouw, R. B., Broadband DC Isolator-Monitors. *Microwave Journal,* 7, November 1964, pp. 75–77.

6-27. McDermott, M. and R. Levy, Very Broadband Coaxial DC Returns Derived by Microwave Filter Synthesis. *Microwave Journal,* 8, February 1965, pp. 33–36.

6-28. Young, L., Microwave Filters-1965. *IEEE Trans. Microwave Theory and Techniques,* MTT-13, September 1965, pp. 489–508.

6-29. Caulton, M. and H. Sobol, Microwave Integrated Circuit Technology—A Survey. *IEEE Jour. Solid-State Circuits,* SC-5, December 1970, pp. 292–303.

6-30. Sobol, H., Applications of Integrated Circuit Technology To Microwave Frequencies. *Proc. IEEE,* 59, August 1971, pp. 1200–1211.

6-31. Keister, F. Z., An Evaluation of Materials and Processes for Integrated Microwave Circuits. *IEEE Trans. Microwave Theory and Techniques,* MTT-16, July 1968, pp. 469–475.

PROBLEMS

6-1. A 50 ohm coaxial line is terminated with a microwave resistor of the type shown in Fig. 6-1. Calculate the SWR at 3 GHz if $R = 55$ ohms and $C = 0.08$ pF. Assume $l \ll 0.05\lambda$. What would be the SWR if the capacitance were negligible?

6–2. Referring to Ex. 6–1 in the text, calculate the SWR at 3 GHz if L' in the lossy-line section is halved and C' is doubled.

6–3. A coaxial line is terminated in a tapered load of the type shown in Fig. 6–3. Calculate the one-way attenuation required for an input SWR of less than 1.004. Assume that the taper itself is reflectionless.

6–4. The length x of an air-insulated shorted line (Fig. 6–6a) with $Z_0 = 75$ ohms is adjusted for an inductive reactance of 90 ohms at 2 GHz. Calculate $dX_{in}/d\omega$ using the smallest possible value of x. Compare the result with that of a lumped inductance having the same reactance at 2 GHz.

6–5. Referring to Fig. 6–6b with $l = \lambda_0$ and $\epsilon_R = 2.25$, show that by varying x, any value of input reactance can be realized.

6–6. Determine the length of a half-wave type dielectric bead (Fig. 6–8a) if $\epsilon_R = 4$ and the design frequency is 5 GHz. Calculate its SWR and insertion loss at 4.6 GHz. Use Fig. 6–9 to verify your answer.

6–7. The thin dielectric bead pictured in Fig. 6–8b behaves like a shunt capacitance when $l \ll \lambda$. Use the admittance transformation to show that with $l \ll \lambda$, the normalized capacitive susceptance $\bar{B}_C = B_C/Y_0 = (\epsilon_R - 1)\phi_0$, where $\phi_0 = 2\pi l/\lambda_0$. Assume that all ϕ_0^2 terms are negligible.

6–8. Design a teflon bead support of the type shown in Fig. 6–8c for a 50 ohm coaxial line. Determine l and a_1 if the outer conductor diameter is 0.8 cm and the highest frequency of interest is 7 GHz.

6–9. A 50 ohm coaxial line contains a dielectric bead support of the type shown in Fig. 6–8c. Its length is $\lambda/4$ at 8 GHz. Calculate its SWR at 4 and 8 GHz if the discontinuity capacitances are 0.10 pF each.

6–10. Design a 16 dB matched Tee attenuator for a 50 ohm coaxial system. What would be the increase in loss if the R_1 resistors were of the correct value but the shunt resistor R_2 was 20 percent less than its correct design value?

6–11. Derive Eqs. (6–12) and (6–14).

6–12. A 50 ohm coaxial system contains a 13 dB lossy-line attenuator. Estimate its lower frequency limit if the resistor element is 10 cm long.

6–13. A matched attenuator is connected to a load having an SWR of 5.7. What value of attenuation will result in an input SWR of 1.12?

6–14. A cutoff attenuator of the type shown in Fig. 6–20 is used as a precision variable attenuator. At 600 MHz, a 2.0 cm increase in length l causes the attenuation to increase from 24 to 40 dB. Calculate the diameter D of the circular guide assuming that the TE_{11} cutoff frequency is much greater than 600 MHz. Having calculated D, justify the assumption.

6–15. Referring to the coaxial phase shifter in Fig. 6–22a, calculate the phase change at 3750 MHz for an increase in length from 8 to 9 cm and from 9 to 10 cm.

6–16. Repeat the above problem if reflections are created by the input and output connectors. Assume $Z_0 = 50$ ohms and that each reflection is due to a series inductive reactance of 10 ohms. (Hint: Use Eq. (4–57) to determine the insertion phase for the 8, 9, and 10 cm lengths.)

6–17. Referring to Fig. 6–24, use the ABCD matrix method to show that with $Z_L = Z_0$, $l_2 = \lambda_0/4$, and $\epsilon_R = 3$, the input SWR $= 9$. Also show that with $l_2 = \lambda_0/2$, the SWR is unity.

6–18. Referring to Fig. 6–32, $Z_{0s} = 2Z_0$, $Z_{01} = 0.8Z_0$, and all lines are $\lambda/4$ long at the design frequency (f_r). Use the Smith chart to verify the data in Fig. 6–31 for $f = f_r$ and $f = 1.3f_r$.

6–19. A dc block for a 50 ohm coaxial line is shown in Fig. 6–34a. With $b_s = 0.40$ cm and $a_s = 0.32$ cm, determine the minimum value of ϵ_R for an input SWR ≤ 1.10 at *half* the design frequency.

6–20. A transmission-line bias circuit (Fig. 6–35b) has been designed for a 50 ohm coaxial system. What is the value of Z_{01} if Z_{02} equals 150 ohms? Calculate the input SWR of the circuit at $0.4f_r$ and $1.6f_r$. Assume that point C is a short circuit at all frequencies of interest.

6–21. For the above problem, prove that the SWR is approximately unity at $0.25f_r$ and $1.75f_r$.

6–22. Design a lowpass lumped-element T filter with $R_k = Z_0 = 75$ ohms and $f_c = 2.5$ GHz. Calculate the insertion loss and SWR at 2.0, 4.0, and 8.0 GHz. Compare the 4.0 and 8.0 GHz loss values with the attenuation values obtained from Eq. (6–28).

6–23. (a) Calculate the input impedance at 4.0 GHz for the filter designed in Prob. 6–22. Assume the output is terminated with a 75 ohm resistor.
   (b) Design a lumped-element π filter having the same values of R_k and f_c as in Prob. 6–22. Compare its input impedance at 4.0 GHz with that of the T filter.

6–24. Calculate the insertion loss at 0.5 and 0.75 GHz for the lumped-element prototype filter described in Ex. 6–4. Using these values and those in Table 6–2, sketch the loss versus frequency characteristic of the filter.

6–25. Design the transmission-line equivalent of the filter described in Prob. 6–22. Assume $f_{max} = 6.0$ GHz. Determine all lengths and diameters for a coaxial version (Fig. 6–38). Note that the low-impedance line section is teflon-filled ($\epsilon_R = 2.1$).

6–26. The last column in Table 6–2 lists the insertion loss for the transmission-line filter described in Ex. 6–4. As explained in the discussion following the example, setting $l_2 = \lambda/8$ rather than $\lambda/4$ at f_{max} results in $l_2 = 0.52$ cm and $Z_{02} = 9.9$ ohms. Calculate the insertion loss at 5.0 GHz for this case and compare it to the corresponding value in Table 6–2.

6–27. Design a π filter having the same specifications as given in Ex. 6–4. Calculate all lengths and diameters. Assume a teflon insulator for the two low-impedance lines.

6–28. The coaxial T filter in Fig. 6–38 has the following line impedances: $Z_0 = 50$ ohms, $Z_{01} = 100$ ohms, and $Z_{02} = 10$ ohms. The lengths are set so that $l_1 = 3\lambda_0/8$ and $l_2 = \lambda/8$ at 6.0 GHz.
   (a) Determine the values of f_c and R_k for the filter.
   (b) Calculate Z_{in} and L_I at 6.0 GHz. Assume $Z_G = Z_L = 50$ ohms.

6–29. Determine all the dimensions for a stripline version of the *m*-derived filter designed in Ex. 6–5. Let $f_{max} = 4.0$ GHz. Assume a stripline system with $\epsilon_R = 1.90$, a ground-plane spacing of 0.30 cm, and a negligible strip thickness.

6–30. Calculate the impedance (Z_{03}) and length (l_3) of the open-circuited stubs described in Ex. 6–5 if f_∞ is set at 1.75 GHz. Assume air-insulated lines.

7

Waveguide Components

For a given operating frequency, waveguide has a greater power handling capacity than coaxial line (Sec. 5–4a). Due to the absence of a center conductor, waveguide components are often less expensive to manufacture than their coaxial counterparts, particularly when investment casting and computerized machining techniques are utilized. This chapter describes a variety of reciprocal waveguide components used in microwave systems. Specifically, the material deals with one- and two-port networks. Multiport components are discussed in Chapter 8, while the special topics of cavities and filters are covered in Chapter 9. Unless otherwise indicated, dominant-mode transmission is assumed (namely, TE_{10} for rectangular guide and TE_{11} for circular guide).

7–1 MODE TRANSDUCERS

It is often necessary to change the mode, direction, or polarization of wave transmission in a microwave system. Some methods of accomplishing these functions are described in this section.

7–1a Coaxial to Waveguide Transitions

Transitions between the TEM coaxial mode and the TE_{10} mode in rectangular guide are frequently used in microwave equipment. In designing such a transition, it is necessary that it be shaped to encourage the desired mode conversion. One such configuration is shown in Fig. 7–1a. The coaxial line is connected to the broad wall of the waveguide with its outer conductor terminating on the wall. The center conductor protrudes into the guide and terminates on the opposite wall. As the TEM

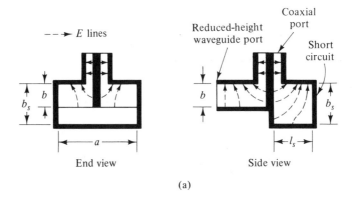

End view Side view

(a)

Note: $\dfrac{Z_{0s}}{Z_{01}} = \dfrac{b_s}{b}$

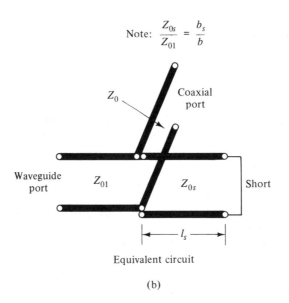

Equivalent circuit

(b)

Figure 7–1 A coaxial-to-waveguide transition and its equivalent circuit. (Note: Reactive effects due to higher-order modes are neglected.)

wave enters the waveguide section, the electric field lines follow along the conducting walls as shown. Also, with the TEM magnetic loops (not shown) in the horizontal plane, they are favorably oriented for conversion to the magnetic loop pattern of the TE_{10} mode. The portion of the TE_{10} wave traveling to the right in the side view is fully reflected by the short ($\Gamma = -1$). With $l_s = \lambda_g/4$ at the design frequency, the reflected wave reinforces the portion traveling to the left resulting in low-loss transmission between the coaxial and waveguide ports.

The equivalent circuit is shown in part b of Fig. 7–1. From a circuit point of view, the function of the quarter-wave shorted stub is to provide an open circuit in shunt with the Z_0 and Z_{01} lines. One might be tempted to obtain the open circuit by leaving the right-hand portion of the rectangular guide unterminated and letting

$l_s = 0$. Unfortunately, this doesn't work since an open-ended waveguide is *not* electrically equivalent to an open circuit. It is, in fact, an excellent radiating antenna.

To achieve a low-SWR transition, the characteristic impedances of the coaxial line (Z_0) and the rectangular waveguide (Z_{01}) must be equal. In most cases, the coax impedance is 50 ohms while that of standard rectangular guide is several times larger.[1] The waveguide impedance can be lowered by either reducing the guide height, as shown in the figure, or by using a ridge waveguide (Ref. 7–12). Keep in mind that, unlike coaxial line, the characteristic impedance of waveguide is a function of frequency. Therefore, the two impedances can only be made equal at the design frequency. If a standard height output guide is desired, a quarter-wave transformer can be incorporated into the waveguide section.

Note that, unlike the output guide, the height of the shorted waveguide section (b_s) has *not* been reduced. With waveguide impedance proportional to height, this means $Z_{0s} > Z_{01}$, which results in a lower *SWR* over the useful waveguide band. The principle involved is the same as discussed in Sec. 6–5a since the equivalent circuits in Figs. 7–1b and 6–30b are the same. However, the SWR performance is not quite as good as indicated in Fig. 6–31 since waveguide impedance is a function of frequency and the variation of λ_g with frequency is greater than that of λ. Problem 7–1 gives an indication of the expected SWR over the useful waveguide band.

Figure 7–2 shows side views of three other coax-to-waveguide transitions. In all cases, the transitions are made directly to standard height waveguide. The one shown in part *a* is useful in high-power applications and is known as a *doorknob* transition. The probe type shown in part *b* is widely used in test equipment. Note that the center conductor of the coaxial line extends part way into the waveguide section. It can be thought of as an antenna that radiates energy into the waveguide.[2] The dielectric sleeve-probe combination provides the required transformation between coaxial and waveguide impedances. This technique results in an SWR of less than 1.25 over the useful frequency range of the rectangular waveguide.

An *in-line* transition is shown in part *c* of Fig. 7–2. In this case, the center conductor of the coaxial line is bent so as to contact the broad wall of the waveguide. The resultant loop encourages energy coupling between the coaxial and waveguide modes. The electric field conversion from TEM to TE_{10} is indicated in the figure. The orientation of the current-carrying loop is such that it converts the vertical plane of the TEM magnetic field (not shown) to a horizontally-polarized magnetic loop as required for the TE_{10} mode. The loop dimensions are adjusted experimentally for best SWR.

The reader should keep in mind that the discontinuity created by a loop or probe produces higher-order modes. For instance, the ones in Fig. 7–2 excite TM modes since they create a component of electric field parallel to the waveguide axis. By proper selection of guide dimensions, these higher-order modes can be attenuated while allowing the dominant mode to propagate. Impedance-matching methods are used to minimize the reactive effects associated with the localized higher modes. Various design configurations are described in Chapter 6 of Ref. 7–10.

---

[1] The various definitions of waveguide impedance are given by Eqs. (5–79) and (5–81).

[2] Since all transitions are reciprocal, it is equally valid to say that the probe acts as a receiving antenna for the TE_{10} energy, thus converting it into the coaxial TEM mode.

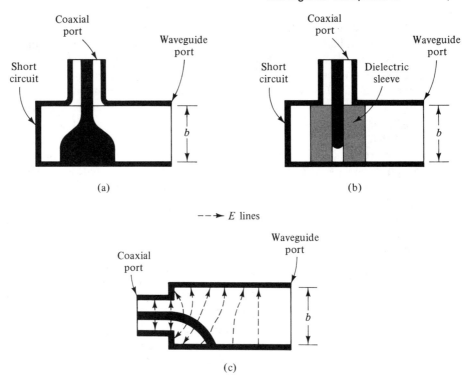

Figure 7–2 Three additional coaxial-to-waveguide transitions (TEM to TE_{10} mode).

Higher-mode transitions. Techniques similar to those described above may be used to convert from the TEM mode to a higher-order waveguide mode. Several examples are given in Ref. 7–10 and Sec. 9.6 of Ref. 7–1. In the absence of specific design procedures, some general comments are in order. When designing a transition between two modes, one should first study the field patterns involved. The transition region should then be configured so as to encourage a smooth conversion of electric and magnetic fields. Actually, one need only visualize one of the two fields since they, in fact, generate each other. For example, in the case of the door-knob transition (Fig. 7–2a), one can see how the shape of the center conductor guides the electric field from its TEM pattern to that required for the TE_{10} waveguide mode. The operation of the loop type (Fig. 7–2c), on the other hand, is more easily visualized in terms of magnetic field conversion. One can picture the current-carrying loop reorienting the TEM magnetic field into the horizontal plane required for TE_{10} wave propagation. These same concepts apply when designing a transition to a higher-order waveguide mode.

Symmetry arguments are also useful in determining which modes are excited by the transition. As an example, consider the transition shown in Fig. 7–3. By extending the coaxial center conductor into the guide as shown, the TM_{11} rectangular waveguide mode is excited. The electric and magnetic field conversion is indicated in the figure. Note that the orientation of the TEM magnetic field is maintained as it

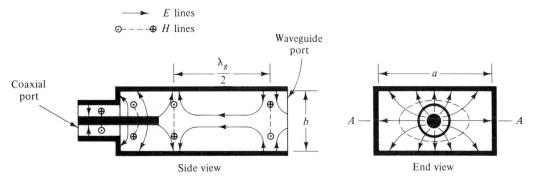

Figure 7–3 A transition between the TEM coaxial mode and the TM_{11} mode in rectangular guide. (Note: If the waveguide were circular, the TM_{01} mode would be excited.)

enters the waveguide, thus exciting the TM_{11} mode. Higher-order TM_{mn} modes are also generated by the transition. The symmetry of the configuration is such that only those with odd values of m and n are possible. The guide dimensions are usually chosen so that the higher-order TM modes are below cutoff and thus only the TM_{11} propagates. A symmetry argument can also be invoked to show that conversion to the dominant TE_{10} mode is impossible. The reasoning is as follows. Since the configuration is symmetrical about the horizontal line A–A (shown in the end view), excitation by the symmetrical TEM mode can only produce waveguide modes that are symmetrical about A–A. The TM_{11} is such a mode, as are all TM_{mn} modes with m and n odd, but the TE_{m0} modes are not.

If the rectangular guide in Fig. 7–3 were replaced by a circular guide, the resultant transition would convert the TEM coaxial mode into the TM_{01} mode in circular guide.

7–1b Waveguide to Waveguide Transitions

Transitions between waveguides of different size, shape, and/or mode are also used in microwave systems. The most common are those in which dominant mode transmission is desired in both guides. Two such transitions are shown in Fig. 7–4. The tapered transition in part a converts the TE_{10} mode in rectangular guide to the TE_{11} mode in circular guide. Reflections are minimized by making the taper length long compared to the operating wavelength. Another version is shown in part b of the figure. The TE_{10} mode enters the circular guide creating a TE_{11} mode pattern. A portion of the TE_{11} mode wave travels directly to the output, while the remainder propagates toward the shorting plate and is fully reflected. By proper adjustment of the length l_s, the two waves arrive at the output port in phase resulting in efficient conversion between the TE_{10} and TE_{11} modes. The higher-order modes generated by the junction can be localized by choosing the guide dimensions so that all but the dominant modes are below cutoff. The reactive effects of the localized modes can be minimized by inserting an impedance-matching network at the input.

Examples of transitions between a dominant mode and a higher-order mode are given in Figs. 7–5 and 7–6. The one described in Fig. 7–5 represents an efficient method of conversion between the TE_{10} rectangular guide mode and the

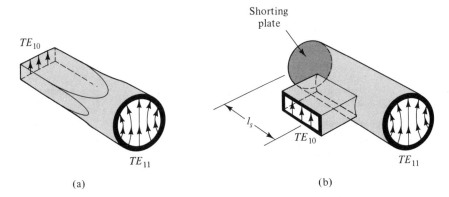

TE_{10}

Shorting plate

l_s

TE_{10}

TE_{11}

TE_{11}

(a)   (b)

Figure 7–4   Two transitions between the TE_{10} mode in rectangular guide and the TE_{11} mode in circular guide.

TM_{01} mode in circular guide. The orientation of the waveguides encourages a smooth transition between the magnetic field patterns of the two modes. This is indicated in the figure. The diameter D of the circular guide must, of course, be sufficiently large to insure that the TM_{01} mode is above cutoff, but the next higher mode is not. The cutoff wavelengths for some circular guide modes are given in Fig. 5–27.

By proper choice of dimensions, the structure in Fig. 7–4b can also be used as a converter between the TE_{10} rectangular guide mode and the TE_{01} mode in circular guide. In this case, the circular guide diameter must be large enough to support TE_{01} mode propagation. The two-dimensional view in Fig. 7–6 describes the mode conversion process. The electric field pattern at the junction is shown on the left. As indicated in the figure, the distorted pattern may be viewed as mainly a combination of the TE_{11} and TE_{01} modes. With the guide sufficiently large to support the TE_{01} mode, it is also capable of propagating the TE_{11} mode. In order to insure TE_{01} mode purity in the circular guide, a mode filter (not shown) is usually inserted to suppress the unwanted TE_{11} mode. The TE_{10} to TM_{01} converter described in Fig. 7–5 also requires a TE_{11} mode suppressor to insure TM_{01} mode purity. In certain cases, the symmetrical nature of the transition prevents the generation of the dominant mode and hence a mode filter is not needed. One such case is the TEM to TM_{11} mode converter shown in Fig. 7–3.

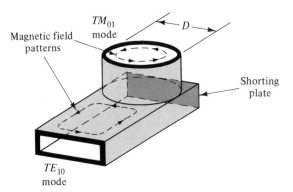

Magnetic field patterns

TM_{01} mode

D

Shorting plate

TE_{10} mode

Figure 7–5   A transition between the TE_{10} mode in rectangular guide and the TM_{01} mode in circular guide.

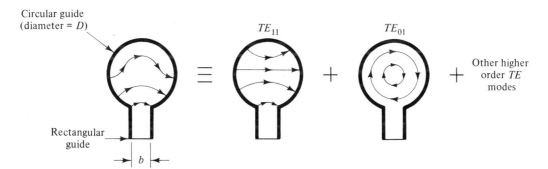

Figure 7–6 Mode conversion from the TE_{10} in rectangular guide to the TE_{01} in circular guide. A mode filter is required to suppress the TE_{11} mode.

Mode filters.   The requirement that only the dominant mode propagate in a waveguide can be satisfied by choosing the guide dimensions so that all other modes are below cutoff in the frequency range of interest. On the other hand, if one wishes to propagate a higher-order mode, mode filters must be used to suppress the modes with lower cutoff frequencies.

One method involves an array of conducting strips or wires inserted in the guide so as to reflect the unwanted modes without interfering with the desired mode. Three examples are shown in Fig. 7–7. In all cases, the mode filters consist of an array of thin metal wires or strips that are everywhere perpendicular to the electric field of the *desired* transmission mode. Modes having all or part of their electric field parallel to the conducting wires or strips are reflected and hence strongly attenuated. Resistive strips are used when absorption rather than reflection of the unwanted modes is desired.

The mode filter shown in part *a* of the figure reflects the TE_{11} mode as well as all TM modes in circular guide. The circularly symmetric TE modes (TE_{01}, TE_{02}, etc.), on the other hand, are unaffected by the presence of the conducting wires or strips. This is the type of mode filter alluded to in the discussion of the TE_{10} to TE_{01} converter (Fig. 7–6).

The filter shown in Fig. 7–7b reflects TE modes without affecting the transmission of the TM_{01} mode. This filter is often used in conjunction with the TE_{10} to TM_{01} converter (Fig. 7–5) to insure TM_{01} mode purity in the circular guide.

The rectangular guide mode filter in Fig. 7–7c consists of an array of conducting strips that are perpendicular to the broad wall of the guide. The principle of

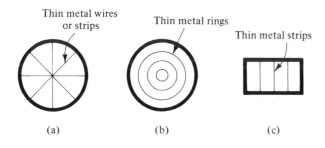

Figure 7–7 Examples of mode filters for (*a*) TE_{01} transmission, (*b*) TM_{01} transmission, and (*c*) TE_{01} rectangular guide transmission.

operation is the same as the others. TE_{01} rectangular mode transmission is unaffected because its horizontally oriented electric field is everywhere perpendicular to the conducting strips. On the other hand, the TE_{10}, TE_{11}, and all the TM modes are reflected since portions of their electric field are parallel to the strips.

Another approach to mode filtering involves cutting an array of thin slots in the walls of the waveguide. The slots are arranged so that the desired mode is unaffected by their presence, while unwanted modes radiate through the slots into an auxiliary guide where they are absorbed. Examples of radiating and nonradiating slots for the TE_{10} mode are shown in Fig. 8–17. These and other examples of mode filtering may be found in Refs. 7–1, 7–10, and 7–11.

7–1c Waveguide Bends and Twists

Bends are used to alter the direction of propagation in a waveguide system. In rectangular guide, the bend can be made in either of two planes as shown in Fig. 7–8. The one in part a is known as an E-plane bend since the orientation of the TE_{10} electric field changes as the wave propagates around the bend. The H-plane bend is so named because only the orientation of the magnetic field changes. The reflection due to the bend is a function of its radius; the larger the radius, the lower the SWR. When the space available is limited, a double-mitred bend (Fig. 7–9) may be preferred. A low SWR is obtained by spacing the two mitred joints $\lambda_g/4$ apart at the design frequency. This technique is applicable to both E- and H-plane bends.

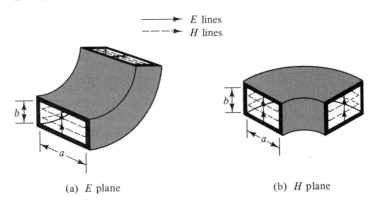

(a) E plane   (b) H plane

Figure 7–8 E- and H-plane bends in rectangular waveguide.

Waveguide bends for higher-order modes present certain difficulties. Mode conversion can occur at the discontinuity created by the bends which results in some of the higher-mode energy being transferred to the dominant mode. To prevent dominant-mode propagation, mode filters must be inserted in the guide. A similar problem occurs in circular waveguide bends wherein TE_{11} propagation around the bend may cause the polarization to change in an unpredictable fashion. A mode filter consisting of an array of metal rods across the waveguide may be used to maintain the polarization of the TE_{11} wave. The orientation of the rods is perpendicular to the desired polarization.

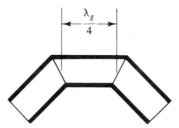

Figure 7–9   A double-mitred bend in rectangular waveguide.

Waveguide twists.   Waveguide twists are used to change the plane of polarization of a propagating wave. Figure 7–10 shows two types that rotate the polarization of the dominant mode by 90 degrees.[3] The gradual twist shown in part *a* changes the plane of polarization in a continuous fashion. Its SWR is typically less than 1.05 when the twist length is greater than a few wavelengths.

Step twists are used when the space available in the propagation direction is limited. The one shown in part *b* of the figure contains a rectangular guide section that is oriented at 45° with respect to the input and output guides. Thus the polarization change takes place in two 45° steps. With its length set equal to $\lambda_g/4$ at the design frequency, a low SWR is obtained over a narrow bandwidth. Multistep twists may be used when broader band performance is required (Ref. 7–15).

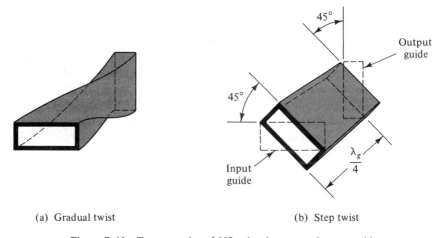

(a) Gradual twist   (b) Step twist

Figure 7–10   Two examples of 90° twists in rectangular waveguide.

7–1d Circular Polarizers

A circular polarizer converts a linearly polarized wave into a circularly polarized wave and vice versa. The equivalence between the two types of waves was explained in Sec. 2–5. A method of converting from linear to circular polarization was described in Ex. 2–5 and Fig. 2–20. A waveguide version is shown in Fig. 7–11a. The vertically polarized input wave is represented by the vector-phasor $\mathbf{E}_{in} = E\underline{/0}$.

---

[3] As explained in Sec. 2–5, the orientation of the electric field defines the plane of polarization of an electromagnetic wave.

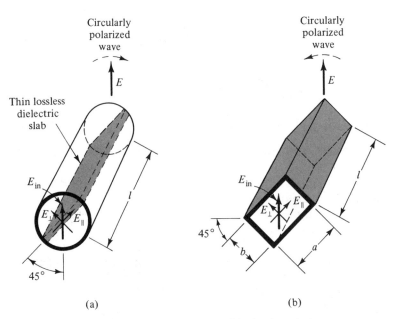

Figure 7–11 Two types of waveguide circular polarizers.

It may be decomposed into two components, one perpendicular to the plane of the thin dielectric slab ($\mathbf{E}_\perp$) and one parallel to the slab ($\mathbf{E}_\parallel$). Since the dielectric is at 45° with respect to the input vector, $\mathbf{E}_\perp = (E/\sqrt{2})\underline{/0}$ and $\mathbf{E}_\parallel = (E/\sqrt{2})\underline{/0}$. Assuming no reflections or loss for either component, the output components are given by

$$\mathbf{E}_{\text{out}_\perp} = (E/\sqrt{2})\underline{/-\beta_\perp l} \quad \text{and} \quad \mathbf{E}_{\text{out}_\parallel} = (E/\sqrt{2})\underline{/-\beta_\parallel l} \qquad (7\text{–}1)$$

where l is the length of the dielectric-loaded waveguide and $\beta_\perp$ and $\beta_\parallel$ are the phase constants for the two components. As explained in Ex. 2–5, the effect of the thin dielectric slab is to slow the parallel component relative to the perpendicular one and hence $\beta_\parallel > \beta_\perp$. By setting the length l so that $\beta_\parallel l - \beta_\perp l = \pi/2$,

$$\mathbf{E}_{\text{out}_\perp} = (E/\sqrt{2})\underline{/-\beta_\perp l} \quad \text{and} \quad \mathbf{E}_{\text{out}_\parallel} = (E/\sqrt{2})\underline{/-\beta_\perp l - \pi/2} \quad (7\text{–}2)$$

These linearly polarized waves are equal in amplitude, perpendicular to each other, and 90° out-of-phase, and hence their sum represents a circularly polarized wave. The direction of polarization is clockwise (looking from the input toward the output) since the parallel component is delayed relative to the perpendicular one. It is interesting to note that if the output wave were reflected, say, by a shorting plate, it would arrive at the input terminals as a horizontally polarized wave. This is because the parallel component experiences a round-trip phase delay of 180° relative to the perpendicular component.

Another form of circular polarizer (Fig. 7–11b) utilizes an air-filled rectangular guide in which the height b is large enough to support the TE_{01} mode. With the guide set at 45° with respect to the input polarization and assuming no reflections, the two output components are given by Eq. (7–1) where, in this case, the perpendicular component represents a TE_{10} wave with $\lambda_c = 2a$ and the parallel component a TE_{01} wave with $\lambda_c = 2b$. For $a > b$, $\beta_\perp > \beta_\parallel$ because the parallel component has the greater phase velocity. Adjusting the length of the guide l for a phase difference of $\pi/2$ yields

$$\frac{l}{\lambda_{g\perp}} - \frac{l}{\lambda_{g\parallel}} = 0.25 \tag{7-3}$$

where λ_g may be computed from Eq. (5–40). With the length so adjusted, the output wave will be a counterclockwise circularly polarized wave. Circular polarizers are also known as *quarter-wave* plates since they require a quarter wavelength difference in electrical length for the two linear waves.

Another form of quarter-wave plate has been described by Simmons (Ref. 7–13). A single-stage version is shown in Fig. 7–12a. The square (or round) waveguide contains a pair of metal septums spaced $\lambda_g/4$ apart in the propagation direction. Septums of this type are discussed in Sec. 7–2. Assuming dominant mode propagation, they have either shunt inductive or capacitive properties depending upon the orientation of the electric field. The circular polarizer operates in the fol-

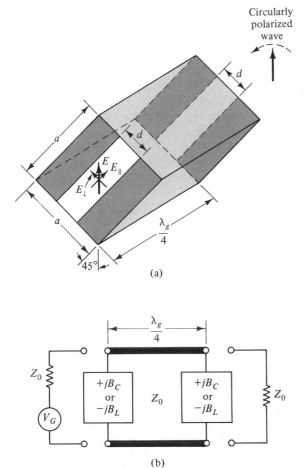

Figure 7–12 A loaded-line quarter-wave plate and its equivalent circuit.

lowing manner. With the square guide set at $45°$ with respect to the vertically polarized incoming wave ($E/\underline{0}$), vector decomposition yields $\mathbf{E}_\perp = (E/\sqrt{2})/\underline{0}$ and $\mathbf{E}_\parallel = (E/\sqrt{2}) /\underline{0}$ as indicated in the figure. The equivalent circuit for the perpendicular component is a pair of shunt capacitors spaced $\lambda_g/4$ apart, while for the parallel component it is a pair of shunt inductors similarly spaced. The analysis that follows shows that the effect of the capacitors is to phase delay the wave, while that of the inductors is to advance it. By setting this phase difference equal to $\pi/2$, the resultant output wave is circularly polarized.

The equivalent circuit for the two linearly polarized waves is shown in part b of Fig. 7–12. For the perpendicular component, the normalized admittance of each shunt element is $+j\bar{B}_C$ while for the parallel component it is $-j\bar{B}_L$. These values are normalized to Y_0, the characteristic admittance of the square guide. Both $\bar{B}_L$ and $\bar{B}_C$ are functions of the dimension d as indicated in Figs. 7–15a and 7–16a, respectively. With the shunt elements spaced $\lambda_g/4$ apart at the design frequency, a normalized ABCD matrix analysis of the circuit for the perpendicular component yields

$$\begin{bmatrix} \bar{A} & \bar{B} \\ \bar{C} & \bar{D} \end{bmatrix} = \begin{bmatrix} 1 & 0 \\ j\bar{B}_C & 1 \end{bmatrix}\begin{bmatrix} 0 & j1 \\ j1 & 0 \end{bmatrix}\begin{bmatrix} 1 & 0 \\ j\bar{B}_C & 1 \end{bmatrix} = \begin{bmatrix} -\bar{B}_C & j1 \\ j(1 - \bar{B}_C^2) & -\bar{B}_C \end{bmatrix} \qquad (7\text{--}4)$$

With the assumption that $Z_G = Z_L = Z_0$, the characteristic impedance of the square guide, Eq. (4–57), may be used to calculate the insertion phase.

For the perpendicular component,

$$\theta_\perp = \arctan\left(\frac{2 - \bar{B}_C^2}{-2\bar{B}_C}\right) = \frac{\pi}{2} + \arctan\left(\frac{2\bar{B}_C}{2 - \bar{B}_C^2}\right) \qquad (7\text{--}5)$$

A similar analysis for the parallel component yields

$$\theta_\parallel = \frac{\pi}{2} - \arctan\left(\frac{2\bar{B}_L}{2 - \bar{B}_L^2}\right) \qquad (7\text{--}6)$$

As explained below Eq. (4–54), a positive value of θ indicates a phase delay, and hence the $\pi/2$ value in both equations may be interpreted as the delay due to the $\lambda_g/4$ line. The second term in Eq. (7–5) represents the phase delay of the two shunt capacitances, while the second term in Eq. (7–6) indicates the phase advance associated with the shunt inductances. Both $\bar{B}_C$ and $\bar{B}_L$ increase with increasing septum penetration. Thus the dimension d can be set so that $\theta_\perp - \theta_\parallel = \pi/2$, the required condition for circular polarization. Problem 7–4 shows that if $\bar{B}_L = \bar{B}_C$, the insertion loss is the same for both waves. This means that the output amplitudes of the perpendicular and parallel waves are equal, which is a necessary condition for circular polarization. Since the septum spacing can only be $\lambda_g/4$ at the design frequency, this circular polarizer is inherently narrowband. Simmons (Ref. 7–13) has shown that increased bandwidth can be achieved by using a multisection design.

It is important that circular polarizers be well matched at both input and output terminals. Failure to do this results in multiple reflections which may cause the output to be elliptically polarized.

The half-wave plate is another useful microwave component. Its configuration is the same as the quarter-wave plate except that $\theta_\perp - \theta_\parallel$ equals π. It has the following properties:

1. If the input wave is circularly polarized, the output wave will also be circularly polarized. The direction of rotation of the output wave will be opposite that of the input wave.

2. If the input wave is linearly polarized, the output wave will also be linearly polarized. A special case of interest is when the input polarization plane is such that the amplitudes of the perpendicular and parallel components are equal. In this situation, the plane of polarization of the output wave will be rotated 90° with respect to that of the input wave.

Both quarter-wave and half-wave plates find application in many microwave components and systems. These include phase shifters, filters, antennas, and ferrite devices.

7–2 WAVEGUIDE DISCONTINUITIES

Any interruption in the uniformity of a transmission line is defined as a discontinuity. This section describes and analyzes several waveguide type discontinuities. In particular, their effect on the dominant mode of propagation is considered.

To satisfy the electromagnetic boundary conditions at a discontinuity usually requires the presence of higher-order modes. When the waveguide dimensions are such that the higher modes are below cutoff, they will be confined to the region near the discontinuity. Since these localized modes represent stored electric and magnetic energy, their effect can be modeled by a reactive network. Once the network has been determined, transmission-line and ac circuit methods may be utilized to calculate the effect of the discontinuity on the propagating dominant mode.[4] An example of this approach was given in Sec. 6–4a, where an abrupt change in the diameter of a coaxial line was modeled as a shunt capacitance. This present section describes the equivalent circuits for some typical waveguide discontinuities.

Change in waveguide height.   An abrupt change in the b dimension of a rectangular guide is shown in Fig. 7–13a. The characteristic impedance of the two waveguides are denoted by Z_{01} and Z_{02}. The change in guide height causes a distortion of the TE_{10} electric field in the vicinity of the step discontinuity. This situation is analogous to the TEM mode distortion caused by a change in the diameter of a coaxial line (Fig. 6–25). Under the assumption that only the dominant TE_{10} mode propagates, the distorted field may be represented by an infinite set of nonpropagating TM modes. Their presence is required because the distorted pattern contains electric field components parallel to the waveguide axis. A symmetry argument can be used to prove that for the symmetrical step shown in the figure, only TM_{1n} modes with n an even integer are generated. Their amplitudes decrease as a function of distance from the step discontinuity, the attenuation constant being given by Eq. (5–42). The analysis in Sec. V of Ref. 7–5 shows that these localized TM modes can be represented by a capacitance shunting the transmission lines. Normalized

---

[4] The mathematical methods required to determine element values of the network are beyond the scope of this text. For details, the reader is referred to Refs. 7–2, 7–3, 7–5, and 7–9.

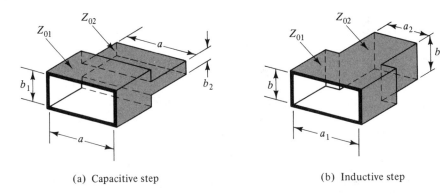

(a) Capacitive step   (b) Inductive step

Figure 7–13 Symmetrical step discontinuities in rectangular waveguide.

capacitive susceptance data is given in Marcuvitz (Ref. 7–6) for both symmetrical and asymmetrical steps. Values for the symmetrical case are repeated in Fig. 7–14a.

The usefulness of the equivalent circuit representation in determining the effect of a discontinuity on the propagating mode is indicated by the following example.

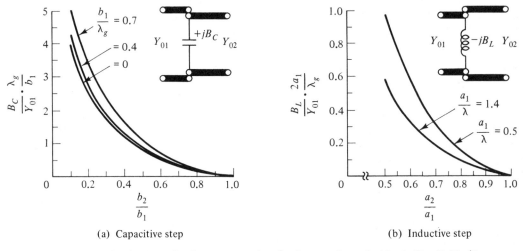

(a) Capacitive step   (b) Inductive step

Figure 7–14 Normalized susceptance data for the step discontinuities in Fig. 7–13. (From N. Marcuvitz, Ref. 7–6.)

Example 7–1:

Referring to Fig. 7–13a, $b_1 = 2$ cm, $b_2 = 1$ cm, and $a = 4$ cm.

(a) Calculate the SWR along the Z_{01} line at 6 GHz if the Z_{02} line is terminated by a matched load. Assume lossless, air-filled waveguides.

(b) A matched 6 GHz generator ($Z_G = Z_{01}$) having an available power $P_A = 1$ W is connected to the Z_{01} line. How much power is delivered to the matched load terminating the Z_{02} line?

Solution: At 6 GHz, $\lambda_0 = 5$ cm. From Eq. (5–40), $\lambda_g = 6.41$ cm, where $\lambda_c = 2(4) = 8$ cm for the TE_{10} mode. From Fig. 7–14a for $b_2/b_1 = 0.5$ and $b_1/\lambda_g = 0.31$,

$$\frac{B_C}{Y_{01}} \approx (0.8)(0.31) = 0.25$$

Thus the normalized admittance terminating the Z_{01} line is

$$\frac{Y_{02}}{Y_{01}} + j\frac{B_C}{Y_{01}} = 2 + j0.25$$

where for a fixed width a, the characteristic impedance of rectangular guide [Eq. (5–43)] is directly proportional to the guide height.
(a) By using Eqs. (3–46) and (3–52),

$$|\Gamma| = 0.34 \qquad \text{and} \qquad SWR = 2.04$$

(b) With $Z_G = Z_{01}$, $P^+ = P_A = 1$ W and the reflected power $P^- = |\Gamma|^2 P^+ = 0.12$ W. Therefore, the power delivered to the Z_{02} load is $1 - 0.12 = 0.88$ W.

Change in waveguide width.   An abrupt change in the a dimension of a rectangular guide (Fig. 7–13b) can be represented by a shunt inductance since only higher-order TE modes are generated by the discontinuity. A symmetry argument shows that for TE_{10} excitation, only TE_{m0} modes with m an odd integer are produced by the symmetrical step shown (Prob. 7–9). Normalized susceptance data is given in Fig. 7–14b and is based upon information found in Marcuvitz (Ref. 7–6). The reference also includes data for the asymmetrical case.

Dielectric discontinuity.   Figure 6–23a illustrates an abrupt dielectric discontinuity in a coaxial line. Its equivalent circuit is shown in part b. The analogous situation in rectangular guide results in the same equivalent circuit if the guide dimensions are identical in both regions and the dielectrics are nonmagnetic. The circuit contains no reactive elements because only a reflected TE_{10} wave is needed to satisfy the boundary conditions at the dielectric interface. Since the equivalent circuit in Fig. 6–23b is valid, the reflection coefficient at the interface for a TE_{10} wave propagating from region 1 toward region 2 is given by Eq. (6–20), where in this case, Z_{01} and Z_{02} are the waveguide impedances are defined by Eq. (5–43). Note, however, that Eqs. (6–21) and (6–22) are not valid for waveguides since the cutoff frequencies are different in the two regions.

Inductive septums.   Three symmetrical inductive discontinuities are shown in Fig. 7–15. For very thin septums ($t < 0.1d$), the equivalent circuit for dominant mode transmission is simply an inductance shunting the waveguide. Approximate expressions for normalized susceptance are given below the figures for two cases. For the restrictions given, the formulas are accurate to within 10 percent over the useful waveguide range. Corrections for finite iris thickness are given in Sec. 6.3 of Ref. 7–7. If the iris thickness is appreciable, series reactive elements must be included in the equivalent circuit. More complete susceptance data for both symmetrical and asymmetrical configurations may be found in Refs. 7–6 and 7–8.

As explained earlier, the equivalent circuit for a particular obstacle derives from the localized higher-order modes generated by it. It may be instructive at this point to use transmission-line concepts to show that the equivalent circuit interpretation is indeed reasonable. As an example, consider the configuration shown in part a of Fig. 7–15. It was explained in Sec. 5–4 that a rectangular waveguide in the TE_{10} mode may be thought of as a two-wire line shunted by an array of quarter-wave

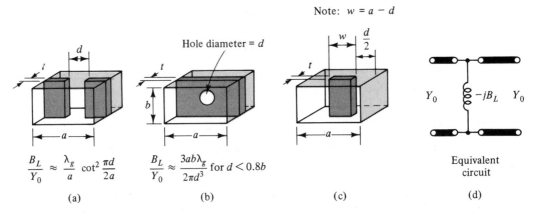

Note: $w = a - d$

$$\frac{B_L}{Y_0} \approx \frac{\lambda_g}{a} \cot^2 \frac{\pi d}{2a}$$

$$\frac{B_L}{Y_0} \approx \frac{3ab\lambda_g}{2\pi d^3} \text{ for } d < 0.8b$$

(a)   (b)   (c)   (d)

Figure 7–15   Three symmetrical inductive irises and their equivalent circuit (for $t < 0.1d$).

shorted stubs (Fig. 5–17). Since the admittance of the stubs is zero, energy propagates down the guide without loss. The configuration in Fig. 7–15a may be similarly viewed, except that at the plane of the septum, the shorted stubs are *less* than a quarter-wavelength long and hence inductive. With the stubs shunting the main line, the equivalent circuit given in the figure is reasonable.

The inductive discontinuities in parts *b* and *c* of Fig. 7–15 are particularly useful when large values of normalized susceptance are desired. The septum shown in part *b* generates both TE and TM higher-order modes. For a round hole, the TE modes are dominant which results in an inductive circuit. An approximate expression for B_L/Y_0 is given in the figure. The thin metal septum shown in part *c* also generates higher-order TE modes and hence behaves like a shunt inductance. The reasonableness of the equivalent circuit can be argued from a circuit point of view. Again referring to Fig. 5–17, some of the TE_{10} *power* current will be shunted through the metal septum. By Ampere's law, magnetic flux is produced which links the conduction current, thus producing an inductive effect. Normalized susceptance data for this configuration is given in Refs. 7–6 and 7–8.

In all cases, the susceptance of the inductive septum decreases with increasing frequency, as might be expected. However, the frequency behavior is not exactly the same as an ideal inductor, wherein B_L is proportional to $1/\omega$.

Capacitive septums.   Two examples of capacitive discontinuities are shown in Fig. 7–16. For TE_{10} transmission, the equivalent circuit is simply a capacitance shunting the waveguide since only TM modes are generated by the discontinuity. Approximate formulas for $t < 0.1b$ are given below the figures. In both cases, B_C increases with increasing frequency. As in the previous case, more exact data may be found in Refs. 7–6 and 7–8.

Capacitive septums are seldom used in high-power applications since the reduced height at the plane of the discontinuity greatly reduces the breakdown voltage of the structure. The inductive septum in Figs. 7–15a is the usual choice in these cases.

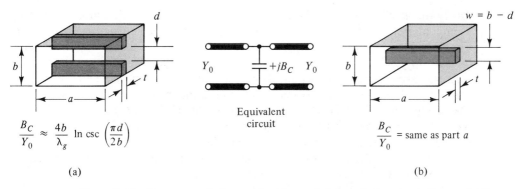

$$\frac{B_C}{Y_0} \approx \frac{4b}{\lambda_g} \ln \csc \left(\frac{\pi d}{2b}\right)$$

$$\frac{B_C}{Y_0} = \text{same as part } a$$

(a)

(b)

Figure 7–16 Two symmetrical capacitive irises and their equivalent circuit (for $t < 0.1b$).

The resonant iris.   Another useful waveguide discontinuity is that shown in Fig. 7–17. It is essentially a combination of the inductive and capacitive irises described in Figs. 7–15a and 7–16a and therefore is equivalent to a parallel L-C circuit. It is known as a resonant iris since, by proper choice of a' and b', it resonates within the useful frequency range of the waveguide. The resonant wavelength (λ_r) may be obtained from the following empirical relationship.

$$\frac{a'}{b'}\sqrt{1 - \left(\frac{\lambda_r}{2a'}\right)^2} = \frac{a}{b}\sqrt{1 - \left(\frac{\lambda_r}{2a}\right)^2} \qquad (7\text{–}7)$$

The reasonableness of this equation is explained by Slater (Ref. 7–4). Solving for the resonant frequency in the case of an air-filled guide yields

$$f_r = \frac{c}{\lambda_r} = \frac{c}{2a}\sqrt{\frac{1 - (b'/b)^2}{(a'/a)^2 - (b'/b)^2}} \qquad (7\text{–}8)$$

The equations are accurate to within 5 percent when the iris thickness is less than $0.1b$.

For a given value of λ_r (or f_r), there are an infinite set of a' and b' values that

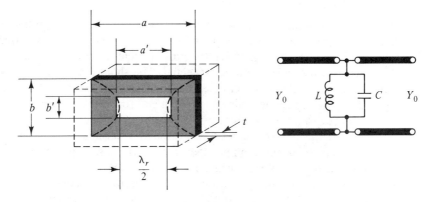

Figure 7–17 A resonant iris and its equivalent circuit (for $t < 0.1b$).

satisfy the above equations, the only restrictions being $2a' > \lambda_r$ and $a'/a > b'/b$. Montgomery (Ref. 7–7) explains that Eq. (7–7) represents a hyperbola with a' and b' as variables. The hyperbola passes through the corners of the rectangular guide as indicated by the dashed lines in Fig. 7–17. Note that the branches are separated by $\lambda_r/2$ at the closest point, which means that a' must be greater than half the resonant wavelength. The size of the opening determines the circuit's frequency sensitivity. The smaller the values of b' and a', the larger the C/L ratio and hence the higher the Q of the circuit. This follows from elementary circuit theory wherein, for parallel resonance, Q is proportional to $\sqrt{C/L}$.

There are many other forms of resonant irises for both rectangular and circular guide. Some of these are described in Refs. 7–1 and 7–7.

Variable height post.   Figure 7–18 shows a round metal rod centrally located in a rectangular guide. By varying the amount of penetration h, a wide range of susceptance values can be realized. One form consists of a metal screw which is threaded through the broadwall of the guide. A commercially available unit, known as a *slide-screw tuner*, uses an adjustable depth probe whose position along the propagation axis is also adjustable. It is very useful in experimental work as an impedance matching device.

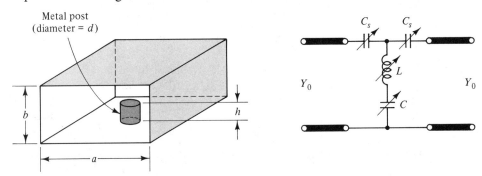

Figure 7–18   A metal tuning post and its approximate equivalent circuit.

The equivalent circuit for the post is shown in the figure. Some experimental data for the element values are given in Ref. 7–6. The configuration has been analyzed by Lewin (Ref. 7–9). When the post diameter is small ($d < 0.15a$), the effect of the series capacitances can be neglected. The shunt portion of the equivalent circuit consists of a series L-C circuit. The inductance is associated with the magnetic flux linking the conduction current in the metal rod, while the capacitance is associated with the displacement current in the remaining length ($b - h$). For $h < 0.5b$, $X_C \gg X_L$ and the adjustable post behaves like a variable capacitance shunting the waveguide. As the penetration h is increased further, X_C approaches X_L. For standard waveguide ($a \approx 2b$), series resonance occurs when h is approximately a quarter-wavelength long. Further penetration causes the net susceptance to become inductive. When $h = b$, $X_C = 0$ and the configuration reduces to an inductance shunting the waveguide. For $d < 0.15a$, the normalized inductive susceptance may be approximated by

$$\frac{B_L}{Y_0} = \frac{2\lambda_g}{a \ln\left(\dfrac{0.22a}{d}\right)} \tag{7-9}$$

More exact data may be found in Refs. 7-6 and 7-8.

Normalized susceptance data for the various irises and posts is easily obtained experimentally. With a matched load connected to the output, one merely measures the input SWR. Since the load and the susceptance are in parallel, $\bar{Y}_{in} = 1 + j\bar{B}$ and hence the magnitude of the reflection coefficient is

$$|\Gamma_{in}| = \frac{|\bar{B}|}{\sqrt{\bar{B}^2 + 4}}$$

By using Eq. (3-52), the following relation is obtained for $|\bar{B}|$.

$$|\bar{B}| = \frac{\text{SWR} - 1}{\sqrt{\text{SWR}}} \tag{7-10}$$

For example, suppose the measured input SWR of an inductive iris, matched-load combination is 5.0. From the above equation, the normalized susceptance of the inductive iris is 1.79.

It must be emphasized that all the equivalent circuits given here assume that only the dominant TE_{10} mode propagates in the guide and hence the higher-order modes are localized to the region near the discontinuity that generated them. The attenuation constant for the higher modes is given by Eq. (5-42). If the guide contains more than one obstacle, their equivalent circuits are valid as long as the higher modes due to one discontinuity are sufficiently attenuated (>20 dB) at the plane of the other discontinuity. The problem of direct coupling between discontinuities is briefly discussed in Sec. 6.9 of Ref. 7-7.

7-3 TERMINATIONS

The matched load, the standard mismatch, and the adjustable short circuit are commonly used transmission-line terminations. Waveguide version are described in this section. Also included is a brief discussion of waveguide flanges.

7-3a Matched Loads

As explained in Sec. 6-1a, a matched load absorbs all the power incident upon it. Most waveguide loads are either of the tapered or stepped variety.

Tapered loads.   Three examples of tapered loads are shown in parts a, b, and c of Fig. 7-19. The first one is used in broadband, low- and medium-power applications. It consists of a tapered conical rod of lossy material that is either fixed in place or movable along the propagation axis. Reflections are quite small (typically, SWR < 1.04) when the taper length l_1 is greater than a few wavelengths at the lowest operating frequency. The length l_2 is chosen so that the combined loss through l_1

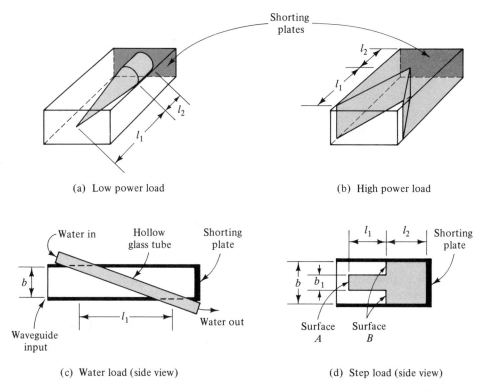

(a) Low power load   (b) High power load

(c) Water load (side view)   (d) Step load (side view)

Figure 7–19   Tapered and step loads in rectangular guide.

and l_2 is greater than 20 dB. The reason for this requirement was explained in Sec. 6–1a.

One application of the matched load is in the measurement of reflections from a two-port network (Prob. 7–15). This is done by placing a matched load at the output port and measuring the input SWR. With a reflectionless load, the measured SWR represents the reflection due to the two-port. If, however, the load is not perfectly matched, there will be some uncertainty regarding the SWR of the two-port itself since the measured input SWR can range from SWR_X/SWR_L to $(SWR_X)(SWR_L)$, where SWR_X and SWR_L represent the values for the two-port and the load, respectively. This is based upon Eq. (4–34) where it is assumed that the two-port is dissipationless and $SWR_X \geq SWR_L$. This uncertainty can be eliminated by using a sliding load and measuring the minimum and maximum values of input SWR. These values are then used to compute the load SWR and the SWR of the two-port network.

The load shown in part b of Fig. 7–19 is useful in high-power applications because most of the lossy element's surface is in intimate contact with the metal waveguide walls. As a result, the heat generated by the absorbed microwave power is conducted to the walls where cooling fins can be used to minimize the temperature rise. In very high-power applications, forced air or water cooling may be required.

Since the load element must withstand high temperatures, ceramic-based absorbing materials are commonly used. Note that the taper begins at the narrow walls of the guide where the electric field is negligible. This minimizes the possibility of arcing at high power levels.

The water load described in part *c* of Fig. 7–19 is used to accurately measure microwave power. It consists of water circulating through a hollow glass tube mounted in the waveguide as shown. The increase in water temperature is proportional to the microwave power absorbed. The tube is inserted at a shallow angle so as to minimize reflections. With l_1 greater than a few wavelengths, the SWR is quite low and essentially all the incident power is dissipated by the circulating water.

Step loads.   The stepped load finds use in applications where there is insufficient room to accommodate a tapered load. Typically it is less than one-half wavelength long and has a useful bandwidth of about 10 percent (compared to 40 percent for a tapered load). The usual configuration is shown in Fig. 7–19d. The l_1 portion of the load acts as a quarter-wave transformer from air-filled waveguide to the fully-loaded lossy guide. Since the quarter-wave section is also lossy, its dimensions are more easily determined experimentally. The b_1 dimension is adjusted so that the reflection at surface *B referred to the input plane* equals the reflection due to surface *A*. That is, $|\Gamma_B|e^{-2\alpha l_1} = |\Gamma_A|$. The length l_1 is adjusted so that the two reflections cancel at the input plane resulting in an input SWR of unity. This means that $Z_{in} = Z_0$, the characteristic impedance of the air-filled guide. As in other type loads, l_2 is chosen so that the combined loss through l_1 and l_2 is greater than 20 dB. As in any quarter-wave transformer, the SWR is a function of the operating frequency. Typically an SWR of less than 1.10 can be maintained over a 10 percent band. Broader-band performance can be achieved with two-or three-section stepped loads.

7–3b Standard Mismatches

Standard mismatches are used to calibrate reflection and impedance measuring equipment. Waveguide versions consist of a well-matched tapered load inserted into a reduced-height waveguide. The configuration and its equivalent circuit is shown in Fig. 7–20, where *b* represents the height of standard waveguide. The desired SWR is obtained by selecting b_1, the height of the reduced-height guide, so that the SWR = b/b_1. This relation assumes that both guide widths are the same, the tapered load is reflectionless, and the discontinuity capacitance C_d is negligible. For these conditions, the SWR is independent of frequency since from Eq. (5–43) the normalized load impedance Z_{01}/Z_0 reduces to b_1/b. Above cutoff, Z_{01} and Z_0 are real and hence, from Eq. (3–55), the input SWR equals b/b_1.

The effect of the discontinuity capacitance C_d is that the SWR is slightly greater than b/b_1 and is frequency dependent. Problem 7–16 shows how b/b_1 can be readjusted to give the desired SWR and that the variation of SWR with frequency is quite small. Standard mismatches having an SWR of 1.50 are widely used in waveguide equipment.

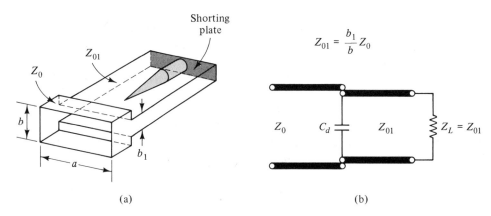

(a)   (b)

Figure 7–20 A standard mismatch and its equivalent circuit.

7–3c Adjustable Short Circuits

The adjustable short circuits shown in Fig. 7–21 are used to provide a variable reactance in waveguide systems. The normalized input impedance is given by

$$\frac{Z_{in}}{Z_0} \equiv \bar{Z}_{in} = j\bar{X}_{in} = j\,\tan\frac{2\pi x}{\lambda_g} \qquad (7\text{--}11)$$

Any value of $\bar{X}_{in}$ can be realized by proper adjustment of the short position x. For example, letting $x = \lambda_g/4$ results in an infinite input reactance, which represents an open circuit. The comments in Sec. 6–1d regarding resistive effects and frequency sensitivity for a coaxial short also apply to the waveguide version. Due to the cutoff effect, the frequency sensitivity of X_{in} is even greater in waveguide.

The contacting-type adjustable waveguide short shown in part a of Fig. 7–21 can suffer from the noisy-contact problem discussed in Sec. 6–1d. This difficulty can be avoided by using a noncontacting adjustable short. One version is shown in part b of the figure. The technique is to insure that ohmic contact occurs at a zero current point, that is, a point of infinite impedance. At the design frequency, both l_1 and l_2

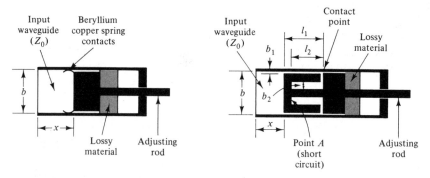

(a) Contacting type (side view)   (b) Noncontacting type (side view)

Figure 7–21 Adjustable short circuits in waveguide. (Note: The lossy material absorbs energy leaking past the contact points.)

are $\lambda_g/4$ long. The l_2 line sections transform the short circuits at points A into open circuits at the contact points. Any resistance R at the contact points is in series with the open and hence the impedance of the combination is infinite, no matter what the value of contact resistance. The l_1 line section transforms the infinite impedance into a short circuit at the front surface of the noncontacting short. Since the short circuit is independent of the contact resistance R, the erratic behavior associated with contacting-type shorts is avoided. The following example illustrates the point.

Example 7-2:

A noncontacting short of the type shown in Fig. 7-21b has the following dimensions:

$$b = 2.0 \text{ cm}, \qquad b_1 = 0.06 \text{ cm}, \qquad b_2 = 0.40 \text{ cm}, \qquad l_1 = l_2 = 1.8 \text{ cm}$$

Calculate the variation in return loss at the design frequency if the normalized resistance (R/Z_0) at the contact point varies from 0 to 0.01. Compare the result with a contacting short having the same variation in R/Z_0.

Solution: The equivalent circuit for the noncontacting short is shown in Fig. 7-22. Assuming a constant guide width throughout,

$$Z_{01}/Z_0 = b_1/b = 0.03 \qquad \text{and} \qquad Z_{02}/Z_0 = b_2/b = 0.20$$

Since l_1 and l_2 are $\lambda_g/4$ at the design frequency, $\lambda_g = 7.2$ cm. The impedance at terminals d-d' (Z_d) is obtained by noting that the Z_{01} line is terminated by the series combination of R and Z_c, where Z_c is the impedance at terminals c-c'. From the impedance transformation equation,

$$Z_d = Z_{01}\frac{(Z_c + R) + jZ_{01}\tan\beta l_1}{Z_{01} + j(Z_c + R)\tan\beta l_1}$$

where $Z_c = jZ_{02}\tan\beta l_2$ and $\beta = 2\pi/\lambda_g$.

The impedance Z at the front surface of the short is $2Z_d$. Making use of the above expressions for Z_c and Z_d, the normalized value of Z becomes

$$\bar{Z} = (-j2\bar{Z}_{01}\cot\beta l)\frac{-(\bar{Z}_{01} + \bar{Z}_{02}) + j\bar{R}\cot\beta l}{(\bar{Z}_{01}\cot^2\beta l - \bar{Z}_{02}) + j\bar{R}\cot\beta l}$$

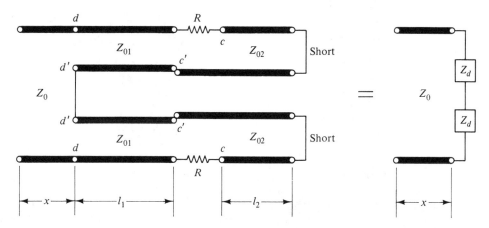

Figure 7-22 Equivalent circuit for the noncontacting short in Fig. 7-21b.

where $l_1 = l_2 = l$ and all impedances have been normalized to Z_0, the characteristic impedance of the input guide. At the design frequency, $\cot \beta l = 0$ and hence $\bar{Z} = 0$, for any value of R, the contact resistance. Thus, independent of R, $|\Gamma| = 1$ and the return loss is 0 dB.

For the contacting-type short, $\bar{Z} = 2\bar{R}$. With $\bar{R} = 0$, $|\Gamma|$ is unity and the return loss is 0 dB. However, for $\bar{R} = 0.01$, $\bar{Z} = 0.02$, and $|\Gamma| = 0.96$, which is equivalent to a return loss of 0.35 dB. This amount of loss variation due to an erratic contact is unacceptable in most applications.

Note that in the above example, the result for the noncontacting short was independent of Z_{01} and Z_{02}. At other than the design frequency, however, Z is a function of both Z_{01} and Z_{02} (and hence b_1 and b_2). In order to maximize the useful bandwidth of the noncontacting short, b_1 should be chosen as small as possible and b_2 as large as possible. The reason is that, at other than the design frequency, $\cot \beta l$ is finite and in order to maintain $Z_c \gg R$, Z_{02} must be as large as possible. By similar reasoning, Z_{01} must be chosen as small as possible in order to maintain a low value of Z_d. This is verified by a numerical example (Prob. 7–18).

7–3d Waveguide Flanges

The connection between waveguides is accomplished with flanges. The types used are the contact or cover flange and the choke flange. Two methods of joining waveguide sections are shown in Fig. 7–23.

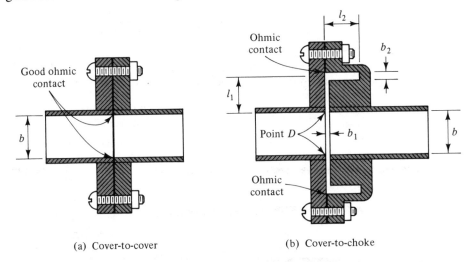

(a) Cover-to-cover   (b) Cover-to-choke

Figure 7–23 Flange coupling between waveguide sections (cross-sectional views).

Contact-type coupling.   The connection between two waveguides may be accomplished with a pair of cover flanges as shown in part *a* of Fig. 7–23. The flanges are bolted together so that the mating surfaces make good ohmic contact, particularly at the points indicated, where longitudinal currents flow along the broad walls of the waveguide. To minimize reflections at the joint, the mating surfaces must be clean and flat. With proper care in aligning the two guides, the SWR due to the joint is typically 1.02 or less and repeatable over many connect/disconnect operations. For this reason, this type coupling is used in most low- and medium-power applications. At high-power levels, an imperfect contact between flanges may interrupt the longitudinal conduction current resulting in arcing at the joint. This problem can be avoided by using a choke flange.

Cover-to-choke flange coupling.   An effective connection between waveguides, particularly in high-power applications, involves the use of one cover flange and one choke flange. A cross-sectional view of the connection is shown in Fig. 7–23b. When the flanges are bolted together, ohmic contact is made at the points indicated since part of the choke flange is undercut (dimension b_1). The flange also contains a circular groove (dimensions b_2 and l_2). The principle of operation is similar to the noncontacting short discussed earlier. With l_2 equal to $\lambda_g/4$, any finite resistance at the contact point is in series with an infinite impedance and hence the current is zero at that point. With the groove diameter chosen so that l_1 also equals $\lambda_g/4$, the impedance at point D is zero, thus providing continuity of longitudinal current flow between waveguides. As in the case of the noncontacting short, broadband performance is achieved by choosing $b_2 \gg b_1$. For example, the standard choke flange for WR-90 guide has $b_1 = 0.030$ cm and $b_2 = 0.25$ cm. SWR values for this type coupling are typically 1.04 or less across the useful waveguide band. Resonances associated with higher modes in the circular groove may restrict the useful range of the choke flange. This problem can be eliminated by filling in the side portions of the circular groove.

The reason that the cover-to-choke flange arrangement is preferred in high-power applications is that the ohmic contact occurs at a minimum current point. Thus, arcing is avoided even if the contact is imperfect and erratic. Choke flanges are also used with circular guides. In designing such a choke, one must be careful to use the appropriate wavelength for the mode excited in the choke section (see Sec. 4.9 of Ref. 7–10).

7–4 ATTENUATORS AND PHASE SHIFTERS

This section describes some of the methods used to control the amplitude and phase of microwave transmission in waveguides. Units designed for use with standard guide sizes are available from most microwave equipment manufacturers.

7–4a Waveguide Attenuators

As explained in Sec. 6–3a, attenuators may be either dissipative or reflective. The waveguide versions described here are all of the dissipative type.

Resistive-card attenuator.   The resistive-card attenuator (Fig. 7–24) may be either fixed or variable. In the fixed version, the card is bonded in place as indicated in part *a* of the figure. The card is tapered at both ends in order to maintain a

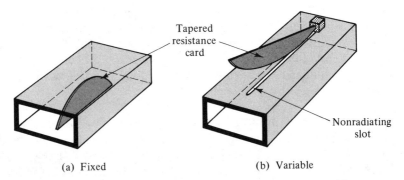

Tapered
resistance
card

Nonradiating
slot

(a) Fixed   (b) Variable

Figure 7–24 Resistive card attenuators in rectangular waveguide.

low input and output SWR over the useful waveguide band. Maximum attenuation
per length is achieved by having the card parallel to the electric field and at the center of the guide where the TE_{10} electric field is maximum. The conductivity and
dimensions of the card are adjusted, usually by trial-and-error, to obtain the desired
attenuation value. The attenuation is a function of frequency, generally becoming
greater with increasing frequency. In high-power versions, ceramic-type absorbing
materials are used instead of the resistive card.

A variable version of this attenuator, known as the *flap attenuator*, is shown in
part *b* of Fig. 7–24. The card enters the waveguide through the slot in the broad
wall, thereby intercepting and absorbing a portion of the TE_{10} wave. The hinge
arrangement allows the card penetration and hence the attenuation to be varied from
zero to some maximum value, typically 30 dB. With the longitudinal slot centered
on the broad wall, none of the TE_{10} wave is radiated.[5]

There are two characteristics of the flap attenuator that make its use awkward
in certain applications. First, the attenuation is frequency sensitive which makes it
inconvenient to use as a calibrated attenuator. Second, the phase of the output signal
is a function of card penetration and hence attenuation. This may result in nulling
difficulties when the attenuator is part of a bridge-type network. Despite these disadvantages, the flap attenuator finds widespread use in microwave equipment.

Rotary-vane attenuator. A variable attenuator that avoids the disadvantages indicated above is the rotary-vane attenuator (Fig. 7–25). The essential parts
consist of three circular waveguide sections, two fixed and one rotatable. Also
included are input and output transitions that provide low-SWR connections to
standard rectangular guide. The attenuation is controlled by rotation of the center
section, minimum loss occurring with $\theta_m = 0$ and maximum loss when $\theta_m = 90°$.
The analysis that follows shows that the attenuation is a function of θ_m only. This is
a significant advantage since the accuracy of the attenuation value merely requires
the precise setting of a mechanical angle. The rotary-vane attenuator is, in fact, so
accurate that it is used as a calibration standard in most microwave laboratories.

The principle of operation is based upon the interaction between plane-polar-
ized waves and thin resistive cards. The effect is analogous to that described in Sec.

[5] Radiating and nonradiating slots are discussed in Sec. 8–3a.

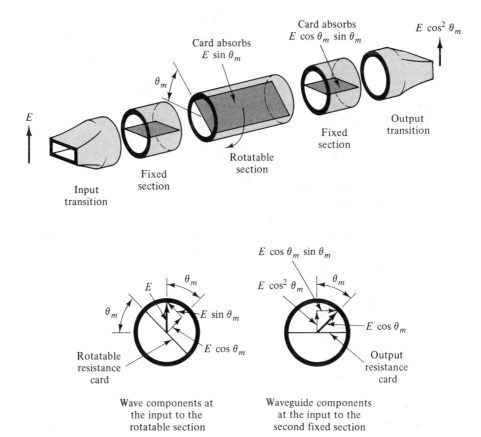

Figure 7–25   The rotary-vane attenuator.

2–5 (Exs. 2–4 and 2–5). Namely, when the electric field of a propagating wave is perpendicular to the plane of the card, the wave attenuation is negligible. On the other hand, when the electric field is parallel to it, the wave is completely absorbed by the resistive card. In practice, each fixed card attenuates the dominant TE_{11} wave by at least 40 dB, while the attenuation of the rotatable card is usually greater than 80 dB.

The input transition shown in the figure converts the TE_{10} wave into a vertically polarized TE_{11} wave in circular guide. The electric field associated with the wave is denoted by E. With the input resistance card perpendicular to the electric field, the wave propagates through the first fixed section without loss. When the card in the rotatable section is horizontal ($\theta_m = 0$), the wave passes through it and the output fixed section without loss. Thus for $\theta_m = 0$, the total loss is 0 dB. For any other angle, the component parallel to the rotatable card ($E \sin \theta_m$) is absorbed and the perpendicular component ($E \cos \theta_m$) arrives at the second fixed section with its polarization at an angle θ_m with respect to the vertical plane. This is indicated in the figure. The portion of the wave that is parallel to the output card ($E \cos \theta_m \sin \theta_m$) is absorbed, while the perpendicular component ($E \cos^2 \theta_m$) proceeds to the ouput port via the circular-to-rectangular transition. With power flow proportional to the square of electric field, the fraction of the incident power that is delivered to a matched load is $\cos^4 \theta_m$. Thus the attenuation (in dB) of the rotary-vane attenuator is

$$A_t = 10 \ \log \left(\frac{1}{\cos^4 \theta_m} \right) \tag{7–12}$$

With a matched generator at the input, the incident power equals the available generator power and therefore the above expression also represents the insertion and transducer loss of the attenuator.

The attenuator can be set for any desired loss value by turning the rotatable section to the appropriate angle θ_m. Its accuracy is determined by the precision with which the angle θ_m can be set (Prob. 7–19). Since A_t is a function of θ_m only, its value is the same at all frequencies within the useful waveguide band, an important advantage in many applications.

Another useful characteristic of the rotary-vane attenuator is that the phase of the output signal is independent of the attenuation setting. The reason is that in order for a signal to arrive at the output, its electric field must have been perpendicular to all three cards. Thus, if $\beta_\perp$ represents the phase constant of the circular guides when the electric field is perpendicular to the card and l the overall length of the three sections, then the total phase delay through the circular sections is $\beta_\perp l$, which is independent of θ_m and hence A_t.

With power proportional to the square of electric field, the fraction of the incident power absorbed by the three resistance cards is 0, $\sin^2 \theta_m$ and $\sin^2 \theta_m \cos^2 \theta_m$, respectively. One might question the necessity of the fixed card at the input side since it absorbs none of the incident wave. It does, however, serve three purposes. First, it helps to absorb waves propagating from the output toward the input (for example due to load reflections). Second, it minimizes multiple reflections due to imperfectly matched resistance cards. The validity of Eq. (7–12) requires that multiple reflection effects are negligible. Finally, the presence of the input card results in a symmetrical configuration, which means that for any value θ_m, the attenuator is matched in both directions. Problem 7–22 shows that if the input card were eliminated, the loss would still be reciprocal but the SWR would not!

The accuracy of the rotary-vane attenuator is limited by imperfectly matched cards and the alignment precision of the resistance cards (Ref. 7–18). Commercially available units are usually calibrated from 0 to 50 dB, with an accuracy of ± 2 percent over the useful waveguide band. Due to the resistance cards, the power-handling capability of these units is typically a few watts or less. For high-power applications, an attenuator configuration that utilizes two power dividers and a phase shifter is often used (Ref. 7–16).

7–4b Waveguide Phase Shifters

The phase delay due to a waveguide section of length l is given by $\beta l = 2\pi l / \lambda_g$, where λ_g is the guide wavelength. When the generator and load are matched to the impedance of the guide, this quantity also represents the insertion phase of the section. The phase can be controlled by varying the guide wavelength. Referring to Eq. (5–40), this may be accomplished by varying either ϵ_R or the guide width a.

Dielectric phase shifter.   Figure 7–26a shows a variable phase shifter that employs a change in the effective dielectric constant of the insulating region within the guide. The configuration is the same as the variable attenuator in Fig. 7–24b except that the resistance card is replaced by a low-loss dielectric. Insertion of the dielectric into the air-filled guide at a point of maximum electric field increases the effective dielectric constant. This causes λ_g to decrease, which increases the phase delay through the fixed-length waveguide section. The tapered configuration minimizes reflections due to the dielectric. The choice of dielectric constant and thickness is governed by two factors. On one hand, increasing either results in a greater amount of phase change. If they are too large, however, higher-mode propagation can occur. The TE_{30} mode can be particularly troublesome. Equation (5–108) may be used to determine the maximum permissible values of thickness and ϵ_R that are consistent with TE_{30} mode suppression.

Another version of the dielectric phase shifter employs a pair of thin rods to move the dielectric slab from a region of minimum electric field to one of maximum field. A description of its operation is given in Sec. 8.16 of Ref. 7–10.

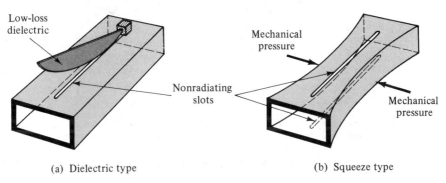

(a) Dielectric type   (b) Squeeze type

Figure 7–26   Variable phase shifters in rectangular waveguide.

Squeeze-type phase shifter.   The phase shifter shown in Fig. 7–26b uses a change in guide width to vary λ_g. The broad walls of the guide contain long nonradiating slots. A clamping arrangement applies pressure to the narrow walls as shown, which reduces the guide width a. This increases λ_g resulting in a decreased phase delay through the waveguide section.

Rotary phase shifter.   Because of its accuracy, the rotary phase shifter is used as a calibration standard in most microwave laboratories. Its operation was first described by Fox (Ref. 7–19). A sketch of the unit is shown in Fig. 7–27. Externally it looks very similar to the rotary-vane attenuator. Like the attenuator, its accuracy depends upon the precise rotation of a circular waveguide. The essential parts of the phase shifter are three circular waveguide sections, two fixed and one rotatable. The fixed sections are quarter-wave plates, while the rotatable one is a half-wave plate. Their properties were explained in Sec. 7–1d. In the discussion that follows, it is assumed that they are of the dielectric type shown in Fig. 7–11a.

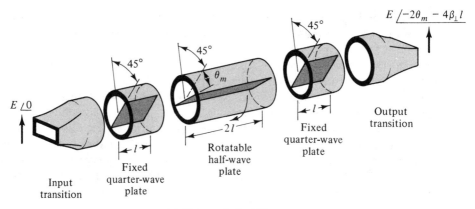

(a) Rotary phase shifter

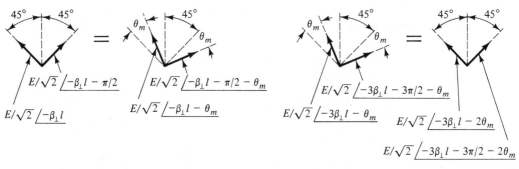

(b) Vector-phasor components at the
input to the rotatable section

(c) Vector-phasor components at the
output of the rotatable section

Figure 7–27   The rotary phase shifter and its vector-phasor representation.

The vector-phasor $E/\underline{0}$ represents the vertically polarized input wave. It may be decomposed into two components, one perpendicular and one parallel to the dielectric slab of the input quarter-wave plate. The value of each component is $E/\sqrt{2}/\underline{0}$. When $\theta_m = 0$, all three dielectric slabs are in line. In this case, the perpendicular and parallel components arrive at the output phase delayed $4\beta_\perp l$ and $4\beta_\perp l + 2\pi$, respectively, where $\beta_\perp$ represents the phase constant of the circular guide for the perpendicular component and $4l$ is the overall length of the three circular sections. With the two components in phase (2π rad being equivalent to zero phase), the output is a vertically polarized wave of value $E/\underline{-4\beta_\perp l}$. The orientation of the output transition is such that the wave is delivered to the output rectangular guide without reflection or loss. The following analysis shows that the output wave experiences an *additional* phase delay of $2\theta_m$ when the halfwave plate is rotated an angle θ_m. Also, the output remains vertically polarized, which means that the phase shifter is lossless and reflectionless for any position of the rotatable section.

As indicated earlier, the input wave may be decomposed into two components. The effect of the input quarter-wave plate is to delay the perpendicular component $\beta_\perp l$ and the parallel component $\beta_\perp l + \pi/2$, which results in a clockwise circularly polarized wave at the input to the rotatable section. When $\theta_m \neq 0$, it is convenient to replace these components by two other components, one perpendicular to the dielectric slab in the rotatable section and one parallel to it. Both sets of components are shown in part b of Fig. 7–27. The equivalence of these two sets was verified in Sec. 2–5 (see Fig. 2–19). With the length of the half-wave plate equal to $2l$, the perpendicular and parallel components are further delayed $2\beta_\perp l$ and $2\beta_\perp l + \pi$, respectively. The resultant components are shown in part c of the figure. This now represents a counterclockwise circularly polarized wave. As before, these components may be replaced by an equivalent set that are perpendicular and parallel to the *output* dielectric slab. Their vector-phasor represenation is shown on the right side of part c in Fig. 7–27. Propagation through the output quarter-wave plate delays these components an additional $\beta_\perp l$ and $\beta_\perp l + \pi/2$, respectively. As a result, the value of the output components are

$$E/\sqrt{2}\underline{/-4\beta_\perp l - 2\theta_m} \qquad \text{and} \qquad E/\sqrt{2}\underline{/-4\beta_\perp l - 2\pi - 2\theta_m}$$

With the waves in phase, vector addition results in a vertically polarized output wave of value $E\underline{/-4\beta_\perp l - 2\theta_m}$, which is shown in part a of Fig. 7–27. Thus, the analysis verifies that rotating the center section an angle θ_m causes the output to be phase delayed an additional $2\theta_m$. Also, with the orientation of the output transition as shown, the vertically polarized wave is delivered to the output port. Since its magnitude equals that of the input wave, the rotary phase shifter is, in principle, lossless for all values of θ_m. Commercial units are available with less than 1.0 dB loss and SWR values below 1.30.

The vector-phasor method of analysis used here is very helpful in understanding the operation of many microwave components. As explained in Sec. 2–5, it makes use of the fact that for sinusoidal signals, the propagating wave can be described by a vector-phasor. In most cases, it represents the transverse electric field of the propagating mode.

Hybrid phase shifter.   The use of the rotary phase shifter is limited to low-power situations, typically a few watts or less. For higher-power applications, the hybrid-type phase shifter is often used. A sketch of the unit is shown in part a of Fig. 7–28. It consists of a 3 dB short-slot coupler and a pair of movable short circuits. The shorts are mechanically connected so that they move as a unit. The operation of the coupler is discussed in Sec. 8–3b (see Fig. 8–30). The sidewall version shown in the figure consists of two rectangular guides, each of width a, having a common narrow wall. Part of the wall is removed to allow energy coupling between the guides. When properly designed, the 3 dB coupler has the following properties. The power associated with a wave entering port 1 splits equally between ports 3 and 4 with no power being delivered to port 2. Also, the wave arriving at port 3 is phase delayed $\pi/2$ relative to the wave at port 4. Due to the coupler's symmetry, injection of a signal at any other port leads to similar effects. For example, a wave entering port 3 splits equally between ports 1 and 2, with the wave at port 1 delayed $\pi/2$ relative to that at port 2. In this case, no power arrives at port 4. With this background, the operation of the hybrid phase shifter is readily explained. The following analysis shows that with the two shorts in the same plane, the phase shifter is loss-

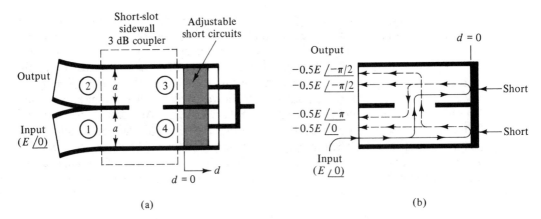

Figure 7–28   The hybrid phase shifter and its phasor representation.

less and that moving the shorts a distance d results in an output phase change of $2\beta d = 4\pi d/\lambda_g$.

Referring to Fig. 7–28a, the TE_{10} wave entering the input port is represented by $E/\underline{0}$. Due to the action of the coupler, waves arriving at ports 4 and 3 equal $E/\sqrt{2}/\underline{0}$ and $E/\sqrt{2}/\underline{-\pi/2}$, respectively. The shorts fully reflect the signals ($\Gamma = -1$), and therefore waves of value $-E/\sqrt{2}/\underline{0}$ and $-E/\sqrt{2}/\underline{-\pi/2}$ reenter ports 4 and 3. Each of these waves experiences a power split as they propagate back through the coupler. The resultant phasor components are shown in part b of Fig. 7–28. Note that the phasor sum of the waves reflected back to the input is zero and hence the input SWR is unity. Also, the sum of the output components is $-E/\underline{-\pi/2} = E/\underline{-3\pi/2}$, which means that the phase shifter is lossless. With the shorts at $d = 0$, the phase delay between the input and output ports is $3\pi/2$ rad.[6]

The phase of the output wave may be varied by moving the two short circuits as a unit. Moving them back a distance d causes the output wave to be delayed an additional $2\beta d$, since the round-trip path of the waves has been increased by $2d$. Thus the phase change in a hybrid phase shifter is

$$\Delta\theta = 2\beta d = \frac{4\pi}{\lambda_g}d \qquad (7\text{–}13)$$

A plot of $\Delta\theta$ as a function of d is show in Fig. 7–29. With an ideal 3 dB coupler and the two shorts in perfect alignment, the curve is linear as indicated by the dashed line. Given the guide width and the operating frequency, one can compute λ_g and hence the slope of the curve. By using a micrometer drive to move the shorts, an accurate phase calibration is easily obtained.

In practice, the $\Delta\theta$ versus d curve exhibits some deviation from linearity as indicated by the solid curve in the figure. In most cases, the nonlinearity is due to the imperfect operation of the 3 dB coupler. A coupler with finite directivity, defined by Eq. (8–15), will leak a portion of the input signal directly to the output of the phase

---

[6] This neglects the phase delay associated with wave propagation from the input to the $d = 0$ plane and back. Since this delay is a constant, it may be ignored when describing the variable phase characteristics. The same holds true for the $3\pi/2$ phase delay.

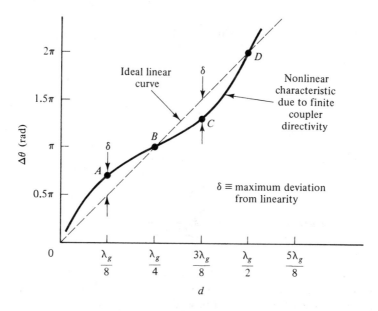

Figure 7–29 Phase change as a function of short position in a hybrid phase shifter.

shifter. When the leakage is small ($E_L \leq 0.1E$), the output signal may be approximated by

$$E_{out} = E \bigg/ {-\frac{3\pi}{2} - 2\beta d} + E_L \underline{/\psi} \qquad (7\text{--}14)$$

where the first term is the output in the absence of leakage and the second represents the leakage. Since the leakage signal does not travel to the short circuits, its phase delay ψ is independent of d.

The effect of leakage on the $\Delta\theta$ characteristic may be determined by a phasor analysis of the output signal. Consider the case where $E_L = 0.1E$. This is equivalent to a coupler with 17 dB directivity since for a 3 dB coupler

$$\text{Directivity} = 20 \log \frac{E}{\sqrt{2}E_L} \qquad (7\text{--}15)$$

A phasor diagram of the two output components described in Eq. (7–14) is shown in Fig. 7–30 for $d = 0$, $\lambda_g/8$, $\lambda_g/4$, and $3\lambda_g/8$. The phase of the leakage signal is arbitrarily chosen as $-\pi/2$. Note that at $d = 0$ and $\lambda_g/4$, the phase of the resultant signal with leakage is the same as without leakage. This corresponds to points B and D in Fig. 7–29. At $d = \lambda_g/8$, the resultant phase delay is slightly greater than that without leakage, while at $d = 3\lambda_g/8$ it is slightly less. These cases represent the points of maximum deviation from linearity and correspond respectively to points A and C on the curve. Referring to the phasor diagram and assuming $E_L \leq 0.1E$, the maximum phase deviation from linearity is given by

$$\delta = \arctan \frac{E_L}{E} = \arctan \frac{1}{\sqrt{2}D} \qquad (7\text{--}16)$$

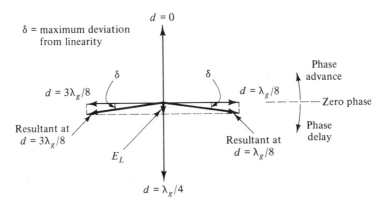

Figure 7–30 Phasor diagram showing the effect of coupler directivity on the phase change of a hybrid phase shifter (for $E_L = 0.10E\underline{/-\pi/2}$).

where $D \equiv$ antilog {Directivity (dB)/20}. For example, in order to insure that δ is less than one degree, the directivity of the 3 dB coupler must be greater than 32 dB, a readily achievable value for a well-designed unit.

The above analysis shows again the usefulness of the vector-phasor representation in explaining the operation of a microwave component.

Fixed phase-shift sections. Microwave bridge circuits often require a waveguide section in which its phase delay differs from that associated with wave propagation through a standard rectangular guide of *equal length*. One method of achieving such a differential phase shift is to alter the guide wavelength by changing the guide width. This causes a differential phase shift

$$\Delta\theta = \left(\frac{2\pi}{\lambda_g} - \frac{2\pi}{\lambda_g'}\right)l \qquad (7\text{–}17)$$

where λ_g and λ_g' are the guide wavelengths for the standard and adjusted guides, respectively. For example, reducing the guide width results in λ_g' being greater than λ_g, which means its phase delay will be less than that of standard guide. Caution must be exercised in using this technique since changing the guide width affects both the cutoff frequency and characteristic impedance. Therefore, some impedance matching of the phase shift section may be required. Also, with guide wavelength being a function of frequency, $\Delta\theta$ will be frequency sensitive. A technique for minimizing this sensitivity is described by Sohon (Ref. 7–17).

Another method of introducing a differential phase change is to insert reactive elements into a waveguide. When properly designed, the reflections due to the elements cancel while their phase effects are additive. The result is a matched differential phase shift section. A typical configuration consists of a pair of either capacitive or inductive septums. The inductive case and its equivalent circuit are shown in Fig. 7–31. The irises are a distances l apart in standard rectangular guide. It is assumed that standard guide is connected to both ends of the phase shift section. For this case, the normalized ABCD matrix of the network is

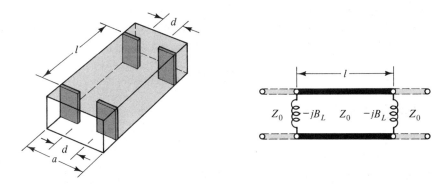

(a) Waveguide structure   (b) Equivalent circuit

Figure 7–31   A differential phase-shift section consisting of a pair of inductive septums in a standard waveguide.

$$\begin{bmatrix} \bar{A} & \bar{B} \\ \bar{C} & \bar{D} \end{bmatrix} = \begin{bmatrix} 1 & 0 \\ -j\bar{B}_L & 1 \end{bmatrix} \begin{bmatrix} \cos \beta l & j \sin \beta l \\ j \sin \beta l & \cos \beta l \end{bmatrix} \begin{bmatrix} 1 & 0 \\ -j\bar{B}_L & 1 \end{bmatrix}$$

$$= \begin{bmatrix} \cos \beta l + \bar{B}_L \sin \beta l & j \sin \beta l \\ j \left\{ \begin{matrix} (1 - \bar{B}_L^2) \sin \beta l \\ - 2\bar{B}_L \cos \beta l \end{matrix} \right\} & \cos \beta l + \bar{B}_L \sin \beta l \end{bmatrix} \qquad (7\text{–}18)$$

where $\beta = 2\pi/\lambda_g$ and $\bar{B}_L \equiv B_L/Y_0$ is the normalized inductive susceptance of each iris. In order for the section to be matched, $\bar{Z}_{in} = (\bar{A} + \bar{B})/(\bar{C} + \bar{D})$ must be unity. Since $\bar{A} = \bar{D}$, this requires that $\bar{B} = \bar{C}$. Using the above matrix expression and simplifying, leads to the condition for a perfect match, namely, $\tan \beta l = -2/\bar{B}_L$. This may be rewritten as

$$\beta l = \frac{\pi}{2} + \arctan \frac{\bar{B}_L}{2} \qquad (7\text{–}19)$$

The insertion phase may be determined with the aid of Eq. (4–57). With some algebraic manipulation and making use of the match condition above Eq. (7–19), the equation for insertion phase reduces to $\theta_I = \arctan (2/\bar{B}_L)$, which may be rewritten as

$$\theta_I = \frac{\pi}{2} - \arctan \frac{\bar{B}_L}{2} \qquad (7\text{–}20)$$

The phase shift in the absence of the irises is βl and therefore the differential phase shift is

$$\Delta\theta_I \equiv \beta l - \theta = 2 \arctan \frac{\bar{B}_L}{2} \qquad (7\text{–}21)$$

Note that $\Delta\theta$ is positive, which means that the effect of the inductive elements is to reduce the overall phase delay. Thus Eq. (7–21) may be interpreted to mean that each shunt inductor introduces a phase *advance* of value $\arctan (\bar{B}_L/2)$.

The capacitive case can be derived by replacing $-j\bar{B}_L$ by $+j\bar{B}_C$, where $\bar{B}_C$ is the normalized capacitive susceptance of each iris. The matrix analysis leads to the following match condition.

$$\beta l = \frac{\pi}{2} - \arctan \frac{\bar{B}_C}{2} \tag{7-22}$$

The differential phase shift is given by

$$\Delta\theta = -2 \arctan \frac{\bar{B}_C}{2} \tag{7-23}$$

Since $\Delta\theta$ is negative, this means that each shunt capacitive element introduces a phase *delay* of value $\arctan (\bar{B}_C/2)$.

This section has described some of the techniques used to obtain variable phase delay and differential phase shift. Other methods are described in Refs. 7–1, 7–10, 7–20, and 7–21.

7–5 ROTARY JOINTS

Some microwave systems require relative motion between electrically connected components. One such example is the connection between a mechanical scanning antenna and a fixed transmitter or receiver. A small amount of motion can usually be accommodated by using either a coaxial cable or a section of flexible waveguide. If, however, continuous rotational motion is required, a rotary joint must be used. A rotary joint is a two-port network that maintains low transmission loss while allowing for complete rotational motion between the ports. This requires that the transmission line linking the two ports be excited by a circularly symmetric mode. Three such modes are the TEM in coax and the TM_{01} and TE_{01} in circular waveguide. Rotary joints that utilize these modes are described here.

Coaxial rotary joints.   In its simplest form, the coaxial rotary joint consists of a pair of axially aligned coaxial lines in which their inner and outer conductors are electrically connected by low-resistance rubbing contacts. For most low-speed, low-power applications, this method provides satisfactory performance despite some variation in contact resistance during rotation. There are situations, however, where even a small resistance change causes an unacceptable variation in transmission loss. This is particularly true in high-power and low-noise applications. To overcome this problem, the coaxial lines are usually coupled by either a capacitive or a choke-type arrangement, the latter being shown in Fig. 7–32. The connection between inner conductors is identical to the dc block described in Fig. 6–34a. Its equivalent circuit is a series-connected, open-circuited stub, indicated by the Z_{03} line in Fig. 7–32b. With the line a quarter wavelength long, the connecting impedance between inner conductors is zero at the design frequency. The outer conductors are joined together by a choke arrangement that operates on the same principle as the noncontacting short (Fig. 7–21b) and the choke flange (Fig. 7–23b). The short circuit is transformed by the quarter wavelength Z_{02} line into an infinite impedance at the bearing

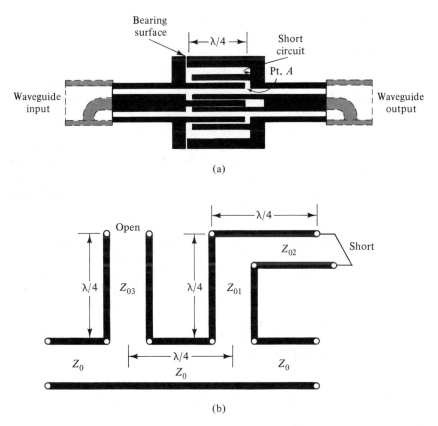

Figure 7–32 A choke-coupled coaxial rotary joint and its equivalent circuit. (The dashed portions in part *a* represent loop type coax-to-waveguide transitions.)

surface. Since the contact resistance at the bearing surface is in series with the infinite impedance, their combined impedance is infinite for all values of contact resistance. The quarter wavelength Z_{01} line transforms the infinite impedance into a short circuit at point *A* resulting in a zero impedance connection between outer conductors. The complete equivalent circuit is shown in part *b* of Fig. 7–32. In order to maintain a low-SWR connection over a broad frequency range, Z_{01} and Z_{03} should be chosen as small as possible and Z_{02} as large as possible. The reasons were explained in Secs. 6–5b and 7–3c.

At other than the design frequency, each stub causes a reflection since it represents a finite reactance in series with the main line. By spacing the stubs a quarter wavelength apart, as indicated in the figure, these reflections tend to cancel, resulting in low loss over a wide frequency band (Prob. 7–28).

The design of a low-friction rotary joint requires good mechanical techniques. For instance, to insure a smooth rotary motion between the bearing surfaces of the outer conductors, a ball-bearing arrangement is usually employed. To prevent contact between the inner conductors during rotation, a low-friction dielectric such as teflon may be necessary. Another important consideration is that both inner and outer conductors be precisely aligned to minimize frictional variations during rotation.

The prevention of loss variations during rotation requires that propagation be restricted to the circularly symmetric TEM mode. Nonsymmetric modes, such as the TE_{11}, can be avoided by choosing the coaxial dimensions in accordance with Eq. (5–18). For a well-designed unit, the variation in loss and SWR during rotation can be made negligible.

The coaxial rotary joint may be converted to a waveguide type by simply incorporating coax-to-waveguide transitions at each end. For example, using transitions of the type shown in Fig. 7–2c, the coaxial unit in Fig. 7–32a becomes an *in-line* waveguide rotary joint. A dashed outline of the transitions is shown in the figure. With this arrangement, the input and output guides have a common propagation axis and rotational motion takes place about the axis.

Waveguide rotary joints. Circular guide may be used to couple the two halves of a waveguide rotary joint. The condition that the transmission characteristics remain constant during rotation requires that the circular waveguide mode be circularly symmetric. In most cases either the TM_{01} or TE_{01} mode is used. The unit shown in Fig. 7–33 makes use of the TM_{01} mode. It consists of two TE_{10} rectangular-to-TM_{01} circular mode transitions that are free to rotate about the propagation axis of the circular guide. The transitions are electrically and mechanically connected by a choke-type arrangement similar to that used in the coaxial rotary joint (Fig. 7–32a).

Since the TM_{01} is not the dominant mode in circular guide, TE_{11} propagation is possible. Therefore, it is critical that the transitions be designed with a high degree of TE_{11} mode suppression. One such transition is shown in Fig. 7–5. TE_{11} mode suppression is obtained by including mode filters of the type shown in Fig. 7–7b. They

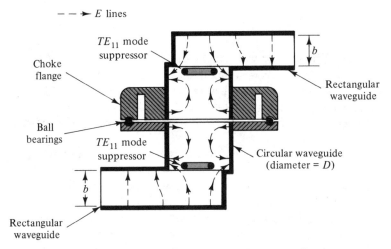

Figure 7–33 A waveguide rotary joint that utilizes the TM_{01} mode in circular waveguide.

are usually placed near the ends of the circular guide section. Shorted stubs may also be used as TE_{11} mode suppressor (Sec. 6.22 of Ref. 7–10).

TE_{11} mode energy in the circular guide section causes variations in loss and SWR during rotation. This can be explained with the aid of Fig. 7–34, where the electric field pattern is shown for two positions of the rotary joint. With the input and output guides positioned as indicated in part a, the TE_{11} energy will be transmitted to the output rectangular guide as shown. However, when the top portion of the joint is rotated 90° (part b), the orientation of the output electric field vector is such that it will be fully reflected. The resultant variation in loss and SWR is a direct consequence of the fact that the TE_{11} mode lacks circular symmetry. With the design techniques indicated above, the amount of TE_{11} power can be kept to less than one percent of the total transmitted power. This translates to a variation in input SWR of less than ±0.02 for 360° rotation.

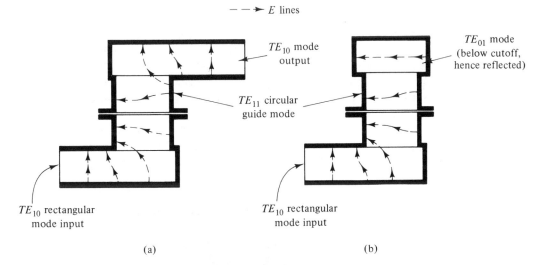

Figure 7–34 An explanation of the variation in transmission characteristics of a waveguide rotary joint due to the presence of the TE_{11} circular guide mode.

Another consideration in the design of a waveguide rotary joint is the length of the circular guide section. The presence of TE_{11} energy causes loss peaks in the transmission characteristics of the rotary joint. The frequencies at which these resonances occur is determined by the length of the circular guide section. This is because TE_{11} reflections at the output propagate back to the input where they are rereflected by the input transition and mode filter. Loss resonances occur when the electrical length of the circular guide is such that the multiple reflections reinforce each other. A detailed analysis of the resonances has been worked out and is described in Sec. 7.4 of Ref. 7–10. Generally, they occur at frequencies where the electrical length of the circular guide is a multiple of a half wavelength. Thus by proper choice of the circular guide length, the resonances can be made to occur outside the frequency band of interest. In broadband applications, TE_{11} mode absorbers are used to suppress the resonances (see Sec. 7.8 of Ref. 7–10).

Waveguide rotary joints can also be designed using the TE_{01} circular guide mode. In this case, the input and output transitions are of the type shown in Fig. 7–6. TE_{01} mode filters (Fig. 7–7a) are usually inserted at each end of the circular guide to insure TE_{01} mode purity.

Another design approach makes use of a circularly polarized TE_{11} wave in the circular guide section. A unit based on this technique is described in Secs. 7.11 to 7.13 of Ref. 7–10. Additional types are described in Refs. 7–22 and 7–23.

All of the waveguide rotary joints described here are capable of handling higher power levels than their coaxial counterparts. In addition, their mechanical design is simpler due to the absence of center conductors and the associated alignment problem.

Multichannel rotary joints.   Certain applications require that waveguides operating in different frequency bands have the capability of being independently rotated about a common axis. This can be accomplished with a multichannel rotary joint. An example of a dual-channel unit is shown in Fig. 7–35. Basically, it consists of a pair of coaxial rotary joints in which the outer conductor of the inner coax is also the inner conductor of the outer coaxial line. Each signal band is coupled to the appropriate waveguide system by well-matched coax-to-waveguide transitions. To maintain TEM mode purity, the coaxial diameters must be chosen to avoid higher-mode propagation, specifically the TE_{11} coaxial mode. Thus the best arrangement is to have the inner coaxial line carry the higher frequency channel.

The mechanical design of multichannel rotary joints can be quite complex. Some practical examples are given in Refs. 7–24 and 7–25.

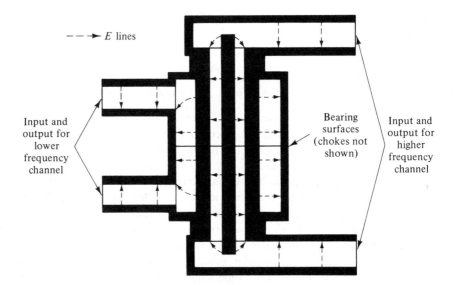

Figure 7–35   A dual-channel rotary joint (cross-sectional view).

7–6 MECHANICAL- AND GAS-TYPE SWITCHES

A variety of techniques are used in designing single-pole single-throw (SPST) and single-pole multithrow (SPMT) switches at microwave frequencies. Depending upon the switching speed and power level involved, the basic switch element can be mechanical, gaseous, ferrite, or semiconductor. The first two types are discussed in this section.

7–6a Mechanical Switches

Mechanical-type switches are particularly suited for applications in which switching speed is not an important consideration. The movement of the switching element may be controlled manually or by means of a solenoid. Due to the mechanical inertia of the element, switching speeds are usually in the order of a fraction of a second. Some typical waveguide configurations are described here. Examples of coaxial units may be found in Sec. 8.21 of Ref. 7–10.

A single-pole single-throw switch is a two-port network that has two possible loss states, one high and one low. In the high-loss state, the incident signal may be either reflected or absorbed. Two examples of reflection-type switches are pictured in Fig. 7–36. The one in part *a* consists of a shorting plate that is capable of being inserted into the waveguide as shown. The separation between the choke and cover flanges is sufficient to allow easy insertion and removal of the metal shorting plate. With the plate across the guide as shown, nearly all of the incident signal is reflected. Isolation values of greater than 90 dB are typical for this type switch. The term *isolation* refers to the insertion loss when the device is in its high-loss state. This terminology is widely used in microwave work.

A narrowband SPST switch is shown in part *b* of Fig. 7–36. It consists of a metal rod that is inserted through the center of the broad wall of the waveguide. The *in* and *out* movement of the rod is usually controlled by a solenoid. With the rod in the *out* position, the loss through the guide is negligible. When the rod is inserted

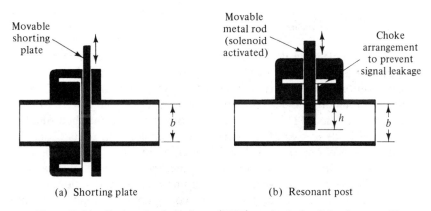

(a) Shorting plate   (b) Resonant post

Figure 7–36   Single-pole single-throw (SPST) mechanical switches in waveguide.

into the guide, the equivalent circuit becomes that shown in Fig. 7–18. The penetration h is adjusted to produce series resonance at the operating frequency, resulting in a high transmission loss. Isolations values of 20 to 30 dB are typical for this type switch. Since the high loss is due to the low shunt impedance at resonance, the unit is inherently narrowband.

Another approach to waveguide switching involves detuning a bandpass filter by inserting a metal rod into the waveguide. A single–pole double–throw (SPDT) switch that utilizes this principle is described in Sec. 9.3 of Ref. 7–1. Also described is a SPDT switch that uses a rotating metal plate and a waveguide tee junction.

The most common form of a mechanical-type waveguide SPDT switch is shown in Fig. 7–37a. Its switching action is based upon the rotation of a 90° bend. Either an E or H-plane bend may be used. For the position shown in the figure, a low-loss connection exists between ports A and B. The loss to the decoupled arm (port C, in this case) is very high, typically 80 to 90 dB. Turning the rotatable section by 90° results in a low-loss connection between ports A and C, with port B being decoupled. This position is indicated by the dashed lines. Noncontacting chokes are usually employed to insure good electrical continuity between the rotatable section and the fixed waveguides.

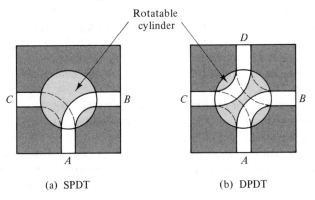

(a) SPDT   (b) DPDT

Figure 7–37 Single-pole double-throw (SPDT) and double-pole double-throw (DPDT) switches in waveguide.

A double-pole double-throw switch based on the same principle is shown in part b of Fig. 7–37. In this case, the rotatable section contains two 90° bends. For the position shown, port A is connected to port B and port C to port D. Ninety degrees rotation of the section results in port A being connected to port C and port B to port D. The DPDT switch can be converted to a matched SPST by connecting matched loads to two of the ports. For example, with ports C and D match terminated, the SWR into ports A and B is unity for either state of the rotatable section. That is, the SPST switch is bilaterally matched in both its high- and low-loss states.

The SWR *during* switching is an important consideration, particularly in high-power applications. In certain SPMT configurations, it can be minimized by controlling the switching sequence. For example, the preferred sequence in many SPDT switches is to connect the decoupled port *before* disconnecting the coupled port. This is known as *make-before-break* switching. The unit shown in Fig. 7–37a does not have this feature and hence its SWR becomes very high during switching. The DPDT switch shown in part *b* also exhibits a high SWR while being switched.

7–6b Gas Discharge Switches

Gas discharge plasmas can also be used to switch microwave energy. This technique is particularly useful in applications requiring the rapid switching of high-power levels. The loss state of the switch can be controlled by a dc or ac voltage or by the microwave power itself. The latter approach is used in the design of TR and ATR duplexers and is discussed in Sec. 8–5b.

The high-loss state is established by applying a control voltage large enough to ionize the gas in the waveguide. Removal of the voltage returns the guide to its low-loss condition. A SPST switch that operates in this manner has been described by Goldie (Ref. 7–26). It represents an extension of the work reported by Hill and Ichiki (Ref. 7–28). A sketch of the unit is shown in Fig. 7–38. It consists of a small section of rectangular guide sealed at both ends by pressure windows. The guide is filled with hydrogen gas at a low pressure, typically 0.1 to 0.3 mm of Hg. In the absence of a control voltage, the insertion loss is less than 0.25 dB. Application of a large dc voltage causes the hydrogen gas in the guide to become ionized. When the free electron density of the plasma exceeds a certain critical value, the microwave signal is attenuated. The interaction between the plasma and the propagating wave can be characterized by the dielectric constant of the plasma which is given by the following relation:

$$\epsilon_R = 1 - \left(\frac{f_p}{f}\right)^2 \tag{7–24}$$

where f is the microwave frequency, $f_p \approx 9000\sqrt{n}$ is the plasma frequency and n is the number of free electrons per cubic centimeter.

The high loss state of the switch is created by causing $f_p \geq f$. This results in a zero or negative value of ϵ_R, which means that wave propagation is forbidden. Thus the incident microwave signal is reflected by the plasma. With n (and hence f_p) a

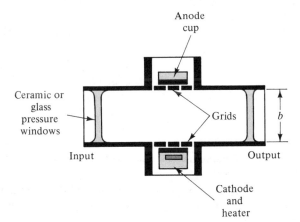

Anode
cup

Ceramic or
glass
pressure
windows

Grids

b

Input

Output

Cathode
and
heater

Figure 7–38   A gas-type SPST switch in rectangular waveguide.

function of the control voltage as well as the gas pressure, the high-loss state can be established by adjusting the voltage. The C-band switch described by Goldie (Ref. 7–26) has an isolation of greater than 30 dB and an arc loss of 0.8 dB. Arc loss is defined as the return loss of the switch in its *fired* state. If the plasma were fully reflective, the arc loss would be zero. This unit is capable of switching hundreds of kilowatts of microwave power in fractions of a microsecond.

Multipactor discharge in a high vacuum can also be employed as a means of switching large amounts of microwave power. An S-band switch based on this principle is described by Forrer and Milazzo (Ref. 7–27). Other techniques include the use of a magnetic field to control the gas discharge. References to this and other switching methods may be found in Refs. 7–27 and 7–28.

REFERENCES

Books

7–1. Southworth, G. C., *Principles and Applications of Waveguide Transmission*, D. Van Nostrand Co., Princeton, NJ, 1950.

7–2. Collin, R. E., *Field Theory of Guided Waves*, McGraw-Hill Book Co., New York, 1960.

7–3. Ghose, R. N., *Microwave Circuit Theory and Analysis*, McGraw-Hill Book Co., New York, 1963.

7–4. Slater, J. C., *Microwave Transmission*, McGraw-Hill Book Co., New York, 1942.

7–5. Schwinger, J. and D. S. Saxon, *Discontinuities in Waveguides*, Gordon and Breach Science Publishers, Inc., New York, 1968.

7–6. Marcuvitz, N., *Waveguide Handbook*, Rad. Lab. Series, vol. 10, McGraw-Hill Book Co., New York, 1951.

7–7. Montgomery, C. G., R. H. Dicke, and E. M. Purcell, *Principles of Microwave Circuits*, Rad. Lab. Series, vol. 8, McGraw-Hill Book Co., New York, 1948.

7–8. Saad, T. S., *Microwave Engineers' Handbook,* vol. 1, Artech House, Dedham, MA, 1971.

7–9. Lewin, L., *Advanced Theory of Waveguides*, Iliffe and Sons, Ltd., London, 1951.

7–10. Ragan, G. L., *Microwave Transmission Circuits*, Rad. Lab. Series, vol. 9, McGraw-Hill Book Co., New York, 1948.

7–11. Ramo, S., J. R. Whinnery, and T. Van Duzer, *Fields and Waves in Communication Electronics,* John Wiley and Sons, Inc., New York, 1965.

Articles

7–12. Cohn, S. B., Design of Simple Broadband Waveguide-to-Coaxial Junctions. *Proc. IRE,* 35, September 1947, pp. 920–926.

7–13. Simmons, A. J., Phase Shift by Periodic Loading of Waveguide and Its Application to Broadband Circular Polarization. *IRE Trans. Microwave Theory and Techniques,* MTT-3, December 1955, pp. 18–21.

7–14. Ayres, W. P., Broad-Band Quarter-Wave Plates. *IRE Trans. Microwave Theory and Techniques*, MTT-5, October 1957, pp. 258–261.

7–15. Wheeler, H. A. and H. Schwiebert, Step-Twist Waveguide Components. *IRE Trans. Microwave Theory and Techniques,* MTT-3, October 1955, pp. 44–52.

7–16. Teeter, W. L. and K. R. Bushmore, A Variable-Ratio Microwave Power Divider and Multiplexer. *IRE Trans. Microwave Theory and Techniques,* MTT-5, October 1957, pp. 227–229.

7–17. Sohon, H., Wide-Band Phase-Delay Circuit. *Proc. IRE,* 41, August 1953, pp. 1050–1052.

7–18. Otoshi, T. Y. and C. T. Stelzried, A Precision Compact Rotary Vane Attenuator. *IEEE Trans. Microwave Theory and Techniques,* MTT-19, November 1971, pp. 843–854.

7–19. Fox, A. G., An Adjustable Waveguide Phase Changer. *Proc. IRE,* 35, December 1947, pp. 1489–1498.

7–20. Vaillancourt, R. M., Errors in a Magic-Tee Phase Changer. *IRE Trans. Microwave Theory and Techniques,* MTT-5, July 1957, pp. 204–207.

7–21. Beatty, R. W., A Differential Microwave Phase Shifter. *IEEE Trans. Microwave Theory and Techniques,* MTT-12, March 1964, pp. 250–251.

7–22. Fromm, W. E., E. G. Fubini, and H. S. Keen, A New Microwave Rotary Joint. *IRE National Convention Record,* March 1958, pp. 78–82.

7–23. Raabe, H. P., A Rotary Joint for Two Microwave Transmission Channels of the Same Frequency Band. *IRE Trans. Microwave Theory and Techniques,* MTT-3, July 1955, pp. 30–41.

7–24. Cohen, M., A Six-Channel Vertically Stacked Coaxial Rotary Joint for the S-, C-, and X-band Region. *Microwave Journal,* 7, November 1964, pp. 71–74.

7–25. Boronski, S., A Multichannel Waveguide Rotating Joint. *Microwave Journal,* 8, June 1965, pp. 102–105.

7–26. Goldie, H., A Fast Broadband High Power Microwave Switch. *Microwave Journal,* 6, August 1963, pp. 76–81.

7–27. Forrer, M. P. and C. Milazzo, Duplexing and Switching with Multipactor Devices. *Proc. IRE,* 50, April 1962, pp. 442–451.

7–28. Hill, R. M. and S. K. Ichiki, Microwave Switching with Low-Pressure Arc Discharge. *IRE Trans. Microwave Theory and Techniques,* MTT-8, November 1960, pp. 628–633.

PROBLEMS

7–1. **(a)** Referring to Fig. 7–1, design a coax-to-waveguide transition with unity SWR at 15 GHz. Calculate l_s and b if $a = 1.58$ cm, $b_s = 0.79$ cm, and $Z_0 = 50$ ohms for the coaxial line. Use the voltage-current definition in Eq. (5–79) to calculate waveguide impedance.

(b) With the values of l_s and b obtained in part a, determine the SWR at 12.4 and 18.0 GHz.

7–2. A circular polarizer of the type shown in Fig. 7–11a has a guide diameter of 3.0 cm. Assume $\beta_\parallel = 1.2\beta_\perp$ and that for $E_\perp$, the effect of the dielectric slab is negligible. Compute l for circular polarization at 8.0 GHz.

7–3. Referring to Fig. 7–11b, $a = 4$ cm and $l = 10$ cm. Calculate the value of b required to convert a 5.0 GHz input vertically polarized wave into a circularly polarized wave at the output. Assume $b < a$ and an air-filled waveguide.

7–4. Derive insertion loss equations for the perpendicular and parallel components of the quarter-wave plate shown in Fig. 7–12. Verify that the two losses are identical when $\bar{B}_L = \bar{B}_C$.

7–5. Referring to Fig. 7–12, determine the values of $\bar{B}_C$ and $\bar{B}_L$ required to satisfy the phase condition for circular polarization when $\bar{B}_L = 2\bar{B}_C$.

7–6. The length of the circular guide section in Fig. 7–11a has been adjusted so that $\beta_\parallel l - \beta_\perp l = \pi$. What is the polarization of the output wave if the input wave is vertically polarized?

7–7. The figure below shows a pair of 3 cm by 6 cm guides separated by a length l of half-height waveguide. At 4 GHz, l is exactly a half wavelength long. All waveguides are air-filled. Assuming symmetrical step discontinuities,
   (a) calculate the input SWR at 4 GHz when the output guide is matched terminated.
   (b) determine the insertion loss at 4 GHz, assuming $Z_G = Z_L = Z_{01}$. Neglect dissipative losses.

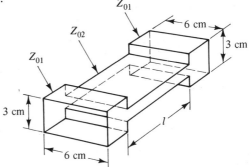

7–8. To solve the previous problem, it was necessary to assume that direct coupling between the step discontinuities was negligible. Verify this by showing that the TM_{12} mode generated at one step is attenuated by more than 20 dB at the plane of the other step.

7–9. Use a symmetry argument to show that the only higher-order modes generated by the symmetrical step discontinuity in Fig. 7–13b are TE_{m0} with m an odd integer. Assume dominant mode excitation.

7–10. A dielectric block ($\epsilon_R = 2.0$) having a 2 cm by 5 cm cross section is inserted into a 2 cm by 5 cm air-filled waveguide that is match terminated. Calculate the input SWR at 4.0 GHz when the dielectric is $\lambda_g/4$ long.

7–11. A 1 cm by 2 cm air-filled waveguide contains an inductive and a capacitive septum spaced 3.40 cm apart along the propagation axis. The inductive iris, of the type shown in Fig. 7–15a, has $d = 1.20$ cm. The capacitive iris, of the type shown in Fig. 7–16a, has $d = 0.40$ cm. Calculate the insertion loss and phase at 10 GHz when $Z_G = Z_L = Z_0$, the characteristic impedance of the waveguide. Assume that $t < 0.1$ cm for both irises.

7–12. A quarter-wave plate of the type shown in Fig. 7–12 has $a = 3.0$ cm and $d = 1.79$ cm.
   (a) Verify that $\theta_\perp - \theta_\parallel \approx 90°$ at 7.5 GHz. Assume negligible iris thicknesses.
   (b) Calculate the insertion losses for the perpendicular and parallel components.

7–13. A 2 cm by 4 cm air-filled waveguide contains a resonant iris of the type shown in Fig. 7–17.
   (a) Calculate the resonant frequency if $a' = 3.0$ cm and $b' = 0.80$ cm.
   (b) To what value must b' be increased in order to maintain the same resonant frequency when a' is increased by 10 percent? Which case has the higher circuit Q?

7–14. A 1.5 cm by 3.0 cm air-filled waveguide contains a thin inductive iris of the type shown in Fig. 7–15c. For an iris width $w = 0.60$ cm, the measured SWR is 20 at 8.0 GHz. Calculate $\bar{B}_L$ and compare your answer with the theoretical data in Marcuvitz (Ref. 7–6).

7–15. The normalized value of a shunt capacitive iris may be determined by placing a matched load at the output and measuring the input SWR.
   (a) Calculate $\bar{B}_C$ if the measured SWR is 4.0 and the load is reflectionless.
   (b) What would be the uncertainty in the calculated value of $\bar{B}_C$ if the SWR of the load had been 1.04?

7–16. A standard mismatch with an SWR = 1.75 is required in a waveguide application. Referring to Fig. 7–20, $b = 1.0$ cm and $a = 2.0$ cm. If it is assumed that $C_d = 0$, the required value of b_1 is 0.571 cm.
   (a) Calculate the actual SWR at 10 GHz when the effect of the discontinuity capacitance is included.
   (b) Readjust the b_1 dimension for an SWR = 1.75 at 10 GHz when the effect of C_d is included.
   (c) Calculate the SWR at 12 GHz using the new value of b_1.
   Use Eqs. (3–46), (3–52), and (3–53) rather than the Smith Chart.

7–17. The adjustable short circuit in Fig. 7–21a uses a rectangular waveguide with an aspect ratio of two and $b = 1.25$ cm.
   (a) Calculate the value of x required for $\bar{Z}_{in} = -j1.5$ at 9 GHz.
   (b) With x fixed, calculate $\bar{Z}_{in}$ at 8 GHz.

7–18. Use the expression for $\bar{Z}$ derived in Ex. 7–2 to calculate the return loss for $\bar{R} = 0$ and 0.01 if $\lambda_g = 10$ cm, $l_1 = l_2 = 1.8$ cm and the values of b_1 and b_2 are those given in the example. Repeat the calculations with $b_1 = b_2 = 0.25$ cm. Which set of values gives better results?

7–19. A rotary-vane attenuator of the type shown in Fig. 7–25 is set for 6.0 dB attenuation.
   (a) Calculate the angular position θ_m of the rotatable section.
   (b) How accurately must θ_m be set so that the attenuation is within ±0.10 dB of the nominal 6.0 dB value?

7–20. A matched microwave source having an available power of 200 mW is connected to the input of a matched rotary-vane attenuator. The attenuator setting is 7.0 dB.
   (a) Calculate the power dissipated in each resistance card when the attenuator is terminated with a matched load ($Z_L = Z_0$).
   (b) Calculate the power dissipated by the input card if $Z_L = 2Z_0$. How much power is reflected back to the source? What is the input SWR in this case?

7–21. The output resistance card of a rotary-vane attenuator (Fig. 7–25) is replaced by a thin metal plate. This causes the horizontal component at the input to the second fixed section to be fully reflected. Calculate the power absorbed by the input card if $\theta_m = 30°$, the input power is 4.0 W, and the ouput is match terminated.

7–22. With a matched generator and load connected to the rotary-vane attenuator shown in Fig. 7–25, both the SWR and loss are reciprocal. Show that if the input resistance card is removed, the loss is reciprocal but the SWR is not.

7–23. Assume that the output quarter-wave plate and circular-to-rectangular transition in Fig. 7–27 are both rotated by 90°. Use a vector-phasor analysis to verify that the phase shifter still works. That is, it is lossless and the phase change equals $2\theta_m$.

7–24. Due to a mechanical malfunction, only the top short in the hybrid phase shifter (Fig. 7–28) can move.
   (a) Calculate the output phase change at 7.5 GHz when the top short moves from $d = 0$ to $d = 0.90$ cm. Assume $a = 3.0$ cm.
   (b) For the above conditions, calculate the change in insertion loss.

7–25. Measurements on a hybrid phase shifter (Fig. 7–28) yield the following results at 6.0 GHz:
   $\theta = 80°$, 125°, and 170° for $d = 2.0$ cm, 2.4 cm, and 2.8 cm, respectively.
   (a) Assuming infinite coupler directivity, calculate the width a of the waveguide.
   (b) Calculate the minimum coupler directivity (in dB) required for the phase versus displacement characteristic to be linear within 2°.

7–26. A matched fixed phase-shift section of the type shown in Fig. 7–31 is required in a 6.0 GHz microwave system.
   (a) Calculate the value of normalized susceptance required for a differential phase shift of 45°.
   (b) Determine l and the iris opening d if the guide width is 4.0 cm. Assume negligible iris thickness and an air-filled waveguide.

7–27. Derive Eqs. (7–22) and (7–23).

7–28. A choke-coupled rotary joint (Fig. 7–32) has been designed for use in a 50 ohm coaxial system. All lines are a quarter wavelength long at the 5.0 GHz design frequency. The diameters have been chosen so that $Z_{01} = Z_{03} = 20$ ohms and $Z_{02} = 40$ ohms.
   (a) Calculate the input SWR at 4.0 GHz.
   (b) Repeat the calculation assuming zero spacing between the stubs. Compare with the results in part a.

8

Reciprocal Multiport Junctions

Multiport junctions find considerable use in microwave systems as power monitors, dividers, and combiners. Because of their unique characteristics, they are also utilized in the design of filters, antennas, switches, mixers, duplexers, and many other components. In this chapter, the properties of reciprocal multiports are discussed in some detail. Most of the units described are of the three- and four-port variety. Both TEM and waveguide type configurations are included. Some typical applications are also given. In most cases, the power wave/scattering matrix formulation is used as an analytical tool. Its properties are reviewed in Appendix D.

8–1 THREE-PORT JUNCTIONS

Reciprocal three-port junctions are characterized by the fact that a change in the terminal conditions at one port affects the conditions at the other ports. This effect is particularly pronounced when the junction is dissipationless. For example, a scattering-matrix analysis shows that it is impossible to match all three ports simultaneously (Ex. D–1 in Appendix D). The reason is that the insertion of a matching network at one port changes the impedance characteristics at the other two ports. This lack of isolation between ports can limit the usefulness of three-port junctions, particularly in power monitoring and divider applications. Subsequent examples will illustrate this point.

8–1a Coaxial and Stripline Configurations

Various methods of dividing and monitoring microwave signals in a coaxial system are described here. The techniques are directly applicable to stripline, microstrip, or any other TEM transmission system. Figure 8–1 shows two coaxial type configura-

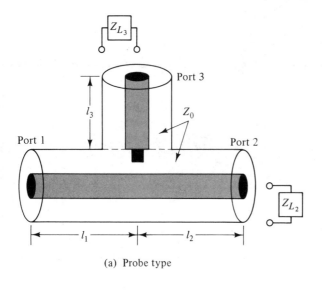

(a) Probe type

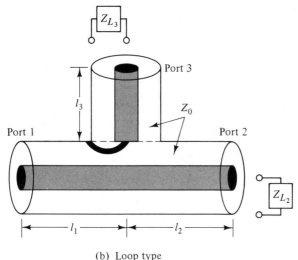

(b) Loop type

Figure 8–1   Coaxial power monitors.

tions that are used to monitor microwave power. The probe type, shown in part *a*, samples signal transmission between ports 1 and 2 by intercepting a portion of the electric field associated with the propagating TEM wave. This produces a signal at port 3 that is proportional to the intercepted field. The loop arrangement, shown in Fig. 8–1*b*, couples power to port 3 by linking some of the magnetic field associated with the TEM wave.

Consider the case of a signal source at port 1 and a matched termination at port 2. The power out of the coupled port (port 3) is directly proportional to the input power. In power-monitoring applications, only a small portion of the incident wave is sampled, typically one percent or less. Therefore, the power delivered to port 2 is approximately equal to the input power. Furthermore, the amount of signal coupled is independent of the length l_2 since port 2 is match terminated. If port 2 were *not* match terminated, the sampled power would depend upon the length l_2 and the terminating impedance since a standing wave would exist on the output line. The following example illustrates the point.

Example 8–1:

A matched generator ($Z_G = Z_0$) with an available power $P_A = 100$ W is connected to port 1 of a probe-type power monitor (Fig. 8–1a). The probe samples 10 percent of the net voltage on the main line. That is, $V_3 = 0.10V_P$, where V_3 is the output voltage at port 3 and V_P is the net voltage on the main line *at the plane of the probe.*

(a) Calculate the power delivered to port 3 if $Z_0 = 50$ ohms and ports 2 and 3 are match terminated.

(b) Determine the variation in output power at port 3 if port 2 is terminated with a 150 ohm resistive load and l_2 is varied from 0.25λ to 0.50λ.

Solution:

(a) With $Z_G = Z_0$, $P_A = P^+ = (V^+)^2/50$, where V^+ and P^+ represent the voltage and power associated with the forward traveling wave. For the given value of P_A, $V^+ = 70.7$ V. Since port 2 is match terminated, there is no reflected wave and hence $V_P = V^+$ (lossless lines are assumed). Therefore, $V_3 = 7.07$ V and $P_3 = V_3^2/50 = 1$ W (that is, one percent of the incident power).

(b) In this case, the reflection coefficient at port 2 is $(150 - 50)/(150 + 50) = 0.50$. With the load resistive and greater than Z_0, a voltage maximum exists at the load and a voltage minimum a quarter wavelength away. For $l_2 = 0.25\lambda$, the net voltage at the plane of the probe is $(1 - |\Gamma|)V^+ = 35.35$ V, while for $l_2 = 0.50\lambda$, the net voltage is $(1 + |\Gamma|)V^+ = 106.05$ V. Thus, V_3 varies from 3.54 to 10.6 V. With port 3 match terminated, $P_3 = V_3^2/50$ and therefore its value varies from 0.25 to 2.25 W.

The reason that P_3 is a function of l_2 is that the probe samples the net voltage on the main line. With l_2 variable, the probe in effect measures the voltage standing-wave pattern. This, in fact, is a commonly used method for measuring SWR. In the above example, the SWR due to the 150 ohm load is 3.0 and therefore the ratio of maximum to minimum power at the probe is $(3.0)^2 = 9.0$. If magnetic field coupling (Fig. 8–1b) had been used, the variation in P_3 would have been exactly the same. However, since the magnetic field pattern is displaced a quarter wavelength relative to the electric field, the maximum value of P_3 would occur when $l_2 = 0.25\lambda$.

The power monitors in Fig. 8–1 use reactive coupling, capacitive for the probe type and inductive for the loop type. Therefore, P_3 is also a function of l_3 when port 3 is *not* match terminated. For example, if l_3 is such that the reflection at port 3 tends to cancel that due to the reactive coupling, P_3 will increase. Conversely, if l_3 is such that the two reflections add, P_3 will decrease. This effect can be minimized by placing some dissipative loss (3 to 6 dB) in the l_3 line.

It is important to understand that the above method of analysis is not valid if a significant portion of the power (say, 10 percent or more) is coupled to port 3. This is because the sampling element distorts the standing wave pattern. For this situation, an equivalent circuit approach should be used (Prob. 8–2).

The coaxial tee junction.   Two forms of coaxial tee junctions are shown in Fig. 8–2. The one in part a is the same as the one in part b except that $Z_{01} = Z_{02} = Z_0$. Consider first Fig. 8–2a, in which port 3 is the input and ports 1 and 2 are terminated by Z_{L_1} and Z_{L_2}, respectively. When both ports 1 and 2 are match terminated, the input power splits equally between them. If $Z_{03} = Z_0$ and the junction is dissipationless, the input SWR is 2.0 and therefore $|\Gamma_3| = (2 - 1)/(2 + 1) = 1/3$. This means that $1/9$ of the incident power is reflected at port 3 and $4/9$ is delivered to each matched termination.

The input reflection can be eliminated with a network that matches the parallel combination of the outputs to the input line. This can be done with a single-section quarter-wave transformer by choosing $Z_{03} = \sqrt{(Z_0)(0.5Z_0)} = 0.707Z_0$ and $l_3 = \lambda/4$ at the design frequency. With port 3 matched, each output receives one-half of the incident power. Note that the inclusion of the matching network increases the SWR into ports 1 and 2 from 2.0 to 3.0 (Prob. 8–3). This is a direct consequence of the

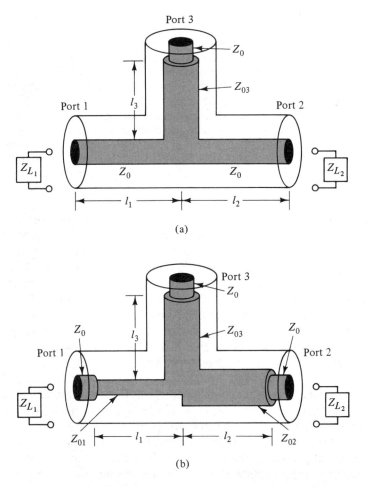

(a)

(b)

Figure 8–2   Coaxial tee junctions.

fact that port 3 is not isolated from the other ports and hence impedance changes in one arm affects all ports.

A more general tee configuration is shown in part b of Fig. 8–2. Figure 8–3 gives the equivalent circuit when ports 1 and 2 are terminated with Z_{L_1} and Z_{L_2}, respectively, and $l_1 = l_2 = l_3 = \lambda/4$. Reactive effects due to localized higher-order modes at the junction have been neglected. This is valid when the transverse dimensions of the coaxial lines are small compared to the operating wavelength. The equivalent circuit also assumes no impedance transformations along the Z_0 lines (that is, their lengths are negligible). Under these conditions, the input impedance at port 3 is given by

$$Z_{\text{in}_3} = \frac{Z_{03}^2}{Z_t} = Z_{03}^2 Y_t = Z_{03}^2 \left(\frac{Z_{L_1}}{Z_{01}^2} + \frac{Z_{L_2}}{Z_{02}^2} \right) \qquad (8\text{–}1)$$

where Z_t is the net impedance terminating the Z_{03} line. Written in normalized form,

$$\bar{Z}_{\text{in}_3} = \bar{Z}_{L_1} \left(\frac{Z_{03}}{Z_{01}} \right)^2 + \bar{Z}_{L_2} \left(\frac{Z_{03}}{Z_{02}} \right)^2 \qquad (8\text{–}2)$$

where the bar notation indicates normalization to Z_0, the characteristic impedance of the connecting lines.

This circuit can be used to either broadband the input SWR or create an uneven division of power between ports 1 and 2. When $Z_{L_1} = Z_{L_2} = Z_0$ and $Z_{01} = Z_{02}$, the power splits equally between the output ports. By setting $Z_{01} = Z_{02} = Z_0(2)^{1/4}$ and $Z_{03} = Z_0/(2)^{1/4}$, the input characteristics are those of a two-section, maximally flat transformer. Thus the input SWR remains low over a wider frequency range than the single-section transformer in Fig. 8–2a.

In order to create an unequal power division when the outputs are match terminated ($Z_{L_1} = Z_{L_2} = Z_0$), the values of Z_{01} and Z_{02} must be different. With the output lines in parallel, the voltage at the junction is common to both lines. Therefore power is proportional to admittance which yields

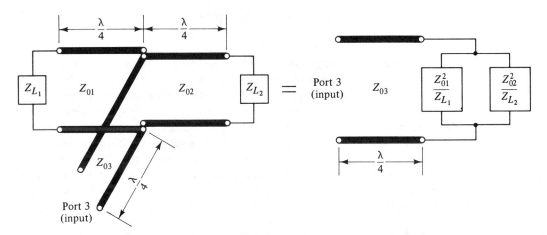

Figure 8–3 Equivalent circuit of the tee junction in Fig. 8–2b when $l_1 = l_2 = l_3 = \lambda/4$.

$$\frac{P_1}{P_2} = \left(\frac{Z_{02}}{Z_{01}}\right)^2 \tag{8-3}$$

where P_1 and P_2 represent the power delivered to ports 1 and 2, respectively. Applying the match condition ($\bar{Z}_{in_3} = 1$) to Eq. (8-2) when $Z_{L_1} = Z_{L_2} = Z_0$ results in

$$Z_{03} = \frac{Z_{02}}{\sqrt{1 + \left(\frac{Z_{02}}{Z_{01}}\right)^2}} = \frac{Z_{02}}{\sqrt{1 + \frac{P_1}{P_2}}} \tag{8-4}$$

For a given power ratio, Eqs. (8-3) and (8-4) contain three unknowns and hence there are an infinite set of solutions for the match condition. One set reduces to the two-section, maximally flat case when $P_1 = P_2$. Namely,

$$Z_{01} = Z_0 \left(2\frac{P_2}{P_1}\right)^{1/4} \quad , \quad Z_{02} = Z_0 \left(2\frac{P_1}{P_2}\right)^{1/4} ,$$

and $\tag{8-5}$

$$Z_{03} = \frac{Z_0}{\left(1 + \frac{P_2}{2P_1} + \frac{P_1}{2P_2}\right)^{1/4}}$$

Problems 8-4 and 8-5 give an indication of the variation of SWR with frequency for this design. It should be noted that except for the case of equal power division, the power ratio P_1/P_2 is also a function of frequency.

As explained earlier, the techniques described here can also be utilized in strip transmission-line systems.

8-1b Waveguide Three-Port Junctions

Typical three-port junctions in rectangular waveguide include the bifurcated guide and the E- and H-plane tees.

The bifurcated waveguide.   A bifurcated guide and its equivalent circuit is shown in Fig. 8-4. It consists of a rectangular guide containing a thin metal plate that effectively converts the guide into two reduced-height waveguides. This configuration is known as an *E-plane bifurcation*. With the plate perpendicular to the TE_{10} electric field, the field and current patterns are unchanged and hence no reflections are created. This assumes that the plate is negligibly thin compared to the guide heights. For this case the input SWR is unity when the outputs are match terminated ($Z_{L_1} = Z_{01}$ and $Z_{L_2} = Z_{02}$). Also, since the electric field is the same in all three guides and voltage is proportional to guide height, $V = Eb$, $V_1 = Eb_1$, $V_2 = Eb_2$, and therefore $V = V_1 + V_2$.[1] The currents in all three guides are the

---

[1] Although voltage is not uniquely defined in a waveguide (Sec. 5-5a), one can still define a voltage that is proportional to the transverse electric field and related to power flow through a surface by $P = V^2/Z$, where Z is the impedance at the surface.

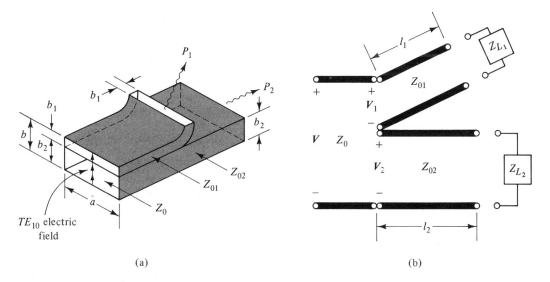

Figure 8–4 The bifurcated waveguide and its equivalent circuit. (Note: For a thin septum, $b_1 + b_2 \approx b$ and therefore $Z_{01} + Z_{02} \approx Z_0$.)

same which means that the E-plane bifurcation represents a series connection. This is indicated in part b of Fig. 8–4.

The power division between output ports when they are match terminated is given by

$$\frac{P_1}{P_2} = \frac{Z_{01}}{Z_{02}} = \frac{b_1}{b_2} \qquad (8\text{–}6)$$

since $P_1 = I^2 Z_{01}$, $P_2 = I^2 Z_{02}$ and characteristic impedance is proportional to guide height. Also, $P_1 + P_2$ equals the input power, assuming lossless guides. In practice, the input is usually standard size rectangular waveguide which means that the reduced-height output guides are not. If the output lines are to be connected to standard size guides, impedance transformers must be used to *step up* the Z_{01} and Z_{02} lines to Z_0. The input SWR and power division are unaffected if the transformers are well matched. As in any three-port junction, the SWR and power division is a function of the terminating impedances as illustrated in the following example problem.

Example 8–2:

A waveguide power divider of the type shown in Fig. 8–4 has $b_1 = b_2$ and $b \approx b_1 + b_2$. Calculate the input SWR and output power ratio under the following conditions.
(a) $Z_{L_1} = Z_{01}$ and $Z_{L_2} = Z_{02}$
(b) $Z_{L_1} = Z_{01}$, $Z_{L_2} = 2Z_{02}$ and $l_2 = \lambda_g/4$
Assume TE_{10} mode propagation.

Solution: The total impedance terminating the input Z_0 line is $Z_1 + Z_2$, where Z_1 and Z_2 are the impedances of the output branches referred to the junction plane. Also, $Z_0 \approx Z_{01} + Z_{02}$ and $Z_{01} = Z_{02}$, since waveguide characteristic impedance is directly proportional to guide height.

(a) With the output lines match terminated, $Z_1 = Z_{01}$ and $Z_2 = Z_{02}$, independent of l_1 and l_2. Therefore, $\bar{Z}_{in} = (Z_1 + Z_2)/Z_0 = (Z_{01} + Z_{02})/Z_0$. Since $Z_0 \approx Z_{01} + Z_{02}$, both $\bar{Z}_{in}$ and the input SWR are unity. Also, the power splits equally between the output ports because with $b_1 = b_2$, $Z_{01} = Z_{02}$. Hence each output signal is 3 dB below the input signal level.

(b) In this case, $Z_1 = Z_{01}$ and $Z_2 = Z_{02}^2/Z_{L_2}$ since $l_2 = \lambda_g/4$. Therefore, $Z_2 = 0.5Z_{02} = 0.5Z_{01}$ and hence $Z_{in} = 1.5Z_{01}$. With $Z_0 = 2Z_{01}$, $\bar{Z}_{in} \equiv Z_{in}/Z_0 = 0.75$ and the input SWR is 1.33.

The power division is no longer equal because $Z_1 \neq Z_2$. With current common in the series circuit, $P_1/P_2 = Z_1/Z_2 = 2.0$. The net input power P_{in} equals $P_1 + P_2$ and therefore $P_1 = 0.67P_{in}$ and $P_2 = 0.33P_{in}$. Using the decibel notation, P_1 and P_2 are respectively 1.76 dB and 4.77 dB below P_{in}. Furthermore, there is a mismatch loss of 0.09 dB due to the 1.33 input SWR. Thus, referenced to the incident power (P_{in}^+), P_1 and P_2 are respectively 1.85 dB and 4.86 dB below P_{in}^+.

The E–plane tee junction.   The three-port waveguide configuration in Fig. 8–5a is known as an *E-plane* or *series* tee. Port 3 is usually designated as the *E* arm and ports 1 and 2 as the coplanar arms. Note that the width of all three guides are the same, which is the most common arrangement. A side view of the tee (part *b*) shows the electric field pattern for a signal input at port 3. Dominant mode propagation is assumed for all arms. Due to the junction's symmetry, the power delivered to ports 1 and 2 are the same. Furthermore, if $l_1 = l_2$, the electric fields at the two outputs are 180° out-of-phase as shown. Associated with the propagating waves are conduction currents on the inner walls of the waveguides. The longitudinal or *power* currents that flow along the broad walls are also indicated. Since a common current flows in the three guides, the *E*-plane tee may be represented by the series connection of three transmission lines. At frequencies where $b \ll \lambda_g$, the equivalent circuit is the same as Fig. 8–4b except that $Z_{01} = Z_{02}$. For standard height guides operating in their useful frequency range, the above inequality does not hold and therefore the equivalent circuit becomes more complicated. The reactive effects associated with

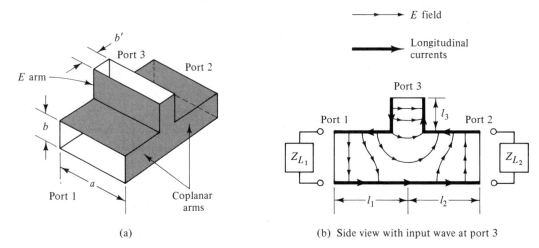

(a)   (b) Side view with input wave at port 3

Figure 8–5   The *E*-plane waveguide tee junction.

the localized higher-order modes may be accounted for by either of the equivalent circuits in Fig. 8–6. Susceptance values as well as values for n_t, d and d' are given by Marcuvitz (Sec. 6.1 of Ref. 8–1). The data is plotted as a function of b'/b and $2b/\lambda_g$.[2]

The circuit in part b of Fig. 8–6 is particularly useful in determining the effect of mismatched terminations on the power division. The data in Marcuvitz indicates that for $b' \leq b$, X is quite small and may be neglected. Because of the impedance transforming properties of transmission lines, it is important that the terminal planes for the equivalent circuits be specified. The line lengths in Fig. 8–6b are necessary to make the terminal planes in both circuits the same. The planes are defined in part c. Marcuvitz's data indicates that the turns ratio n_t of the transformer is frequency sensitive, its value decreasing with increasing frequency. One can also obtain the value of n_t experimentally by simply measuring the SWR into a coplanar arm with the remaining ports match terminated. From the equivalent circuit with $X = 0$, the input impedance at plane A is $Z_A = Z_0 + n_t^2 Z_0'$ and therefore $\bar{Z}_A \equiv Z_A/Z_0 = 1 + (b'/b)n_t^2$. Since $\bar{Z}_A$ is real and greater than one, the SWR into port 1 equals $\bar{Z}_A$. Thus,

$$n_t = \sqrt{(\text{SWR} - 1)\frac{b}{b'}} \qquad (8\text{–}7)$$

Once n_t is known, additional characteristics of the tee junction are easily determined as indicated by the following example and Probs. 8–7, 8–8, and 8–9.

Example 8–3:

An E-plane tee with $b' = b$ has an SWR of 1.64 at port 1 when the other ports are match terminated. Calculate the power delivered to ports 2 and 3 when a matched generator with an available power of one watt is connected to port 1.
Solution: From Eq. (8–7), $n_t = 0.8$. Since current is common in a series connection,

$$P_2 = I^2 Z_0, \qquad P_3 = I^2 n_t^2 Z_0 \qquad \text{and} \qquad P_{\text{in}} = I^2 (1 + n_t^2) Z_0$$

Therefore, $P_2 = 0.61 P_{\text{in}}$ and $P_3 = 0.39 P_{\text{in}}$, where $P_{\text{in}} = \{1 - |\Gamma_1|^2\} P_{\text{in}}^+$ and Γ_1 is the reflection coefficient at port 1. For a matched generator, P_{in}^+ is the available generator power. With $|\Gamma_1| = (1.64 - 1)/(1.64 + 1) = 0.24$,

$$P_{\text{in}} = 0.94 \text{ W}, \qquad P_2 = 0.57 \text{ W} \qquad \text{and} \qquad P_3 = 0.37 \text{ W}.$$

The division of power can be controlled by means of a capacitive or inductive iris in the E arm. Equivalent circuits for certain cases have been obtained by Marcuvitz (Secs. 6.2 and 6.8 of Ref. 8–1). As before, the circuit in Fig. 8–6b (with $X = 0$) may be used for SWR and power calculations. The turns ratio can be determined experimentally with the aid of Eq. (8–7). For example, if $b' = b$ and the measured SWR into the coplanar arm is 3.0, the turns ratio $n_t = \sqrt{2}$. Note that this case corresponds to a perfect match into port 3, the E arm.

---

[2] The parameter b/λ_g in Figs. 6.1–4 through 6.1–14 of Marcuvitz should read $2b/\lambda_g$.

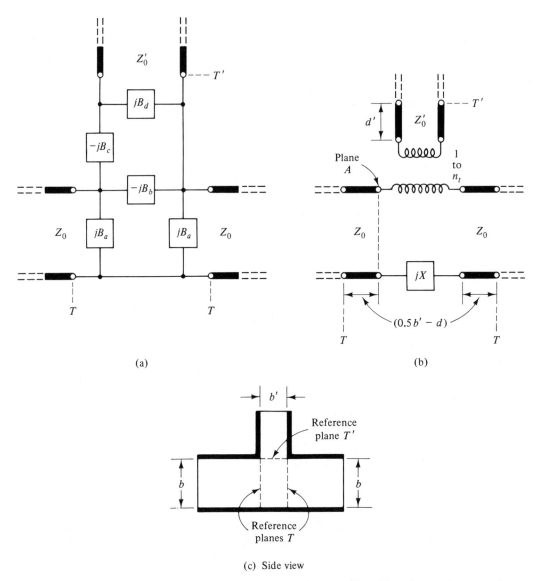

(a)   (b)

(c) Side view

Figure 8–6 Two equivalent circuits for the E-plane tee. (Note: The reference planes associated with terminals T and T' are indicated in part c.)

The H-plane tee junction. The three-port junction in Fig. 8–7 is known as an *H-plane* or *shunt* tee. Port 3 is usually designated as the H arm and ports 1 and 2 as the coplanar arms. Also shown is the flow of power currents for a signal into the H arm. The current division indicates that the three waveguides are connected in shunt. A signal into the H arm splits equally between the coplanar arms. The TE_{10} electric field pattern is indicated in part b of the figure. Note that if $l_1 = l_2$, the output electric fields are in phase. Matching into the H arm is usually accomplished with an inductive iris or post located as shown.

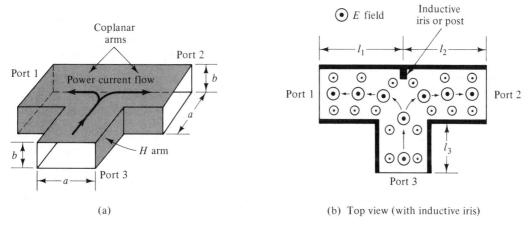

(a)

(b) Top view (with inductive iris)

Figure 8–7   The H-plane waveguide tee junction.

The H-plane tee can also be represented by an equivalent circuit consisting of several reactances. Element values may be found in Marcuvitz (Sec. 6.5 of Ref. 8–1). A simplified equivalent circuit is shown in Fig. 8–8. Values of d, d', and X are given in Marcuvitz. Over the useful frequency range of the guides, $|X| > 2Z_0$ and may be neglected for approximate calculations. As before, the transformer turns ratio may be determined experimentally. Neglecting the shunt reactance X, the normalized admittance at plane A with the other ports match terminated is $\bar{Y}_A = 1 + n_t^2$. With $\bar{Y}_A$ real and greater than unity, the SWR into port 1 is $1 + n_t^2$ and therefore

$$n_t = \sqrt{\text{SWR} - 1} \qquad (8\text{–}8)$$

This simplified equivalent circuit is most accurate when $|X| \gg Z_0$, a condition that holds at the center of the waveguide's frequency band.

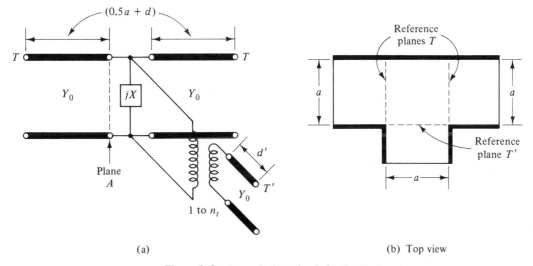

(a)

(b) Top view

Figure 8–8   An equivalent circuit for the H-plane tee.

Examples 8–1 and 8–2 showed that for any three-port network the division of power between output ports is a function of the terminating impedances. Because this effect is often undesirable, the use of hybrid junctions and directional couplers is preferred in most applications. These are discussed in the following sections.

8–2 HYBRID JUNCTIONS

A hybrid junction is a four-port network in which a signal incident on any one of the ports divides between two output ports with the remaining port being isolated. The assumption is that all output ports are terminated in a perfect match. Under these conditions, the input to any port is perfectly matched. This section describes hybrids in which the power is divided equally between the output ports. These are known as 3 dB hybrids. Both waveguide and TEM structures are analyzed.

The hybrid (magic) tee.   One method of realizing a 3 dB hybrid is to combine an E-plane and H-plane tee as shown in Fig. 8–9a. By virtue of the orientation of the E and H arms, ports 3 and 4 are decoupled. For example, assuming single-mode propagation, a TE_{10} wave entering the E arm cannot couple to the H arm since the electric field vector would be parallel to the broadwall of the H arm and hence below cutoff. The following argument shows that if the E and H arms are matched, the four-port junction forms a 3 dB hybrid, commonly known as a *magic tee*.

Consider two equal amplitude signals entering ports 3 and 4. The one entering the H arm splits equally between the coplanar arms. If the length of the coplanar arms are equal, the output signals at ports 1 and 2 are in phase. The vector-phasors are indicated by the solid arrows in part b of Fig. 8–9. The signal entering the E arm also splits equally, but in this case the output signals at ports 1 and 2 are 180° out-of-phase, as indicated by the dashed arrows. If both the E and H arms are matched, the amplitude of all four output phasors are the same. Thus assuming equal

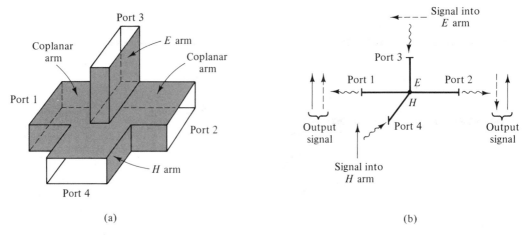

(a)   (b)

Figure 8–9 The hybrid (magic) tee junction. (Note: All waveguides are identical with characteristic impedance Z_0.)

line lengths, the signals add at port 1 and cancel at port 2. Since the network is reciprocal, this means that an input signal at port 1 splits equally between the E and H arms with none appearing at the other coplanar arm. Furthermore, there is no reflection at port 1 because all of the input power is delivered to ports 3 and 4 (assuming matched terminations). In a similar manner, an input signal at port 2 splits equally between ports 3 and 4 with none appearing at port 1. Therefore with the E and H arms matched and decoupled, the coplanar arms are also matched and decoupled. Power into any port produces an equal power split between the two output ports, which means that the magic tee is a 3 dB hybrid.

Although the above explanation assumed arms of equal length, the results are valid even when they are unequal. The only condition required is that $|\Delta\phi_H - \Delta\phi_E| = \pi$, where $\Delta\phi_H$ is the difference in output phases when the H arm is the input and $\Delta\phi_E$ is the output phase difference when the E arm is the input.

As in the case of the E and H-plane tee junctions, an equivalent circuit involving several reactive elements can be developed for the magic tee (Sec. 7.8 of Ref. 8–1). An alternate equivalent circuit that is very useful in analyzing power division is shown in Fig. 8–10. It is a combination of the equivalent circuits in Figs. 8–6b and 8–8a except that the line lengths and the reactive elements have been neglected. This circuit, in fact, resembles the hybrid coil arrangement once used in telephone equipment. With the E and H arms matched, the turns ratios of the ideal transformers are $n_E = n_H = \sqrt{2}$.

Let us now summarize the properties of an ideal magic tee with matched terminations.

1. All ports are perfectly matched.

2. The E and H arms are decoupled as are the two coplanar arms.

3. A signal into a coplanar arm splits equally between the E and H arms. For each output signal, $P_{\text{out}} = P_{\text{in}}/2$ and therefore $V_{\text{out}} = V_{\text{in}}/\sqrt{2}$.

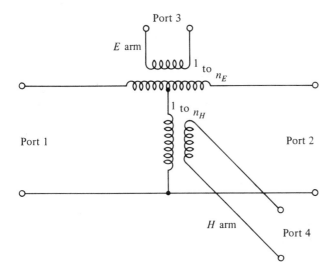

Figure 8–10 An equivalent circuit for the magic tee junction. (Note: When $n_E = n_H = \sqrt{2}$, all ports are matched.)

4. A signal into the H arm splits equally between the coplanar arms, the outputs being in phase equidistant from the junction.

5. A signal into the E arm splits equally between the coplanar arms, the outputs being 180° out-of-phase equidistant from the junction.

6. For signals into both coplanar arms,
   (a) The signal output from the E arm equals $1/\sqrt{2}$ times the phasor difference of the input signals.
   (b) The signal output from the H arm equals $1/\sqrt{2}$ times the phasor sum of the input signals.

Because of the last property, the E arm is often called the *difference* arm and the H arm the *sum* arm. The reason for the $1/\sqrt{2}$ factor is that with all impedances equal, an equal power split reduces voltages (and currents) by $1/\sqrt{2}$.

Based on the properties described here, the scattering matrix for an ideal hybrid tee may be stated in the following form.

$$[S] = \begin{bmatrix} 0 & 0 & 1/\sqrt{2} & 1/\sqrt{2} \\ 0 & 0 & -1/\sqrt{2} & 1/\sqrt{2} \\ 1/\sqrt{2} & -1/\sqrt{2} & 0 & 0 \\ 1/\sqrt{2} & 1/\sqrt{2} & 0 & 0 \end{bmatrix} \tag{8-9}$$

This assumes that the length of the coplanar arms are the same and the length of the E and H arms have been chosen so that S_{13} and S_{14} are positive real. Without these assumptions, phase angles would have to be added to the non-zero scattering elements.

One of the main advantages of a magic tee, or in fact any hybrid, is that the power delivered to one port is independent of the termination at the other output port. This is because the two output ports are decoupled. For this reason, hybrid junctions are much more widely used than three-port junctions. This is illustrated by the example that follows. The solution makes use of the incident and scattered waves defined in Appendix D. Some of their properties are summarized here.

1. A wave incident on port k is indicated by $\boldsymbol{a}_k$, while a wave exiting from port k is indicated by $\boldsymbol{b}_k$.

2. Both a_k and b_k are rms-phasor quantities such that $P_k^+ = a_k^2$ and $P_k^- = b_k^2$ are respectively the average power associated with waves propagating toward and away from port k. (Note, a_k and b_k represent the rms values of $\boldsymbol{a}_k$ and $\boldsymbol{b}_k$.)

3. The *net* power into port k is $P_{in} = a_k^2 - b_k^2$, while the net power delivered to a termination Z_L at port k is given by $P_L = b_k^2 - a_k^2 = b_k^2\{1 - |\Gamma_L|^2\}$.

Example 8–4:

A matched generator ($Z_G = Z_0$) with an available power of one watt is connected to the H arm of a magic tee. The E arm is match terminated and the length of the coplanar arms are the same. Calculate the power delivered to the terminations at ports 1, 2, and 3 and the power reflected at port 4 when
(a) Ports 1 and 2 are match terminated ($Z_{L_1} = Z_{L_2} = Z_0$).

(b) Ports 1 and 2 are terminated in the following manner,

$$Z_{L_1} = 2.4Z_0 \quad \text{and} \quad Z_{L_2} = 0.6Z_0.$$

Also calculate the SWR into the H arm for both cases.

Solution: With the generator matched to the line, the incident power

$$P_4^+ = P_A = 1 \text{ W}.$$

(a) In this case, all ports are matched and therefore the input SWR is unity. Since no power is reflected, $P_{\text{in}} = P_4^+ = 1$ W which splits equally between ports 1 and 2. Thus $P_1 = P_2 = 0.5$ W and $P_3 = 0$.

(b) The incident and scattered waves in a magic tee are indicated in Fig. 8–11, where port 4 is the input and ports 1, 2, and 3 are terminated with Z_{L_1}, Z_{L_2} and Z_{L_3}, respectively. With the E arm match terminated, $Z_{L_3} = Z_0$ and therefore $a_3 = 0$. Thus by virtue of the magic tee's properties, $b_1 = a_4/\sqrt{2}$ and $b_2 = a_4/\sqrt{2}$. The reflected waves in the coplanar arms are $a_1 = \Gamma_1 a_4/\sqrt{2}$ and $a_2 = \Gamma_2 a_4/\sqrt{2}$. Note that Γ_1 and Γ_2 are the reflection coefficients associated with Z_{L_1} and Z_{L_2} *referenced to the junction plane*. The waves emanating from the difference and sum arms (that is, the E and H arms) are respectively, $b_3 = 0.5a_4(\Gamma_1 - \Gamma_2)$ and $b_4 = 0.5a_4(\Gamma_1 + \Gamma_2)$. With the E arm match terminated, the net power delivered to the load terminations are

$$P_1 = b_1^2 - a_1^2 = 0.5(1 - |\Gamma_1|^2)a_4^2,$$

$$P_2 = 0.5(1 - |\Gamma_2|^2)a_4^2$$

and

$$P_3 = 0.25|\Gamma_1 - \Gamma_2|^2 a_4^2,$$

where $a_4^2 = P_4^+$, the incident power at the input port. The reflected power at the input is

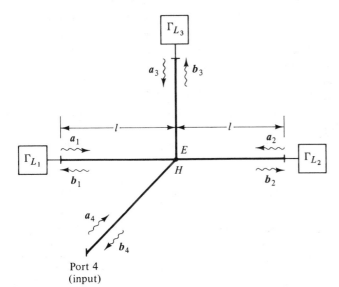

Port 4
(input)

Figure 8–11   Incident and scattered waves in a magic tee. (See Ex. 8–4.)

$$P_4^- = b_4^2 = 0.25|\,\Gamma_1 + \Gamma_2\,|^2 P_4^+$$

and therefore the net input power

$$P_{\text{in}} = P_4^+ - P_4^- = (1 - |\,\Gamma_{\text{in}}\,|^2)P_4^+,$$

where $|\,\Gamma_{\text{in}}\,| = b_4/a_4$.

In this example, $\Gamma_{L_1} = (2.4 - 1)/(2.4 + 1) = 0.41$ and $\Gamma_{L_2} = (0.6 - 1)/(0.6 + 1) = -0.25$. Therefore, $\Gamma_1 = 0.41\underline{/-2\beta l}$ and $\Gamma_2 = -0.25\underline{/-2\beta l}$, where l is the length of each coplanar arm. Calculation of the various powers when $P_4^+ = 1.0$ W yields

$$P_1 = 0.416 \text{ W} \qquad P_2 = 0.469 \text{ W}$$

$$P_3 = 0.109 \text{ W} \qquad P_4^- = 0.006 \text{ W}$$

Note that their sum equals P_4^+, the incident power at the input. As long as the two line lengths are the same, the results are independent of l. With $|\,\Gamma_{\text{in}}\,| = \sqrt{0.006} = 0.078$, the input SWR = 1.17 and the net input power $P_4 = (1 - |\,\Gamma_{\text{in}}\,|^2)P_4^+ = 0.994$ W, which is the sum of P_1, P_2, and P_3. This is expected since the magic tee itself is lossless.

A few comments are in order regarding the above example. First, with $\Gamma_3 = 0$, ports 1 and 2 are completely isolated and therefore P_1 is independent of Z_{L_2} and P_2 is independent of Z_{L_1}. This is not true for a three-port power divider (Prob. 8–11). Second, if the generator is not matched to the line, the incident power at the input is *not* equal to P_A, the available generator power. To calculate P_{in}^+ (P_4^+ in the example), Eq. (3–69) must be used. It is repeated here for convenience.

$$P_{\text{in}}^+ = P_A \frac{1 - |\,\Gamma_G\,|^2}{|\,1 - \Gamma_G\Gamma_{\text{in}}\,|^2} \qquad (8\text{–}10)$$

where Γ_G and Γ_{in} are the generator and input reflection coefficients. Note that Γ_G must be referenced to the same terminal plane as Γ_{in}, namely, the input. If both Γ_G and Γ_{in} are non-zero, their magnitude and phase values are needed to calculate P_{in}^+. Finally, if $\Gamma_3 \neq 0$ or the magic tee is not perfectly matched, ports 1 and 2 are no longer decoupled. This situation involves multiple reflections of the incident and scattered waves. As a result, b_k now represents the phasor sum of the scattered waves at port k as indicated by Eq. (D–8) in Appendix D.[3] The use of flow graphs as described in Sec. 4–4 is helpful in solving such problems.

The hybrid ring. One of the first hybrid junctions developed for microwave use was the hybrid ring, sometimes referred to as the *rat race*. Six basic configura-

---

[3] The reader is cautioned to avoid a common error in power calculations. Assuming correlated signals, one *cannot* determine the total output power at a port by simply summing the powers of the individual waves leaving that port. The principle of superposition requires that one first find the phasor sum of the individual waves and *then* square its rms value. For example, if $b_k = 3\underline{/\phi_1} + 5\underline{/\phi_2}$, the output power is *not necessarily* $3^2 + 5^2 = 34$ W. To determine the actual value of $P_k^- = b_k^2$, ϕ_1 and ϕ_2 (or their difference) must be known. For instance, if the waves are *in phase*, the rms value of b_k is 8, while if they are 180° out-of-phase, b_k is 2. Thus, depending upon the relative phase of the two waves, the value of P_k^- can be anywhere between 4 and 64 W. This comment also applies to the calculation of $P_k^+ = a_k^2$, where a_k is the rms value of the phasor sum of the waves entering port k.

tions have been described by Tyrrell (Ref. 8–11). Two versions, a shunt-connected TEM type and a series-connected waveguide type, are shown in Fig. 8–12. The center conductor pattern for a shunt-connected stripline or coaxial unit is shown in Fig. 8–12a. It consists of a transmission-line ring having a mean circumference of $3\lambda/2$ and a characteristic impedance equal to $\sqrt{2}Z_0$, where Z_0 is the characteristic impedance of the connecting lines.

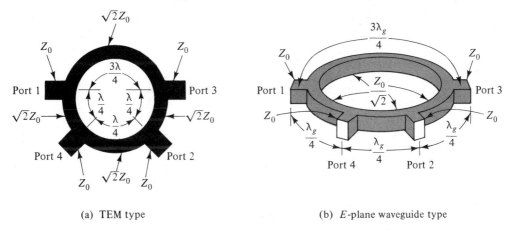

(a) TEM type   (b) E-plane waveguide type

Figure 8–12   TEM and waveguide type hybrid rings.

The hybrid ring has the following properties. Ports 1 and 2 are decoupled as are ports 3 and 4. Also, a signal entering either port 1 or 2 splits equally between ports 3 and 4, while a signal input at port 3 or 4 splits equally between ports 1 and 2. To understand its operation, consider a signal input at port 1. The signal splits and propagates in both directions, thereby creating a standing wave pattern in the transmission-line ring. A voltage null exists at port 2 since the difference in path lengths causes the two waves to arrive at that point 180° out-of-phase. When ports 3 and 4 are match terminated, the null is unaffected by their presence. This is because the phase shift and loss experienced by the wave as it passes port 3 is identical to that of the wave propagating past port 4. Identical voltage maximums exist at ports 3 and 4 and therefore equal power is available at each port. With the characteristic impedance of the ring equal to $\sqrt{2}\,Z_0$, port 1 is matched when ports 3 and 4 are matched terminated. This can be understood with the aid of Fig. 8–13. Since a voltage null exists at port 2, shorting its terminals does not affect the standing wave pattern in the ring. The quarter-wave lines transform the short circuit into open circuits at ports 3 and 4 resulting in the equivalent circuit shown in the figure. The $\lambda/4$ and $3\lambda/4$ lines transform each load impedance (Z_0) to $2Z_0$ as seen at port 1. Because of the parallel connection at port 1, the total input impedance is Z_0 and hence the input is matched.

The following conclusions may be drawn from this analysis. With port 1 as the input and the remaining ports match terminated,

1. Port 1 is matched.
2. Ports 1 and 2 are decoupled.

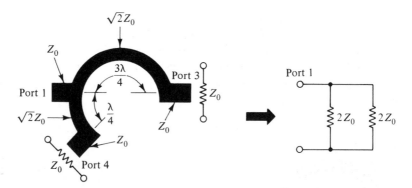

Figure 8–13 The input impedance of a hybrid ring with output ports match terminated.

3. The input power splits equally between ports 3 and 4. Thus $b_3 = ja_1/\sqrt{2}$ and $b_4 = -ja_1/\sqrt{2}$, where a_1 is the incident wave at port 1. The 180° phase difference between b_3 and b_4 is due to the $\lambda/2$ difference in the path lengths. By symmetry, a signal entering port 3 splits equally between ports 1 and 2 with port 4 isolated. A similar argument can be used to show that power into port 2 splits equally between ports 3 and 4 with port 1 isolated. That is, $b_1 = 0$, $b_3 = -ja_2/\sqrt{2}$, and $b_4 = -ja_2/\sqrt{2}$, where a_2 is the incident wave at port 2. Note that in this case, b_3 and b_4 are in phase.

Based on the above properties, the scattering matrix for the hybrid ring may be written in the following form.

$$[S] = \begin{bmatrix} 0 & 0 & j/\sqrt{2} & -j/\sqrt{2} \\ 0 & 0 & -j/\sqrt{2} & -j/\sqrt{2} \\ j/\sqrt{2} & -j/\sqrt{2} & 0 & 0 \\ -j/\sqrt{2} & -j/\sqrt{2} & 0 & 0 \end{bmatrix} \qquad (8\text{–}11)$$

The isolation between ports 1 and 2 and between ports 3 and 4 is a result of the $\lambda/2$ difference in path length for the two waves in the ring. This condition only exists at the design frequency and therefore the useful bandwidth of the hybrid ring is limited. Both Albanese and Peyser (Ref. 8–12) and Reed and Wheeler (Ref. 8–13) have analyzed its frequency response. The latter article employs even-odd mode theory, which is widely used in analyzing hybrids and directional couplers. For most applications, the useful bandwidth of the hybrid ring is about 20 percent. Techniques for increasing the bandwidth significantly have been developed (Ref. 8–12).

A waveguide hybrid ring is shown in Fig. 8–12b. The ports are connected to the waveguide ring via E-plane tee junctions. Since the E-plane junction is essentially a series connection, this version is the dual of the one shown in part a. Therefore, the match condition requires that the characteristic impedance of the ring be $Z_0/\sqrt{2}$, where Z_0 is the impedance of the connecting guides. This follows from the principle of duality.

The Wilkinson power divider.   An N-port power divider that splits an input signal into N equiphase–equiamplitude outputs has been described by Wilkinson (Ref. 8–14). The structure is a hybrid junction because internally connected resistors provide isolation between output ports. This concept was utilized by Parad and Moynihan (Ref. 8–15) to design a two-way power divider capable of unequal power division. The most widely used arrangement, however, is that of equal power division. The center conductor pattern for a stripline version is shown in Fig. 8–14. The principle of operation described below applies equally to a coaxial version.[4]

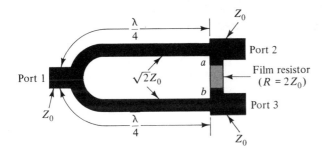

Figure 8–14 A two-way power divider with equal power division.

When a signal enters port 1, it splits into equal-amplitude equiphase output signals at ports 2 and 3. Since each end of the resistor (points a and b) is at the same potential, no current flows through it and therefore the resistor is decoupled from the input. With the characteristic impedance of the quarter-wave lines equal to $\sqrt{2}Z_0$, the input is matched when ports 2 and 3 are match terminated.

Next, consider a signal input at port 2. In this case, it splits equally between port 1 and the resistor R with none appearing at port 3. The resistor thus serves the important function of decoupling ports 2 and 3. These characteristics can be verified by an even-odd mode analysis. The basis of the even-odd mode analysis is that a voltage source (V_G) connected to port 2 can be treated as the superposition of two *pairs* of sources at ports 2 and 3 each having a value $V_G/2$. Their polarities are indicated in part a of Fig. 8–15. Note that the matched termination at port 1 is represented by two terminations in parallel, each of value $2Z_0$. Also the resistor R is represented by two resistors in series, each of value $R/2$. This allows us to define a symmetry plane indicated in the figure by the dashed horizontal line.

Consider first the even-mode pair. Symmetry requires that points a and b be at the same potential, which means that no current flows through the resistor R. Thus point d can be open circuited without affecting even-mode operation. Similarly, point c can be open circuited and therefore either half of the circuit may be represented by the equivalent circuit in part b. By choosing the characteristic impedance of the quarter-wave line as $\sqrt{2}Z_0$, the circuit is matched at port 2 (and also at port 3). That is, $Z_{2e} = Z_0$, where the subscript e is used to denote even-mode quantities. Therefore $V_{2e} = V_G/4$ and by virtue of the quarter-wave line, $V_{1e} = -j\sqrt{2}V_G/4$. Because of the even-order symmetry, $V_{3e} = V_G/4$.

[4]A magic tee with its E arm match terminated represents a waveguide version, the H arm being equivalent to port 1 in Fig. 8–14.

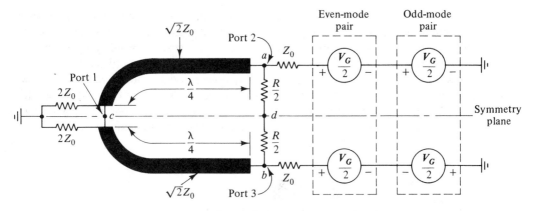

(a) Replacement of a matched generator at port 2 and
a matched load at port 3 by even and odd-mode pairs

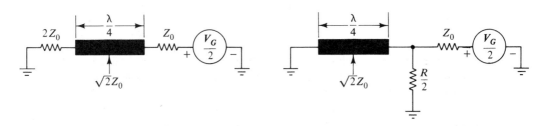

(b) Even-mode equivalent circuit   (c) Odd-mode equivalent circuit

Figure 8–15   Even-odd mode analysis of the two-way power divider in Fig. 8–14.

Next, consider the odd-mode pair. The opposite polarity generators create zero
potential values at points c and d, which means that these points can be shorted to
ground. Thus the equivalent circuit for the top half becomes that shown in part c of
Fig. 8–15. The bottom half has the same equivalent circuit except that the polarity
of the generator is reversed. The input to the quarter-wave shorted line is an open
circuit and hence the input impedance for the odd mode is $Z_{2o} = R/2$, where the
subscript o denotes odd-mode quantities. For a perfect match, the resistor value is
made equal to $2Z_0$. This is indicated in Fig. 8–14. For $R = 2Z_0$, the odd-mode
voltage at port 2 is $V_{2o} = V_G/4$ and by virtue of odd-order symmetry, $V_{3o} = -V_G/4$.
Also, with point c at ground potential, $V_{1o} = 0$.

Superimposing the even- and odd-mode solutions for an input at port 2 results
in the following voltages.

$$V_2 = V_{2e} + V_{2o} = V_G/2 \quad , \quad V_3 = V_{3e} + V_{3o} = 0$$
$$V_1 = V_{1e} + V_{1o} = -jV_2/\sqrt{2} \quad , \quad V_{ab} = V_2 - V_3 = V_2$$

where V_{ab} is the voltage across the resistor. Furthermore, with port 2 matched for
both modes the input impedance at port 2 is Z_0. Thus the even-odd mode analysis
verifies that the circuit in Fig. 8–14 is a hybrid since all ports are matched and ports

2 and 3 are decoupled. Note that for a signal input at either port 2 or 3, half the power is dissipated in the resistor and half is delivered to port 1.

The quarter-wave lines limit the useful bandwidth of the hybrid to about 40 percent. Parad and Moynihan (Ref. 8–15) describe a compensated version with improved bandwidth. Theoretical performance characteristics as well as design data for both equal and unequal power division are given in Ref. 8–3. Significant improvement in bandwidth can be achieved by using a multisection design. This problem has been analyzed by Cohn (Ref. 8–16) for the case of equal power division. Detailed design information and performance characteristics are given. For example, a four-section hybrid provides excellent performance over a 4 to 1 frequency band, while a seven-section version has a 10 to 1 frequency capability.

It should be noted that any of the 3 dB hybrids described here can also be used as a power combiner. In the case of the Wilkinson divider, for instance, signals injected at ports 2 and 3 produce a voltage at port 1 that is proportional to their phasor sum. That is, $V_1 = -j(V_2 + V_3)/\sqrt{2}$. The power associated with the phasor *difference* is dissipated in the resistor, thus preserving the isolation between ports 2 and 3.

The hybrid junctions discussed thus far are of the 0–180° variety. That is, in all cases, the output signals are either in phase or 180° out-of-phase. The next section describes another class, known as 90° or *quadrature* hybrids, in which the output signals are 90° out-of-phase, two examples being the branch-line and short-slot couplers.

The wide variety of hybrid configurations is a tribute to the ingenuity of microwave engineers. Among these are the Alford hybrid (Ref. 8–17), the Kahn E-plane hybrid (Ref. 8–18), and the Lange coupler (Ref. 8–19). Hybrids based on quasi-optical techniques have also been developed for use at millimeter wavelengths. Some examples are given by Levy (Ref. 8–4).

Before concluding this section, special mention should be made of the remarkable hybrid circuits described by Ruthroff (Ref. 8–51). The ultra wideband performance of these hybrids is achieved through the use of transmission-line transformers that are wound on miniature ferrite toroids. Bandwidths spanning several octaves are readily obtained.

8–3 DIRECTIONAL COUPLERS

A directional coupler is a four-port network in which portions of the forward and reverse traveling waves on a line are *separately* coupled to two of the ports.[5] As a result, it finds extensive use in systems that require the measurement of amplitude and phase for a traveling wave. In its most common form, the directional coupler consists of two transmission lines and a mechanism for coupling signals between them. Its properties may be summarized with the aid of the schematic diagram in Fig. 8–16. With matched terminations on the output ports,

---

[5] This contrasts with the three-port networks discussed earlier in which the signal output at a port is proportional to the phasor combination of the waves traveling between the remaining two ports. For example, port 3 in Fig. 8–1 samples the standing wave pattern due to the traveling waves between ports 1 and 2.

1. A portion of the wave traveling from port 1 to port 4 is coupled to port 3, but *not* to port 2.
2. A portion of the wave traveling from port 4 to port 1 is coupled to port 2, but *not* port 3.
3. Similar comments apply to waves propagating between ports 2 and 3. That is, a portion of the wave incident on port 2 is coupled to port 4, but *not* port 1; and a portion of the wave incident on port 3 is coupled to port 1, but *not* port 4.

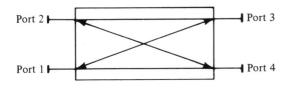

Port 2   Port 3

Port 1   Port 4

Figure 8–16 Schematic diagram of a directional coupler. (Note: The port numbering may differ in other texts and articles.)

The above conditions are indicated in the schematic diagram by the pair of double-sided arrows. Note that in an ideal directional coupler, ports 1 and 2 are decoupled as are ports 3 and 4. Hybrid junctions have this same property and therefore are examples of directional couplers. Since the ones discussed in the previous section involved equal power division, they are classified as 3 dB couplers. As is the case for an ideal hybrid junction, all ports of an ideal directional coupler are perfectly matched. That is, with any three ports match terminated, the remaining port has an input SWR of unity. In fact, it can be shown that if all ports of a dissipationless, reciprocal, four-port network are matched, it is a directional coupler (see Sec. 12.17 of Ref. 8–2). Thus, the scattering matrix for an ideal directional coupler reduces to

$$[S] = \begin{bmatrix} 0 & 0 & S_{13} & S_{14} \\ 0 & 0 & S_{23} & S_{24} \\ S_{13} & S_{23} & 0 & 0 \\ S_{14} & S_{24} & 0 & 0 \end{bmatrix} \tag{8-12}$$

where the port numbering is given in Fig. 8–16.[6] This can be further reduced since, with the aid of Eq. (D–11), it can be shown (Prob. 8–18) that $|S_{13}| = |S_{24}| \equiv k_c$ and $|S_{14}| = |S_{23}| \equiv k_t$, where k_c is the coupling coefficient and $k_t = \sqrt{1 - k_c^2}$ is the transmission coefficient. By proper selection of the terminal planes and the use of Eq. (D–12), the scattering matrix is further simplified. Namely,

$$[S] = \begin{bmatrix} 0 & 0 & jk_c & k_t \\ 0 & 0 & k_t & jk_c \\ jk_c & k_t & 0 & 0 \\ k_t & jk_c & 0 & 0 \end{bmatrix} \tag{8-13}$$

where $k_t = \sqrt{1 - k_c^2}$. For small or loose coupling ($k_c \leq 0.1$), $k_t \approx 1.0$.

Note the similarity between this matrix and that of the hybrids discussed earlier [Eqs. (8–9) and (8–11)]. This is because the coupler is, in fact, a hybrid junc-

---

[6] Since the port numbering in other texts may differ from that used here, care must be exercised in comparing results.

tion. Because the two output signals are 90° out-of-phase, the coupler is classified as a 90° or *quadrature*-type hybrid.

The performance of a directional coupler is usually described in terms of its coupling and directivity, which are defined in the following manner.

$$\text{Coupling (dB)} \equiv 10 \log \frac{P_i}{P_c} = 10 \log K_c \qquad (8–14)$$

$$\text{Directivity (dB)} \equiv 10 \log \frac{P_c}{P_d} = 10 \log K_d \qquad (8–15)$$

where P_i is the incident power at the input port, $P_c = k_c^2 P_i$ is the coupled power, and P_d is the power out of the decoupled port. Note that, as defined, the coupling factor $K_c = 1/k_c^2$ and the directivity factor K_d are greater than unity. These definitions assume that all output ports are match terminated. Thus, *coupling* is a measure of how much of the incident power is being sampled, while *directivity* is a measure of how well the coupler distinguishes between forward and reverse traveling waves. For an ideal directional coupler, $P_d = 0$ and therefore the directivity is infinite. Most applications require directivity values in excess of 30 dB.

The term *isolation* is sometimes used to describe the directive properties of a coupler. It is defined as

$$\text{Isolation (dB)} \equiv 10 \log \frac{P_i}{P_d} = 10 \log K_c K_d \qquad (8–16)$$

Note that *isolation* (dB) equals *coupling* plus *directivity*.

To clarify the meaning of these terms, consider a coupler (Fig. 8–16) with port 1 as the input. If, for example, it is a 20 dB coupler with 30 dB directivity, $K_c = 100$ and $K_d = 1000$. Therefore, $P_3 = 10^{-2} P_1^+$ and $P_2 = 10^{-3} P_3 = 10^{-5} P_1^+$, which means that P_3 and P_2 are respectively 20 dB and 50 dB below the incident signal level. Also, with the input matched, $P_1^- = 0$ and hence $P_4 = P_1^+ - P_2 - P_3$. In this example, $P_4 \approx P_1^+$ and thus $k_t \approx 1$, which is always true for loose coupling.

Many types of directional couplers have been developed over the years as evidenced by the hundreds of published articles on the subject. Generally, they can be classified as either discrete-type or distributed-type couplers. Some of the more widely used are discussed in this section. For an even more detailed treatment, the reader is referred to Levy (Ref. 8–4). This reference also contains an excellent bibliography as does Ref. 8–20. Much design information is given in Refs. 8–3, 8–5, and 8–6.

8–3a Discrete Coupling

The usual configuration for a directional coupler consists of two transmission lines with some mechanism for coupling between them. Techniques for coupling microwave energy include inductive loops, capacitive probes, and sections of line connected directly between the two transmission lines. In waveguide, the most common method is to use apertures in the guide walls, usually in the form of thin slots or circular holes. Figure 8–17 shows several possible aperture positions in a rectan-

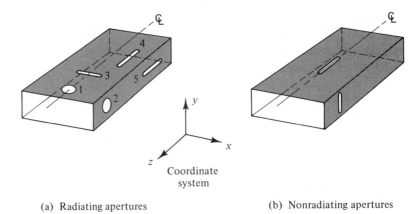

(a) Radiating apertures   (b) Nonradiating apertures

Figure 8–17 Radiating and nonradiating apertures in rectangular guide (TE_{10} mode propagation).

gular guide. Assuming TE_{10} mode propagation, the ones in part *a* radiate a portion of the wave energy and therefore find use in antenna arrays as well as directional couplers. In the coupler case, one or more apertures are placed in the common wall between two adjacent waveguides, thereby allowing the transfer of electromagnetic energy between them.

The theory of radiation through small apertures was developed by Bethe (Ref. 8–21) and is described in various texts (for example, Refs. 8–4 and 8–7). Both electric and magnetic coupling effects are analyzed. Figure 8–18 shows the field patterns for a circular hole centrally located in the broad wall of a waveguide. For the case of electric coupling, the electric field (E_y) in the main guide extends into the secondary guide as shown. Bethe showed that the coupled waves could be described

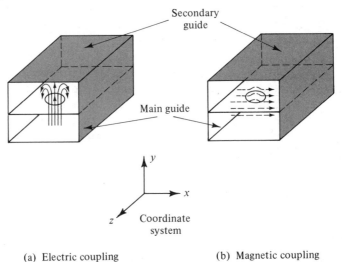

(a) Electric coupling   (b) Magnetic coupling

Figure 8–18 Examples of electric and magnetic coupling through a waveguide aperture.

in terms of an electric dipole perpendicular to the plane of the aperture. Magnetic coupling is shown in part b of Fig. 8–18, where in this case H_x couples into the secondary guide. The associated waves can be described in terms of a magnetic dipole parallel to the aperture plane. Thus, referring to Fig. 8–17, radiation through aperture 1 is due to both electric and magnetic coupling. On the other hand, coupling through the circular hole in the narrow wall (aperture 2) is magnetic since only H_z exists at the aperture. For the case of thin slots, the coupling is essentially magnetic. Only the magnetic field component parallel to the long dimension of the slot couples through the aperture.

An alternate way of viewing magnetic coupling is in terms of the interruption of conduction currents. Conduction currents for the TE_{10} mode were described in Sec. 5–4a (Fig. 5–19). For instance, the slot indicated by aperture 3 radiates energy since the longitudinal conduction current (J_z) is interrupted, thus creating a displacement current which produces a coupled magnetic field (H_x in this case). Apertures 4 and 5 interrupt transverse conduction currents (J_x and J_y, respectively), thereby producing magnetic coupling via the H_z component. The thin slots shown in Fig. 8–17b, on the other hand, are nonradiating since they do not interrupt any conduction currents. If the one on the broad wall is offset from the center line, it becomes a radiating slot, as indicated by aperture 4 in Fig. 8–17a. The apertures shown in part a of the figure provide the coupling mechanism for most discrete-type waveguide couplers.

Multihole waveguide couplers.   This type directional coupler is usually used in applications that require loose coupling (say, 15 dB or greater). Two- and three-hole versions are shown in Fig. 8–19. They consist of two identical rectangular waveguides having a common wall. As explained above, the apertures in the common wall produce wave coupling between the guides. The waveguides may be arranged so that they have either a common narrow wall or broad wall. In both cases, the distance l between apertures is $\lambda_g/4$ at the design frequency.

The operation of the two-hole version may be understood with the aid of Fig. 8–19a, where it is assumed that port 1 is the input. Both apertures are of identical shape and size. The wave incident on port 1 is designated by the phasor $\mathbf{a}_1 = a_1/\underline{0}$. As it propagates toward port 4, some of the wave is coupled through the first hole with portions propagating toward ports 3 and 2. These waves are denoted by $k_f a_1/\underline{-\beta l}$ and $k_r a_1/\underline{0}$, where k_f and k_r are the forward and reverse *aperture coupling coefficients*. Note that the wave at port 3 is phase delayed βl, where $\beta = 2\pi/\lambda_g$ is the phase constant for the waveguide. For loose coupling both k_f and k_r are less than 0.1. Therefore, the second aperture couples the same amount producing forward and reverse waves of value $k_f a_1/\underline{-\beta l}$ and $k_r a_1/\underline{-2\beta l}$, respectively. The coupled waves are shown in the figure. By phasor addition, the waves emerging from ports 3 and 2 are $\mathbf{b}_3 = 2k_f a_1/\underline{-\beta l}$ and $\mathbf{b}_2 = k_r a_1(1/\underline{0} + 1/\underline{-2\beta l})$. The power associated with these waves is given by b_3^2 and b_2^2. Thus, $P_3 = 4k_f^2 a_1^2$ and $P_2 = k_r^2 a_1^2 |(1 + \cos 2\beta l) - j \sin 2\beta l|^2 = 4k_r^2 a_1^2 \cos^2 \beta l$. Since a_1^2 is the incident power at the input port, the coupling and directivity, defined by Eqs. (8–14) and (8–15), for a two-hole coupler are given by

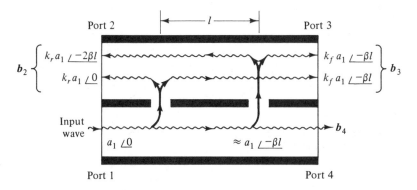

(a) Two-hole coupler

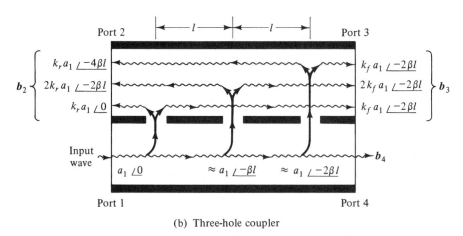

(b) Three-hole coupler

Figure 8–19 Two- and three-hole directional couplers. (Note: In part b, the center hole is larger than the end ones. The waves shown are for a 1-2-1 binomial coupler.)

$$\text{Coupling} = 10 \log\left(\frac{1}{4k_f^2}\right) \quad \text{and} \quad \text{Directivity} = 10 \log\frac{(k_f/k_r)^2}{\cos^2 \beta l}$$

Coupling is a function of k_f only, which depends upon aperture size and the operating frequency. For a small circular hole of diameter d, k_f is proportional to d^3 (Ref. 8–4). When the hole is located in the narrow guide wall, k_f increases as the operating frequency decreases. This is because the interrupted transverse currents, which produce the coupling, become larger when the operating frequency approaches cutoff. The frequency response for a hole located in the broad wall is more complex and is discussed in Ref. 8–4. For either location, the coupling for the two-hole coupler is independent of βl because the path lengths traversed by the two waves arriving at port 3 are identical.

Directivity, on the other hand, is a function of βl since the phase difference for the two waves arriving at port 2 is $2\beta l$. The above expression for directivity of a two-hole coupler may be rewritten as

$$\text{Directivity} = 10 \log \left(\frac{k_f}{k_r}\right)^2 + 10 \log \left(\frac{1}{\cos \beta l}\right)^2 \qquad (8\text{-}17)$$

The first term represents the inherent directivity of the apertures, while the second is associated with the wave cancellation produced by the $2\beta l$ phase difference. Except for aperture 1, the ones shown in Fig. 8-17a are nondirective (that is, $k_f = k_r$). At the design frequency, $l = \lambda_g/4$ and therefore the coupler has infinite directivity. This is because the two waves arriving at port 2 are 180° out-of-phase. For a fixed length l, directivity is a function of β and therefore frequency. Figure 8-20 shows directivity as a function of βl for the two-hole coupler. Curves are also given for three- and four-hole binomial couplers. These will be discussed shortly. In all cases it is assumed that $k_f = k_r$.

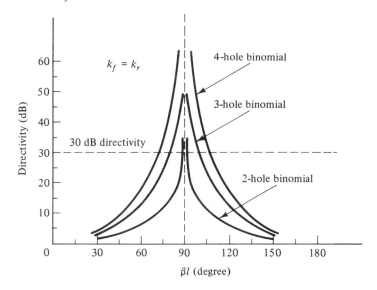

Figure 8-20 Directivity of multihole binomial couplers as a function of βl. (Note: $k_f = k_r$ is assumed.)

Directivity is a measure of the coupler's ability to separately sample forward and reverse traveling waves. Referring to Fig. 8-16, this means that the forward wave (port 1 to 4) and the reverse wave (port 4 to 1) are sampled at ports 3 and 2, respectively. If, however, the coupler has finite directivity some of the reverse wave appears at port 3 and some of the forward wave appears at port 2. These represent unwanted signals. Many applications require greater than 30 dB directivity which means that the ratio of wanted-to-unwanted signal at the coupled port must exceed 1000. With this requirement, the useful frequency band of the two-hole coupler is only a few percent (Prob. 8-19).

It was shown in Sec 4-2b that the useful bandwidth of quarter-wave transformers could be substantially increased by using multisection techniques, the most common being the Butterworth and Tchebyscheff designs. These methods may also be used to broadband the directivity of a directional coupler. In the Butterworth case, the distribution of aperture coupling coefficients is in accordance with the binomial coefficients. To illustrate, consider the three-hole (two-section) coupler shown in Fig. 8–19b. The binomial coefficients are 1, 2, and 1, which means that the coupling for the center aperture is twice that of the end apertures. For an incident wave at port 1, the phasor sum of the three waves at port 3 is given by $b_3 = 4k_f a_1 \underline{/-2\beta l}$ and thus $P_3 = b_3^2 = 16k_f^2 a_1^2$. The phasor sum of the waves at port 2 is

$$b_2 = k_r a_1 \underline{/0} + 2k_r a_1 \underline{/-2\beta l} + k_r a_1 \underline{/-4\beta l}$$

$$= 2k_r a_1 (1 + \cos 2\beta l) \underline{/-2\beta l} = 4k_r a_1 \cos^2 \beta l \underline{/-2\beta l}$$

Therefore $P_2 = b_2^2 = 16\, k_r^2 a_1^2 \cos^4 \beta l$, and the directivity of a three-hole binomial coupler becomes

$$\text{Directivity} = 10 \log \left(\frac{k_f}{k_r}\right)^2 + 10 \log \left(\frac{1}{\cos \beta l}\right)^4 \qquad (8\text{–}18)$$

This is plotted in Fig. 8–20 for $k_f = k_r$. The directivity of a four-hole (3-section) binomial version is also shown. Note the improvement in bandwidth as the number of sections is increased. The above analysis can be extended to an n-section $(n + 1$ holes) binomial coupler resulting in a directivity given by

$$\text{Directivity} = 10 \log \left(\frac{k_f}{k_r}\right)^2 + 10 \log \left(\frac{1}{\cos \beta l}\right)^{2n} \qquad (8\text{–}19)$$

Precise test equipment applications usually require couplers with greater than 50 dB directivity over the useful waveguide band. This means that for a binomial design at least ten sections are needed (Prob. 8–21).

Another approach to broadbanding the directivity of a multihole directional coupler is to select the aperture coupling values in accordance with the Tchebyscheff polynomials. This leads to an equal-ripple directivity characteristic. The response for a three-section design is shown in Fig. 8–21. For $k_f = k_r$, the minimum directivity over the useful bandwidth for an n-section directional coupler is given by

$$\text{Min. Directivity} = 10 \log \left\{ T_n \left(\frac{1}{\cos \phi_0}\right) \right\}^2 \qquad (8\text{–}20)$$

where the useful bandwidth is defined by $\phi_0 \le \beta l \le \pi - \phi_0$ and T_n denotes the Tchebyscheff polynomial of degree n. Closed expressions for $T_n(x)$ are given in Eq. (4–25). For comparison purposes, the response of a three-section binomial design (dashed curve) is also shown in Fig. 8–21. For a given minimum directivity, the bandwidth of a Tchebyscheff design is greater than that of a Butterworth design having the same number of sections (Prob. 8–22). Tchebyscheff element values for up to seven sections are given by Levy (Ref. 8–4). Additional design data may be found in Refs. 8–5, 8–6, and 8–22. Most of this data is based on a loose-coupling approximation. Levy (Ref. 8–24) uses even-odd mode theory to overcome this restriction. A synthesis procedure based upon a distributed lowpass prototype filter is also presented. A further refinement of the technique is described in a later paper (Ref. 8–25).

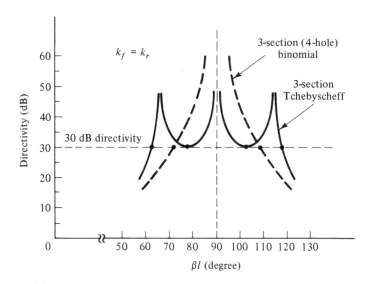

Figure 8-21 Directivity characteristics of a three-section Tchebyscheff directional coupler with 30 dB minimum directivity. From Eq. (8-20), $\phi_0 = 61.8°$ for 30 dB minimum directivity.

There are many other types of directional couplers that utilize discrete coupling. Four waveguide versions are shown in Fig. 8-22. All of them use aperture coupling through a common waveguide wall.

The Riblet-Saad coupler.   The coupler shown in part a of Fig. 8-22 was developed by Riblet and Saad and is described in Ref. 8-26. It achieves high directivity by combining the coupling characteristics of two thin slots located at the same position along the propagation axis. The longitudinal slot couples H_z producing in-phase waves in the secondary guide indicated by $k_z a_1$. The transverse slot couples H_x producing out-of-phase waves indicated by $+k_x a_1$ and $-k_x a_1$. If the slot dimensions are the same, the ratio k_x/k_z is proportional to the ratio H_x/H_z in the main guide. For $\lambda_g = 2a$, which occurs at about the middle of the waveguide's useful frequency band, $k_x/k_z = 1$ and hence perfect wave cancellation ($b_2 = 0$) occurs at port 2. The signal coupled to port 3 is $b_3 = 2k_x a_1$. Levy (Ref. 8-4) explains that the change of coupling with frequency is minimum when $\lambda_g = 2a$. Thus maximum directivity and flattest coupling occur at the same frequency. The directivity is frequency sensitive since the ratio k_x/k_z is proportional to frequency. For a single pair of slots, greater than 20 dB directivity can be achieved over a 25 percent frequency band. For greater bandwidth and directivity, multisection designs are necessary.

The Schwinger reversed-phase coupler.   A sketch of this coupler is shown in Fig. 8-22b. It consists of two thin slots spaced $\lambda_g/4$ apart at the design frequency. The propagation direction of the coupled wave is opposite to that of the wave in the main guide. Note that the port numbering is consistent with the schematic diagram in Fig. 8-16. With the offset slots located on opposite sides of the

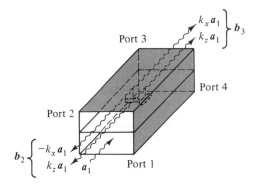

(a) Riblet-Saad coupler

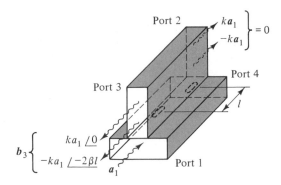

(b) Schwinger reverse-phase coupler

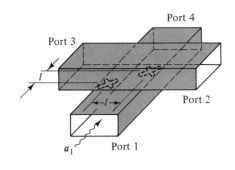

(c) Moreno cross-guide coupler

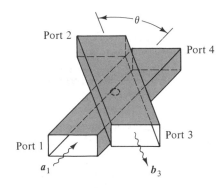

(d) Bethe-hole coupler

Figure 8–22   Four waveguide directional couplers that utilize aperture coupling.

broad wall, out-of-phase H_z components are coupled into the secondary guide. The portions of the waves that propagate toward port 2 travel the same distance and therefore cancel (that is, $b_2 = 0$). On the other hand, the waves traveling to port 3 add since $2\beta l = 180°$ at the design frequency. Note that for this type coupler, coupling is a function of βl but directivity is not. Thus its characteristics are just the opposite of the two-hole coupler described in Fig. 8–19a.

The Moreno cross-guide coupler.   The operation of this coupler (Fig. 8–22c) is similar to the Schwinger coupler in that it utilizes the inherent 180° phase difference of H_z on opposite sides of the waveguide's broad wall. The wave in the main guide is magnetically coupled through the two cross slots into the secondary guide. Since the path lengths for the components traveling toward port 2 are the same, the waves are 180° out-of-phase and therefore cancel. Thus the directivity of a Moreno coupler is independent of βl. At the design frequency, $l = \lambda_g/4$ and therefore the waves coupled to port 3 add. These waves are in phase because the 180° phase delay due to $2\beta l$ offsets the inherent 180° phase reversal associated with the magnetic coupling of H_z. Referring to Fig. 8–22c, this means that a signal incident on port 1 couples to port 3 but not to port 2. As is true for the Schwinger coupler, the coupling of a Moreno cross-guide coupler is a function of βl and therefore is frequency sensitive.

Design data for the Moreno, Schwinger, and Riblet-Saad directional couplers may be found in Refs. 8-5 and 8-27.

The Bethe-hole coupler.

One version of this coupler is shown in Fig. 8-22d. The coupling aperture is a circular hole centrally located in the broad wall of the waveguide. By proper adjustment of θ, the angle between the axes of the guides, a high degree of directivity can be achieved. Unlike the couplers discussed thus far, the Bethe-hole coupler requires both electric and magnetic coupling between the guides. These coupling mechanisms were described earlier (see Fig. 8-18). For an input signal at port 1, electric coupling produces in-phase waves of equal amplitude propagating toward ports 2 and 3. Magnetic coupling (H_x) produces out-of-phase waves, also of equal amplitude, in the secondary guide. By proper adjustment of the angle θ, the electric and magnetic coupling coefficients can be made equal resulting in complete wave cancellation at port 2. Thus, as indicated in the figure, power into port 1 couples to port 3 but not to port 2. The angle that yields maximum directivity is obtained from the following equation.

$$\cos \theta = \frac{1}{2}\left(\frac{\lambda_g}{\lambda_0}\right)^2 \tag{8-21}$$

Note that for $\lambda_0 = \sqrt{2}\,a$, $\theta = 0$ which means that it behaves like a reverse coupler. The reason for this is that the electric field vector is inverted as it couples into the secondary guide but the magnetic vector is not (see Fig. 8-18).

Expressions for coupling and directivity are given in Sec. 6.4 of Ref. 8-7. Another version of the Bethe-hole coupler that utilizes an offset circular hole is also described.

The branch-line coupler.

The coupling mechanism used in this directional coupler is two or more quarter-wave line sections (or branches) connected between a pair of transmission lines. Any form of transmission line may be used. Figure 8-23 shows a two-branch coaxial version. The characteristic impedance and length of the coupling lines are denoted by Z_{02} and l. The spacing between the branches is also equal to l as shown. At the design frequency, $l = \lambda/4$. The characteristic impedance of the lines between the branches is denoted by Z_{01} while that of the input and output lines is denoted by Z_0.

The operation of the branch-line coupler and many other couplers may be analyzed using even-odd mode theory. The method of analysis is described in Refs. 8-4 and 8-13 and summarized here. Referring to Fig. 8-23, an incident wave a_1 at port 1 may be expressed in terms of even and odd pairs of incident waves at ports 1 and 2. The four waves, each of amplitude $a_1/2$, are shown at the left of the figure. For the even-mode pair, a voltage maximum and current zero exists at all points along the symmetry line (indicated by the dashed line). The odd-mode pair creates a voltage zero and current maximum everywhere along the symmetry line. For a given mode pair, the transmission and reflection characteristics of the top and

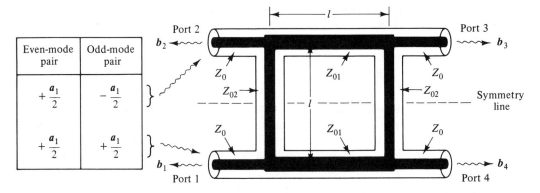

Figure 8–23 A coaxial branch-line directional coupler. (Note: $l = \lambda/4$ at the design frequency.)

bottom halves of the symmetrical four-port are the same. Therefore by superposition, the resultant signals out of each port can be expressed in terms of transmission and reflection coefficients. Namely,

$$b_1 = \frac{a_1}{2}(\Gamma_e + \Gamma_o) \quad , \quad b_2 = \frac{a_1}{2}(\Gamma_e - \Gamma_o)$$

$$b_3 = \frac{a_1}{2}(T_e - T_o) \quad , \quad b_4 = \frac{a_1}{2}(T_e + T_o)$$

(8–22)

where Γ_e and Γ_o are respectively the even- and odd-mode reflection coefficients and T_e and T_o are respectively the even- and odd-mode transmission coefficients. These coefficients may be expressed in terms of the normalized ABCD matrix elements. From Eqs. (C–10) and (C–15),

$$\Gamma = \frac{\bar{A} + \bar{B} - \bar{C} - \bar{D}}{\bar{A} + \bar{B} + \bar{C} + \bar{D}} \quad \text{and} \quad T = \frac{2}{\bar{A} + \bar{B} + \bar{C} + \bar{D}} \quad (8–23)$$

where it is assumed that the output ports are match terminated.

The branch-line coupler shown in Fig. 8–23 may be analyzed using the even-odd mode theory described here. Equivalent two-port networks for even and odd-mode excitation are given in Fig. 8–24, where the open and short circuits represent the impedance of the symmetry plane for the two modes. At the design frequency $l = \lambda/4$ ($\beta l = 90°$) and therefore the network reduces to a pair of $\lambda/8$ stubs separated by a quarter-wave line. The normalized ABCD matrix for even- and odd-mode excitation is given by

$$\begin{bmatrix} \bar{A} & \bar{B} \\ \bar{C} & \bar{D} \end{bmatrix} = \begin{bmatrix} 1 & 0 \\ \pm j/\bar{Z}_{02} & 1 \end{bmatrix} \begin{bmatrix} 0 & j\bar{Z}_{01} \\ j/\bar{Z}_{01} & 0 \end{bmatrix} \begin{bmatrix} 1 & 0 \\ \pm j/\bar{Z}_{02} & 1 \end{bmatrix}$$

$$= \begin{bmatrix} \mp \bar{Z}_{01}/\bar{Z}_{02} & j\bar{Z}_{01} \\ j\bar{Z}_{01}\left(\dfrac{1}{\bar{Z}_{01}^2} - \dfrac{1}{\bar{Z}_{02}^2}\right) & \mp \bar{Z}_{01}/\bar{Z}_{02} \end{bmatrix}$$

(8–24)

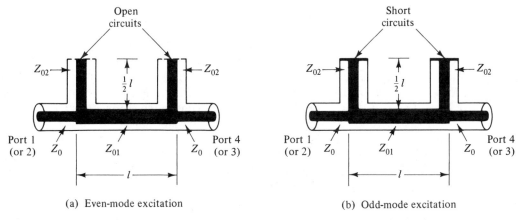

(a) Even-mode excitation   (b) Odd-mode excitation

Figure 8–24 Equivalent circuits for even- and odd-mode excitation of the branch-line coupler in Fig. 8–23.

where $\bar{Z}_{01} \equiv Z_{01}/Z_0$ and $\bar{Z}_{02} \equiv Z_{02}/Z_0$. Some of the matrix elements have a double sign. The upper sign is associated with the even-mode solution and the lower sign with the odd-mode.

By setting both Γ_e and Γ_o equal to zero, port 1 is matched ($b_1 = 0$) and port 2 is decoupled ($b_2 = 0$). Due to the coupler's symmetry, this means that all ports are matched and ports 3 and 4 are also decoupled. Thus, the branch-line coupler is an ideal directional coupler when $\Gamma_e = \Gamma_o = 0$. From Eq. (8–23), this requires that $\bar{A} + \bar{B} = \bar{C} + \bar{D}$. Since $\bar{A} = \bar{D}$ for both even and odd modes, the condition reduces to $\bar{B} = \bar{C}$ or

$$\bar{Z}_{02} = \frac{\bar{Z}_{01}}{\sqrt{1 - \bar{Z}_{01}^2}} \tag{8–25}$$

For this condition, the transmission coefficient becomes $T = (\bar{A} + \bar{B})^{-1}$. By using Eq. (8–24) and the above condition, the even- and odd-mode values become

$$T_e = -\sqrt{1 - \bar{Z}_{01}^2} - j\bar{Z}_{01} \quad \text{and} \quad T_o = \sqrt{1 - \bar{Z}_{01}^2} - j\bar{Z}_{01} \tag{8–26}$$

Thus from Eq. (8–22), the output waves at ports 4 and 3 are

$$b_4 = -j\bar{Z}_{01}a_1 \quad \text{and} \quad b_3 = -\sqrt{1 - \bar{Z}_{01}^2}\, a_1 \tag{8–27}$$

where a_1 is the incident wave at port 1. Note that b_3 is phase delayed 90° relative to b_4 and $b_3^2 + b_4^2 = a_1^2$, the incident power at port 1. From Eq. (8–14), the coupling at the design frequency for the two-branch coupler in Fig. 8–23 becomes

$$\text{Coupling} = 10 \log \left(\frac{1}{1 - \bar{Z}_{01}^2} \right) \tag{8–28}$$

since $P_3 = b_3^2$ and $P_1 = a_1^2$.

With Z_0 given and $l = \lambda/4$ at the design frequency, the branch-line coupler design reduces to solving Eqs. (8–28) and (8–25) for Z_{01} and Z_{02}. Note that for any finite value of coupling, Z_{01} is less than Z_0. For example, a 3 dB coupler (that is, equal power division) requires $Z_{01} = Z_0/\sqrt{2}$ and $Z_{02} = Z_0$. Since the outputs are 90° out-of-phase, this represents a quadrature hybrid.

Branch-line couplers are easily constructed in strip transmission lines because the impedance of the various lines can be adjusted by controlling the strip width. Waveguide versions are usually constructed with E-plane (series) junctions and as such are the dual of the coaxial version described here. In the waveguide case, the required line impedances are easily realized by controlling the guide height.

It is sometimes convenient to maintain a constant impedance in the main and secondary lines (that is, $Z_{01} = Z_0$). A two-branch coupler cannot be realized with this constraint, three or more coupling branches are required. The three-branch case has been analyzed using the even-odd mode procedure (Ref. 8–13). Waveguide versions are described in Ref. 8–28.

Branch-line couplers are particularly useful in applications that require strong coupling (< 10 dB). Since the connecting lines can only be a quarter wavelength long at the design frequency, both coupling and directivity are frequency sensitive. Some indication of the useful bandwidth for two- and three-branch couplers is found in Ref. 8–13 and Prob. 8–27. To obtain high directivity over a wide frequency band (say, 30 dB over a 30 percent band), more than three branches are required. One design approach is to keep the impedance of the main and secondary lines constant ($Z_{01} = Z_0$) and choose the branch-line impedances to produce the desired directivity response. Designs based on Butterworth and Tchebyscheff responses are given in Ref. 8–29 and 8–30. Patterson (Ref. 8–32) used lowpass-highpass filter theory to obtain similar results.

Further improvements in bandwidth can be achieved by varying the main and secondary line impedances. This case, with useful design data for up to nine branches, has been presented by Young (Ref. 8–31). A synthesis procedure for symmetrical branch-line couplers has been developed by Levy and Lind (Ref. 8–33). Values for Butterworth and Tchebyscheff designs up to nine branches are given in tabular form. A later paper by Levy (Ref. 8–34) presents a design theory based upon the use of Zolotarev functions. This technique is particularly useful for couplers requiring tight coupling values.

8-3b Distributed Coupling

The term *distributed* or *continuous* coupling refers to the situation wherein the coupling between a pair of transmission lines occurs over an extended length in the propagation direction. The coupling mechanism can be either uniform or nonuniform over the length of the interaction region. Examples of TEM and waveguide directional couplers that utilize uniform coupling are presented here. The symmetrical configurations described are analyzed using even-odd mode theory. An alternative analytical approach involves the solution of a pair of coupled linear differential equations. This method has been described in detail by Miller (Ref. 8–35).

The coplanar stripline coupler. In 1954, Oliver (Ref. 8–36) and Firestone (Ref. 8–37) showed that the natural electric and magnetic coupling between a pair of TEM transmission lines produces directional coupling. Specifically, a wave propagating along one line induces a wave on the other line that propagates in the opposite direction. This effect is the basis for many wideband TEM directional couplers (for example, Refs. 8–38 and 8–39).

Cross-sectional views of various coupled-line pairs are shown in Fig. 8–25. The ones in parts b and c are used when tight coupling (10 dB or less) is required. The configuration shown in part b is often used in the design of 3 dB couplers. For loose coupling, the coplanar stripline arrangement shown in part a is the most common. The center conductor patterns for single- and three-section directional couplers are shown in Fig. 8–26. Note that the ports are numbered to reflect the reverse nature of the coupling mechanism. Thus, ports 1 and 3 are coupled as are ports 2 and 4.

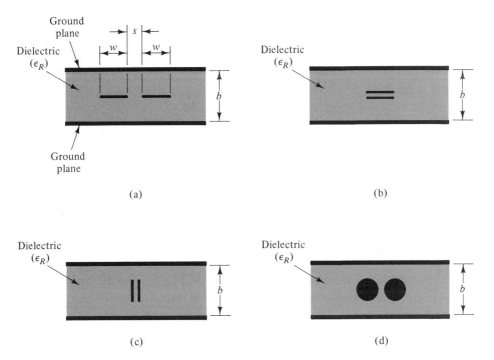

Figure 8–25   Cross-sectional views of four widely used TEM coupled-line pairs.

The single-section, coplanar stripline coupler (Fig. 8–26a) will now be analyzed using even-odd mode theory. The results obtained may be applied to any of the coupled-line configurations shown in Fig. 8–25. Consider a signal $\mathbf{a}_1 = a_1/\underline{0}$ incident on port 1. It can be viewed as the superposition of even and odd signal pairs at ports 1 and 3 as indicated in parts a and b of Fig. 8–27. Their transverse electromagnetic field patterns are shown in parts c and d, respectively. Since both are TEM modes in a uniform dielectric, their velocities are equal. For even-mode excitation, the two center strips are at the same potential and their currents are equal and in the same direction. This produces the electromagnetic field pattern shown in part c of the figure. Thus the symmetry plane indicated by the dashed vertical line represents

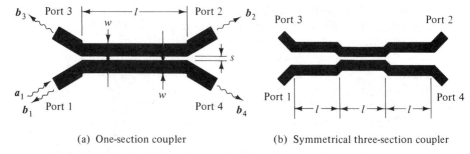

(a) One-section coupler   (b) Symmetrical three-section coupler

Figure 8–26   Conductor patterns for one- and three-section coplanar, stripline couplers.

a magnetic wall, which is defined as a boundary having zero tangential magnetic field. From an impedance point of view, it represents an open circuit. For odd-mode excitation, the center strips are of opposite polarity and have oppositely directed currents, which results in the field pattern shown in part d of Fig. 8–27. In this case, the symmetry plane represents an electric wall or a short circuit (that is, tangential E is zero).

Because of the dissimilar field patterns, the characteristic impedance of the even and odd modes (denoted by Z_{0e} and Z_{0o}) are unequal. Z_{0e} and Z_{0o} represent, respectively, the characteristic impedance of each half of the coupled-line pair for even- and odd-mode excitation. For the coplanar stripline structure, Z_{0o} is less than

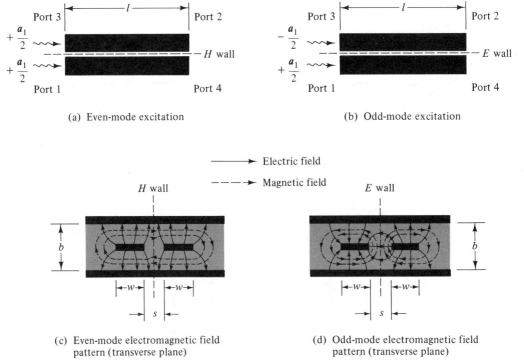

(a) Even-mode excitation   (b) Odd-mode excitation

(c) Even-mode electromagnetic field pattern (transverse plane)   (d) Odd-mode electromagnetic field pattern (transverse plane)

Figure 8–27   Even- and odd-mode excitation for the coplanar coupler shown in Fig. 8–26a.

Z_{0e} since the capacitance per length associated with the odd-mode field pattern is greater than that of the even mode. Static field methods have been used to calculate Z_{0e} and Z_{0o} for various coupled-line pairs. The ones in parts a, b, and c of Fig. 8–25 have been computed by Cohn (Refs. 8–40 and 8–41). His nomogram for the coplanar case is reproduced in Fig. 8–28. Approximate impedance formulas for the coupled-line pair in Fig. 8–25d is given in Ref. 8–4. Additional data for coupled striplines may be found in Ref. 8–3.

The even-odd mode analysis for a coupled-line directional coupler reduces to finding the transmission and reflection coefficients of two equal-length transmission lines having impedances Z_{0e} and Z_{0o}. The normalized ABCD matrices for the two cases are

$$
\begin{bmatrix} \cos \beta l & j\bar{Z}_{0e} \sin \beta l \\ j\dfrac{\sin \beta l}{\bar{Z}_{0e}} & \cos \beta l \end{bmatrix}_{\substack{\text{even} \\ \text{mode}}} \quad \text{and} \quad \begin{bmatrix} \cos \beta l & j\bar{Z}_{0o} \sin \beta l \\ j\dfrac{\sin \beta l}{\bar{Z}_{0o}} & \cos \beta l \end{bmatrix}_{\substack{\text{odd} \\ \text{mode}}} \tag{8–29}
$$

where $\bar{Z}_{0e} \equiv Z_{0e}/Z_0$, $\bar{Z}_{0o} \equiv Z_{0o}/Z_0$ and Z_0 is the characteristic impedance of the connecting lines. Since the velocities are equal, $\beta_e = \beta_o = \beta = 2\pi/\lambda$. From Eq. (8–23), the even-mode coefficients are

$$
\Gamma_e = \frac{j(\bar{Z}_{0e} - \bar{Z}_{0e}^{-1}) \sin \beta l}{2 \cos \beta l + j(\bar{Z}_{0e} + \bar{Z}_{0e}^{-1}) \sin \beta l} \tag{8–30}
$$

and

$$
T_e = \frac{2}{2 \cos \beta l + j(\bar{Z}_{0e} + \bar{Z}_{0e}^{-1}) \sin \beta l} \tag{8–31}
$$

The expressions for Γ_o and T_o are the same as above except that $\bar{Z}_{0e}$ is replaced by $\bar{Z}_{0o}$. For a perfect input match, $b_1 = 0$, which from Eq. (8–22) requires that $\Gamma_e + \Gamma_o = 0$. This condition is satisfied when

$$
\bar{Z}_{0e} \bar{Z}_{0o} = 1 \quad \text{or} \quad Z_{0e} Z_{0o} = Z_0^2 \tag{8–32}
$$

Because of symmetry, all four ports of the coupled-line pair are matched when this condition is satisfied. Thus the resultant four-port network is an ideal directional coupler. This may be verified by showing that the impedance condition of Eq. (8–32) results in infinite directivity ($b_2 = 0$). In this case $b_2 = (T_e - T_o)a_1/2$.[7] Note that match and directivity are independent of βl (assuming $\beta_e = \beta_o$). This is an important conclusion since with Z_{0e} and Z_{0o} being frequency insensitive, the coupled-line pair is, in principle, an ideal directional coupler at all frequencies.

The output waves from ports 3 and 4 are given by the following expressions.

$$
b_3 = (\Gamma_e - \Gamma_o)\frac{a_1}{2} = \frac{ja_1(\bar{Z}_{0e} - \bar{Z}_{0o}) \sin \beta l}{2 \cos \beta l + j(\bar{Z}_{0e} + \bar{Z}_{0o}) \sin \beta l} \tag{8–33}
$$

---

[7] Since port numbers 2 and 3 have been interchanged relative to the branch-line coupler, the expressions for b_2 and b_3 in Eq. (8–22) must be interchanged.

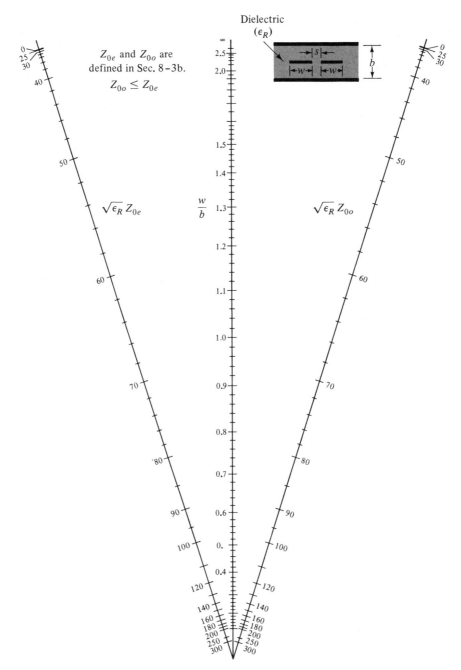

(a) Nomogram giving w/b as a function of Z_{0e} and Z_{0o}

Figure 8–28 Nomograms for obtaining w/b and s/b for the coplanar stripline pair shown in Fig. 8–25a. (From S. Cohn, Ref. 8–40; ©1955 IRE now IEEE, with permission.)

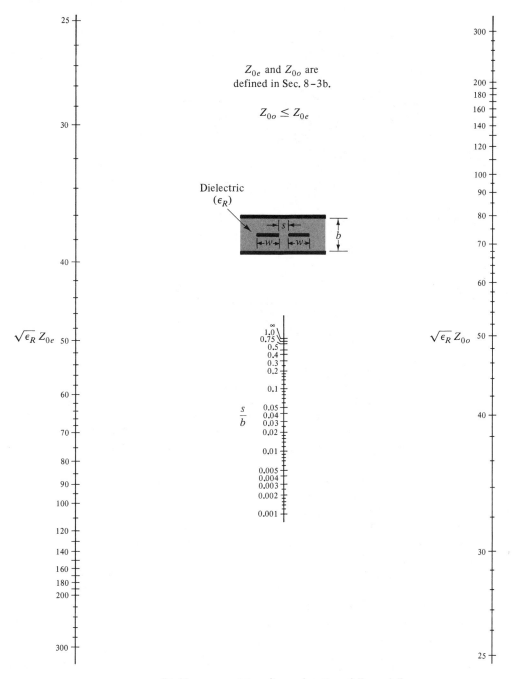

Z_{0e} and Z_{0o} are
defined in Sec. 8–3b.

$$Z_{0o} \leq Z_{0e}$$

(b) Nomogram giving s/b as a function of Z_{0e} and Z_{0o}

Figure 8–28 (*cont.*)

and

$$b_4 = (T_e + T_o)\frac{a_1}{2} = \frac{2a_1}{2\cos\beta l + j(\bar{Z}_{0e} + \bar{Z}_{0o})\sin\beta l} \tag{8-34}$$

where use has been made of Eq. (8–32). Note that the outputs are 90° out-of-phase for all values of βl. Also, the coupled wave (b_3) propagates opposite to the incident wave (a_1). Thus, the coupled-line pair is a *reverse* directional coupler.

Coupling is frequency sensitive since it is a function of βl. The coupling coefficient $k_c = b_3/a_1$ is maximum when $l = \lambda/4$. At this frequency it is given by

$$k_c = \frac{\bar{Z}_{0e} - \bar{Z}_{0o}}{\bar{Z}_{0e} + \bar{Z}_{0o}} = \frac{Z_{0e} - Z_{0o}}{Z_{0e} + Z_{0o}} \tag{8-35}$$

From Eq. (8–14), coupling is $10 \log K_c$, where $K_c = 1/k_c^2$ and therefore

$$\text{Maximum Coupling} = 10 \log \left(\frac{Z_{0e} + Z_{0o}}{Z_{0e} - Z_{0o}}\right)^2 \tag{8-36}$$

Equations (8–32) and (8–35) may be combined to provide a pair of design equations for a one-section, quarter-wave, coupled-line directional coupler. Namely,

$$Z_{0e} = Z_0\sqrt{\frac{1 + k_c}{1 - k_c}} \quad\text{and}\quad Z_{0o} = Z_0\sqrt{\frac{1 - k_c}{1 + k_c}} \tag{8-37}$$

Thus, given Z_0 and the desired maximum coupling, Z_{0e} and Z_{0o} values may be obtained, from which the transverse dimensions of the coupled-line pair are determined. For the coplanar stripline version (Fig. 8–26), Cohn's nomograms in Fig. 8–28 may be used to determine the strip width w and the spacing s. The ground-plane spacing b should be chosen to avoid higher-mode propagation (see Sec. 5–3). Impedance data for other types of coupled lines may be found in Refs. 8–3, 8–42, and 8–43.

Since the length l can only equal $\lambda/4$ at the design frequency, the coupling value is frequency sensitive. Figure 8–29 gives an indication of the coupling variation for a single-section, quarter-wave coupler. For most applications, its useful bandwidth is about an octave (≈ 66 percent). Larger bandwidths require the cascading of quarter-wave sections. The coupling variation for a symmetrical three-section unit based on an equal-ripple design is shown in Fig. 8–29. Note that the coupling is within ± 0.40 dB of the nominal value over a four-to-one frequency range. A coplanar stripline version of the coupler is shown in Fig. 8–26b. For even larger bandwidth and less coupling variation, additional sections are necessary. For example, Ref. 8–44 describes a nine-section 3 dB coupler with a coupling variation of only ± 0.10 dB over an eight-to-one frequency range.

A review of multisection directional coupler theory is given in Ref. 8–4. Design data for both symmetric and asymmetric cases have been tabulated in Refs. 8–44, 8–45, and 8–46.

Some design difficulties are encountered if the coupled-line pair has unequal phase velocities for the even and odd modes (for example, in microstrip). Levy (Ref. 8–4) shows that this results in a directivity that is frequency dependent. He also describes the effect of impedance tolerances on the coupler's directivity. A third factor

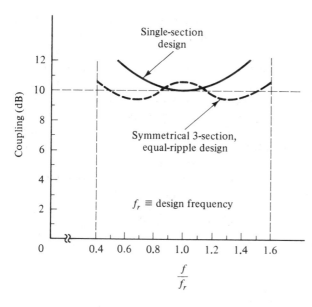

Figure 8–29 Coupling variation for a 10 dB single-section, quarter-wave coupled-line directional coupler. Also shown (dashed curve) is the variation of a symmetrical three-section coupler (Fig. 8–26b).

that affects directivity is reflections due to discontinuities between the coupled lines and the output lines. This limitation is common to all directional couplers. The following example illustrates the point.

Example 8–5:

The directional coupler in Fig. 8–26a has $Z_0 = 50$ ohms, $Z_{0e} = 62.5$ ohms, $Z_{0o} = 40$ ohms, and $l = \lambda/4$ at the design frequency.
(a) Calculate the coupling (in dB) at the design frequency.
(b) With port 1 as the input, calculate the directivity (in dB) if the discontinuity at port 3 has a reflection coefficient $\Gamma_3 = 0.05$. Assume that reflections due to the discontinuities at the other three ports have been eliminated.

Solution:
(a) With $l = \lambda/4$, the coupling is maximum and is given by Eq. (8–36). Thus,

$$\text{Coupling} = 10 \log (4.56)^2 = 13.2 \, \text{dB}.$$

(b) Let $\boldsymbol{a}_1 = a_1\underline{/0}$ represent the input wave at port 1. From Eqs. (8–33) and (8–34) for $\beta l = 90°$, $\boldsymbol{b}_3 = 0.22a_1\underline{/0}$, and $\boldsymbol{b}_4 = 0.98a_1\underline{/-90°}$.
The reflected wave at port 3 is $0.22\Gamma_3 a_1\underline{/0}$ and the portion that arrives at port 2 is $\boldsymbol{b}_2 = 0.98(0.22\Gamma_3 a_1\underline{/-90°})$. The power delivered to a matched load at port 2 is b_2^2, while the net power out of port 3 is $\{1 - |\Gamma_3|^2\}b_3^2 \approx b_3^2$. Thus the directivity reduces to

$$\text{Directivity} \approx 10 \log \left(\frac{b_3}{b_2}\right)^2 = 10 \log \left|\frac{1}{0.98\Gamma_3}\right|^2 \approx 10 \log \left|\frac{1}{\Gamma_3}\right|^2.$$

For $|\Gamma_3| = 0.05$, the directivity is 26 dB.

Note that the directivity is approximately equal to the return loss of the mismatch at the coupled port. Reflections at the output port (port 4) also affect the coupler's directivity (Prob. 8–32). In some cases, a reflection is purposely introduced at the coupled port to cancel the leakage signal due to finite coupler directivity.

The Riblet short-slot coupler.   There are many directional couplers that utilize continuous coupling between waveguides. Miller (Ref. 8–35) has developed a comprehensive theory based on an integral formulation. He also describes several coupler configurations. Other waveguide versions have been developed by Riblet (Ref. 8–47), Hadge (Ref. 8–48), Cook (Ref. 8–49), and Tomiyasu and Cohn (Ref. 8–50). A sketch of the widely used Riblet short-slot coupler is shown in Fig. 8–30. Although it can be designed for practically any value of coupling, its most common version is as a 3 dB coupler. In fact, when one refers to the Riblet short-slot coupler, a 3 dB coupling value is implied. The coupler consists of two adjacent rectangular waveguides in which a portion of the common narrow wall has been removed to provide wave coupling between the guides. The width and length of the coupling region are $2a$ and l, respectively. Its operation can be explained in terms of even and odd modes.

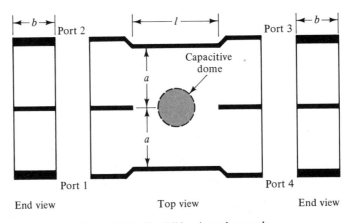

Figure 8–30   The Riblet short-slot coupler.

A TE_{10} wave incident on port 1 can be considered as the sum of even- and odd-mode signal pairs at ports 1 and 2. The even-mode pair consists of in-phase TE_{10} signals as indicated in Fig. 8–31a. By symmetry, this produces a TE_{10} wave in the coupling region which results in output waves at ports 3 and 4 that are delayed $\beta_e l$, where $\beta_e = 2\pi/\lambda_{g_e}$. Since the guide width is $2a$, $\lambda_c = 4a$ for the even mode and therefore

$$\lambda_{g_e} = \frac{\lambda}{\sqrt{1 - \left(\dfrac{\lambda}{4a}\right)^2}} \tag{8–38}$$

The odd-mode pair (Fig. 8–31b) creates a TE_{20} wave in the coupling region since the input signals are 180° out-of-phase. The associated output waves are delayed $\beta_o l$ as indicated, where $\beta_o = 2\pi/\lambda_{g_o}$. For the TE_{20} mode, λ_c equals the guide width in the coupling region and hence

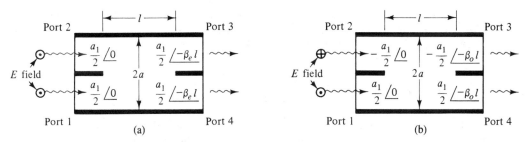

Figure 8-31 Even- and odd-mode analysis of the Riblet short-slot coupler shown in Fig. 8-30.

$$\lambda_{g_o} = \frac{\lambda}{\sqrt{1 - \left(\dfrac{\lambda}{2a}\right)^2}} \tag{8-39}$$

The phase difference between the even and odd modes is

$$\Delta\theta = \beta_e l - \beta_o l = 2\pi\left(\frac{l}{\lambda_{g_e}} - \frac{l}{\lambda_{g_o}}\right) \tag{8-40}$$

By superposition, an input wave a_1 at port 1 produces output waves

$$b_3 = \frac{a_1}{2}\underline{/-\beta_e l} - \frac{a_1}{2}\underline{/-\beta_o l} = -ja_1 \sin\frac{\Delta\theta}{2}\underline{/-\beta_o l - \frac{\Delta\theta}{2}} \tag{8-41}$$

$$b_4 = \frac{a_1}{2}\underline{/-\beta_e l} + \frac{a_1}{2}\underline{/-\beta_o l} = a_1 \cos\frac{\Delta\theta}{2}\underline{/-\beta_o l - \frac{\Delta\theta}{2}}$$

where use has been made of Eq. (8-40) and some trigonometric identities. Note that the output signals are 90° out-of-phase and their amplitudes are periodic functions of half the even-odd mode phase difference. Also note that power is conserved since the sum of the output powers $P_3 + P_4 = b_3^2 + b_4^2 = a_1^2$. For 3 dB coupling, $\Delta\theta$ must equal $\pi/2$. This results in $b_3 = b_4 = a_1/\sqrt{2}$ or $P_3 = P_4 = a_1^2/2$, where a_1^2 is the incident power at port 1. As opposed to the magic tee, the Riblet 3 dB coupler is a 90° or quadrature hybrid since the equal amplitude output signals are 90° out-of-phase. Several interesting applications utilize a combination of the two hybrid types.

The above analysis assumed that the even and odd modes are perfectly matched (that is, Γ_e and $\Gamma_o = 0$). Reflections due to the common wall discontinuities at the ends of the coupling region adversely affect the match and directivity of the coupler. The odd-mode reflection is negligible because its electric field (the TE_{20} mode) is zero along the center line. In practice, a capacitive dome (pictured in Fig. 8-30) is used to cancel the even-mode reflection and establish the $\pi/2$ phase difference required by Eq. (8-40).

Another version of the 3 dB waveguide coupler has been described by Hadge (Ref. 8-48), wherein coupling occurs through a common broad wall. In this case, the coupled wave b_3 leads b_4 by 90°.

Some of the more commonly used directional couplers have been discussed in this section. Many others have been developed over the years as evidenced by the bibliographies in Refs. 8–4, 8–20, and 8–52.

8–4 OTHER MULTIPORT JUNCTIONS

This section describes two other reciprocal multiports that are sometimes used in microwave systems. They are the dual-mode transducer and the turnstile junction.

Dual-mode transducers. Two examples of dual-mode transducers are shown in Fig. 8–32. Both configurations consist of two rectangular guides and a circular waveguide. Dominant-mode propagation is assumed in all guides. The circular guide is a dual-mode transmission system since it is capable of propagating two orthogonal TE_{11} modes. Associated with each of these modes is a port and hence the dual-mode transducer is a four-port junction. For example, ports 3 and 4 in Fig. 8–32 correspond to the linearly polarized TE_{11} waves E_h and E_v, respectively.

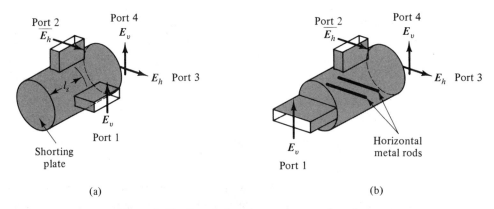

(a)   (b)

Figure 8–32   Two dual-mode transducer configurations.

Consider the configuration in part *a* of the figure. It is similar to the mode transducer described in Fig. 7–4b, except that it has two rectangular waveguides. The two guides are perpendicular to each other and therefore ports 1 and 2 are decoupled. The conversion mechanism between the TE_{10} in rectangular guide and the TE_{11} in circular guide was described in Sec. 7–1b. As explained, transmission between the rectangular and circular guides can be maximized by proper adjustment of the length l_s. Thus, incident TE_{10} waves at ports 1 and 2 are transmitted to the circular guide as linearly polarized TE_{11} waves (E_v and E_h, respectively). Since the dual-mode transducer is reciprocal, it can also function as a mode separator. That is, if a TE_{11} wave of arbitrary polarization is incident on the circular guide, its E_v component will be delivered to port 1 and its E_h component to port 2. It is this property of mode separation that makes the dual-mode transducer so useful in ferrite, duplexer, and antenna applications.

Another version of the dual-mode transducer is shown in part *b* of Fig. 8–32. Because of its symmetrical configuration, it has better isolation between the rectangular guide ports than its counterpart in part *a*. As in the previous case, a vertically polarized TE_{11} wave (E_v) at the circular guide is transmitted to port 1 while a horizontally polarized wave (E_h) is delivered to port 2. For the E_h wave, the horizontal metal rods serve the same function as the short circuit in part *a*. Thus, by proper positioning of these reflective elements, transmission between ports 2 and 3 is maximized. A vertically polarized wave, on the other hand, is unaffected by their presence since the electric field E_v is perpendicular to the rods. To maximize transmission between ports 1 and 4, a matching transformer (not shown) is usually inserted between port 1 and the circular guide.

Still another version is shown in Fig. 8–33. In this case, the dual-mode guide has a square cross section. Typical performance is 30 dB isolation between the rectangular guides and an SWR into all ports of less than 1.15 over a 15 percent frequency band.

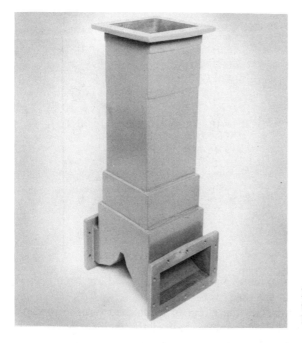

Figure 8–33   A broadband dual-mode transducer. (Courtesy of Atlantic Microwave Corp., Bolton, Mass.)

The turnstile junction.   The most common configuration of a turnstile junction is shown in Fig. 8–34. It is a reciprocal six-port that consists of four rectangular guides in the form of an *H*-plane cross and a circular guide whose axis is perpendicular to the plane of the cross. As indicated in the figure, ports 5 and 6 represent two orthogonal linearly polarized TE_{11} waves in the circular guide and thus are decoupled. Also, the orientation of the guides are such that port 5 is decoupled from ports 2 and 4, and port 6 is decoupled from ports 1 and 3. Due to its symmetrical form, the turnstile junction is best analyzed using symmetry operators (Refs. 8–2 and 8–8). The analysis shows that when all ports are matched, ports 1 and 3 are decoupled as are ports 2 and 4. By assuming a dissipationless structure, the scattering matrix for a matched turnstile junction reduces to

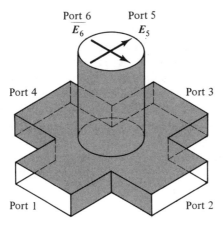

Port 6   Port 5
E_6 E_5

Port 4   Port 3

Port 1   Port 2   **Figure 8–34**   A turnstile junction.

$$[S] = \begin{bmatrix} 0 & \dfrac{1}{2} & 0 & \dfrac{1}{2} & \dfrac{1}{\sqrt{2}} & 0 \\[2mm] \dfrac{1}{2} & 0 & \dfrac{1}{2} & 0 & 0 & \dfrac{1}{\sqrt{2}} \\[2mm] 0 & \dfrac{1}{2} & 0 & \dfrac{1}{2} & -\dfrac{1}{\sqrt{2}} & 0 \\[2mm] \dfrac{1}{2} & 0 & \dfrac{1}{2} & 0 & 0 & -\dfrac{1}{\sqrt{2}} \\[2mm] \dfrac{1}{\sqrt{2}} & 0 & -\dfrac{1}{\sqrt{2}} & 0 & 0 & 0 \\[2mm] 0 & \dfrac{1}{\sqrt{2}} & 0 & -\dfrac{1}{\sqrt{2}} & 0 & 0 \end{bmatrix} \qquad (8\text{–}42)$$

It has been assumed that the reference planes of the ports have been chosen so that all the scattering coefficients are real.

Equation (8–42) summarizes the effect of a matched turnstile on an incident wave at any of the six ports. For example, half of the incident power at port 1 is transmitted to port 5 ($S_{51} = 1/\sqrt{2}$) while the remaining half splits equally between ports 2 and 4 ($S_{21} = S_{41} = 1/2$). Note that no power appears at ports 3 and 6. Due to the turnstile's symmetry, a similar power distribution occurs for an incident signal at any other rectangular port. The power distribution for an incident TE_{11} wave at the circular guide depends upon its polarization. For any polarization state, the wave can be decomposed into two orthogonal linearly polarized waves. These are represented in the figure by the vector-phasors E_5 and E_6. The E_5 component splits equally between ports 1 and 3 with the output signals being 180° out-of-phase ($S_{15} = 1/\sqrt{2}$ and $S_{35} = -1/\sqrt{2}$). In a similar manner, the E_6 component divides equally between ports 2 and 4. In both cases, the electric field behavior is similar to an E-plane tee in that the output waves are 180° out-of-phase at points equidistant from the junction.

The matched turnstile junction has several useful applications. Some of these are described and analyzed by Altman (Sec. 8.4 of Ref. 8–8). For example, by placing movable shorts in a decoupled pair of rectangular guides (for example, ports 2 and 4), the turnstile can function as a polarization analyzer and synthesizer. Altman shows that if the position of the two shorts from the junction differ by $\lambda_g/4$, an incident signal at a rectangular guide port (for example, port 1) is completely transmitted to the circular guide. Furthermore, depending upon the position of the shorts, linear or circular polarization can be produced in the circular guide.

8–5 SOME MULTIPORT APPLICATIONS

The uses of multiport junctions are many and varied. A few applications that show their versatility are given in this section. The examples also indicate how microwave circuits of this type can be analyzed.

8–5a Directional Coupler Applications

A directional coupler is capable of separately sampling the forward and reverse waves on a transmission line. Two applications that make use of this characteristic are the power monitor and the reflectometer. The effect of coupler directivity on the accuracy of both systems is analyzed.

The power monitor.   Part *a* of Fig. 8–35 shows a typical power-monitoring application, wherein the meter at port 3 samples a portion of the transmitter power. The transmitter and the main load, in this case a radiating antenna, are connected to ports 1 and 4, respectively. The coupler value is chosen so that practically all the transmitter power is delivered to the antenna port. Typical values are in the 30 to 60 dB range. For example, a 30 dB coupler samples only 0.10 percent of the forward power ($P_1^+ = a_1^2$) and therefore 99.9 percent appears at the antenna port. With a perfectly matched load at port 2 and infinite coupler directivity, the meter reading is *independent* of antenna reflections when $Z_G = Z_0$.[8] Even shorting the antenna port ($|\Gamma_4| = 1$) has no effect on the reading since any reflection at port 4 is absorbed by the internal impedance of the transmitter (Z_G) and the matched load at port 2.

Part *b* of Fig. 8–35 defines the incident and scattered waves at all four ports. Note that $a_2 = \Gamma_2 b_2$ and $a_4 = \Gamma_4 b_4$, where Γ_2 and Γ_4 are the reflection coefficients of the terminations at ports 2 and 4, respectively. It is assumed that the power meter is perfectly matched ($\Gamma_3 = 0$) and the transmitter is matched to the line. With $Z_G = Z_0$, $a_1^2 = P_T$, the available transmitter power.

Let us first consider the ideal case of infinite directivity and $\Gamma_2 = 0$. The coupled signal $b_3 = S_{31} a_1$, where S_{31} is the scattering coefficient associated with transmission from port 1 to port 3. The reflected wave at the antenna is $\Gamma_4 b_4$. A portion of it is coupled to port 2 and therefore $b_2 = S_{24} \Gamma_4 b_4$. With $\Gamma_2 = 0$, this wave is

---

[8] When $Z_G \neq Z_0$, the meter reading *is* a function of Γ_4 since port 3 samples P_1^+, and P_1^+ is related to Γ_4 via Eq. (8–10). Note that for loose coupling, $|\Gamma_{in}| \approx |\Gamma_4|$.

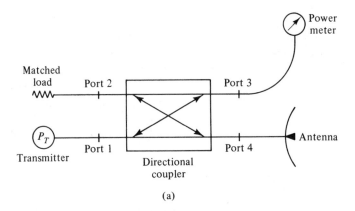

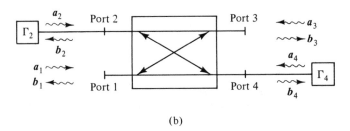

(b)

Figure 8–35 Power monitoring with a directional coupler.

completely absorbed by the matched load at port 2 and thus

$$P_2 = b_2^2 = |S_{24}|^2 |\Gamma_4|^2 b_4^2 = \frac{|\Gamma_4|^2}{K_c} b_4^2 \qquad (8\text{–}43)$$

where $K_c = |S_{24}|^{-2} = |S_{31}|^{-2}$ is the coupling factor as defined in Eq. (8–14). For loose coupling (>20 dB), $b_4 \approx a_1$ and therefore $P_2 \approx |\Gamma_4|^2 P_T/K_c$.

With $\Gamma_2 = 0$ and the directivity infinite, the only signal arriving at port 3 is $S_{31}a_1$ and thus the power delivered to the matched power meter is $P_3 = |S_{31}|^2 a_1^2 = P_T/K_c$. For example, if the coupling value is 30 dB, $K_c = 1000$, $P_3 = P_T/1000$, and $P_2 = |\Gamma_4|^2 P_T/1000$. For the situation described here, the power meter reading is independent of Γ_4 and is directly proportional to P_T. Thus, if the coupling factor is precisely known and the meter is accurate, P_T may be determined from the power meter reading (namely, $P_T = K_c P_3$).

Let us next consider the case where Γ_2 and the directivity are finite. With $a_3 = 0$, the output wave at port 3 is given by

$$b_3 = S_{31}a_1 + S_{34}\Gamma_4 b_4 + S_{32}\Gamma_2 S_{21}a_1 + S_{32}\Gamma_2 S_{24}\Gamma_4 b_4$$

where the ports and waves are all defined in Fig. 8–35b. The first term is the normal coupled signal between ports 1 and 3, while the second and third terms are associated with the fact that finite coupler directivity produces leakage signals between ports 3 and 4 and between ports 1 and 2. The fourth term represents that portion of the reflected wave at port 4 that is coupled to port 2 and then reflected by the termination at port 2. The above expression may be rewritten in terms of the coupling and directivity factors. For loose coupling, $b_4 \approx a_1 = \sqrt{P_T}$ and $|S_{32}| = |S_{41}| \approx 1$. Thus,

$$b_3 = \sqrt{\frac{P_T}{K_c}} \left[1\underline{/\phi_1} + \frac{|\Gamma_4|}{\sqrt{K_d}} \underline{/\phi_2} + \frac{|\Gamma_2|}{\sqrt{K_d}} \underline{/\phi_3} + |\Gamma_2 \Gamma_4|\underline{/\phi_4} \right] \qquad (8-44)$$

where ϕ_1 to ϕ_4 represent the phase of the individual waves arriving at port 3. Note that $|S_{21}| = |S_{34}| = 1\sqrt{K_c K_d}$.

In most cases, only the magnitude of the reflection and scattering coefficients are known. This means that only the extreme values of $P_3 = b_3^2$ can be determined. Namely,

$$P_{3\substack{max \\ min}} = \frac{P_T}{K_c} \left[1 \pm \left(\frac{|\Gamma_2| + |\Gamma_4|}{\sqrt{K_d}} + |\Gamma_2 \Gamma_4| \right) \right]^2 \qquad (8-45)$$

where it has been assumed that the first term (the direct coupled wave) is much greater than the others. The plus sign of the $\pm$ symbol is for the case when all the waves arrive at port 3 in phase, while the minus sign represents the case when ϕ_2, ϕ_3, and ϕ_4 are 180° out-of-phase with respect to ϕ_1.

The assumption of loose coupling simplifies the above analysis. Without this assumption, the problem is more complex but still solvable. The flow graph approach described in Sec. 4-4 provides a logical and systematic method for solving such problems.

Under ideal conditions, the value of transmitter power (P_T) would be determined by merely multiplying the meter reading at port 3 (P_3) by the coupling factor (K_c). Equation (8-45) shows that such a calculation is in error when the directivity is finite and/or Γ_2 is non-zero. For example, suppose the coupler directivity is 26 dB ($K_d = 400$) and $|\Gamma_2| = 0.10$. Then if $|\Gamma_4| = 0.30$, the meter reading can be anywhere from $0.9P_T/K_c$ to $1.1P_T/K_c$. Thus, calculating P_T from $K_c P_3$ would result in an uncertainty of ± 10 percent or ± 0.4 dB [note: $10 \log (1 \pm 0.1) = \pm 0.4$].

Two special cases of interest are $\Gamma_2 = 0$ with finite directivity and vice versa. For $\Gamma_2 = 0$, Eq. (8-45) reduces to

$$P_{3\substack{max \\ min}} = \frac{P_T}{K_c} \left[1 \pm \frac{|\Gamma_4|}{\sqrt{K_d}} \right]^2 \qquad (8-46)$$

By reasoning similar to that used to obtain Eq. (8-45), the maximum and minimum values of power delivered to port 2 can also be obtained. With $\Gamma_2 = 0$,

$$P_{2\substack{max \\ min}} = \frac{P_T}{K_c} \left[|\Gamma_4| \pm \frac{1}{\sqrt{K_d}} \right]^2 \qquad (8-47)$$

For the case of infinite directivity but finite Γ_2, Eq. (8-45) reduces to

$$P_{3\substack{max \\ min}} = \frac{P_T}{K_c} \left[1 \pm |\Gamma_2 \Gamma_4| \right]^2 \qquad (8-48)$$

Note that the measurement uncertainty created by a mismatch at port 2 is similar to that associated with finite coupler directivity.

The reflectometer.   Another useful application of the directional coupler involves measuring the reflection coefficient magnitude (and hence SWR) of an unknown load impedance. The circuit arrangement is shown in Fig. 8–36 where the signal source is connected to port 1 and the load impedance Z_L to port 4. Detectors D_f and D_r at ports 3 and 2 sample the forward and reflected waves, respectively. The dc voltages out of the detectors are directly proportional to the microwave power levels at each port. Assuming identical detector sensitivities, $V_r/V_f = P_2/P_3 = (b_2/b_3)^2$. The ratio meter takes the square root of V_r/V_f and displays it on the meter. The scale ranges from zero to unity since it represents the reflection coefficient magnitude of the load Z_L. In some cases, the meter face has a corresponding SWR scale, ranging from one to infinity. The meter reading is an accurate indication of load reflections if the coupler has infinite directivity and the detectors are perfectly matched. As before, finite directivity produces some uncertainty in the measured value of $|\Gamma_L|$. The analysis closely follows that of the power monitor. From Eqs. (8–47) and (8–46),

$$P_{2\substack{max \\ min}} = \frac{|\Gamma_L|^2 P_T}{K_c}\left[1 \pm \frac{1}{|\Gamma_L|\sqrt{K_d}}\right]^2$$

and   (8–49)

$$P_{3\substack{max \\ min}} = \frac{P_T}{K_c}\left[1 \pm \frac{|\Gamma_L|}{\sqrt{K_d}}\right]^2$$

where, as before, loose coupling is assumed.

Assuming identical detectors, the ratio-meter reading M will equal $\sqrt{P_2/P_3}$. The maximum possible reading occurs when P_2 is largest and P_3 smallest (vice versa for the minimum reading). Therefore,

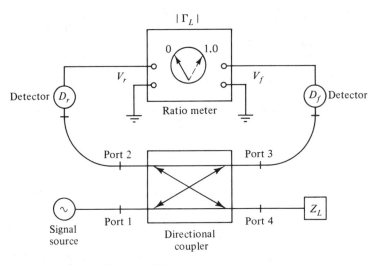

Figure 8–36   A reflectometer circuit.

$$M_{\substack{max \\ min}} = |\Gamma_L| \left[\frac{1 \pm \dfrac{1}{|\Gamma_L|\sqrt{K_d}}}{1 \mp \dfrac{|\Gamma_L|}{\sqrt{K_d}}} \right] \tag{8–50}$$

Note that with infinite directivity, M is exactly equal to $|\Gamma_L|$. For the case of finite directivity, there will be some error when the meter reading is used as a measure of $|\Gamma_L|$. For example, if $|\Gamma_L| = 0.5$ and $K_d = 400$ (26 dB directivity), the meter could read anywhere between 0.439 and 0.564. To improve the accuracy of the measurement, higher coupler directivity is required. In precise systems, couplers with greater than 50 dB directivity are used.

The method of analysis described here can also be used to determine the uncertainty associated with a phase measurement (Prob. 8–42).

8–5b 3 dB Hybrid Applications

Two components that utilize 3 dB hybrids were described in earlier chapters. The E-H tuner (Sec. 4–1d) made use of the magic-tee configuration, while the hybrid phase shifter (Sec. 7–4b) utilized the Riblet short-slot coupler. Several other applications are described in Chapter 8 of Ref. 8–8. One important use for these hybrids is in the design of microwave duplexers. A duplexer is a three-port network that allows a single antenna to both transmit and receive electromagnetic signals while protecting the sensitive receiver from the transmitter power. The duplexing function is indicated in Fig. 8–37. In an ideal duplexer, all of the transmitter power is radiated by the antenna and none appears at the receiver. Also, all of the received signal is delivered to the receiver port. In most cases, the received signal represents transmitter energy that has been reflected by a target and intercepted by the antenna. For a continuous wave (cw) system, transmission and reception occur simultaneously, while in a pulsed system they occur sequentially. In the pulsed case, the antenna is essentially time-shared. Both systems are described here.

Passive duplexers for cw systems.   Any 3 dB hybrid can serve as a passive duplexer. It is not, however, an ideal one since half the transmitter power is wasted as is half of the received signal. This 6 dB loss can be avoided by using

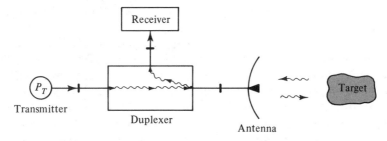

Figure 8–37 A schematic diagram of a duplexer.

either a circulator-type duplexer (Ref. 8–57) or a circularly polarized system (discussed in this section). Despite its shortcomings, it is instructive to analyze the operation of the hybrid-type passive duplexer.

For purposes of discussion, a magic-tee version is shown in Fig. 8–38 with a cw source connected to port 4 (the H arm) and a receiver to port 3 (the E arm). The common antenna is connected to one coplanar arm and a matched load to the other. The operation of the duplexer follows directly from the properties of a 3 dB hybrid. A cw signal into port 4 splits equally between ports 1 and 2. Assuming perfectly matched components, half the source power is absorbed by the matched load and half is radiated by the antenna. Note that none of the transmitter power appears at the receiver port. This is important since even a few milliwatts of power can permanently damage the receiver. At the same time, any signal received by the antenna splits between the E and H arms with half being delivered to the receiver at port 2. The other half is absorbed by the internal resistance of the cw source. Thus this system allows the simultaneous transmission and reception of microwave signals via a single antenna.

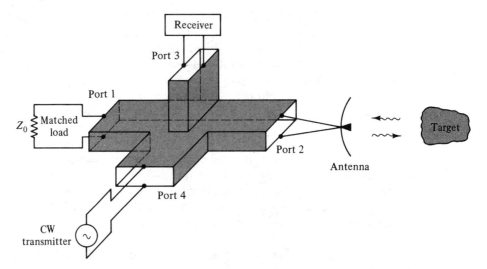

Figure 8–38   The magic-tee hybrid as a cw duplexer.

Suppose now that the antenna is mismatched, thus causing a portion of the transmitter power to be reflected by the antenna. Half of this signal is delivered to the receiver port and hence the transmitter and receiver are no longer isolated from each other. Reflections from an imperfect load termination produce the same effect. This circuit arrangement was analyzed in Sec. 8–2 (Ex. 8–4). For this application, Γ_1 and Γ_2 represent the reflection coefficients of the load termination and the antenna, respectively. From the results in part b of the example, $P_3 = 0.25 |\Gamma_1 - \Gamma_2|^2 a_4^2$, where a_4^2 is the transmitter power incident at port 4 and P_3 is the portion that appears at the receiver port. The worst case (maximum P_3) occurs when the two reflected signals arrive at port 3 in phase resulting in $P_{3_{max}} = 0.25\{|\Gamma_1| + |\Gamma_2|\}^2 a_4^2$. For example, if the antenna SWR is 1.50 and the load SWR is 1.20, $P_{3_{max}} = 0.021a_4^2$ or 2.1 percent of the transmitter power. Most applications require that this leakage signal be much smaller, which means that the SWR of both the antenna and the load must be extremely low (Prob. 8–43). In some cases, the required SWR values are so low that they cannot be achieved in practice. This means that the passive duplexer method is no longer feasible and a dual-antenna system must be used instead.

Another method of passive duplexing is described in Fig. 8–39. It consists of a dual-mode transducer and a quarter-wave plate. The transmitter, receiver, and antenna are connected as shown. By virtue of the transducer's properties, the transmitter and receiver are isolated from each other. The action of the quarter-wave plate is to convert the linearly polarized transmitter signal into a circularly polarized wave that is radiated by the antenna. For purposes of illustration, a dielectric-type quarter-wave plate is shown in figure. The radiated wave in this case will have a counterclockwise sense as indicated. Depending upon the shape and surface of the target, the reflected signal may contain both clockwise and counterclockwise components. Generally, the component having the same sense as the transmitted signal (counterclockwise, in this case) will be the larger one. The quarter-wave plate converts this signal into a horizontally polarized wave that is delivered to the receiver. Note that in this duplexer system there is no signal loss on either transmission or reception, thus avoiding the 6 dB loss associated with the hybrid-type duplexer. This method, however, requires that the radiated signal be circularly polarized.

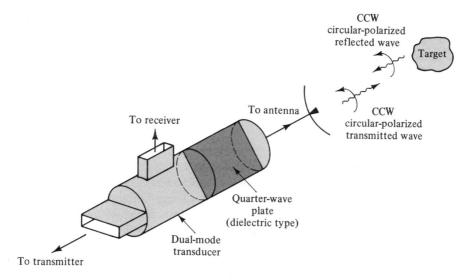

Figure 8–39   A circular polarized system for passive duplexing.

Duplexers for pulsed systems.   For systems that use a pulsed transmitter, the duplexer usually consists of a switching arrangement that allows a common antenna to be time-shared by the transmitter and receiver. Branched and balanced types are shown in Figs. 8–41 and 8–43. In most high-power applications, the switching action is provided by gas-filled TR (transmit-receive) and ATR (anti-transmit-receive) tubes. Waveguide versions are shown in Fig. 8–40.

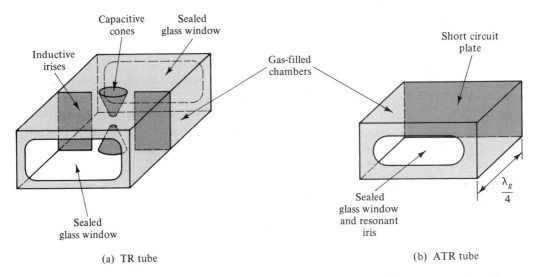

(a) TR tube   (b) ATR tube

Figure 8–40 Waveguide versions of gas-type TR and ATR tubes. (Flanges are not shown.)

The TR tube is a two-port device that exhibits low loss for low-level signals and high reflection loss for high-level signals. It is in essence a single-pole single-throw switch that is activated by a high-level microwave signal. A single-stage version is shown in part *a* of Fig. 8–40. Incorporated into each flange is a hermetically sealed glass window that prevents the gas, usually argon, from leaking out of the waveguide chamber. The capacitive cones and inductive irises form a parallel resonant circuit across the waveguide. The following qualitative explanation describes the operation of the TR tube.

When a high-power microwave signal impinges on the gas-filled tube, the electric field between the cones is sufficiently large to ionize the gas. In this state, the impedance between the cones is quite low, practically a short circuit. As a result, most of the incident signal is reflected and hence little is transmitted through the TR tube. For a low-level signal, the electric field is insufficient to ionize the gas and therefore the TR tube behaves like a bandpass filter. Its center frequency and bandwidth is a function of the gap between the cones and the dimensions of the irises. The useful bandwidth for a single-stage version is only a few percent. Multistage units are usually required to obtain larger bandwidths (Ref. 8–9).

The ATR tube is a one-port device whose input impedance is a function of the incident microwave power level. A waveguide version is shown in part *b* of Fig. 8–40. It consists of a resonant iris at the input flange followed by a quarter-wavelength shorted line. The iris opening has a glass seal for containing the gas within the waveguide chamber. When a high-level signal impinges upon the ATR tube, the gas becomes ionized and a virtual short circuit exists at the plane of the resonant iris. A low-level signal, on the other hand, is transmitted past the resonant circuit and reflected by the short. Since the short is located a quarter-wavelength away, the impedance at the plane of the iris is infinite, an open circuit. Thus in both high- and low-level states the magnitude of the reflection coefficients are unity, but their angles differ by 180°.

An example of a duplexer that utilizes TR and ATR tubes is shown in Fig. 8–41. It is known as a *branched duplexer* and finds use in narrow bandwidth applications. A typical waveguide version is shown in part *a* of the figure. It consists of two *E*-plane tee junctions separated by $\lambda_g/4$, a TR tube and an ATR tube. The antenna, receiver, and transmitter ports are as shown. In the discussion that follows, it is assumed that when the gas is ionized by the high-level microwave signal, the resonant iris in the ATR tube and the capacitive cones in the TR tube are shorted.[9] When the pulsed transmitter is on, the gas in each tube becomes ionized. This situation can be represented in the equivalent circuit (part *b* of the figure) by closing the two single-pole single-throw switches. Because of the $\lambda_g/2$ line, a zero impedance condition exists all along the broadwall of the waveguide. As a result, the transmitter power is fully delivered to the antenna. When the transmitter pulse ends, the gas in the tubes becomes deionized. This situation can be represented by opening the two switches in the equivalent circuit. The ATR now produces an open circuit at the broadwall of the main guide, thus effectively disconnecting the transmitter from the antenna. This open circuit is transformed into a short circuit at the plane of the *E*-plane tee containing the TR tube. Since the TR tube is now in its low-loss state, the received signal at the antenna port is delivered to the receiver port.

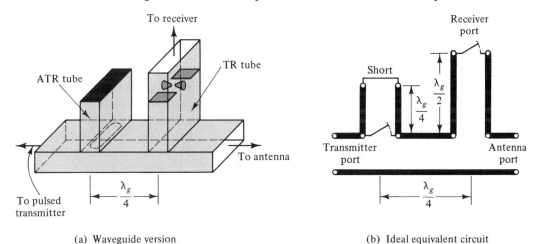

(a) Waveguide version   (b) Ideal equivalent circuit

Figure 8–41   A branched duplexer system utilizing TR and ATR tubes. (Note: In part *b*, the switches are open for low-level signals and closed for high-level signals.)

A sketch of receiver power as a function of time for a typical gas-type duplexer system is shown in Fig. 8–42. The pulsed transmiter is *on* during the interval $0 \leq t \leq \tau$ and *off* during the interval $\tau < t < T_r$, where τ is the pulse width and $f_r = 1/T_r$ is the pulse repetition rate. The fraction of the time that the transmitter is *on* is known as the *duty cycle* and is given by $\tau/T_r = \tau f_r$. For example, a transmitter with a $1.0\mu s$ pulse width and a repetition rate of 1000 Hz has a duty cycle of 0.001 or 0.10 percent. The large energy spike at the beginning of the transmitter pulse is

---

[9] The electrical properties of the ionized gas are, in fact, quite complex. For further information, the reader is referred to Refs. 8–9, 8–53, 8–54, and 8–55.

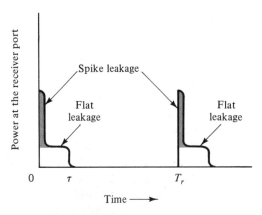

Figure 8–42 Receiver power as a function of time for a typical gas-type duplexer system. The width of the transmitter pulse is τ and its pulse repetition rate is $f_r = 1/T_r$.

due to the fact that it takes a finite time for the gas in the TR tube to ionize. This means that some of the transmitter energy will appear at the receiver port before the TR tube has had a chance to fully reach its high-loss state. For most well-designed systems, the spike leakage (represented by the shaded area) is much less than one erg. After the initial spike, the gas is fully ionized and the leakage power is essentially constant for the remainder of the pulse interval. This *flat leakage* is associated with the fact that the ionized gas column of the TR tube is not a short circuit but has a finite impedance. For example, if the transmitter-to-receiver loss is 70 dB and the transmitter peak power is 100 kW, the flat leakage will be 10 mW, an acceptable level for most radar receivers.

In practice, the ionized gas columns of both the TR and ATR tubes have finite values of resistance. The factors affecting these values have been studied by several workers (for example, Refs. 8–54, 8–55, and 8–56). As a result of this resistance, some of the transmitter pulse is absorbed by the ionized gas. This is known as the *arc loss* and is an important consideration in the design of high-power duplexers. There are many factors that must be considered in designing a duplexer for a pulsed system. These include bandwidth, arc loss, the low-level insertion loss, power-handling capability, spike and flat leakage, recovery time, and tube life. Design compromises are often necessary since the improvement of one characteristic may lead to the deterioration of another. For example, spike leakage can be reduced by introducing a *keep-alive* in the TR tube. This, however, usually results in higher insertion loss and shorter tube life. More complete discussions of the design considerations in gas-type duplexers may be found in Refs. 8–9, 8–10, 8–54, and 8–57.

A duplexer configuration that is widely used in broadband systems is shown in Fig. 8–43. It is known as a *balanced duplexer* and consists of a pair of 3 dB hybrids and a dual TR section. For purposes of explanation, a waveguide version that utilizes two Riblet short-slot couplers is indicated. With the pulse transmitter *on*, the amplitude and phase of the signals arriving at the TR tubes are as shown. These high-power signals *fire* the TR tubes producing a low impedance (ideally a short) across the guides, which in turn causes the signals to be reflected. With the TR tubes located in the same transverse plane, the reflected signals arrive at the antenna port in phase. This is indicated in the figure. Ideally, all of the transmitter pulse ($\mathbf{a}_1$) will be

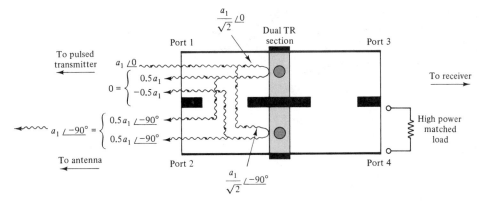

(a) Transmitter on. TR tubes are highly reflective

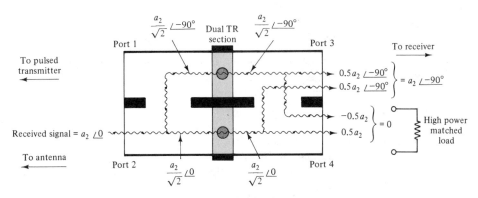

(b) Transmitter off. TR tubes are in their low-loss state

Figure 8–43   A balanced duplexer system utilizing a pair of Riblet short-slot couplers and a dual TR section.

delivered to the antenna. Once the transmitter pulse ends, the TR tubes deionize and return to their low-loss state. A signal received by the antenna ($\mathbf{a_2}$) is fully delivered to the receiver port. The hybrids cause the received signal to split and then recombine as shown in the figure.

The balanced duplexer has three advantages. Unlike the branched type, it requires no critical line lengths and hence is capable of wider band performance. Second, it can handle higher power since each TR tube only interacts with half the transmitter power. Finally, the action of the hybrids produces a significant reduction (typically 5 to 10 dB) in the spike and flat leakage to the receiver. In fact, if the leakage through each TR tube were identical, *none* would appear at the receiver. This is because the two leakage signals arrive at the receiver port 180° out-of-phase, while those at port 4 are in phase and absorbed by the matched load.

Balanced duplexers that utilize dual ATR tubes have also been developed and have the same advantages described here (Refs. 8–54 and 8–57). There are other type duplexer systems that combine hybrid junctions with ferrite and/or semiconductor devices. One method utilizes a ferrite circulator in combination with a TR tube or a semiconductor limiter. An excellent review of various active and passive duplexers is given by Harvey (Ref. 8–57). Also included is a comprehensive bibliography on the subject.

REFERENCES

Books

8–1. Marcuvitz, N., *Waveguide Handbook,* Rad. Lab. Series, vol. 10, McGraw-Hill Book Co., New York, 1951.

8–2. Montgomery, C. G., R. H. Dicke, and E. M. Purcell, *Principles of Microwave Circuits,* Rad. Lab. Series, vol. 8, McGraw-Hill Book Co., New York, 1948.

8–3. Howe, Jr., H., *Stripline Circuit Design,* Artech House, Dedham, MA, 1974.

8–4. Levy, R., Directional Couplers, in *Advances in Microwaves,* vol. 1, L. Young, ed., Academic Press, New York, 1966.

8–5. Saad, T.S., *Microwave Engineers' Handbook,* vol. 2, Artech House, Dedham, MA, 1971.

8–6. Matthaei, G. L., L. Young, and E. M. T. Jones, *Microwave Filters, Impedance Matching Networks and Coupling Structures,* McGraw-Hill Book Co., New York, 1964.

8–7. Collin, R. E., *Foundations for Microwave Engineering,* McGraw-Hill Book Co., New York, 1966.

8–8. Altman, J. L., *Microwave Circuits,* D. Van Nostrand Co., New York, 1964.

8–9. Smullin, L. D. and C. G. Montgomery, *Microwave Duplexers,* Rad. Lab. Series, vol. 14, McGraw-Hill Book Co., New York, 1948.

8–10. Kraszewski, A., *Microwave Gas Discharge Devices,* Iliffe Books, Ltd., London, 1967.

Articles

8–11. Tyrell, W. A., Hybrid Circuits for Microwaves. *Proc. IRE,* 35, November 1947, pp. 1294–1306.

8–12. Albanese, V. J. and W. P. Peyser, An Analysis of a Broadband Coaxial Hybrid Ring. *IRE Trans. Microwave Theory and Techniques,* MTT-6, October 1958, pp. 369–373.

8–13. Reed, J. and G. Wheeler, A Method of Analysis of Symmetrical Four-Port Networks. *IRE Trans. Microwave Theory and Techniques,* MTT-4, October 1956, pp. 246–252.

8–14. Wilkinson, E. J., An *N*-Way Hybrid Power Divider. *IRE Trans. Microwave Theory and Techniques,* MTT-8, January 1960, pp. 116–118.

8–15. Parad, L. I. and R. L. Moynihan, Split-Tee Power Divider. *IEEE Trans. Microwave Theory and Techniques,* MTT-13, January 1965, pp. 91–95.

8–16. Cohn, S. B., A Class of Broadband Three-Port TEM-Mode Hybrids. *IEEE Trans. Microwave Theory and Techniques,* MTT-16, February 1968, pp. 110–116.

8–17. Alford, A. and C. W. Watts, A Wideband Coaxial Hybrid. *IRE Nat. Conv. Record,* 4 (part 1), 1956, pp. 171–179.

8–18. Kahn, W. K., *E*-Plane Forked Hybrid-*T* Junction. *IRE Trans. Microwave Theory and Techniques,* MTT-3, December 1955, pp. 52–58.

8–19. Lange, J., Interdigitated Stripline Quadrature Hybrid. *IEEE Trans. Microwave Theory and Techniques,* MTT-17, December 1969, pp. 1150–1151.

8–20. Caswell, W. E. and R. F. Schwartz, The Directional Coupler-1966. *IEEE Trans. Microwave Theory and Techniques,* MTT-15, February 1967, pp. 120–123.

8–21. Bethe, H. A., Theory of Diffraction by Small Holes. *Phys. Rev.,* 66, October 1944, pp. 163–182.

8–22. Shelton, W., Compact Multihole Waveguide Directional Couplers. *Microwave Journal,* 4, July 1961, pp. 89–92.

8–23. Riblet, H. J., A Mathematical Theory of Directional Couplers. *Proc. IRE,* 35, November 1947, pp. 1307–1313.

8–24. Levy, R., Analysis and Synthesis of Waveguide Multiaperture Directional Couplers. *IEEE Trans. Microwave Theory and Techniques,* MTT-16, December 1968, pp. 995–1006.

8–25. Levy, R., Improved Single and Multiaperture Waveguide Coupling Theory, Including Explanation of Mutual Interaction. *IEEE Trans. Microwave Theory and Techniques,* MTT-28, April 1980, pp. 331–338.

8–26. Riblet, H. J. and T. S. Saad, A New Type of Waveguide Directional Coupler. *Proc. IRE,* 36, January 1948, pp. 61–64.

8–27. Anderson, T. N., Directional Coupler Design Nomograms. *Microwave Journal,* 2, May 1959, pp. 34–38.

8–28. Reed, J., The Multiple Branch Waveguide Coupler. *IRE Trans. Microwave Theory and Techniques,* MTT-6, October 1958, pp. 398–403.

8–29. Lomer, P.D. and J. W. Crompton, A New Form of Hybrid Junction for Microwave Frequencies. *Proc. IEE* (London), B104, May 1957, pp. 261–264.

8–30. Levy, R., A Guide to the Practical Application of Chebyshev Functions to the Design of Microwave Components. *Proc. IEE* (London), C106, June 1959, pp. 193–199.

8–31. Young, L., Synchronous Branch-Guide Directional Couplers for Low and High Power Applications. *IRE Trans. Microwave Theory and Techniques,* MTT-10, November 1962, pp. 459–475.

8–32. Patterson, K. G., A Method for Accurate Design of a Broadband Multibranch Waveguide Coupler. *IRE Trans. Microwave Theory and Techniques,* MTT-7, October 1959, pp. 466–473.

8–33. Levy, R. and L. F. Lind, Synthesis of Symmetrical Branch-Guide Directional Couplers. *IEEE Trans. Microwave Theory and Techniques,* MTT-16, February 1968, pp. 80–89.

8–34. Levy, R., Zolotarev Branch-Guide Couplers. *IEEE Trans. Microwave Theory and Techniques,* MTT-21, February 1973, pp. 95–99.

8–35. Miller, S. E., Coupled Wave Theory and Waveguide Applications. *Bell System Tech. J.,* 33, May 1954, pp. 661–719.

8–36. Oliver, B. M., Directional Electromagnetic Couplers. *Proc. IRE,* 42, November 1954, pp. 1686–1692.

8–37. Firestone, W. L., Analysis of Transmission Line Directional Couplers. *Proc. IRE,* 42, October 1954, pp. 1529–1538.

8–38. Jones, E.M.T. and J. T. Bolljahn, Coupled Strip Transmission Line Filters and Directional Couplers. *IRE Trans. Microwave Theory and Techniques,* MTT-4, April 1956, pp. 75–81.

8–39. Shimizu, J. K. and E.M.T. Jones, Coupled-Transmission Line Directional Couplers. *IRE Trans. Microwave Theory and Techniques,* MTT-6, October 1958, pp. 403–410.

8–40. Cohn, S. B., Shielded Coupled-Strip Transmission Line. *IRE Trans. Microwave Theory and Techniques,* MTT-3, October 1955, pp. 29–38.

8–41. Cohn, S. B., Characteristic Impedance of Broadside-Coupled Strip Transmission Line. *IRE Trans. Microwave Theory and Techniques,* MTT-8, November 1960, pp. 633–637.

8–42. Getsinger, W. J., Coupled Rectangular Bars Between Parallel Plates. *IRE Trans. Microwave Theory and Techniques,* MTT-10, January 1962, pp. 65–72.

8–43. Shelton, P. J., Impedances of Offset Parallel Coupled Strip Transmission Line. *IEEE Trans. Microwave Theory and Techniques,* MTT-14, January 1966, pp. 7–15.

8–44. Cristal, E. G. and L. Young, Theory and Tables of Optimum Symmetrical TEM-Mode Coupled Transmission-Line Directional Couplers. *IEEE Trans. Microwave Theory and Techniques,* MTT-13, September 1965, pp. 544–558.

8–45. Toulios, P. P. and A. C. Todd, Synthesis of Symmetrical TEM-Mode Directional Couplers. *IEEE Trans. Microwave Theory and Techniques,* MTT-13, September 1965, pp. 536–544.

8–46. Levy, R., Tables for Asymmetric Multi-Element Coupled Transmission Line Directional Couplers. *IEEE Trans. Microwave Theory and Techniques,* MTT-12, May 1964, pp. 275–279.

8–47. Riblet, H. J., The Short-Slot Hybrid Junction. *Proc. IRE,* 40, February 1952, pp. 180–184.

8–48. Hadge, E., Compact Top-Wall Hybrid Junction. *IRE Trans. Microwave Theory and Techniques,* MTT-1, March 1953, pp. 29–30.

8–49. Cook, J. S., Tapered Velocity Couplers. *Bell System Tech. J.,* 34, July 1955, pp. 807–822.

8–50. Tomiyasu, K. and S. B. Cohn, The Transvar Directional Coupler. *Proc. IRE,* 41, July 1953, pp. 922–926.

8–51. Ruthroff, C. L., Some Broad-Band Transformers. *Proc. IRE,* 47, August 1959, pp. 1337–1342.

8–52. A 28 Year Index of the *IEEE Trans. Microwave Theory and Techniques,* MTT-29, June 1981.

8–53. Brown, S. C., The Interaction of Microwaves with Gas-Discharge Plasmas. *IRE Trans. Microwave Theory and Techniques,* MTT-7, January 1959, pp. 69–72.

8–54. Muehe, Jr., C. E., High-Power Duplexers. *IRE Trans. Microwave Theory and Techniques,* MTT-9, November 1961, pp. 506–512.

8–55. Margenau, H., Conduction and Dispersion of Ionized Gases at High Frequencies. *Phys. Rev.,* 69, May 1946, pp. 508–513.

8–56. Ward, C. S., F. A. Jellison, N. J. Brown, and L. Gould, The Arc Loss of Multi-megawatt Gas Discharge Duplexers. *IEEE Trans. Microwave Theory and Techniques,* MTT-13, November 1965, pp. 801–805.

8–57. Harvey, A. F., Duplexing Systems at Microwave Frequencies. *IRE Trans. Microwave Theory and Techniques,* MTT-8, July 1960, pp. 415–430.

PROBLEMS

8–1. A matched generator with an available power of 80 W is connected to port 1 of a loop-type power monitor (Fig. 8–1b). With all ports match terminated, 0.40 W is delivered to port 3. Calculate the variation in output power at port 3 when the reflection coefficient magnitude due to Z_{L_2} is 0.25 and l_2 is varied over a half wavelength.

8–2. If the probe shown in Fig. 8–1a is connected directly to the center conductor of the main line, the equivalent circuit becomes that shown in Fig. 8–2a with $Z_{03} = Z_0$. Given $Z_{L_2} = 150$ ohms, $Z_G = Z_0 = 50$ ohms and a 100 W source at port 1, calculate the power delivered to a matched load at port 3 for $l_2 = 0.25\lambda$ and 0.50λ.

8–3. The coaxial tee in Fig. 8–2a has $l_3 = \lambda/4$. Z_{03} has been chosen so that with ports 1 and 2 match terminated, the input SWR at port 3 is unity. Prove that with match terminations at ports 2 and 3, the SWR into port 1 is 3.0.

8–4. Referring to the coaxial tee in Fig. 8–2b with port 3 as the input,
(a) Calculate Z_{01}, Z_{02}, and Z_{03} for equal power division and unity input SWR. Assume $Z_0 = 50$ ohms and $l_1 = l_2 = l_3 = \lambda/4$ at the design frequency.
(b) Calculate the SWR into port 3 at *half* the design frequency.

8–5. Repeat Prob. 8–4 for the case $P_1 = 3P_2$. Also calculate the output power ratio (P_1/P_2) at half the design frequency.

8–6. Referring to Fig. 8–4, $b_2 = 3b_1$, $l_1 = \lambda_g/4$, $l_2 = \lambda_g/2$, $Z_{L_1} = 0.8Z_{01}$ and $Z_{L_2} = 1.2Z_{02}$. Determine the input SWR and the output power ratio (P_1/P_2). Assume $b_1 + b_2 \approx b$.

8–7. For the E-plane tee described in Ex. 8–3,
(a) Calculate the SWR into port 3 when ports 1 and 2 are match terminated.
(b) Determine the insertion loss (in dB) between ports 1 and 3 when port 2 is match terminated and a matched generator is connected to port 3.

8–8. An E-plane tee (Fig. 8–5) has $b = 2.0$ cm, and $b' = 1.0$ cm. With ports 2 and 3 match terminated, the SWR into port 1 is 1.25. Calculate the power delivered to ports 2 and 3 when a 5.0 W matched generator is connected to port 1.

8–9. Port 3 of an E-plane tee is terminated in a short circuit. The length l_3 has been adjusted for zero transmission between ports 1 and 2. Find l_3 if $\lambda_g = 6.0$ cm and referring to Fig. 8–6b, $n_t = 0.6$, $X = 0$, and $d' = 0.30$ cm.

8–10. The equivalent circuit of an H-plane tee is given in Fig. 8–8. Calculate the input SWR at ports 1, 2, and 3 if $n_t = 0.6$ and $X = 5Z_0$. Assume all output ports are match terminated.

8–11. With ports 1 and 2 of an H-plane tee match terminated, the SWR into port 3 is unity. Assuming $|X| \gg 2Z_0$,
(a) Calculate the transformer turns ratio n_t.
(b) With a one watt matched generator connected to port 3, calculate the power delivered to load impedances at ports 1 and 2 when $Z_{L_1} = 2.4Z_0$ and $Z_{L_2} = 0.6Z_0$. Neglect line lengths (that is, $0.5a + d \approx 0$). Compare the results with those in part *b* of Ex. 8–4.

8–12. Consider the equivalent circuit in Fig. 8–10 with $n_E = n_H = \sqrt{2}$ and ports 2, 3, and 4 terminated in Z_0. Prove that with a generator at port 1, no power is delivered to port 2. Also verify that the input impedance at port 1 is Z_0.

8–13. Repeat part *b* of Ex. 8–4 for the case where the lengths of the coplanar arms are $l_1 = 0.50\lambda_g$ and $l_2 = 0.25\lambda_g$, respectively.

8–14. A microwave generator with $P_A = 2$ W and $Z_G = 0.5Z_0$ is connected to the H arm of a matched magic tee. Identical loads are connected to equal length coplanar arms. Referring to Fig. 8–11, $\Gamma_{L_3} = 0$ and $|\Gamma_{L_1}| = |\Gamma_{L_2}| = 0.40$. Calculate the maximum and minimum power that can be absorbed by the load at port 1.

8–15. Referring to Fig. 8–14, calculate the SWR into port 2 if the length of the $\sqrt{2}Z_0$ lines are 0.20λ rather than $\lambda/4$. Use even-odd mode analysis and assume ports 1 and 3 are match terminated.

8–16. The Wilkinson divider in Fig. 8–14 can be used as a power combiner by injecting coherent signals into ports 2 and 3. Calculate the power delivered to a matched termination at port 1 if the voltages at ports 2 and 3 are 30° out-of-phase and each has an rms value of 12 V. Assume $Z_0 = 50$ ohms. Also calculate the power dissipated in the resistor.

8–17. A 25 W signal is incident on port 2 of a directional coupler (Fig. 8–16). Calculate the power out of ports 1 and 4 if the coupling is 33 dB and the directivity is 24 dB.

8–18. The scattering matrix for an ideal directional coupler is given by Eq. (8–12). Use Eq. (D–11) to prove that $|S_{13}| = |S_{24}|$ and $|S_{14}| = |S_{23}|$.

8–19. A two-hole waveguide directional coupler (Fig. 8–19a) has $k_f = k_r$ and $l = 1.50$ cm. The width of each air-filled guide is 4.00 cm.
  (a) Determine the frequency at which the directivity is infinite.
  (b) Determine the frequency range over which the directivity is greater than 30 dB.

8–20. Verify that the directivity of a four-hole binomial coupler is $10 \log (1/\cos \beta l)^6$ when $k_r = k_f$. Assume loose coupling (that is, $b_4 \approx a_1$).

8–21. A ten-section binomial coupler has been designed in WR-90 waveguide. The spacing between apertures is 0.97 cm. Assuming $k_r = k_f$, calculate the directivity at 8.2, 10.0, and 12.4 GHz.

8–22. A four-section Tchebyscheff coupler has been designed with a minimum directivity of 36 dB. Calculate the useful bandwidth (in MHz) if the width of each air-filled guide is 4.00 cm and the spacing between apertures is 1.50 cm. Assume $k_f = k_r$. Note that $T_4(x) = 8x^4 - 8x^2 + 1$. Compare the bandwidth to that of a four-section binomial design.

8–23. Prove that for a Bethe-hole coupler, maximum directivity occurs at $\theta = 0$ when $\lambda_0 = \sqrt{2}\,a$.

8–24. Design a coaxial two-branch coupler (Fig. 8–23) with 6.0 dB coupling. The design frequency is 3.0 GHz, $\lambda = 0.8\lambda_0$ and $Z_0 = 50$ ohms. Calculate l, Z_{01} and Z_{02}.

8–25. Consider a three-branch coaxial coupler in which the impedance of the main lines (Z_{01}) equals that of the connecting lines (Z_0). The line impedance of the end branches (Z_{02}) are identical and that of the center branch is Z_{03}. Derive the condition for a perfect match (and hence infinite directivity). Assume all line lengths are $\lambda/4$. Also derive an equation for coupling as a function of Z_{03}.

8–26. Use the results of the previous problem to show that with $Z_{01} = Z_{02} = Z_{03} = Z_0$, the coupling is 0 dB.

8–27. With $Z_{01} = Z_0/\sqrt{2}$ and $Z_{02} = Z_0$, the two-branch coupler in Fig. 8–23 has infinite directivity and 3.0 dB coupling at the design frequency (f_r). Calculate the coupling and directivity of $0.9f_r$. (Hint: Use Eqs. (8–22) and (8–23) and the appropriate normalized ABCD matrices.)

8–28. Prove that for a matched, coupled-line directional coupler (Fig. 8–26a), the sum of the output powers ($b_3^2 + b_4^2$) equals the input power (a_1^2) for all values of βl.

8–29. Design a one-section, quarter-wave coupler using coplanar stripline (Fig. 8–26a). The design frequency is 2.5 GHz and the desired maximum coupling is 18 dB. Calculate s, w, and l if $b = 0.60$ cm, $\epsilon_R = 2.25$ and $Z_0 = 50$ ohms.

8–30. For the coplanar, stripline coupler shown in Fig. 8–26a, $l = \lambda/4$ at 5.0 GHz, $Z_0 = 50$ ohms, $Z_{0e} = 62.5$ ohms, and $Z_{0o} = 40$ ohms. Calculate the coupling (in dB) at 4.0, 5.0, and 6.0 GHz.

8–31. For the previous problem, calculate the coupling and directivity at 5.0 GHz if $Z_{0e} = 57$ ohms instead of 62.5 ohms. Note that in this case $Z_{0e} Z_{0o} \neq Z_0^2$.

8–32. Repeat part b of Ex. 8–5 if the reflection coefficients due to discontinuities at ports 3 and 4 are $0.05\underline{/0}$ and $0.05\underline{/90°}$ respectively.

8–33. A Riblet short-slot coupler is shown in Fig. 8–30.
    (a) Calculate l for 3.0 dB coupling at 9.0 GHz when $a = 2.00$ cm. Assume air-filled guides.
    (b) For the above value of l, calculate the coupling at 10.0 GHz.

8–34. A Riblet short-slot coupler (Fig. 8–30) has $a = 2.00$ cm. Calculate the slot length l for 0 dB coupling at 9.0 GHz.

8–35. A 5.0 W wave is incident on port 1 of a 3.0 dB short-slot coupler (Fig. 8–30). The remaining ports are terminated with loads having the following reflection coefficients: $\Gamma_2 = 0$, $\Gamma_3 = 0.20$, and $\Gamma_4 = 0.30\underline{/45°}$. Calculate the power dissipated in each load.

8–36. The power meter shown in Fig. 8–35a reads 8.0 mW when $P_T = 2.0$ W and $Z_G = Z_0$.
    (a) Calculate the coupling value (in dB) if the antenna SWR is unity.
    (b) How much power is dissipated by the matched load if the antenna SWR is 2.0 and the coupler directivity is infinite.
    (c) Determine the largest possible increase and decrease in the power meter reading (in dB) if the coupler directivity is 20 dB.

8–37. The antenna in a power monitor circuit (Fig. 8–35) has an SWR of 3.0. The load at port 2 has an SWR of 1.25, and the directivity of the 30 dB coupler is 26 dB. What is the range of P_T values that could result in a meter reading of 7.0 mW? Assume $Z_G = Z_0$.

8–38. For the power monitor described in the previous problem, determine the power meter reading if $P_T = 7.0$ W, $\phi_1 = \phi_2 = 0$, $\phi_3 = 90°$, and $\phi_4 = -90°$. Note: ϕ_1 to ϕ_4 are defined below Eq. (8–44).

8–39. Consider a coupler circuit similar to that in Fig. 8–35 except that the matched power meter is connected to port 2, and port 3 is terminated by an adjustable load having $\Gamma_3 = 0.1\underline{/\phi}$. Given $P_T = 9.0$ W, $Z_G = Z_0$, $\Gamma_4 = 0.3\underline{/0}$, and a coupler value of 20 dB, determine the power meter reading when ϕ is adjusted for a minimum reading. Assume infinite directivity and loose coupling.

8–40. The reflectometer circuit shown in Fig. 8–36 is terminated by a purely reactive load. Specify the minimum directivity required in order for the meter to read $|\Gamma_L|$ to within one percent of its true value.

8–41. Determine the maximum meter reading for the reflectometer circuit (Fig. 8–36) if $Z_L = 3Z_0$ and the SWR of the detectors D_r and D_f are 1.10 and 1.20, respectively. Assume infinite directivity. Consider only first-order reflections from D_r and D_f.

8–42. Referring to the coupler diagram in Fig. 8–35b, $b_3 = 0.10\ a_1\underline{/0}$ when Γ_2 and Γ_4 are zero. Calculate the range of phase values for b_3 if $|\Gamma_2| = |\Gamma_4| = 0.20$. Assume infinite directivity and loose coupling.

8–43. A 20 W transmitter is connected to port 4 of the magic-tee duplexer shown in

Fig. 8–38. The termination at port 1 has an SWR of 1.04. What is the maximum allowable antenna SWR in order to insure that the receiver power is less than 10 mW?

8–44. A pulsed transmitter with a duty cycle of 0.001 and a pulse power of 100 kW is connected to port 1 of a balanced duplexer (Fig. 8–43). In the *fired* state, the shunt resistance of each TR tube is 5 ohms. Calculate the arc loss (in dB) and the average power absorbed by each TR tube. Assume $Z_0 = 400$ ohms and all ports are match terminated.

8–45. For the duplexer described in the previous problem,
   (a) Calculate the flat leakage for the given values of shunt resistances.
   (b) What is the flat leakage if the resistance of one TR tube is 4 ohms and that of the other is 2 ohms?

9

Microwave Resonators and Filters

Resonant circuits are used extensively in electronic networks and systems. They serve as key elements in oscillators, tuned amplifiers, frequency meters, phase equalizers, as well as in bandpass and bandstop filters. Below roughly 300 MHz, resonant circuits usually consist of lumped inductors and capacitors. At microwave frequencies, however, these lumped-element circuits have certain disadvantages. First of all, the requirement that the element dimensions be much smaller than the operating wavelength limits their voltage and current capabilities. Second, the inductance and capacitance values required at the higher microwave frequencies are sometimes difficult to realize, particularly with high Q. Furthermore, the circuit tends to radiate at high frequencies resulting in substantial losses, especially at the resonant frequency. To overcome these limitations, transmission-line techniques are widely used to realize high-Q microwave resonant circuits. In general, distributed-type circuits can be made with lower dissipative losses and hence higher Q. One example is the cavity resonator. Its properties are analyzed in detail and its use in bandpass and bandstop filters are discussed. Other transmission-line resonators and filters are also described.

As a preliminary to the discussion of microwave resonators and filters, the properties of lumped-element resonant circuits are reviewed.

9–1 A REVIEW OF RESONANT CIRCUITS

Figure 9–1 shows series and parallel resonant L-C circuits connected to ac sources. For the series case (part a), a constant voltage generator V_G in series with R_G is used to represent the ac source, while for the parallel case (part b), the equivalent constant-current representation is used. Thus $I_G = V_G/R_G$ and $G_G = 1/R_G$.

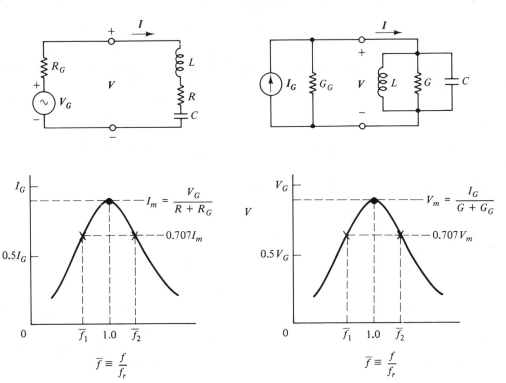

Figure 9-1 Series and parallel resonant L-C circuits. (Note: $R_G = 1/G_G$, $I_G = V_G/R_G$, and $V_G = I_G/G_G$.)

The series and parallel resonant circuits are represented by the R-L-C and G-L-C elements, respectively. R and G are associated with the losses in the resonant circuits themselves and *not* with external losses. Note that $G \neq 1/R$. For both circuits, the resonant frequency is given by

$$\omega_r = 2\pi f_r = \frac{1}{\sqrt{LC}} \tag{9-1}$$

In the series case, the impedance of the resonant circuit is $Z = R + j(\omega L - 1/\omega C)$. At resonance, $Z = R$ since $\omega_r L = 1/\omega_r C$. For the parallel case, the admittance is $Y = G + j(\omega C - 1/\omega L)$ and at resonance $Y = G$. The frequency response of current and voltage for the two cases are shown below the respective circuits.

One can define a quality factor Q for any resonance phenomena in terms of its stored energy and energy loss at the resonant frequency.[1] That is,

---

[1] Q can also be defined in terms of the circuit's transient response. See, for example, Sec. 1.04 of Ref. 9-1.

$$Q \equiv 2\pi \frac{\text{Energy Stored}}{\text{Energy Loss per Cycle}}\bigg|_{\text{at } \omega_r} = \omega_r \frac{\text{Energy Stored}}{\text{Average Power Loss}}\bigg|_{\text{at } \omega_r} \quad (9\text{--}2)$$

There are actually three Q factors that can be defined, depending upon which loss is being considered. Namely,

$$\text{Unloaded } Q\text{: } Q_U \equiv \omega_r \frac{\text{Energy Stored in the Resonant Circuit}}{\text{Power Loss in the Resonant Circuit}}\bigg|_{\text{at } \omega_r}$$

$$\text{External } Q\text{: } Q_E \equiv \omega_r \frac{\text{Energy Stored in the Resonant Circuit}}{\text{Power Loss in the External Circuit}}\bigg|_{\text{at } \omega_r} \quad (9\text{--}3)$$

$$\text{Loaded } Q\text{: } Q_L \equiv \omega_r \frac{\text{Energy Stored in the Resonant Circuit}}{\text{Total Power Loss}}\bigg|_{\text{at } \omega_r}$$

The unloaded Q is a measure of the quality of the resonant circuit itself. Infinite Q_U means that the resonant circuit is dissipationless. External Q is a measure of the degree to which the resonant circuit is coupled to the external circuitry. Methods of controlling this coupling are described in the next section. A relationship among the three Q factors can be deduced from Eq. (9–3). Namely,

$$\frac{1}{Q_L} = \frac{1}{Q_E} + \frac{1}{Q_U} \quad (9\text{--}4)$$

Therefore, the loaded Q is also a function of the degree of coupling between the resonant circuit and the external circuitry. For the single-tuned versions shown here, it describes the frequency response of the network. Referring to Fig. 9–1,

$$Q_L = \frac{f_r}{f_2 - f_1} = \frac{\text{Resonant Frequency}}{\text{3 dB Bandwidth}} \quad (9\text{--}5)$$

where f_1 and f_2 are the frequencies at which the current (or voltage) is 0.707 of its value at resonance. Note that these correspond to the half-power points on a power versus frequency curve. Also note that $f_r = \sqrt{f_1 f_2}$.
For the series resonant circuit in Fig. 9–1a,

$$Q_U = \frac{\omega_r L}{R}, \qquad Q_E = \frac{\omega_r L}{R_G}, \qquad \text{and} \qquad Q_L = \frac{\omega_r L}{R + R_G} \quad (9\text{--}6)$$

For the parallel resonant circuit in Fig. 9–1b,

$$Q_U = \frac{\omega_r C}{G}, \qquad Q_E = \frac{\omega_r C}{G_G}, \qquad \text{and} \qquad Q_L = \frac{\omega_r C}{G + G_G} \quad (9\text{--}7)$$

Since the frequency sensitivity of the resonant circuit is a function of Q_L, it follows that loaded Q must be related to the change in reactance (or susceptance) per deviation in source frequency from resonance ($\Delta\omega = \omega - \omega_r$). The relationships are as follows:

For series resonance,

$$Q_L = \frac{1}{2} \lim_{\Delta\omega \to 0} \left[\frac{\dfrac{\Delta X}{R_G + R}}{\dfrac{\Delta\omega}{\omega_r}} \right] = \frac{\omega_r}{2(R_G + R)} \frac{dX}{d\omega} \bigg|_{\omega = \omega_r} \tag{9-8}$$

For parallel resonance,

$$Q_L = \frac{1}{2} \lim_{\Delta\omega \to 0} \left[\frac{\dfrac{\Delta B}{G_G + G}}{\dfrac{\Delta\omega}{\omega_r}} \right] = \frac{\omega_r}{2(G_G + G)} \frac{dB}{d\omega} \bigg|_{\omega = \omega_r} \tag{9-9}$$

If the unloaded Q is infinite ($R = G = 0$), these expressions define the external Q. Conversely, with R_G and G_G equal to zero, they represent the unloaded Q of the circuit. Thus, the various Q's for series and parallel resonant circuits may be defined in terms of their reactance and susceptance slopes at resonance. These expressions are particularly useful in analyzing transmission-line resonant circuits.

The impedance and admittance of the circuits in Fig. 9–1 are $Z = R + j(\omega L - 1/\omega C)$ for the series case and $Y = G + j(\omega C - 1/\omega L)$ for the parallel case. In transmission-line circuits, the resistive and reactive elements are not lumped but distributed. As a result, these expressions for Z and Y are not very useful. It would be helpful to rewrite them in terms of quantities that are easily measured at microwave frequencies. This can be done by making use of Eqs. (9–1), (9–6), and (9–7).

For series resonance,

$$Z = (R_G + R)\left[\frac{Q_L}{Q_U} + jQ_L\epsilon_f \right] \approx R_G\left[\frac{Q_L}{Q_U} + jQ_L\epsilon_f \right] \tag{9-10}$$

For parallel resonance,

$$Y = (G_G + G)\left[\frac{Q_L}{Q_U} + jQ_L\epsilon_f \right] \approx G_G\left[\frac{Q_L}{Q_U} + jQ_L\epsilon_f \right] \tag{9-11}$$

The approximate expressions are valid for $Q_U \gg Q_E$. For a matched source, $R_G = Z_0$ and $G_G = Y_0$, where $Z_0 = 1/Y_0$ is the characteristic impedance of the connecting lines. In both equations, ϵ_f is defined as

$$\epsilon_f \equiv \frac{\omega}{\omega_r} - \frac{\omega_r}{\omega} = \frac{f}{f_r} - \frac{f_r}{f} \tag{9-12}$$

where ω and f represent the frequency of the ac source. Thus ϵ_f is an alternate way of expressing the frequency variable. This form is convenient when dealing with resonant circuits since $\epsilon_f = 0$ when the source frequency equals the resonant frequency of the circuit. For $f > f_r$, ϵ_f is positive, while for $f < f_r$, ϵ_f is negative. For example, if the source frequency is 10 percent below f_r, then $f/f_r = 0.90$ and $\epsilon_f = -0.21$. With Δf defined as the deviation in source frequency from resonance,

$$\epsilon_f = \left(1 + \frac{\Delta f}{f_r}\right) - \left(1 + \frac{\Delta f}{f_r}\right)^{-1} \approx 2\frac{\Delta f}{f_r} \tag{9-13}$$

where $\Delta f \equiv f - f_r$. The approximate form is accurate to within 10 percent for $|\Delta f| \le 0.20 f_r$.

The various expressions developed here will now be used to obtain insertion loss and phase equations for single-stage bandpass and bandstop filters.

The single-stage bandpass filter. A bandpass filter can be constructed using either a series-connected series resonant circuit or a shunt-connected parallel resonant circuit. The filter will be analyzed using the series representation, shown in Fig. 9–2. The resultant expressions for loss and phase are valid for either representation. The assumption is made that $R_G = R_L = Z_0$, the characteristic impedance of the input and output connecting lines. As a result, the insertion and transducer losses are identical, since the power delivered to the load *before* inserting the filter is the available generator power, namely, $P_A = V_G^2/4Z_0$. With the filter inserted as shown, the load power is

$$P_L = \left|\frac{V_G}{R + 2Z_0 + j(\omega L - 1/\omega C)}\right|^2 Z_0 = \frac{V_G^2/4Z_0}{\left(\dfrac{R + 2Z_0}{2Z_0}\right)^2 \left[1 + \left(\dfrac{\omega_r L \epsilon_f}{R + 2Z_0}\right)^2\right]} \tag{9-14}$$

where ω_r and ϵ_f are defined in Eqs. (9–1) and (9–12). For the series resonant circuit, $Q_U = \omega_r L/R$ as indicated in Eq. (9–6). The equations for Q_E and Q_L, however, must be modified to account for the fact that the resonant circuit is being *loaded* by $R_G + R_L = 2Z_0$. Thus $Q_E = \omega_r L/2Z_0$ and $Q_L = \omega_r L/(R + 2Z_0)$. With the aid of these expressions and Eq. (9–14), the insertion and transducer loss (in dB) for the single-stage bandpass filter becomes

$$L_I = L_T = 10 \log\left[\frac{1 + (Q_L \epsilon_f)^2}{\left(1 - \dfrac{Q_L}{Q_U}\right)^2}\right]$$

or   (9-15)

$$L_I = L_T = 10 \log\left(1 - \frac{Q_L}{Q_U}\right)^{-2} + 10 \log[1 + (Q_L \epsilon_f)^2]$$

where use has been made of Eq. (9–4).

The first term represents the loss at resonance since with $\epsilon_f = 0$ the second term of the equation is zero. Note that when Q_U is infinite ($R = 0$), there is no loss at resonance. The second term of the loss expression represents the frequency sensitive portion which, of course, produces the bandpass filter characteristic. It is important to note that for high-quality resonators ($Q_U \gg Q_L$) the second term represents a *mismatch loss*. That is, the filter characteristic is produced by reflections rather than dissipation. A sketch of the loss versus frequency characteristic is shown in Fig.

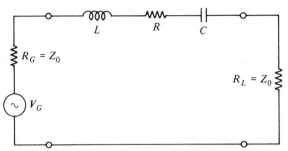

Figure 9–2 Series representation of a single-stage bandpass filter.

9–3. For convenience, both f/f_r and ϵ_f scales are included. Note that in this case f_1 and f_2 are defined as the frequencies at which the loss is 3 dB *greater* than the loss at resonance. The 3 dB bandwidth and Q_L are related by Eq. (9–5).

Also shown in Fig. 9–3 is the insertion phase versus frequency characteristic of the filter. Insertion phase is defined in Eq. (4–54). For the single-stage bandpass filter,

$$\theta_I = \arctan\left(\frac{\omega L - 1/\omega C}{R + 2Z_0}\right) = \arctan(Q_L\epsilon_f) \qquad (9–16)$$

Note that, as defined, a positive value of θ_I indicates phase *delay* while a negative value denotes phase *advance*.

The loss and phase equations could have been derived using the ABCD matrix method (Sec. 4-3a). The next case is analyzed in this manner.

The single-stage bandstop filter.   A bandstop filter can be constructed with either a shunt-connected series resonant circuit or a series-connected parallel resonant circuit. The latter arrangement is shown in Fig. 9–4. Again using the assumption that $R_G = R_L = Z_0$, the normalized ABCD matrix becomes

$$\begin{bmatrix} \bar{A} & \bar{B} \\ \bar{C} & \bar{D} \end{bmatrix} = \begin{bmatrix} 1 & 1/\bar{Y} \\ 0 & 1 \end{bmatrix}$$

where $\bar{Y} \equiv Y/Y_0$ and $Y = G + j(\omega C - 1/\omega L) = G(1 + jQ_U\epsilon_f)$. The insertion and transducer loss is given by

$$L_I = L_T = 10 \log\left|1 + \frac{Y_0}{2Y}\right|^2 \qquad (9–17)$$

For the connection shown in the figure, the resonant circuit is loaded by an external admittance of value $(2Z_0)^{-1} = 0.5Y_0$. Therefore, $Q_E = \omega_r C/0.5Y_0$ and $Q_L = \omega_r C/(G + 0.5Y_0)$. Making use of these expressions, the following loss equation can be derived for a single-stage bandstop filter.

$$L_I = L_T = 10 \log\left[\frac{1 + (Q_L\epsilon_f)^2}{\left(\dfrac{Q_L}{Q_U}\right)^2 + (Q_L\epsilon_f)^2}\right] = 10 \log\left[\left(\frac{Q_U}{Q_L}\right)^2 \frac{1 + (Q_L\epsilon_f)^2}{1 + (Q_U\epsilon_f)^2}\right] \qquad (9–18)$$

A sketch of the loss versus frequency characteristic is shown in Fig. 9–5. At resonance, the loss equals $10 \log (Q_U/Q_L)^2$. For infinite unloaded Q, the loss

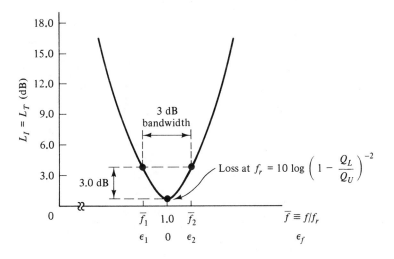

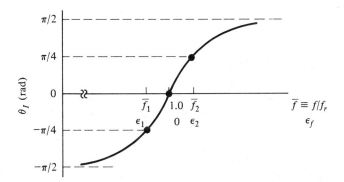

Figure 9–3 Loss and phase characteristics of a single-stage bandpass filter.

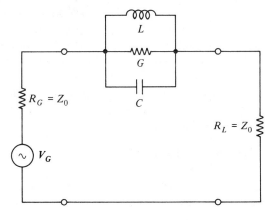

Figure 9–4 Series representation of a single-stage bandstop filter.

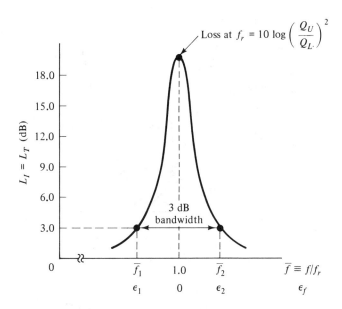

Figure 9–5   Loss characteristic of a single-stage bandstop filter.

becomes infinite. This is because the impedance of a dissipationless parallel resonant circuit is infinite at resonance. For $Q_U \gg Q_L$, Eq. (9–5) gives the relationship between the 3 dB bandwidth and Q_L, where the 3 dB bandwidth is defined in Fig. 9–5.

The insertion phase of the single-stage bandstop filter is given by

$$\theta_I = \arctan Q_L \epsilon_f - \arctan Q_U \epsilon_f \qquad (9–19)$$

The expressions derived in this section are useful in approximating the loss and phase characteristics of bandpass and bandstop filters that utilize microwave resonators.

9–2 PRINCIPLES OF MICROWAVE RESONATORS

This section describes the basic principles of microwave resonators. In most cases, the resonator consists of two reflective boundaries separated by a section of low-loss transmission line. The line may be coaxial, stripline, hollow waveguide, or any other guided transmission structure. This point of view was alluded to at the end of Sec. 3–2 where multiple reflections due to a mismatched load and generator created an oscillatory effect. The characteristics of the resonators discussed in this section are derived from transmission-line theory and the relationships in Sec. 9–1. These characteristics may also be deduced from an analysis of the electromagnetic field in the resonator. This approach is used in Sec. 9–3 to determine the resonant frequencies and unloaded Q of certain waveguide resonators.

9–2a Open and Shorted TEM Lines as Resonators

In Sec. 3–4 it was shown that for certain values of βl, the input characteristics of shorted and open-circuited transmission lines are very similar to those of ordinary series and parallel resonant circuits. For example, Fig. 3–14 shows that in the vicinity of $\beta l = n\pi$ ($n = 1, 2, 3, \ldots$), the input impedance of a shorted line approximates that of a series L-C circuit at resonance. At frequencies where the length of the shorted line is an odd-multiple of a quarter wavelength (that is, $\beta l = (2n - 1)\,\pi/2$), its input characteristics approximate that of a parallel L-C circuit at resonance. Similar comments apply to an open-circuited line (see Fig. 3–16) or, in fact, any reactively terminated line. Unlike simple L-C circuits, the input impedance and admittance of transmission-line sections have an infinite number of poles and zeros. To account for this fact, the equivalent circuit representation becomes more complex. For example, Fig. 9–6 shows the equivalent circuit for a shorted transmission-line section. The L_0 element provides the dc impedance zero, while the L_n-C_n series circuits provide the impedance zeros at $\beta l = n\pi$. In accordance with Foster's reactance theorem, the circuit has alternate poles and zeros of impedance as a function of frequency. The impedance poles (admittance zeros) are associated with the frequencies at which $\beta l = (2n - 1)\,\pi/2$. A more complete discussion of this circuit representation is given in Secs. 11.13 and 11.14 of Ref. 9–1. In this section we will focus on the circuit characteristics near the *primary* resonant frequency (the narrowband case). Hence the simple L-C circuit models of Fig. 9–1 will be used to describe the transmission-line resonators.

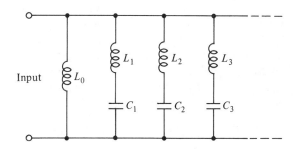

Figure 9–6 Equivalent circuit of a shorted transmission-line section (lossless case). L_0 represents the dc inductance of the shorted line. The series L-C circuits account for the impedance zeros that occur when $\beta l = n\pi$ ($n = 1, 2, 3, \ldots$).

Unloaded Q of transmission-line resonators.   Figure 9–7 shows a short-circuited transmission line connected to a microwave generator. Its input impedance is given by Eq. (3–81) with $Z_L = 0$ and $d = l$. That is, $Z_{\text{in}} = Z_{01} \tanh (\alpha l + j\beta l)$. For low-loss lines ($\alpha l < 0.1$ Np), $\tanh \alpha l \approx \alpha l$ and therefore the input impedance may be approximated by

$$Z_{\text{in}} \approx \frac{Z_{01}\alpha l(1 + \tan^2 \beta l) + jZ_{01}\tan \beta l}{1 + (\alpha l \tan \beta l)^2} \tag{9–20}$$

In the vicinity of $\beta l = n\pi$, the impedance behavior is similar to that of a series resonant circuit. For this condition, $\tan^2 \beta l \ll 1$ and the input impedance reduces to

$$Z_{\text{in}} \approx Z_{01}\,\alpha l + jZ_{01} \tan \beta l \tag{9–21}$$

An expression for the unloaded Q of the shorted-line resonator can be obtained from this equation. With $R_G = 0$, Eq. (9–8) becomes the *unloaded Q* of a series resonant circuit, where R represents its input impedance at resonance. For the shorted line at series resonance, $\beta l = n\pi$ and therefore the input impedance is simply $Z_{01}\,\alpha l$. Thus,

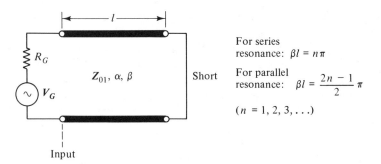

Figure 9–7 A short-circuited transmission-line resonator.

$$Q_U = \frac{\omega_r}{2Z_{01}\,\alpha l}\,\frac{d(Z_{01}\tan\beta l)}{d\omega}\bigg|_{\omega=\omega_r} \qquad (9\text{–}22)$$

where $\omega_r = n\pi v/l$. Since $\beta = \omega/v$ and it is assumed that Z_{01} and v are independent of frequency (a valid assumption for TEM lines), the above expression reduces to

$$Q_U = \frac{\beta}{2\alpha} = \frac{\pi}{\alpha\lambda_r} \qquad (9\text{–}23)$$

where α and β are the attenuation and phase constants of the transmission line at resonance and λ_r is the resonant wavelength. This expression is equally valid for open-circuited resonators (Prob. 9–3). For example, if an air-insulated TEM line with $\alpha = 0.005$ Np/m is used to construct a resonator with $f_r = 2$ GHz ($\lambda_r = 0.15$ m), one can expect an unloaded Q of about 4000, a fairly good value at microwave frequencies. This calculation is slightly optimistic since it does not include the losses associated with an imperfect short or open.

Q_E and Q_L limitation of direct-coupled resonators. Most microwave generators are matched to the characteristic impedance of their output line. For instance, commercially available coaxial generators usually employ a 50 ohm output line and hence $R_G = 50$ ohms. It will now be shown that when a shorted or open line resonator is connected directly to such a source, its external Q and hence its loaded Q is quite small. This means that narrowband resonant circuits cannot be realized by such an arrangement.

Consider the shorted line shown in Fig. 9–7. Its input admittance is $Y_{01}\coth(\alpha l + j\beta l)$. At frequencies in the vicinity of $\beta l = (2n-1)\pi/2$, its input characteristics are that of a parallel resonant circuit. For the case of a lossless line,

$$Y_{in} = -jY_{01}\cot\beta l \qquad (9\text{–}24)$$

An expression for the external Q may be obtained from Eq. (9–9) by setting $G = 0$. With $G_G = 1/R_G$ and $Y_{01} = 1/Z_{01}$, the result is

$$Q_E = \frac{\omega_r l}{2v}\frac{R_G}{Z_{01}} = \frac{(2n-1)\pi}{4}\frac{R_G}{Z_{01}} \qquad (9\text{–}25)$$

For high-quality resonators ($Q_U \gg Q_E$), this expression also represents the loaded Q

of the parallel resonance. Suppose now a quarter-wave shorted line ($n = 1$) with $Z_{01} = 40$ ohms is connected to a 50 ohm source. This would result in a Q_E (and Q_L) value of about unity. Since many applications required Q_L values in excess of ten (that is, 3 dB bandwidth $< 10\%$), it is apparent that connecting a shorted- or open-line resonator directly to the generator will not do the job!

From Eq. (9–25), it would appear that high Q_E values can only be achieved by decreasing Z_{01}, or increasing R_G and n. None of these alternatives are practical. A substantial decrease in Z_{01} increases line losses (see Sec. 5–2) which, in turn, degrades the unloaded Q of the circuit. R_G, on the other hand, cannot be increased substantially since for a matched generator this requires that the characteristic impedance of its output line also be increased. For microwave transmission lines, Z_0 values much in excess of 150 ohms are impractical. Finally, increasing n is also impractical since the physical length of the resonator becomes large and the associated line losses increase. For example, if $n = 100$, the circuit is 199 times longer than a quarter-wavelength resonator. A further and more significant disadvantage is that the circuit will have spurious responses very close to the desired resonant frequency. With $n = 100$, for instance, the two nearest ones ($n = 99$ and 101) are only one percent from the desired response. The close proximity of these spurious responses is unacceptable in practically all applications. Similar comments apply to open-circuited transmission-line resonators.

Another method of increasing Q_E for the shorted-line resonator is to connect many of them in parallel. For m such resonators, Q_E is m times the value obtained from Eq. (9–25). This is because the circuit susceptance B and hence $dB/d\omega$ is multiplied by m. This approach is illustrated in Fig. 9–8a for $m = 6$. Part b shows the limiting case for large m. The resultant hollow cylindrical enclosure is one example of a whole class of microwave structures known as *cavity resonators*. A discussion of their properties is given in Sec. 9–2b and in Sec. 9–3.

There are many techniques for controlling the external Q of a resonator without changing the generator impedance. The most widely used involves reactive coupling between the generator and the resonator. Two examples are analyzed here.

The inductively coupled shorted-line resonator.   Figure 9–9 shows a shorted line connected to a matched generator via a shunt inductance L. The following analysis shows that by a suitable choice of L, one can realize any desired value of Q_E and Q_L. Since the shorted line and the inductance are connected in parallel, the total input susceptance is

$$B_{in} = -\frac{1}{\omega L} - Y_{01} \cot \beta l = -B_L - Y_{01} \cot \beta l$$

From Eq. (9–9) with $G = 0$ and $G_G = 1/R_G = Y_0$,

$$Q_E = -\frac{\omega_r}{2Y_0} \frac{d}{d\omega}\left(\frac{1}{\omega L} + Y_{01} \cot \frac{\omega l}{v}\right)\Bigg|_{\omega = \omega_r} \tag{9–26}$$

where for parallel resonance, ω_r is defined as the frequency at which $B_{in} = 0$. That is,

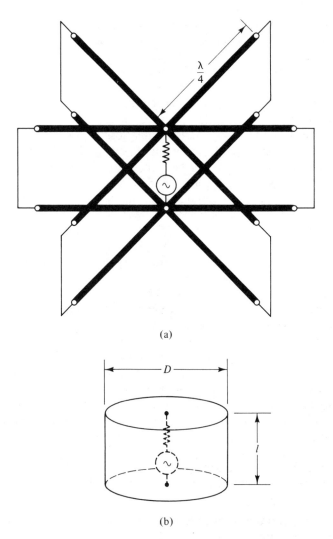

(a)

(b)

Figure 9–8 The cylindrical cavity resonator viewed as the parallel connection of shorted quarter-wave line sections. An exact field analysis (Sec. 9–3b) shows that for the resultant TM_{010} mode, the resonant wavelength $\lambda_r = 1.306\, D$.

$$\frac{1}{\omega_r L} = -Y_{01} \cot \frac{\omega_r l}{v} \qquad (9\text{--}27)$$

Performing the differentiation in Eq. (9–26) and using the above resonance condition yields

$$Q_E = \frac{Z_0}{2Z_{01}} \left[\bar{B}_{L_r} + \phi_r (1 + \bar{B}_{L_r}^2) \right] \qquad (9\text{--}28)$$

where $B_{L_r} \equiv 1/\omega_r L$, $\bar{B}_{L_r} \equiv B_{L_r}/Y_{01}$ and $\phi_r \equiv \omega_r l/v = 2\pi l/\lambda_r$.

The derivation of the above expression makes use of the trigonometric identity $\csc^2 \beta l = 1 + \cot^2 \beta l$ and assumes that Z_{01} and v are independent of frequency. The

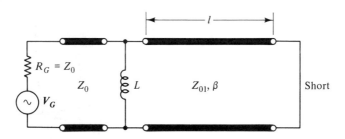

Figure 9–9   An inductively coupled shorted-line resonator.

resonance condition of Eq. (9–27) may be rewritten as

$$\phi_r = \frac{2\pi l}{\lambda_r} = \arctan\left(\frac{-1}{\overline{B}_{L_r}}\right) = \pi - \arctan\left(\frac{1}{\overline{B}_{L_r}}\right) \qquad (9\text{--}29)$$

Equations (9–28) and (9–29) represent the design relations for the inductively coupled, shorted resonator. The following example problem illustrates the calculation.

Example 9–1:

An inductively coupled resonator circuit (Fig. 9–9) is connected to a 50 ohm matched generator. Determine the values of L and l required for a 3 dB bandwidth of 60 MHz centered at 3000 MHz. Assume an air-insulated TEM line with $Z_{01} = 75$ ohms.

Solution: From Eq. (9–5), $Q_L = 3000/60 = 50$. Assuming $Q_U \gg Q_L$, Q_E is also 50. For a high-Q circuit, $\overline{B}_{L_r}$ is large and hence $\phi_r \approx \pi$. With this assumption, Eq. (9–28) yields

$$\overline{B}_{L_r}^2 + 0.318\overline{B}_{L_r} - 46.7 = 0 \qquad \text{or} \qquad \overline{B}_{L_r} = 6.68$$

Therefore, $B_{L_r} = 6.68/75$ and $L = 1/\omega_r B_{L_r} = 0.596$ nH. From Eq. (9–29),

$$\phi_r = \pi - \arctan(1/6.68) = 2.99 \text{ rad}$$

or $l = (2.99)10/2\pi = 4.759\,\text{cm}$

Note that in calculating $\overline{B}_{L_r}$, the assumption that $\phi_r \approx \pi$ results in only a 2 percent error.

This example shows that B_{L_r} can be chosen to produce any required value of Q_E (and hence Q_L). Once B_{L_r} has been determined, the exact length of the shorted-line section may be obtained from the resonance equation. Note that the circuit produces a parallel resonant condition since the input admittance is zero at ω_r. A series resonant characteristic can be obtained by simply adding a quarter-wave line at the input to the resonator.

Figure 9–10 shows the graphical solution of Eq. (9–27). The intersections of the B_L curve and the susceptance curve for the shorted line represent resonant conditions. Solutions exist in the vicinity of ϕ_r equal to π, 2π, etc. Since $\phi_r \approx \pi$ is the desired resonance, the first spurious response is nearly an octave away in frequency. The graphical solution also shows why a large value of B_L produces a large value of Q_E as indicated by Eq. (9–28). From Eq. (9–9), it is clear that a large susceptance

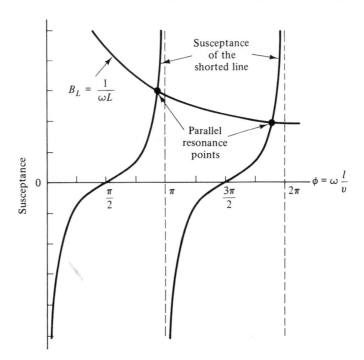

Figure 9–10 Graphical solution of Eq. (9–27), the resonance condition for the inductively coupled shorted-line resonator.

slope ($dB/d\omega$) at the resonant frequency is the key to high Q. By increasing B_L, the resonant point moves toward $\phi = \pi$ where the susceptance slope is very steep. This, in turn, results in high Q_E values. Thus B_L may be chosen to yield any desired value of Q_E.

It should be noted that the resonant condition [Eq. (9–29)] is a function of coupling (that is, B_L). As a result, any change in coupling will cause a slight shift in the resonant frequency. This is true for many resonator configurations.

The capacitively coupled open-line resonator.   It is often convenient in strip transmission-line systems to use an open-circuited line as a resonator. The center conductor pattern and its equivalent circuit is shown in Fig. 9–11. The open-line resonator is coupled to the matched generator via a series capacitance C. The derivation of Q_E and the resonant condition is similar to that of the inductively coupled shorted line. In this case, the total input reactance is

$$X_{\text{in}} = -\frac{1}{\omega C} - Z_{01} \cot \beta l = -X_C - Z_{01} \cot \frac{\omega l}{v}$$

Series resonance occurs when $X_{\text{in}} = 0$ and hence the resonant condition is given by

$$\phi_r \equiv \frac{2\pi l}{\lambda_r} = \arctan\left(\frac{-1}{\overline{X}_{C_r}}\right) = \pi - \arctan\left(\frac{1}{\overline{X}_{C_r}}\right) \tag{9–30}$$

where $X_{C_r} \equiv 1/\omega_r C$, $\overline{X}_{C_r} \equiv X_{C_r}/Z_{01}$, and λ_r is the resonant wavelength. The

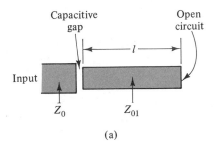

(a)

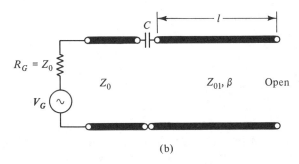

(b)

Figure 9–11   A capacitively coupled open-circuited stripline resonator and its equivalent circuit.

external Q may be derived from Eq. (9–8) with $R = 0$. For a matched generator, $R_G = Z_0$ and therefore

$$Q_E = \frac{Z_{01}}{2Z_0}[\bar{X}_{C_r} + \phi_r(1 + \bar{X}_{C_r}^2)] \tag{9–31}$$

where $\phi_r \approx \pi$ for high values of Q_E ($\bar{X}_{C_r} \gg 1$). The capacitance value and hence X_{C_r} may be chosen to obtain any desired value of Q_E. Note that this configuration produces a series resonance since the input impedance is zero at ω_r.

The tapped half-wave resonator.   A high-Q parallel resonant circuit can be realized by properly tapping into a half-wave, transmission-line resonator. Figure 9–12 illustrates the center conductor pattern for a strip transmission-line version that is open circuited at both ends. The following analysis verifies that the resonator is a half wavelength long at the resonant frequency ($l = \lambda_r/2$). Other resonances occur at frequencies where l is a multiple of a half wavelength.

The input line, denoted by Z_0, is connected directly to the resonator as shown. The total input susceptance of the two shunt-connected, open-circuited line sections is

$$B_{in} = Y_{01} \tan \frac{\omega l_1}{v} + Y_{01} \tan \frac{\omega(l - l_1)}{v} \tag{9–32}$$

The condition for parallel resonance is $B_{in} = 0$. This requires that $\omega_r(l - l_1)/v = n\pi - \omega_r l_1/v$ or $l = n\lambda_r/2$, the primary resonance being with $n = 1$. Thus the resonant frequency is determined by the *overall length* of the open-circuited resonator. Equation (9–9) with $G = 0$ may be used to derive its external Q. By assuming a matched generator ($G_G = Y_0$), the result is

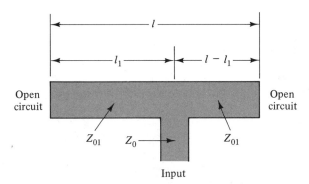

Note: $l = \dfrac{\lambda_r}{2}$ and therefore $f_r = \dfrac{c}{2l\sqrt{\mu_R \epsilon_R}}$

Figure 9–12 The center conductor pattern for a tapped open-circuited half-wave stripline resonator.

$$Q_E = \frac{\pi}{2} \frac{Z_0}{Z_{01}} \sec^2\left(\frac{2\pi l_1}{\lambda_r}\right) \qquad (9\text{--}33)$$

The derivation is left to the reader (Prob. 9–7). Note that Q_E (and hence Q_L) may be controlled by l_1, the position at which the input line is connected to the resonator. For example, if $Z_0 = 50$ ohms, $Z_{01} = 25$ ohms, and $l_1 = 0.30\lambda_r$, $Q_E \approx 33$. Larger values of Q_E can be realized by choosing l_1 closer to $0.25\lambda_r$. In practice, however, values much in excess of 100 are impractical since the unloaded Q of the resonator deteriorates rapidly as l_1 approaches a quarter wavelength. This is because line losses cannot be neglected when l_1 (and $l - l_1$) is nearly $0.25\lambda_r$. The difficulty is similar to that described in Ex. 3–5 of Sec. 3–6b.

In addition to having a zero of input admittance at f_r, this resonator also has *impedance* zeros at frequencies above and below f_r. Assuming l_1 is the longer of the two lines, the lower frequency one occurs when $l_1 = \lambda/4$ while the higher one occurs when $l - l_1 = \lambda/4$. These impedance zeros are useful in improving the loss characteristics of narrowband filters (see Sec. 9–4).

Other transmission-line resonators.   The resonators described thus far represent but a few of the many possible configurations utilized at microwave frequencies. In the majority of cases, the resonator structure consists of two reflective boundaries separated by a length of low-loss transmission line.[2] For the inductively coupled, shorted-line resonator (Fig. 9–9), the reflections are created by the shunt inductance and the short circuit, while for the one shown in Fig. 9–11, they are created by the series capacitance and the open circuit. In the case of the resonator described in Fig. 9–12, the reflections are due to the open circuits at each end of the Z_{01} line. This same principle applies to the three coaxial resonators shown in Fig. 9–13. The one in part a consists of a coaxial line that is shorted at both ends. It exhibits resonant behavior at frequencies where l is a multiple of a half wavelength. The electric field pattern for $n = 1$ is as indicated. It resembles the displacement pattern of a vibrating string pinned at its end points. As in the case of the vibrating

---

[2] Optical resonators also make use of this principle. One example is the Fabry-Perot resonator (see Sec. 7.5 of Ref. 9–2).

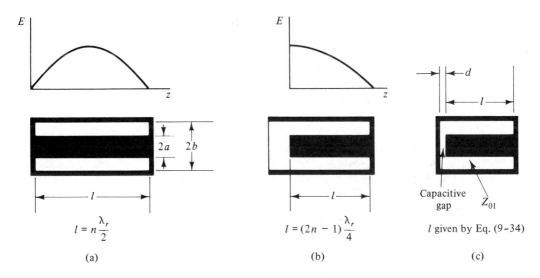

Figure 9–13   Three coaxial resonator configurations.

string, the coaxial resonator has other natural frequencies corresponding to $n = 2$, $3, \ldots$. For the TEM field pattern, the resonant frequencies of the higher modes are harmonically related to that of the fundamental mode ($n = 1$). Unlike the vibrating string, however, the coaxial resonator has additional resonant frequencies that are not necessarily harmonics of the fundamental. These are associated with higher-order mode propagation along the coaxial line, one example being the TE_{11} mode described in Sec. 5–2. Because of the cutoff effect, their resonant frequencies are *not* harmonically related, even though the resonance condition still requires that l be a multiple of a half wavelength (Prob. 9–10).

The coaxial resonator illustrated in Fig. 9–13b is open circuited at one end and shorted at the other end. Resonance occurs at frequencies where the line length is an odd multiple of a quarter wavelength. As before, resonant behavior is produced by two large reflections separated by a length of low-loss line. A sketch of the electric field pattern for the primary resonance is shown in the figure. For this resonator, the first spurious response occurs at *three* times the primary resonant frequency.

The configuration described in Fig. 9–13c is known as a *foreshortened coaxial-line resonator* and finds use in microwave oscillators and amplifiers. By making the spacing d very small, the electric field in the gap can be made quite large, which results in strong interaction between an electron beam passing through the gap and the microwave field (see Sec. 5.2 of Ref. 9–5). Parallel resonance occurs when the total circuit susceptance is zero. Thus the resonant equation is

$$\omega_r C = Y_{01} \cot \frac{\omega_r l}{v} \tag{9–34}$$

where C is the capacitance of the gap region.

The external and loaded Q of any resonator depends upon the degree of coupling to the source and load. For the methods described in Figs. 9–9, 9–11, and 9–12, the value of Q_E can be computed from Eqs. (9–28), (9–31), and (9–33), respectively. In many other cases, the coupling is adjusted experimentally to give the desired Q_E value. For instance, probes and loops are commonly used to couple into and out of both coaxial and waveguide resonators (see Fig. 9–18). The degree of coupling is controlled by varying their dimensions, position, and orientation.

9-2b Cavity Resonators

Hollow metallic enclosures exhibit resonance behavior when excited by an electromagnetic field. These enclosures are known as *cavity resonators*[3] and are widely used at frequencies above 3 GHz. The energy stored by the resonator is associated with the electromagnetic field within the volume of the cavity. The quality of these resonators can be quite high at microwave frequencies with typical values of unloaded Q ranging from 5,000 to 50,000. As with the transmission-line resonators described earlier, cavity resonators have an infinite number of resonant modes. These can be determined by solving Maxwell's equations and applying the boundary conditions appropriate to the particular configuration. This approach is used in the next section to analyze rectangular and cylindrical resonators. In this section, the same structures are studied from a transmission-line point of view.

The rectangular cavity.   A rectangular cavity resonator may be viewed as a rectangular waveguide shorted at both ends. This is illustrated in Fig. 9–14 for a guide of height b, width a, and length l. The cavity is resonant at frequencies where l is a multiple of $\lambda_g/2$, λ_g being the guide wavelength for propagation along the z axis. To verify this, consider the rectangular guide to be shorted at $z = l$ but *not* at $z = 0$. A signal entering at $z = 0$ produces a standing wave pattern. Because of the short, the transverse electric field (E_x and E_y) is zero at $z = l$ and at multiples of a half wavelength along the z axis. At certain frequencies an electric field null occurs at $z = 0$. Two such cases are shown in Fig. 9–15. Since the transverse electric field is zero at the $z = 0$ plane, the introduction of a shorting plate at that point does not

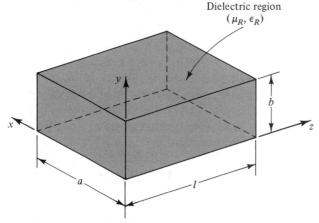

Dielectric region
(μ_R, ϵ_R)

Figure 9–14   A rectangular cavity resonator.

---

[3] The cavity resonator was originally proposed by W. W. Hansen (Ref. 9–17).

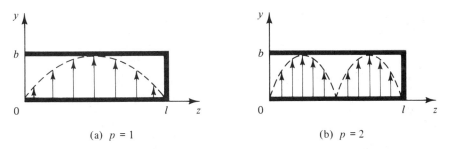

(a) $p = 1$ (b) $p = 2$

Figure 9–15   Electric standing-wave patterns for a shorted rectangular waveguide.

disturb the standing wave pattern. Thus at selected frequencies, the electromagnetic field within the cavity can be sustained even in the absence of a signal source. The resonant condition is therefore given by

$$l = p\frac{\lambda_g}{2} \tag{9–35}$$

where $p = 1, 2, 3, \ldots$. With p being any positive integer, the cavity has an infinite number of resonances.[4] Actually, the number is triply infinite since the standing-wave argument can be applied to any of the TE_{mn} and TM_{mn} modes associated with rectangular waveguide transmission.

A standing wave is the result of two oppositely directed traveling waves. Therefore, the fields in the cavity may be viewed as a z-directed TE or TM wave that is reflected back and forth between the $z = 0$ and $z = l$ boundaries. For $\beta l = p\pi$, the waves add constructively thus producing the resonant condition. Of course, one could equally well argue that the reflecting surfaces are at $x = 0$ and $x = a$ or at $y = 0$ and $y = b$, and that the wave propagates in the x or y direction. All three views lead to the same conclusions. By convention, the x-y plane is usually defined as the transverse plane. Hence the resonant modes of the cavity are denoted by TE_{mnp} and TM_{mnp} where TE_{mn} and TM_{mn} indicate waveguide modes with the x-y plane being transverse. As explained in Sec. 5–4a, m represents the number of half-sine variations in the x direction and n the number of half-sine variations in the y direction. For example, a rectangular cavity operating in the TE_{102} mode may be viewed as a TE_{10} standing wave pattern having *two* half-wave variations in the z direction.

Because of the cutoff effect, the guide wavelength is given by

$$\lambda_g = \frac{\lambda}{\sqrt{1 - \left(\dfrac{\lambda}{\lambda_c}\right)^2}} \tag{9–36}$$

where λ_c is the cutoff wavelength of the guide and λ is the operating wavelength. This equation was derived in Sec. 5–5a. Substitution into Eq. (9–35) results in the following condition for resonance

---

[4] Note that the integer p, rather than n, is used to indicate the number of half wavelengths in the z direction. This is because n is used as one of the subscripts to denote the particular waveguide mode.

$$l = \frac{p}{2} \frac{\lambda_r}{\sqrt{1 - \left(\frac{\lambda_r}{\lambda_c}\right)^2}} \tag{9-37}$$

where λ_r is the resonant wavelength. Solving for λ_r yields

$$\lambda_r = \frac{1}{\sqrt{\left(\frac{1}{\lambda_c}\right)^2 + \left(\frac{p}{2l}\right)^2}} \tag{9-38}$$

Since $f_r \lambda_r = c/\sqrt{\mu_R \epsilon_R}$, the resonant frequencies for the various cavity modes are given by

$$f_r = \frac{c}{\sqrt{\mu_R \epsilon_R}} \sqrt{\left(\frac{1}{\lambda_c}\right)^2 + \left(\frac{p}{2l}\right)^2} \tag{9-39}$$

where μ_R and ϵ_R are the material constants of the dielectric filling the cavity. For rectangular waveguide, λ_c is given by Eq. (5–67) and thus the resonant equation for a rectangular cavity becomes

$$\lambda_r = \frac{1}{\sqrt{\left(\frac{m}{2a}\right)^2 + \left(\frac{n}{2b}\right)^2 + \left(\frac{p}{2l}\right)^2}} \tag{9-40}$$

where a, b, and l are defined in Fig. 9–14. In general, the resonant frequencies are different for the various modes. The one having the *lowest* frequency is usually called the *primary* mode. For a rectangular cavity with b as the smallest dimension, it is the TE_{101} mode. Note that for a given set of mode numbers, the resonant wavelength (and frequency) for TE and TM cavity modes are identical. Modes having the same resonant frequency are said to be *degenerate*.

Electromagnetic field patterns for some of the TE and TM rectangular cavity modes are given in Sec. 9–3 (Figs. 9–22 and 9–23). The fields represent the stored energy within the cavity. Some energy is also dissipated by the cavity. This loss is associated with the finite conductivity of the metallic walls and the dielectric loss of the insulating region. For an air-filled cavity, the dielectric loss is negligible. In general, Q_U is proportional to the volume-to-surface area ratio. This is because the energy is stored volumetrically, while power loss is mainly due to the conduction currents on the inner cavity walls. Q_U values in excess of 10,000 are easily realized in well-designed rectangular cavities. In practice, the inner cavity walls are usually silver plated to lower surface resistivity and polished to reduce surface roughness.

In order to determine the unloaded Q for a particular mode, the stored energy and power loss must be calculated from the field equations. This procedure is described in Sec. 9–3. The result for a rectangular cavity operating in the TE_{10p} mode is given in Eqs. (9–63) and (9–64). Equation (9–67) can be used to account for dielectric losses.

The cylindrical cavity. A circular waveguide shorted at both ends is widely used as a cavity resonator. The configuration is shown in Fig. 9–16 where D is the diameter of the cylindrical cavity and l is its length. As before, resonance occurs at

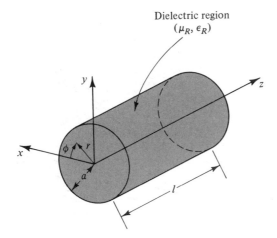

Figure 9–16 A cylindrical cavity resonator. (Diameter $D = 2a$.)

frequencies where l is a multiple of $\lambda_g/2$. Since circular guide is capable of propagating TE_{nm} and TM_{nm} waves, the cavity modes are denoted by TE_{nmp} and TM_{nmp}, where p represents the number of half wavelengths in the z direction. Field patterns for some of the cavity modes are shown in Figs. 9–24 and 9–25. The resonant wavelength for a particular mode is given by Eq. (9–38). As explained in Sec. 5–5b, the cutoff wavelength in circular waveguide is proportional to the guide diameter. Letting $\lambda_c = K_{nm}D$, the resonant wavelength for a cylindrical cavity becomes

$$\lambda_r = \frac{1}{\sqrt{\left(\dfrac{1}{K_{nm}D}\right)^2 + \left(\dfrac{p}{2l}\right)^2}} \tag{9–41}$$

The resonant frequency is calculated from $f_r\lambda_r = c/\sqrt{\mu_R\,\epsilon_R}$. Values of K_{nm} for several TE and TM modes are given in Table 5–3 and repeated in Table 9–1.

TABLE 9–1   Values of K_{nm} for Several TE and TM Modes in Circular Guide ($\lambda_c = K_{nm}D$)

K_{nm} for TE Modes			K_{nm} for TM Modes		
n \\ m	1	2	n \\ m	1	2
0	0.820	0.448	0	1.306	0.569
1	1.706	0.589	1	0.820	0.448
2	1.029	0.468	2	0.612	0.373

Figure 9–17 shows the relationship between resonant frequency and the diameter-to-length ratio for several modes. It is known as a *cavity mode chart* and finds use in resonator design. Note that except for TM modes with $p = 0$, increasing the cavity length l causes the resonant frequency to decrease. Thus an adjustable length cylindrical cavity provides a simple way of realizing a tunable resonant circuit

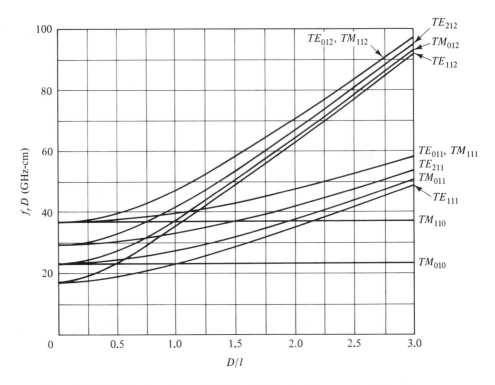

Figure 9–17 Mode chart for an air-filled cylindrical cavity resonator. (Note: These mode characteristics all become straight lines if one plots $[f_r D]^2$ versus $[D/l]^2$)

at microwaves. The mode chart is useful in determining the frequency range over which a particular resonance can be tuned without interference from other modes (Prob. 9–14).

TM modes with $p = 0$ are interesting in that the resonant condition is independent of cavity length. These modes are not easily described in terms of two reflections separated by a length of transmission line. In the case of the TM_{010} mode, the cavity may be viewed as a large number of shorted two-wire lines in parallel. This concept was illustrated earlier in Fig. 9–8.

The TE_{01p} set of cylindrical cavity modes are of considerable engineering interest. For an air-filled cavity, the unloaded Q of these modes can be quite high, typically 20,000 to 60,000. This can be understood with the aid of Eq. (9–23), which indicates that the key to high unloaded Q is a low-attenuation transmission line between the two reflecting surfaces.[5] Since the TE_{01} mode in circular guide has extremely low loss (see Fig. 5–28), one might expect that its use would result in a very high-quality resonator. The field analysis in the next section supports this expectation. Expressions for the unloaded Q of an air-filled cylindrical cavity operating in the TE_{01p} mode are given by Eqs. (9–83) and (9–84). Note that for a given D/l

---

[5] The calculation of Q_U from Eq. (9–23) produces an optimistic value since the losses due to the end plates are neglected.

ratio, Q_U decreases with increasing frequency since R_s is proportional to $\sqrt{f_r}$. Problem 9–13 gives an indication of realizable Q values at microwave frequencies.

The length of a cylindrical cavity, and hence its resonant frequency, can be made variable by simply replacing one of its end walls with a movable shorting plate. Due to the absence of *axial* conduction currents, the TE_{01p} modes are easily tunable. This is because no current flows between the cylindrical wall and the movable end wall, and hence the problems usually associated with an erratic ohmic contact are avoided.

Coupling to cavity resonators.   The discussion thus far has dealt only with *uncoupled* cavity resonators. In order to excite a particular mode, the cavity must be properly coupled to an external source. Four examples of cavity coupling are illustrated in Fig. 9–18. Part *a* shows a cylindrical cavity that is loop-coupled to a coaxial line. The conduction current in the loop produces a linking magnetic field as indicated. Thus the loop is capable of exciting any cavity mode that has a portion of its magnetic field linking the loop. For the cylindrical cavity shown, this includes all the TM_{nmp} modes. Part *b* illustrates probe-type coupling between a coaxial line and a cavity. The electric field pattern created by the probe is similar to that of the coaxial-to-waveguide transition described in Fig. 7–3. Any cavity mode having an electric field component parallel to the probe can be excited by this coupling arrangement. For the cylindrical cavity shown, this includes all the TE_{nmp} modes except those with $n = 0$. With the probe centrally located along the l dimension (as indicated in the figure), only TE modes with p an odd integer can be excited since those with p even have an electric field null at the plane of the probe. To produce TE modes with p even, the probe must be offset along the l dimension. When the probe is inserted into one of the end walls of the cavity, a z component of electric field is created, which results in TM modes being excited. These same methods can be used to couple into rectangular cavities as well as the coaxial resonators described in Fig. 9–13.

Parts *c* and *d* of Fig. 9–18 illustrate aperture coupling between a cylindrical cavity and a rectangular waveguide operating in the dominant TE_{10} mode. As explained in Sec. 8–3a, a magnetic field component that is parallel to the long dimension of the slot will be coupled through the aperture. For the slot orientation shown in part *c*, the TE_{10} magnetic field couples into the cavity as shown, which allows any of the TM resonant modes to be excited. This coupling arrangement is often used to generate the TM_{01p} modes. When the waveguide and aperture are centrally located along the l dimension, p must be even since, for odd values, the magnetic field is zero at that point in the cavity. The coupling arrangement in part *d* of Fig. 9–18 is similar to that in part *c* except that the rectangular guide and aperture have been rotated 90°. This allows the TE modes to be excited in the cavity, the TE_{11p} and the TE_{01p} being the most commonly used. Shown in the figure is the electric field pattern for the TE_{01p} modes. The process by which the TE_{10} electric field is converted to a TE cavity mode is similar to that described in Fig. 7–6.

Although many modes can be excited by the coupling mechanisms described here, only those having a resonant frequency equal to or nearly equal to the source

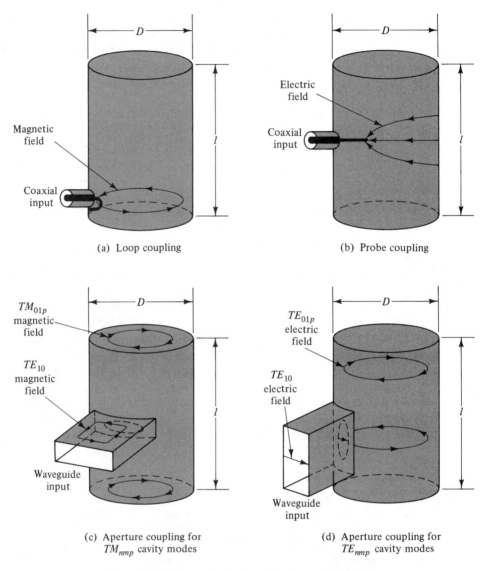

(a) Loop coupling   (b) Probe coupling

(c) Aperture coupling for   (d) Aperture coupling for
   TM_{nmp} cavity modes TE_{nmp} cavity modes

Figure 9–18 Four methods of coupling to a cavity resonator.

frequency will be excited to any appreciable degree. For degenerate modes (that is, those with identical resonant frequencies), the coupling must be arranged to favor the desired mode while minimizing the excitation of the unwanted one. In most cases, this is easily achieved since a particular coupling method usually excites either TE or TM modes, but not both.

These same methods may be used to couple the cavity to an external load. For example, the addition of a coupling loop directly opposite that shown in Fig. 9–18a results in a two-port cavity with bandpass filter characteristics. The filter will also have spurious responses that are associated with other cavity modes. The mode chart is useful in determining the frequencies of the nearest ones.

For the techniques described in Fig. 9-18, the degree of coupling is usually adjusted experimentally to produce the desired value of Q_E. One method of cavity coupling that is amenable to transmission-line analysis is the inductively coupled rectangular cavity shown in Fig. 9-19. The cavity is connected to the source via a rectangular waveguide (Z_0) operating in the TE_{10} mode. Energy is coupled to the cavity through a shunt inductive iris. Although the figure shows a circular aperture of the type described in Fig. 7-15b, any shunt inductive obstacle may be used. The characteristic impedance of the rectangular guide that forms the cavity is denoted by Z_{01}. Since the propagating mode of the input waveguide is TE_{10}, the symmetrical iris is capable of exciting any of the TE_{mnp} modes with m odd and n zero or even. However, only the TE_{10p} modes will be considered here. The equivalent circuit for the waveguide configuration in Fig. 9-19 is the same as that given in Fig. 9-9. Therefore, the derivation of the resonant condition and Q_E is similar to that described earlier. Letting B_L denote the inductive susceptance of the coupling iris, resonance occurs when

$$\phi_r \equiv \frac{2\pi l}{\lambda_{g_r}} = p\pi - \arctan\left(\frac{1}{\bar{B}_{L_r}}\right) \qquad (9\text{-}42)$$

where $\bar{B}_{L_r} \equiv B_{L_r}/Y_{01}$ and B_{L_r} denotes the inductive susceptance at the resonant frequency. With two exceptions, this equation is the same as Eq. (9-29). First of all, it has been generalized to any integer value of p. Second, λ_r has been replaced by λ_{g_r}, the TE_{10} guide wavelength at resonance for the Z_{01} waveguide section. Its value can be computed from Eq. (9-36), where $\lambda_c = 2a_1$ and $\lambda = \lambda_r = c/f_r\sqrt{\mu_R \epsilon_R}$. Note that because of the inductive coupling, resonance does *not* occur when $l = p\lambda_g/2$. This condition is approached only when $\bar{B}_{L_r}$ is large, which is the case when the coupling aperture is very small.

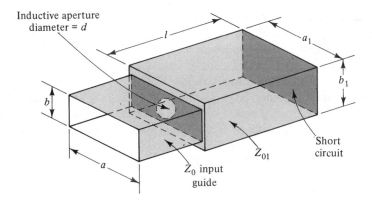

Figure 9-19   An inductively coupled rectangular cavity.

In waveguide systems, it is common practice to define Q in terms of reciprocal guide wavelength rather than ω. Thus for parallel resonance, Eq. (9-9) is replaced by

$$Q_L = \frac{1/\lambda_g}{2(G_G + G)} \frac{dB}{d(1/\lambda_g)}\bigg|_{\lambda_g = \lambda_{g_r}} \qquad (9\text{-}43)$$

For $G = 0$, the above expression represents the external Q.[6] Using this definition and the fact that for a waveguide inductive iris, B_L is proportional to λ_g, one can derive an expression for the external Q of the inductively coupled rectangular cavity (Fig. 9–19). The result is *exactly* the same as Eq. (9–28), where in this case $\phi_r = 2\pi l/\lambda_{g_r}$ and Z_0 and Z_{01} are defined by Eq. (5–43). Note that when $a = a_1$, $Z_0/Z_{01} = b/b_1$.

Equations (9–28) and (9–42) represent the design relations for an inductively coupled rectangular cavity connected to a matched waveguide source. For a given pair of 3 dB points, Q_L may be computed from Eq. (9–91) when the guide dimensions are known. In many applications, $Q_U \gg Q_L$ and therefore $Q_L \approx Q_E$. Thus for a given cavity mode, $\bar{B}_{L_r}$ may be computed from Eq. (9–28) by assuming $\phi_r \approx p\pi$. Once $\bar{B}_{L_r}$ is known, the exact length of the cavity can be determined from Eq. (9–42). When both guide sections are the same (that is, $a_1 = a$ and $b_1 = b$), the dimension of the inductive iris may be calculated from the appropriate equation in Fig. 7–15.

In the vicinity of the resonant frequency, the input admittance at the plane of the inductive aperture approximates that of a parallel resonant circuit. From Eq. (9–11), the input admittance at resonance ($\epsilon_f = 0$) is real. That is, $Y_{in} = (G_G + G)Q_L/Q_U = G_G Q_E/Q_U$. For a matched generator, $G_G = Y_0$ and therefore

$$\bar{Y}_{in} \equiv \frac{Y_{in}}{Y_0} = \frac{Q_E}{Q_U} \tag{9–44}$$

If the aperture size is adjusted so that $Q_E = Q_U$, then $\bar{Y}_{in} = 1.0$ and the cavity is matched to the input line. This condition is known as *critical coupling* and results in all the incident power being absorbed by the cavity. For smaller apertures, $\bar{B}_L$ and hence Q_E is larger. This situation, namely, $Q_E > Q_U$ and $\bar{Y}_{in} > 1$, is known as the *undercoupled* case. Conversely for larger apertures, $Q_E < Q_U$ and $\bar{Y}_{in} < 1$, which represents the case of an *overcoupled* cavity.

It is important to understand that the representation of a cavity resonator by a single R-L-C equivalent circuit is only valid at frequencies near the resonant frequency of the particular cavity mode. Unlike the R-L-C model, the cavity has a multiplicity of admittance poles and zeroes. Therefore, an analysis of its behavior well away from resonance must include the effect of the nearest poles and zeros. Furthermore, for a given coupling arrangement, the degree of coupling for each mode is different. The reader is referred to Sec. 11.14 of Ref. 9–1 and Sec. 7.4 of Ref. 9–2 for further discussions on these points.

Other microwave resonators. In addition to the resonators discussed thus far, there are many other ways of realizing a resonant circuit at microwave frequencies. Some are described in Refs. 9–4, 9–6, and 9–18. Two of these are illustrated in Fig. 9–20. The one in part a is known as a *dielectric resonator* and consists of a

---

[6] When using this definition of Q, the frequency variable ϵ_f defined in Eq. (9–12) is replaced by ϵ_λ which is defined in Eq. (9–90). Furthermore, the relationship between Q_L and bandwidth is given by Eq. (9–91).

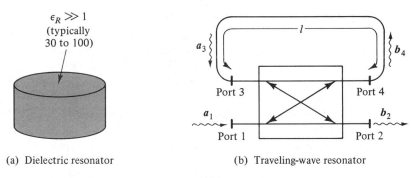

(a) Dielectric resonator   (b) Traveling-wave resonator

Figure 9–20   Two additional types of microwave resonators.

solid, low-loss, dielectric disc having a large value of ϵ_R, typically 30 to 100. The resonance behavior is associated with the large reflections at the air-dielectric interfaces. A field analysis of this configuration is given in Refs. 9–18 and 9–19. Dielectric resonators find use in bandpass and bandstop filters (Refs. 9–19 and 9–20). In principle, the resonator losses are due solely to the finite loss tangent of the dielectric. With most of the fields confined to the dielectric region, the unloaded Q of the resonator is simply the reciprocal of the dielectric loss tangent. In practice, however, skin-effect losses associated with the input and output coupling structures can be comparable to the dielectric losses. This is particularly true for high-quality dielectric materials.

The coupler arrangement shown in part b of Fig. 9–20 is known as a *traveling-wave resonator*. It consists of a high-directivity directional coupler in which the coupled port (port 4) is connected to the decoupled port (port 3) by a length l of low-loss line. A portion of the input signal at port 1 couples to port 4 and then proceeds around the ring to port 3 where it combines with more of the coupled signal. A steady-state analysis of this feedback circuit (Ref. 9–4) shows that the wave entering the ring is given by

$$b_4 = \frac{jk_c}{1 - \sqrt{1 - k_c^2}\, e^{-(A_t + j\phi)}}\, a_1 \qquad (9\text{--}45)$$

where k_c is the coupling coefficient of the directional coupler. A_t and ϕ are respectively the total attenuation and phase delay around the ring. Resonance occurs when $\phi = 2n\pi$, where $n = 1, 2, 3, \ldots$. At resonance, the fields in the ring can be considerably larger than those in the main line.[7] For this reason, the traveling-wave resonator is sometimes used to test the peak-power capability of microwave circuits.

Other phenomena can also be utilized to create microwave resonant circuits. For example, the ferrimagnetic resonance effect in single-crystal YIG spheres is widely used in the design of electronically tunable resonators (Refs. 9–21 and 9–22). The resonant frequency is controlled by the applied magnetic field. Typical values of Q_U range from 1000 to 6000.

---

[7] This is true for any resonant circuit. For instance, the field strength at certain points in a cavity is much greater than its input value. See Sec. 9–6b and Eq. (9–141).

★ 9–3 FIELD ANALYSIS OF CAVITY RESONATORS

In this section, rectangular and cylindrical cavity resonators are analyzed using Maxwell's equations. Expressions for the resonant frequency and the electromagnetic field of the various modes are developed. These are used to derive equations for the unloaded Q of certain modes.

In developing the field equations, it is assumed that the cavity walls are perfect conductors and the region inside the cavity is a perfect insulator. Therefore conduction currents exist only on the inner wall surfaces. The phasor form of Maxwell's equations is used since sinusoidal excitation is assumed. The resultant vector wave equations for a charge-free region are given by Eqs. (5–54) and (5–55) and are repeated here.

$$\nabla^2 \vec{E} + \omega^2 \mu \epsilon \vec{E} = 0 \qquad (9\text{–}46)$$

$$\nabla^2 \vec{H} + \omega^2 \mu \epsilon \vec{H} = 0 \qquad (9\text{–}47)$$

where $\mu \equiv \mu_R \mu_0$ and $\epsilon \equiv \epsilon_R \epsilon_0$ define the electric and magnetic properties of the insulating region.

9–3a The Rectangular Cavity Resonator

Figure 9–14 describes a rectangular cavity of width a, height b, and length l. For the rectangular coordinate system shown, Eqs. (9–46) and (9–47) become

$$\frac{\partial^2 \vec{E}}{\partial x^2} + \frac{\partial^2 \vec{E}}{\partial y^2} + \frac{\partial^2 \vec{E}}{\partial z^2} = -\omega^2 \mu \epsilon \vec{E} \qquad (9\text{–}48)$$

and

$$\frac{\partial^2 \vec{H}}{\partial x^2} + \frac{\partial^2 \vec{H}}{\partial y^2} + \frac{\partial^2 \vec{H}}{\partial z^2} = -\omega^2 \mu \epsilon \vec{H} \qquad (9\text{–}49)$$

As is the case for waveguide transmission, cavity modes can also be divided into two sets, namely, TE and TM.

TE$_{mnp}$ modes ($E_z \equiv 0$, $H_z \neq 0$). By equating the z components of Eq. (9–49), the following phasor equation is obtained.

$$\frac{\partial^2 H_z}{\partial x^2} + \frac{\partial^2 H_z}{\partial y^2} + \frac{\partial^2 H_z}{\partial z^2} = -\omega^2 \mu \epsilon H_z \qquad (9\text{–}50)$$

This equation can be solved by a separation of variables technique similar to that described in Sec. 5–5. Let

$$H_z = XYZ \qquad (9\text{–}51)$$

where X, Y, and Z are functions of only x, y, and z, respectively. Therefore,

$$\frac{1}{X}\frac{d^2 X}{dx^2} + \frac{1}{Y}\frac{d^2 Y}{dy^2} + \frac{1}{Z}\frac{d^2 Z}{dz^2} = -\omega^2 \mu \epsilon$$

Equating each term to a constant yields

$$\frac{1}{X}\frac{d^2 X}{dx^2} = -k_x^2 \qquad \frac{1}{Y}\frac{d^2 Y}{dy^2} = -k_y^2 \qquad \text{and} \qquad \frac{1}{Z}\frac{d^2 Z}{dz^2} = -k_z^2$$

where $k_x^2 + k_y^2 + k_z^2 = \omega^2 \mu\epsilon$.

The solution to these equations can be written as the sum of sine and cosine terms. For example,

$$X = A \cos k_x x + B \sin k_x x$$

The expressions for Y and Z are similar. The boundary conditions at a conducting surface are given by Eqs. (2–42) and (2–43). For the six cavity walls, they are

$$\frac{\partial H_z}{\partial x} = 0 \text{ at } x = 0 \text{ and } x = a, \qquad \frac{\partial H_z}{\partial y} = 0 \text{ at } y = 0 \text{ and } y = b,$$

and $H_z = 0$ at $z = 0$ and $z = l$.

The application of these six conditions to Eq. (9–51) results in

$$H_z = H_0 \cos\frac{m\pi}{a}x \cos\frac{n\pi}{b}y \sin\frac{p\pi}{l}z \qquad (9\text{–}52)$$

where $k_x = m\pi/a$, $k_y = n\pi/b$, $k_z = p\pi/l$, and m, n, and p are positive integers. H_0 is a constant that is proportional to the signal level. Substituting the above expression for H_z into Eq. (9–50) yields

$$\left\{ \left(\frac{m\pi}{a}\right)^2 + \left(\frac{n\pi}{b}\right)^2 + \left(\frac{p\pi}{l}\right)^2 - \omega^2 \mu\epsilon \right\} H_z = 0$$

If $H_z = 0$, all field components are zero and the solution is trivial. In order for the field to be non-zero, the following condition must hold

$$\left(\frac{m\pi}{a}\right)^2 + \left(\frac{n\pi}{b}\right)^2 + \left(\frac{p\pi}{l}\right)^2 = \omega_r^2 \mu\epsilon = \left(\frac{2\pi}{\lambda_r}\right)^2$$

where $\omega_r = 2\pi f_r$ is the resonant frequency and λ_r the resonant wavelength of the rectangular cavity. Solving for λ_r and f_r,

$$\lambda_r = \frac{1}{\sqrt{\left(\dfrac{m}{2a}\right)^2 + \left(\dfrac{n}{2b}\right)^2 + \left(\dfrac{p}{2l}\right)^2}} \qquad (9\text{–}53)$$

and

$$f_r = \frac{c}{\sqrt{\mu_R \epsilon_R}} \sqrt{\left(\frac{m}{2a}\right)^2 + \left(\frac{n}{2b}\right)^2 + \left(\frac{p}{2l}\right)^2} \qquad (9\text{–}54)$$

Thus, an electromagnetic field can exist in the lossless cavity only when the excitation frequency is exactly f_r. This is clearly a resonance behavior. Since m, n, and p can take on all integer values, the cavity has an infinite number of resonant modes each with its own resonant frequency. The one with the *lowest* resonant frequency is known as the *primary* mode. For a rectangular cavity, it is the TE_{101} mode, where b is assumed to be the smallest of the three dimensions.

As explained in the previous section, the standing wave pattern in the cavity may be viewed as a pair of equal amplitude TE or TM waves propagating in the plus and minus z directions. This can be seen by rewriting Eq. (9–52) as follows:

$$H_z = H_0 \cos k_x x \cos k_y y \left\{ \frac{e^{jk_z z} - e^{-jk_z z}}{2j} \right\} \tag{9–55}$$

The first exponential term represents a wave traveling in the minus z direction with a phase constant k_z, while the second term represents a plus z-directed wave. Using this form of H_z and the relations in Eq. (5–62), the remaining field components may be derived. For example,

$$E_x = j\omega_r \mu \frac{k_y}{k_c^2} H_0 \cos k_x x \sin k_y y \left\{ \frac{e^{jk_z z} - e^{-jk_z z}}{2j} \right\}$$

$$= j\omega_r \mu \frac{k_y}{k_c^2} H_0 \cos k_x x \sin k_y y \sin k_z z$$

Expressions for E_y, H_x and H_y may be similarly derived.[8] Thus, the field components for the TE_{mnp} cavity modes become

$$E_x = j\omega_r \mu \frac{k_y}{k_c^2} H_0 \cos k_x x \sin k_y y \sin k_z z$$

$$E_y = -j\omega_r \mu \frac{k_x}{k_c^2} H_0 \sin k_x x \cos k_y y \sin k_z z$$

$$E_z = 0 \tag{9–56}$$

$$H_x = -\frac{k_x k_z}{k_c^2} H_0 \sin k_x x \cos k_y y \cos k_z z$$

$$H_y = -\frac{k_y k_z}{k_c^2} H_0 \cos k_x x \sin k_y y \cos k_z z$$

$$H_z = H_0 \cos k_x x \cos k_y y \sin k_z z$$

where $k_x = m\pi/a$, $k_y = n\pi/b$, $k_z = p\pi/l$, and $k_c^2 = k_x^2 + k_y^2$. These equations can be used to sketch the field patterns of the various cavity modes. For example, the electric and magnetic field for the TE_{101} mode at four instants in time is shown in Fig. 9–21. The time-dependent form of the field components are obtained from Eqs. (9–56) by reinserting the $\sqrt{2}e^{j\omega_r t}$ factor and taking the real part thereof. For $m = 1$, $n = 0$, and $p = 1$,

$$\mathscr{E}_y = 2\sqrt{2} \frac{a}{\lambda_r} \eta H_0 \sin \frac{\pi}{a} x \sin \frac{\pi}{l} z \sin \omega_r t$$

---

[8] In deriving H_x and H_y, note the double signs in Eq. (5–62). Use the upper ones for the plus z-directed wave and the lower ones for the minus z-directed wave.

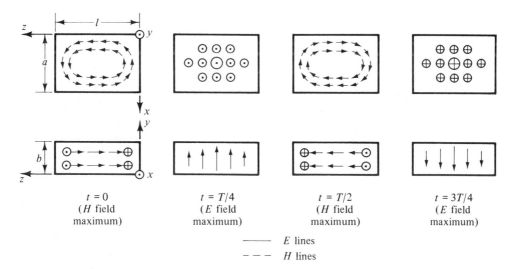

Figure 9–21 The electromagnetic field pattern for the TE_{101} mode in a rectangular cavity at four instants in time. (T = period of the ac signal.)

$$\mathcal{H}_x = -\sqrt{2}\, H_0 \frac{a}{l} \sin \frac{\pi}{a} x \cos \frac{\pi}{l} z \cos \omega_r t$$

$$\mathcal{H}_z = \sqrt{2}\, H_0 \cos \frac{\pi}{a} x \sin \frac{\pi}{l} z \cos \omega_r t \qquad\qquad (9\text{–}57)$$

$$\mathcal{E}_x = 0, \ \mathcal{E}_z = 0, \text{ and } \mathcal{H}_y = 0$$

where $\eta = \sqrt{\mu/\epsilon} = 377\sqrt{\mu_R/\epsilon_R}$ ohms. Two views of the electromagnetic field at $t = 0, T/4, T/2,$ and $3T/4$ are shown in the figure. Note that the field components have no variations in the y direction and one half-wave variation in the x and z directions since $n = 0$ and $m = p = 1$. Generalizing, the subscripts m, n, and p of the TE_{mnp} mode represent the half-wave periodicity of the field in the x, y, and z directions, respectively.

Referring to Fig. 9–21 and Eq. (9–57), it is apparent that the electric and magnetic fields are 90° out-of-phase. When the magnetic field is maximum, the electric field is zero and vice versa. This means that no real power flow occurs, the cavity merely stores the energy associated with the field. As is the case with an ordinary L-C circuit, the stored energy changes from magnetic to electric and back to magnetic every half cycle.

The field patterns for various TE cavity modes are shown in Fig. 9–22. Expressions for their resonant wavelength and frequency are given by Eqs. (9–53) and (9–54). The mode numbering is for the coordinate system shown in the figure, and therefore an interchange of axes would require a corresponding change in mode numbers.

TM$_{mnp}$ modes ($H_z \equiv 0$, $E_z \neq 0$). By equating the z components of Eq. (9–48), the following phasor equation is obtained.

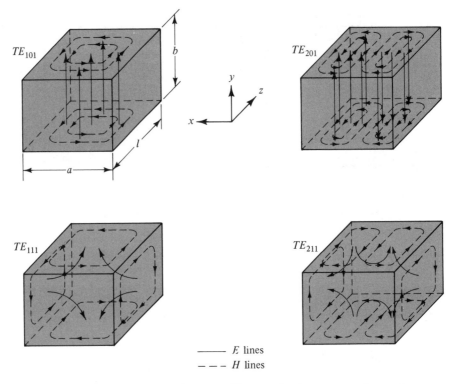

Figure 9–22 Field patterns for some TE_{mnp} modes in a rectangular cavity. (Note: The E and H fields are 90° out-of-phase.)

$$\frac{\partial^2 \boldsymbol{E}_z}{\partial x^2} + \frac{\partial^2 \boldsymbol{E}_z}{\partial y^2} + \frac{\partial^2 \boldsymbol{E}_z}{\partial z^2} = -\omega^2 \mu\epsilon \boldsymbol{E}_z \qquad (9\text{–}58)$$

Again, using the separation of variables technique results in

$$\boldsymbol{E}_z = E_0 \sin\frac{m\pi}{a}x \, \sin\frac{n\pi}{b}y \, \cos\frac{p\pi}{l}z \qquad (9\text{–}59)$$

where, as before, m, n, and p represent the half-wave periodicity of the field in the x, y, and z directions, respectively. The boundary conditions used to solve for $\boldsymbol{E}_z$ are

$$\boldsymbol{E}_z = 0 \qquad \text{at} \qquad x = 0,\, x = a,\, y = 0,\, \text{and } y = b$$

and

$$\frac{\partial \boldsymbol{E}_z}{\partial z} = 0 \qquad \text{at} \qquad z = 0 \text{ and } z = l$$

Substituting Eq. (9–59) into Eq. (9–58) leads to the same resonance condition as for the TE modes. Thus Eqs. (9–53) and (9–54) are also valid for the TM modes in a rectangular cavity. The remaining field components are obtained by substituting the expression for $\boldsymbol{E}_z$ into Eq. (5–84). For the TM_{mnp} cavity modes, they are

$$E_x = -\frac{k_x k_z}{k_c^2} E_0 \cos k_x x \sin k_y y \sin k_z z$$

$$E_y = -\frac{k_y k_z}{k_c^2} E_0 \sin k_x x \cos k_y y \sin k_z z$$

(9–60)

$$E_z = E_0 \sin k_x x \sin k_y y \cos k_z z$$

$$H_x = j\omega_r \epsilon \frac{k_y}{k_c^2} E_0 \sin k_x x \cos k_y y \cos k_z z$$

$$H_y = -j\omega_r \epsilon \frac{k_x}{k_c^2} E_0 \cos k_x x \sin k_y y \cos k_z z$$

$$H_z = 0$$

where $k_x = m\pi/a$, $k_y = n\pi/b$, $k_z = p\pi/l$, and $k_c^2 = k_x^2 + k_y^2$.

The field patterns for the TM cavity modes may be deduced from the above relations. Some are pictured in Fig. 9–23. Note that if the coordinate axes are interchanged ($x \to z$, $z \to y$, and $y \to x$), the TM_{110} mode pattern is exactly the same as the TE_{101}. On the other hand, the TM_{111} mode cannot be viewed as a TE mode since all three components of the electric field exist.

It should be emphasized that the expressions for λ_r and f_r of the TM modes are the same as the TE modes. Thus for a given set of mode numbers, the TE and TM modes are *degenerate*, that is, they resonate at the same frequency. In order for a rectangular cavity mode to exist, at least two of its integer values must be non-zero. For a given set of m, n, and p values, the existence of a mode may be determined by substituting the values into the expressions for the field components. For instance, TM_{mn0} modes can exist in the cavity, but TE_{mn0} modes cannot!

Unloaded Q of the TE_{10p} cavity mode.   The preceding analysis assumed perfectly conducting cavity walls and a perfect insulator within. This, of course, leads to a resonant device that is dissipationless. The fact of the matter is that all resonant circuits have some loss. A measure of the circuit's quality is its unloaded Q (Q_U) and is defined in Eq. (9–3). This definition can be used to determine the unloaded Q for any mode of the cavity. To illustrate the method, an expression for Q_U will now be derived for the TE_{10p} mode in a rectangular guide.

Although the energy stored in a cavity is continually shifting from electric to magnetic and back, the total energy is a constant. Its value may be determined by simply evaluating the electric energy at the instant when $\mathscr{E}$ is maximum and $\mathscr{H}$ is zero. From Eq. (2–7),

$$U_E = U = \frac{1}{2} \int_{\text{vol}} \vec{D} \cdot \vec{E} \, dV = \frac{\epsilon}{2} \int_0^l \int_0^a \int_0^b (\sqrt{2} E_y)^2 \, dy \, dx \, dz$$

where E_y is the rms value of E_y given in Eq. (9–56). With $m = 1$ and $n = 0$, $k_x = \pi/a$, $k_y = 0$, $k_c = k_x$, and therefore

$$U = 4\epsilon \eta^2 H_0^2 b \left(\frac{a}{\lambda_r}\right)^2 \int_0^l \int_0^a \sin^2 \frac{\pi}{a} x \sin^2 \frac{p\pi}{l} z \, dx \, dz$$

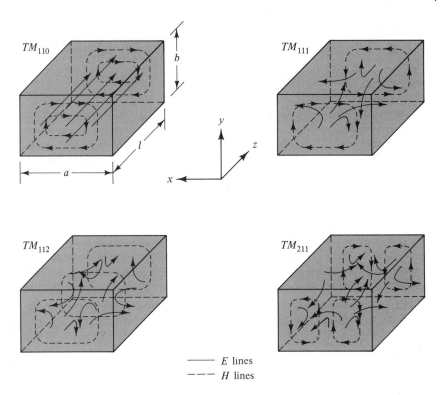

——— E lines
– – – H lines

Figure 9–23 Field patterns for some TM_{mnp} modes in a rectangular cavity. (Note: The E and H fields are 90° out-of-phase.)

or

$$U = \frac{\eta}{v}\left(\frac{a}{\lambda_r}\right)^2 H_0^2 \, abl \tag{9–61}$$

where $\eta = \sqrt{\mu/\epsilon}$, $v = 1/\sqrt{\mu\epsilon}$, and abl represents the volume of the cavity.

In order to determine the power dissipated by the cavity, it will be assumed that, despite the finite surface resistivity of the walls, Eq. (9–56) accurately describes the electromagnetic field within the cavity. This assumption is very good for all metals commonly used in the construction of cavity resonators.

The ohmic losses due to the walls of the cavity is given by

$$P_d = R_s \int_S K^2 dS = R_s \int_S H_T^2 dS$$

where R_s, the surface resistivity, is defined in Eq. (2–78) and H_T represents the rms value of the tangential magnetic field at the conducting surfaces. The integration is over the complete inner surface area S. Note that use has been made of Eq. (2–42). Integrating over three of the cavity walls and doubling the result yields the dissipated power. Namely,

$$P_d = 2R_s \left[\int_0^a \int_0^b H_{x_1}^2 \, dy \, dx + \int_0^l \int_0^b H_{z_1}^2 \, dy \, dz + \int_0^l \int_0^a (H_{x_2}^2 + H_{z_2}^2) dx \, dz \right]$$

where H_{x_1} is the rms value of H_x at the $z = 0$ wall, H_{z_1} is the rms value of H_z at the $x = 0$ wall, and the subscript 2 denotes H values at the $y = 0$ wall. Using the rms values of the field components in Eq. (9–56) for $m = 1$ and $n = 0$ yields

$$P_d = R_s H_0^2 \left[ab\left(\frac{pa}{l}\right)^2 + bl + \frac{al}{2}\left(\frac{pa}{l}\right)^2 + \frac{al}{2} \right] \qquad (9\text{–}62)$$

Since the unloaded Q is defined as $\omega_r U/P_d$ and $\omega_r/v = 2\pi/\lambda_r$, Eqs. (9–61) and (9–62) produce the following expression for Q_U of a rectangular cavity in the TE_{10p} mode.

$$Q_U = \frac{\pi\eta}{\lambda_r R_s} \frac{abl\{1 + (pa/l)^2\}}{2bl\{1 + (a/l)(pa/l)^2\} + al\{1 + (pa/l)^2\}} \qquad (9\text{–}63)$$

An alternate form is obtained by substituting λ_r from Eq. (9–53).

$$Q_U = \frac{\pi\eta}{2R_s} \frac{\{1 + (pa/l)^2\}^{3/2}}{2\{1 + (a/l)(pa/l)^2\} + (a/b)\{1 + (pa/l)^2\}} \qquad (9\text{–}64)$$

For a square cavity ($l = a$), the above expressions reduce to

$$Q_U = \frac{2\pi}{\lambda_r} \frac{\eta}{R_s}\left(\frac{\text{Volume}}{\text{Surface Area}}\right) = \frac{\pi}{2} \frac{\eta}{R_s} \frac{\sqrt{1 + p^2}}{2 + a/b} \qquad (9\text{–}65)$$

Note that for a given resonant frequency, Q_U is maximized by maximizing the volume-to-surface area of the cavity. This is generally true, since energy is stored volumetrically, while power loss is a surface effect. The optimum shape for a rectangular cavity is cubic. For a cubic cavity operating in the TE_{10p} mode,

$$Q_U = \sqrt{1 + p^2} \frac{\pi}{6} \frac{\eta}{R_s} \qquad (9\text{–}66)$$

Note that Q_U decreases with increasing frequency since R_s is proportional to $\sqrt{f_r}$. This holds true for any size or shape cavity.

In general, larger Q_U values can be achieved by using a higher-order cavity mode. The reason is that for a given resonant frequency, higher-order modes require a larger cavity size which, in turn, increases the volume-to-surface area ratio. This is verified by Eqs. (9–65) and (9–66) where larger values of p produce higher Q_U values. This approach, however, has its drawbacks since the number of modes having resonant frequencies near that of the wanted mode increases. The close proximity of these unwanted modes can, for example, lead to difficulties in tuning the resonator. The mode chart in Fig. 9–17 illustrates this point.

Equations (9–63) to (9–66) are valid for computing the unloaded Q of a rectangular cavity filled with air or a very low-loss dielectric. The actual values realized in practice are slightly lower since the roughness of the metal walls increases the effective surface area of the cavity. Silver plating to several skin depths and polishing the inner surfaces to reduce roughness is a widely used technique for improving Q_U.

On occasion, the cavity is dielectric-filled to reduce its size while maintaining the desired resonant frequency. Since all dielectrics have a finite loss tangent (tan δ), the unloaded Q is adversely affected. The power dissipated by the dielectric is $\omega_r U \tan \delta$. When combined with the conductor losses (P_d), the unloaded Q for the dielectric-filled cavity becomes

$$Q_U = \frac{\omega_r U}{P_d + \omega_r U \tan \delta} = \frac{1}{\dfrac{1}{Q_c} + \tan \delta} \qquad (9\text{–}67)$$

where Q_c is the unloaded Q *due to conductor losses alone* [namely, Eqs. (9–63) to (9–66)]. The above expression assumes that the field pattern is not appreciably altered by the presence of losses. This is a valid assumption for practically all microwave cavities.

Example 9–2:

A cubic-shaped copper cavity is required to resonate at 7500 MHz in the TE_{101} mode. Calculate its dimensions and unloaded Q if

(a) the cavity is air-filled.
(b) the cavity is dielectric-filled with $\epsilon_R = 5.0$ and tan δ = 0.0004.

Solution: For a cubic cavity operating in the TE_{101} mode,

$$f_r = \frac{c}{\sqrt{2}\, a \sqrt{\epsilon_R}} \qquad \text{or} \qquad a = \frac{c}{\sqrt{2} f_r \sqrt{\epsilon_R}}$$

where at 7500 MHz, $c/f_r = 4.0$ cm.

(a) Since $\epsilon_R = 1$, $a = 2.828$ cm.
  From Eq. (9–66) with $p = 1$,

$$Q_U = \sqrt{2}\, \frac{\pi}{6}\, \frac{\eta}{R_s} = 0.74 \frac{\eta}{R_s}$$

For air, $\eta = 377$ ohms. R_s may be calculated from Eq. (2–78). For a copper surface, $R_s = 0.023$ ohm per square at 7500 MHz and therefore $Q_U = 12{,}130$.

(b) With $\epsilon_R = 5$, $a = 1.265$ cm.
  For $\epsilon_R = 5$, $\eta = 377/\sqrt{5} = 168.6$ ohms. Since R_s is unchanged, $Q_c = 5425$.
  From Eq. (9–67) with tan δ = 0.0004,

$$Q_U = \frac{10{,}000}{1.84 + 4.0} = 1712$$

This example problem clearly shows the disadvantage of a dielectric-filled resonator. Although the size of the cavity has been reduced, the deterioration in Q_U is substantial. This is due to both the losses *in* the imperfect dielectric and the increased conductor losses due to the *presence* of the dielectric. Note that even if the dielectric were perfect, the value of Q_U would be less than half that of the air-filled cavity.

9-3b The Cylindrical Cavity Resonator

Figure 9–16 describes a circular cylindrical resonator of diameter D and length l. The field analysis that follows is similar to that of the rectangular cavity. As before, equating the z components in Eqs. (9–46) and (9–47) yields

$$\frac{1}{r}\frac{\partial}{\partial r}\left(r\frac{\partial E_z}{\partial r}\right) + \frac{1}{r^2}\frac{\partial^2 E_z}{\partial \phi^2} + \frac{\partial^2 E_z}{\partial z^2} = -\omega^2 \mu\epsilon E_z \tag{9–68}$$

and

$$\frac{1}{r}\frac{\partial}{\partial r}\left(r\frac{\partial H_z}{\partial r}\right) + \frac{1}{r^2}\frac{\partial^2 H_z}{\partial \phi^2} + \frac{\partial^2 H_z}{\partial z^2} = -\omega^2 \mu\epsilon H_z \tag{9–69}$$

where the r, ϕ, and z coordinates are shown in Fig. 9–16.

TE_{nmp} modes ($E_z \equiv 0$, $H_z \neq 0$).   Equation (9–69) can be solved by separation of variables. Let

$$H_z = R\Phi Z \tag{9–70}$$

where R, Φ, and Z are functions of only r, ϕ, and z, respectively. Therefore,

$$\frac{1}{rR}\frac{\partial}{\partial r}\left(r\frac{\partial R}{\partial r}\right) + \frac{1}{r^2\Phi}\frac{\partial^2 \Phi}{\partial \phi^2} + \frac{1}{Z}\frac{\partial^2 Z}{\partial z^2} = -\omega^2 \mu\epsilon$$

Equating the first two terms to $-k_c^2$ and the third to $-k_z^2$ results in the following equations

$$\frac{r}{R}\frac{\partial}{\partial r}\left(r\frac{\partial R}{\partial r}\right) + k_c^2 r^2 = -\frac{1}{\Phi}\frac{\partial^2 \Phi}{\partial \phi^2} \tag{9–71}$$

$$\frac{d^2 Z}{dz^2} + k_z^2 Z = 0 \tag{9–72}$$

where $k_z^2 + k_c^2 = \omega^2 \mu\epsilon$. Equating both sides of Eq. (9–71) to a constant (n^2) yields

$$\frac{d^2 \Phi}{d\phi^2} + n^2 \Phi = 0 \tag{9–73}$$

and

$$\frac{d^2 R}{dr^2} + \frac{1}{r}\frac{dR}{dr} + \left(k_c^2 - \frac{n^2}{r^2}\right)R = 0 \tag{9–74}$$

The solutions to Eqs. (9–72), (9–73), and (9–74) are

$$Z = A \cos k_z z + B \sin k_z z$$

$$\Phi = C \cos n\phi + D \sin n\phi \tag{9–75}$$

$$R = F J_n(k_c r) + G N_n(k_c r)$$

where A, B, C, D, F, and G are constants and J_n and N_n represent the nth order Bessel functions of the first and second kind. These functions are plotted in Fig. 5–31 for some integer values of n.

The resonator configuration requires that the field be finite at $r = 0$. This means that the N_n solution is unacceptable since it becomes infinite when $r = 0$. Thus G must equal zero. Also, since the field and hence Φ must be single-valued, n is restricted to integer values. Finally, one of the two Φ solutions may be discarded. Both lead to the same field patterns except that one is displaced 90° in the ϕ direction. Setting $D = 0$ leaves only the cos $n\phi$ term.[9] Substituting Eq. (9–75) into Eq. (9–70) yields

$$\boldsymbol{H_z} = CF \cos n\phi (A \cos k_z z + B \sin k_z z) J_n(k_c r)$$

The boundary conditions at $z = 0$ and $z = l$ require that $\boldsymbol{H_z} = 0$. Therefore $A = 0$ and $k_z l = p\pi$, where p is a positive integer. The $\boldsymbol{H_z}$ solution reduces to

$$\boldsymbol{H_z} = H_0 J_n(k_c r)\cos n\phi \sin k_z z \tag{9–76}$$

where $k_z = p\pi/l$ and H_0 is an arbitrary amplitude constant. At $r = a$, the boundary condition requires that $d\boldsymbol{H_z}/dr = 0$, which means $J_n'(k_c a) = 0$. The prime denotes differentiation with respect to $k_c r$. The mth root of this equation is designated by q_{nm}' and thus $q_{nm}' = k_c a$. Several root values are listed in Table 5–2.

The resonance condition is obtained by substituting Eq. (9–76) into Eq. (9–69). This yields

$$\omega_r^2 \mu\epsilon = k_c^2 + k_z^2 = \left(\frac{q_{nm}'}{a}\right)^2 + \left(\frac{p\pi}{l}\right)^2$$

Thus the resonant frequency and wavelength for TE_{nmp} modes in a cylindrical cavity are

$$f_r = \frac{c}{\sqrt{\mu_R \epsilon_R}} \sqrt{\left(\frac{1}{K_{nm}D}\right)^2 + \left(\frac{p}{2l}\right)^2} \tag{9–77}$$

and

$$\lambda_r = \frac{1}{\sqrt{\left(\frac{1}{K_{nm}D}\right)^2 + \left(\frac{p}{2l}\right)^2}} \tag{9–78}$$

where $D = 2a$ and $K_{nm} \equiv \pi/q_{nm}'$. Values of K_{nm} for some of the TE modes are given in Table 9–1.

The integer p has the same meaning as in the rectangular cavity, namely, the number of half-wave field variations in the z direction. On the other hand, n indicates the number of *full-cycle* variations in the circumferential or ϕ direction. Since the $J_n'(q_{nm}'r/a)$ function behaves like a damped sinusoid, m describes the radial varia-

---

[9] With this choice, the electric field for the dominant TE_{111} mode will, for the most part, be directed along the y axis.

tions of the field.[10] By convention, n is the first of the three subscripts used to denote a cylindrical cavity mode.

The remaining field components can be derived from the relations in Eq. (5–94) by a procedure similar to that used for the rectangular cavity. For the TE_{nmp} modes in a cylindrical cavity resonator,

$$E_r = j\omega_r\mu\frac{H_0}{k_c^2}\frac{n}{r}J_n(k_cr)\sin n\phi \sin k_zz$$

$$E_\phi = j\omega_r\mu\frac{H_0}{k_c}J_n'(k_cr)\cos n\phi \sin k_zz$$

$$E_z = 0 \tag{9–79}$$

$$H_r = \frac{k_z}{k_c}H_0J_n'(k_cr)\cos n\phi \cos k_zz$$

$$H_\phi = -\frac{k_z}{k_c^2}\frac{n}{r}H_0J_n(k_cr)\sin n\phi \cos k_zz$$

$$H_z = H_0J_n(k_cr)\cos n\phi \sin k_zz$$

where $k_z = p\pi/l$, $k_c = q_{nm}'/a$, and J_n' is the derivative of J_n with respect to k_cr. The field patterns for some of the TE cavity modes are shown in Fig. 9–24.

TM_{nmp} modes ($H_z \equiv 0$, $E_z \neq 0$).   Equation (9–68) can be solved for E_z by employing the same method used to obtain H_z for the TE modes. The boundary conditions in this case are $E_z = 0$ at $r = a$ and $\partial E_z/\partial z = 0$ at $z = 0$ and $z = l$. The remaining field components are obtained by substituting the resultant expression for E_z into Eq. (5–104). This yields the following field equations for the TM_{nmp} modes in a cylindrical cavity.

$$E_r = -\frac{k_z}{k_c}E_0J_n'(k_cr)\cos n\phi \sin k_zz$$

$$E_\phi = \frac{k_z}{k_c^2}\frac{n}{r}E_0J_n(k_cr)\sin n\phi \sin k_zz$$

$$E_z = E_0J_n(k_cr)\cos n\phi \cos k_zz \tag{9–80}$$

$$H_r = -j\omega_r\epsilon\frac{E_0}{k_c^2}\frac{n}{r}J_n(k_cr)\sin n\phi \cos k_zz$$

$$H_\phi = -j\omega_r\epsilon\frac{E_0}{k_c}J_n'(k_cr)\cos n\phi \cos k_zz$$

$$H_z = 0$$

where $k_z = p\pi/l$, $k_c = q_{nm}/a$, and q_{nm} is the mth root of J_n.

[10] Some Bessel functions are plotted in Fig. 5–31. Additional properties are described in Sec. 3.26 of Ref. 9–1 and Appendix II of Ref. 9–2.

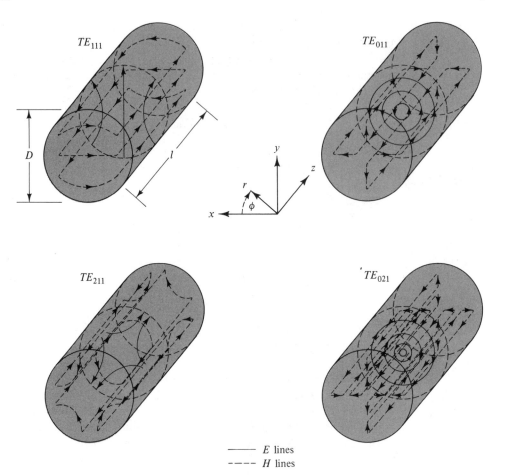

Figure 9–24 Field patterns for some TE_{nmp} modes in a cylindrical cavity. (Note: The E and H fields are 90° out-of-phase.)

The field patterns for some of the TM cavity modes are shown in Fig. 9–25. Their resonant frequencies and wavelengths are given by Eqs. (9–77) and (9–78) *except that $K_{nm} \equiv \pi/q_{nm}$*. Values of K_{nm} for some of the TM modes are given in Table 9–1. In most cases, these values are different than those of the TE modes. As a result, most TE and TM modes are *not* degenerate, an exception being the TE_{0mp} and the TM_{1mp} modes. This follows from the fact that $J_0'(x) = -J_1(x)$. For $D/l \leq 0.99$, the primary mode in a cylindrical cavity is the TE_{111}; otherwise it's the TM_{010}. The TE_{111} mode is analogous to the TE_{101} mode in a rectangular cavity. There are other mode similarities. For instance, the TM_{01p} modes are analogous to the TM_{11p} rectangular cavity modes.

Figure 9–17 illustrates the functional relationship between the resonant frequency and the D/l ratio of the cylindrical cavity. It is known as a *mode chart* and is discussed in Sec. 9–2b. Note that for TM modes with $p = 0$, the resonant frequency is independent of cavity length. TE modes with $p = 0$ do not exist.

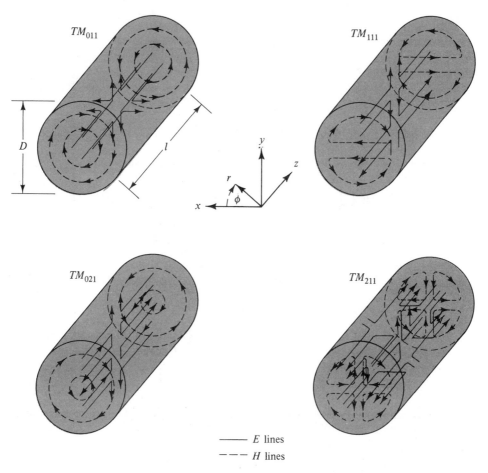

Figure 9–25   Field patterns for some TM_{nmp} modes in a cylindrical cavity. (Note: The E and H fields are 90° out-of-phase.)

Unloaded Q of the TE_{01p} cavity modes.   As discussed in Sec. 9–2b, the TE_{01p} modes in a cylindrical cavity are of considerable engineering interest because of their ease of tuning and high Q_U values. An expression for Q_U will now be derived using the field expressions in Eq. (9–79). The analysis proceeds in essentially the same manner as that used for the TE_{10p} rectangular cavity modes.

With $n = 0$ and $m = 1$, $E = E_\phi$ and its peak value is given by

$$E_\phi = \sqrt{2}\ H_0 \frac{\omega_r \mu}{k_c} J_0'(k_c r)\sin\frac{p\pi}{l}z$$

where $k_c = 3.832/a$. Therefore,

$$U = \epsilon\left(\frac{\omega_r \mu}{k_c}\right)^2 H_0^2 \int_0^a \int_0^l \int_0^{2\pi} \left\{J_0'(k_c r)\sin\frac{p\pi}{l}z\right\}^2 r\, d\phi\, dz\, dr\, \Bigg]$$

or

$$U = \frac{\pi}{2} \frac{\eta}{v} H_0^2 \left(\frac{2\pi}{\lambda_r}\right)^2 \frac{a^2 l}{k_c^2} \{-J_0(q'_{01}) J_2(q'_{01})\} \tag{9-81}$$

where $k_c a = q'_{01} = 3.832$ and $J_1(q'_{01}) = 0$. To solve the above integral, use has been made of the following relations:

$$J_0'(u) = -J_1(u) \qquad \text{and} \qquad \int u J_1^2(u)\, du = \frac{u^2}{2}\{J_1^2(u) - J_0(u) J_2(u)\}$$

The power dissipated by the cylindrical wall and the two end walls is given by

$$P_d = H_0^2 R_s \left[\int_0^l \int_0^{2\pi} \{J_0(k_c a)\sin k_z z\}^2 a\, d\phi\, dz + 2\int_0^a \int_0^{2\pi} \left\{\frac{k_z}{k_c} J_0'(k_c r)\right\}^2 r\, d\phi\, dr\right]$$

where $k_z = p\pi/l$ and the second integral is evaluated at the $z = 0$ end wall. Again using the above relations,

$$P_d = \pi H_0^2 R_s \left[a l J_0^2(q'_{01}) + 2a^2 \left(\frac{p\pi}{l} \frac{a}{q'_{01}}\right)^2 \{-J_0(q'_{01}) J_2(q'_{01})\}\right] \tag{9-82}$$

Since $Q_U \equiv \omega_r U/P_d$ and $\omega_r/v = 2\pi/\lambda_r$,

$$Q_U = \frac{\dfrac{\eta}{2R_s} \dfrac{a^2 l}{k_c^2} \left(\dfrac{2\pi}{\lambda_r}\right)^3}{a l + 2a^2 \left(\dfrac{p\pi}{l} \dfrac{a}{q'_{01}}\right)^2}$$

where use has been made of the fact that $J_2(q'_{01}) = -J_0(q'_{01})$. Using the resonance condition of Eq. (9–78) yields

$$Q_U = \frac{\eta}{2R_s} \frac{\left\{(q'_{01})^2 + \left(\dfrac{p\pi D}{2l}\right)^2\right\}^{3/2}}{(q'_{01})^2 + \dfrac{D}{l}\left(\dfrac{p\pi D}{2l}\right)^2} \tag{9-83}$$

where the cylinder diameter $D = 2a$ and $q'_{01} = 3.832$. An alternate form is

$$Q_U = \frac{\lambda_r}{2\pi\delta_s} \frac{\left\{(q'_{01})^2 + \left(\dfrac{p\pi D}{2l}\right)^2\right\}^{3/2}}{(q'_{01})^2 + \dfrac{D}{l}\left(\dfrac{p\pi D}{2l}\right)^2} \tag{9-84}$$

where δ_s is the skin depth at f_r. Note that like the rectangular cavity, Q_U decreases with increasing frequency since R_s is proportional to $\sqrt{f_r}$. Larger Q_U values may be obtained by using higher values of p. Finally, when the cavity is dielectric-filled, the overall unloaded Q is obtained from Eq. (9–67) where Q_c is the value calculated from either of the above expressions for Q_U.

There are many other forms of microwave resonators. These include radial, conical, and spherical shaped cavities. Some are described in Refs. (9–1) and (9–6).

Cavity perturbations.   On occasion, it is necessary to determine the shift in resonant frequency due to a small perturbation in a cavity having a known field distribution. For example, Fig. 9–26 shows a cylindrical cavity with a small indentation of volume ΔV in the top wall. Slater (Ref. 9–7) has developed the following expression for the change in resonant frequency Δf_r due to a volume perturbation ΔV.

$$\frac{\Delta f_r}{f_r} = \frac{\int_{\Delta V}(\mu H^2 - \epsilon E^2)dV}{\int_V(\mu H^2 + \epsilon E^2)dV} = \frac{\int_{\Delta V}(\mu H^2 - \epsilon E^2)dV}{4U} \tag{9–85}$$

This expression is based on the fact that, in general, a small perturbation in a cavity wall will affect one type of energy more than the other. As a result, the resonant frequency must shift so as to again equalize the electric and magnetic energies. For instance, if the cylindrical cavity in Fig. 9–26 is operated in the TM_{010} mode and perturbed as shown, the electric energy will decrease. Since the magnetic field at that point is negligible, the change in stored energy is

$$-\frac{\epsilon}{4}\int_{\Delta V}(E_{\text{peak}})^2\,dV \approx -\frac{\epsilon(\sqrt{2}E_0)^2}{4}\Delta V \tag{9–86}$$

since from Eq. (9–80) with $n = 0$ and $p = 0$, $E_z = E_0$ along the $r = 0$ axis. The total energy in the unperturbed cavity may be computed when the electric energy is maximum and the magnetic energy is zero. Thus for the TM_{010} mode,

$$U = \frac{\epsilon}{2}\int_0^a\int_0^l\int_0^{2\pi}\{\sqrt{2}\,E_0 J_0(k_c r)\}^2\,r\,d\phi\,dz\,dr = 2\pi l\epsilon E_0^2\int_0^a r\,J_0^2(k_c r)dr$$

or

$$U = \pi a^2 l\epsilon E_0^2 J_1^2(q_{01}) = V\epsilon E_0^2 J_1^2(q_{01}) \tag{9–87}$$

where $\pi a^2 l = V$, the volume of the unperturbed cavity. Therefore, the fractional change in resonant frequency for the TM_{010} mode is

$$\frac{\Delta f_r}{f_r} = -\frac{\Delta V/V}{2J_1^2(2.405)} = -1.85\frac{\Delta V}{V} \tag{9–88}$$

since $q_{01} = 2.405$ and $J_1(2.405) = 0.52$. If, for example, $\Delta V/V$ is 1.0 percent, the indentation results in a 1.85 percent *decrease* in the resonant frequency.

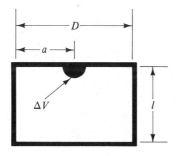

Figure 9–26 Cylindrical cavity resonator with a small indentation (of volume ΔV) in the top wall.

9–4 NARROWBAND MICROWAVE FILTERS

Microwave systems often require means for suppressing unwanted signals and/or separating signals having different frequencies. These functions are performed by electric filters. Filters are usually categorized by their frequency characteristics, namely, lowpass, highpass, bandstop, and bandpass. The first two types were discussed briefly in Sec. 6–6. Techniques for realizing bandstop and bandpass filters at microwave frequencies are described in this and the following section. Narrowband versions are covered in this section, while moderate and wideband designs are reviewed in Sec. 9–5. The term *narrowband* is generally used to indicate bandwidths of less than 10 percent, while *wideband* usually denotes bandwidths in excess of 40 percent.

Over the years, literally hundreds of microwave filter configurations have been developed. Only a handful of the more common types will be described here. For additional information, the reader is referred to the extensive bibliographies in Refs. 9–8, 9–12, 9–16, and 9–23. Detailed design procedures for many filters are given in Refs. 9–8, 9–9, and 9–10.

9–4a Bandstop Filters

The equivalent circuit of a single-stage bandstop filter is shown in Fig. 9–4. With $R_G = R_L = Z_0$, the expressions for insertion loss and phase are given by Eqs. (9–18) and (9–19), respectively. The loss characteristic is shown in Fig. 9–5. An alternate equivalent circuit for the bandstop filter is given in Fig. 9–27. Equations (9–18) and (9–19) are also valid for this circuit, where in the case $Q_U = \omega_r L / R$ and $Q_L = \omega_r L / (R + 0.5 Z_0)$.

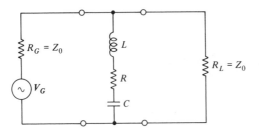

Figure 9–27 Alternate equivalent circuit for a single-stage bandstop filter.

Microwave versions of these two circuits can be constructed using either open or shorted quarter-wave lines. For example, an open-circuited line that is a quarter wavelength long at f_r may be used to approximate the series resonant circuit in Fig. 9–27. As explained in Sec. 9–2a, this technique has two disadvantages. First, the quarter-wave line has a multiplicity of impedance zeros while the series R-L-C circuit has only one. For narrowband filters, this is not a serious drawback. Second, the realizable values of Q_E and Q_L are quite small when a quarter-wave line is connected directly to a source and load [see Eq. (9–25)]. This, of course, presents a problem when designing narrowband filters since bandwidth is inversely proportional to Q_L. As a result, narrowband designs generally use inductively coupled shorted-line resonators (Fig. 9–9) and capacitively coupled open-line resonators (Fig. 9–11).

Two bandstop filters that employ these resonators will now be analyzed. One is a TEM version while the other utilizes rectangular waveguide.

A TEM bandstop filter.   Figure 9–28a shows a capacitively coupled, open-circuited resonator shunt connected to a transmission line (Z_0). Both the generator and load are matched to the main line. The center conductor pattern for a strip transmission-line version is shown in part b of the figure. The following analysis assumes TEM lines.

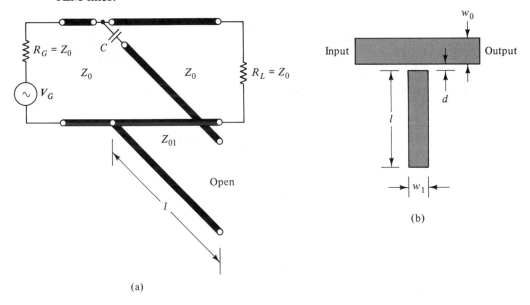

Figure 9–28   A bandstop filter that utilizes a capacitively coupled open-line resonator. (Part b shows the center conductor pattern for a strip transmission-line version.)

Since the resonator has an impedance zero at its resonant frequency (f_r), the main line is effectively shorted at f_r and thus no power is delivered to the load R_L. In other words, the insertion (and transducer) loss is infinite. At frequencies very close to f_r, the impedance of the resonator is small and hence the loss remains high. The circuit has a bandstop characteristic since the loss decreases as the frequency deviates from f_r. For a given 3 dB bandwidth and center frequency, the filter design proceeds in the following manner. Use Eq. (9–5) to calculate Q_L. For high-quality resonators ($Q_U \gg Q_L$), $Q_E \approx Q_L$. The value of capacitive coupling ($X_{C_r} \equiv 1/\omega_r C$) is obtained from the following equation.[11]

$$Q_E = \frac{Z_{01}}{Z_{01}}[\overline{X}_{C_r} + \phi_r(1 + \overline{X}_{C_r}^2)] \tag{9–89}$$

where $\overline{X}_{C_r} \equiv X_{C_r}/Z_0$ and for narrowband filters, it is assumed that $\phi_r \approx \pi$. Although

---

[11] Note that this equation differs from Eq. (9–31) by a factor of two. In this case (Fig. 9–27) the resonator is *loaded down* by two resistors in parallel. Thus the Thevenin resistance of the generator-load combination is $0.5Z_0$ rather than Z_0.

the value of Z_{01} is arbitrary, it is usually chosen between $0.5Z_0$ and $2Z_0$. Once $\overline{X}_{C_r}$ (and hence C) is obtained from Eq. (9–89), the exact length of the resonator l may be determined from Eq. (9–30), where $\lambda_r = v/f_r$ and v is the wave velocity in the Z_{01} line. Note that in determining the exact value of l, the assumption $\phi_r \approx \pi$ is *not* used.

For the strip transmission-line version shown in Fig. 9–28b, the strip widths w_0 and w_1 are determined from the given values of Z_0 and Z_{01}. The gap width d is chosen to give the required value of capitance C. Design curves that relate gap width to capacitance may be found in Refs. 9–8 and 9–12.

Figure 9–29 shows the loss characteristics of a bandstop filter with $f_r = 1.00$ GHz and $Q_E = 20$. The dashed curve is based on Eq. (9–18) which assumes that the impedance behavior of the capacitively coupled resonator is the same as the series R-L-C circuit in Fig. 9–27. An infinite unloaded Q (that is, $R = 0$) is also assumed. To obtain the exact loss characteristics of the filter (the solid curve), one must calculate L_l for the two-port network in Fig. 9–28a (Prob. 9–20). Note that the approximate results are only accurate in the vicinity of f_r since Eq. (9–18) does not take into account the additional impedance zeros of the transmission-line resonator (the second zero occurs at about 2.03 GHz).

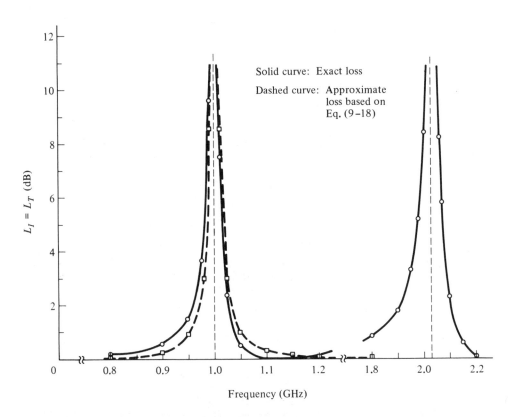

Figure 9–29 Typical loss characteristics for the bandstop filter shown in Fig. 9–28. The parameter values are $Z_0 = Z_{01} = 50$ ohms, $C = 1.36$ pF, $f_r = 1.00$ GHz, and $l = 0.436\lambda_r$.

Finally, it is important to realize that the loss characteristics of the filter are due to *reflection* rather than absorption. The assumption of infinite unloaded Q means that the resonator itself absorbs no power and hence the filter loss represents a mismatch loss. For example, at the 3 dB points, $|\Gamma_{in}| = 0.707$ and hence the input SWR is 5.83. Unless otherwise noted, filters described here are of the reflective type.

Waveguide bandstop filter.   Bandstop filters can also be realized using a series-connected parallel resonant circuit. A lumped-element version is shown in Fig. 9–4. The expressions for insertion loss and phase are given by Eqs. (9–18) and (9–19), respectively. A transmission-line version is shown in Fig. 9–30a. The resonator is the inductively coupled, shorted-line type described in Sec. 9–2a. This filter arrangement is often used in waveguide systems. A waveguide version is shown in part b of the figure, where TE_{10} waveguide transmission and a TE_{10p} rectangular cavity mode is assumed. In most practical filters, $p = 1$. Note that the guide width is the same for both the Z_0 and the Z_{01} lines. Since characteristic impedance is directly proportional to guide height, $Z_{01}/Z_0 = b_1/b$. A symmetrical inductive iris is shown in the figure, although any of the types described in Fig. 7–15 may be used.

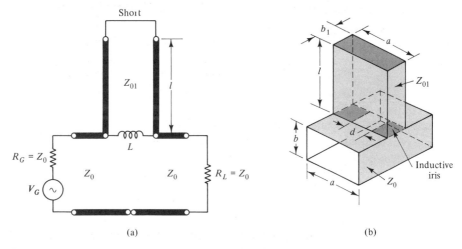

(a)   (b)

Figure 9–30   A bandstop filter that utilizes an inductively coupled, shorted-line resonator. Part b shows a waveguide version of the filter.

As in the previous case, the loss is very high at the resonant frequency f_r and decreases as the signal frequency f deviates from f_r. The loss expression for the waveguide version is the same as Eq. (9–18) except that ϵ_f is replaced by ϵ_λ which is defined as

$$\epsilon_\lambda \equiv 2\left(\frac{\lambda_{g_r}}{\lambda_g} - 1\right) \tag{9–90}$$

where λ_g is the guide wavelength and λ_{g_r} is its value at the resonant frequency f_r. The relationship between Q_L and the 3 dB points must also be modified to account for the cutoff effect in waveguide. Assuming $Q_U \gg Q_L$, the 3 dB points are defined by $Q_L \epsilon_\lambda = \pm 1$, which leads to the following expression,

$$Q_L = \frac{\lambda_{g_r}^{-1}}{\lambda_{g_2}^{-1} - \lambda_{g_1}^{-1}} \qquad \text{where} \qquad \lambda_{g_r} \equiv \frac{2\lambda_{g_1}\lambda_{g_2}}{\lambda_{g_1} + \lambda_{g_2}} \qquad (9\text{--}91)$$

λ_{g_1} and λ_{g_2} represent the guide wavelengths at f_1 and f_2, the frequencies at the 3 dB loss points. Note that in the absence of a cutoff effect, $\lambda_g = \lambda = v/f$ and the above expression for Q_L reduces to Eq. (9–5).

Values of Q_U for the TE_{10p} rectangular cavity modes are quite high and can be calculated from Eq. (9–64). In general $Q_U \gg Q_L$ and therefore $Q_E \approx Q_L$. The amount of inductive coupling ($\bar{B}_{L_r}$) needed to produce the required value of Q_L is obtained from the following relation.

$$Q_E = \frac{Z_0}{Z_{01}}[\bar{B}_{L_r} + \phi_r(1 + \bar{B}_{L_r}^2)] \approx \frac{Z_0}{Z_{01}}(\bar{B}_{L_r} + p\pi\bar{B}_{L_r}^2) \qquad (9\text{--}92)$$

where the approximate expression is valid for narrowband filters (namely, $\bar{B}_{L_r} > 3$). This equation differs from Eq. (9–28) since in this case the Thevenin impedance seen by the resonator is $2Z_0$ rather than Z_0. Furthermore, for the waveguide version, $\phi_r \equiv 2\pi l/\lambda_{g_r}$ rather than $2\pi l/\lambda_r$.

The resonance condition is *similar* to Eq. (9–29). Namely,

$$\phi_r \equiv \frac{2\pi l}{\lambda_{g_r}} = \arctan\left(\frac{-1}{\bar{B}_{L_r}}\right) = p\pi - \arctan\left(\frac{1}{\bar{B}_{L_r}}\right) \qquad (9\text{-}93)$$

In most cases, $p = 1$ is the desired resonance and $p = 2, 3$, etc. represent spurious responses associated with the higher-order TE_{10p} cavity modes. As in the previous case, it should be noted that Eq. (9–18) does not include the effect of these spurious responses. Problem 9–25 compares loss values obtained from Eq. (9–18) with those obtained from an exact transmission-line analysis.

Multisection bandstop filters. Many bandstop applications require a higher *skirt selectivity* than can be obtained with a single-section filter. Skirt selectivity is defined as the slope of the loss versus frequency characteristic in the vicinity of the 3 dB points. For a given bandwidth, this slope can be increased by increasing the number of filter sections. A lumped-element version of a multisection bandstop filter is shown in Fig. 9–31a. Typical loss curves for one and two-section filters are given in part *b* of the figure. Note that the lumped-element prototype contains both series and shunt-connected resonant circuits. It is usually difficult and inconvenient to incorporate both types into a transmission-line filter. This difficulty can be overcome by using quarter-wave inverters. For example, consider the two-section stripline filter shown in Fig. 9–32a. As explained earlier, the capacitively coupled, open-circuited resonators approximate series resonant circuits. Thus the one on the left can be represented by the L_1-C_1 combination in part *a* of Fig. 9–31. The one on the right also approximates a shunt-connected, series resonant circuit. The quarter-wave line transforms its impedance properties into that of a series-connected parallel resonant circuit *at the plane of the first resonator*, thus simulating the L_2-C_2 combi-

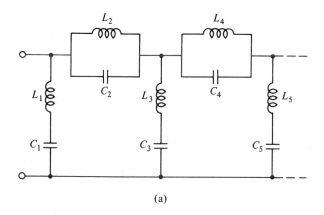

(a)

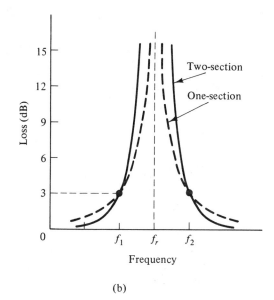

(b)

Figure 9–31 A multisection bandstop filter network.

nation in Fig. 9–31a. This conversion is a direct result of the impedance transformation equation [see Eq. (3–85)].

The waveguide version in part *b* of Fig. 9–32 uses a pair of series-connected inductively coupled resonators spaced $\lambda_{g_r}/4$ apart. The one on the left approximates a parallel resonant circuit connected in series, while the quarter-wave line transforms the second resonator into a shunt-connected series resonant circuit at the plane of the first resonator.

The following analysis verifies that a significant increase in skirt selectivity can be achieved with a two-section filter. Consider the transmission-line circuit in Fig. 9–33. It approximates the stripline bandstop filter shown in Fig. 9–32a. For the narrowband case, the frequency sensitivity of the resonant circuits is much greater than that of the quarter-wave line. Thus it is reasonable to assume $\beta l \approx 90°$ at all frequencies of interest. With this assumption, the normalized ABCD matrix of the

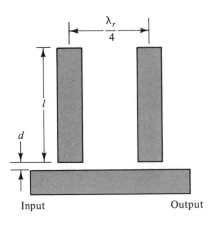

Note: l is obtained from
Eq. (9–30)

(a) Stripline version

Note: l is obtained from
Eq. (9–93)

(b) Waveguide version

Figure 9–32 Stripline and waveguide versions of a two-section bandstop filter. (Note: In the waveguide version, the spacing is usually $3\lambda_{g_r}/4$, since at $\lambda_{g_r}/4$ the sections may overlap.)

two-port network becomes

$$\begin{bmatrix} \overline{A} & \overline{B} \\ \overline{C} & \overline{D} \end{bmatrix} = \begin{bmatrix} 1 & 0 \\ -j/\overline{X} & 1 \end{bmatrix}\begin{bmatrix} 0 & j \\ j & 0 \end{bmatrix}\begin{bmatrix} 1 & 0 \\ -j/\overline{X} & 1 \end{bmatrix} = \begin{bmatrix} 1/\overline{X} & j \\ j\dfrac{\overline{X}^2 - 1}{\overline{X}^2} & 1/\overline{X} \end{bmatrix}$$

where $\overline{X} \equiv X/Z_0 = \left(\omega L - \dfrac{1}{\omega C}\right)\Big/Z_0 = \dfrac{\omega_r L}{Z_0}\left(\dfrac{\omega}{\omega_r} - \dfrac{\omega_r}{\omega}\right) = \dfrac{1}{2}Q_E\,\epsilon_f.$

The reader is reminded that ϵ_f is defined by Eq. (9–12) and that $Q_E = \omega_r L/0.5Z_0$ is the external Q of the individual resonators when connected to a Z_0 source and a Z_0 load. For the capacitively coupled open-circuited resonator, Q_E is given by Eq. (9–89).

The insertion and transducer loss of the filter can be derived with the aid of Eq. (6–31), which results in

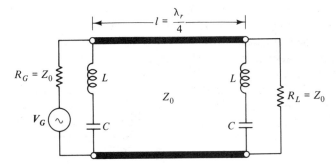

Figure 9–33 Approximate equivalent circuit of the stripline filter pictured in Fig. 9–32a.

$$L_I = L_T = 10 \log \left[1 + \frac{4}{(Q_E \, \epsilon_f)^4} \right] = 10 \log \left[1 + \frac{1}{(Q_T \, \epsilon_f)^4} \right] \quad (9-94)$$

where $Q_T = Q_E/\sqrt{2}$. Note that $Q_T = f_r/(3 \text{ dB bandwidth})$ since the 3 dB points occur when $Q_T \, \epsilon_f = \pm 1$. Thus Q_T, the total Q of the filter, is the reciprocal of the *fractional* 3 dB bandwidth.

The analysis for the two-section waveguide version in Fig. 9–32b produces the same loss equation except that ϵ_f is replaced by ϵ_λ, Q_E is for the inductively coupled resonator [Eq. (9–92)] and the relationship between Q_T and 3 dB bandwidth is given by Eq. (9–91) where Q_T replaces Q_L.

It should be noted that the above analysis neglects resonator losses since an infinite unloaded Q has been assumed. By applying this same assumption to Eq. (9–18), the loss expression for the single-section bandstop filter becomes

$$L_I = L_T = 10 \log \left[1 + \frac{1}{(Q_E \, \epsilon_f)^2} \right] \quad (9-95)$$

A comparison of the loss characteristics for the one- and two-section filters is given in Fig. 9–31b. It is apparent that in the vicinity of the 3 dB points, the two-section filter has the greater loss slope. This can be understood by referring to Eqs. (9–94) and (9–95). In the single-section case, the bandstop region is defined by $|Q_E \, \epsilon_f| < 1$, while for the two-section filter, it is defined by $|Q_T \, \epsilon_f| < 1$. Since raising a number *less* than unity to the fourth power is *smaller* than squaring it, the two-section filter produces higher loss *in the stopband*. Conversely, *outside* the stopband, $|Q_E \, \epsilon_f|$ and $|Q_T \, \epsilon_f|$ are *greater* than unity and therefore raising its value to the fourth power is greater than squaring it. Thus the loss outside the stopband is *less* for the two-section filter.

The higher skirt selectivity of the two-section filter may also be explained in terms of wave reflections. Within the stopband, the reflection due to each resonator is very large since its shunt reactance (series susceptance in the waveguide version) is very low. The quarter-wavelength spacing causes these large reflections to be additive as seen at the input. This can be verified by a Smith chart analysis. Thus in the stopband, the overall reflection loss of the two-section filter is much greater than that of the single-section unit. Conversely, *outside* the stopband, the individual reflections are small and for the quarter-wavelength spacing, these reflections tend to cancel.[12] Thus, outside the stopband, the overall loss for the two-section filter is *less* than that of the individual sections and also less than that of the single-section filter. This can also be verified by a Smith chart analysis.

This technique can be extended to filters with more than two sections. One might surmise that by increasing the number of sections, a more idealized bandstop characteristic can be achieved. When properly designed, this is indeed the case. In general, the loss characteristic for an n-section filter is similar to Eq. (9–94) except that the exponent for the $(Q_T \, \epsilon_f)$ term is $2n$. The theory and design of multisection bandstop filters is discussed in Chapter 12 of Ref. 9–8 and Chapter 2 of Ref. 9–9.

---

[12] For the narrowband case, we assume that the spacing is approximately a quarter wavelength at all frequencies of interest.

Other bandstop filters.   There are many other forms of narrowband bandstop filters. For example, a TEM version that employs capacitively coupled shorted-line resonators is described in Sec. 12.05 of Ref. 9–8. In this case, the primary resonance occurs when the line is slightly less than a quarter wavelength long. As a result, the frequency of the first spurious response is roughly three times that of the primary resonance. This represents an improvement over the open-line case wherein the first spurious response occurs at approximately $2f_r$ (see Fig. 9–29).

A variation of the waveguide filter shown in Fig. 9–32b is described in Ref. 9–24. It is known as the *E-plane cutoff* filter since the width of the shorted waveguide section is less than that of the main guide. A three-element version is pictured on the front cover of the book. This arrangement is particularly useful when one is interested in low loss at frequencies below the stopband region. Since the shorted waveguide sections are below cutoff at these frequencies, the E-plane cutoff filter has high power-handling capability in the passband.

The waveguide cutoff effect can also be combined with hybrid circuits to produce similar filter characteristics. These are described in Refs. 9–24 and 9–25. Other high-power filters are discussed in Ref. 9–23 and Chapter 15 of Ref. 9–8. Some of these (for example, the waffle-iron filter) provide excellent bandstop characteristics not only for TE_{10} mode transmission but for higher-order TE modes as well.

9–4b Bandpass Filters

The equivalent circuit of a single-stage bandpass filter is shown in Fig. 9–2. With $R_G = R_L = Z_0$, the expressions for insertion loss and phase are given by Eqs. (9–15) and (9–16), respectively. These characteristics are plotted as a function of frequency in Fig. 9–3. In principle, microwave bandpass filters can be constructed using the one-port resonators described in the previous section. For example, a strip transmission line version might use a series-connected, capacitively coupled open-circuited resonator. This series connection, however, is not easily realized in a TEM transmission line. Similarly, shunt connecting an inductively coupled shorted resonator in a waveguide system results in a cumbersome structure. These practical limitations are easily overcome by using two-port resonators. Stripline and waveguide versions are described in Figs. 9–34 and 9–36.

Stripline bandpass filter.   The bandpass filter shown in Fig. 9–34 utilizes a two-port capacitively coupled stripline resonator. The gap spacing d is chosen to produce the required value of series capacitance. As is common practice, the characteristic impedance of the resonator section (Z_{01}) equals that of the connecting lines.

Before proceeding with a circuit analysis, it is useful to understand the operation of the filter in terms of wave reflections. The capacitance value C is chosen to produce a large value of reactance (X_C) in the frequency range of interest. As will be shown, the larger the value of X_C, the narrower the filter bandwidth. Since the reactance is in series, a large value of X_C produces a large wave reflection. The total reflection due to a *pair* of such capacitors depends upon the spacing between them. By proper choice of l, the individual reflections can be made to cancel each other at

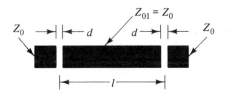

(a) Center conductor pattern

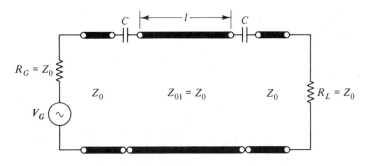

(b) Equivalent transmission-line circuit

Figure 9–34   A capacitively coupled stripline bandpass filter.

a specified frequency (f_r) resulting in full transmission through the filter. This is readily verified by a Smith chart analysis which shows that for $X_C \gg Z_0$, the proper spacing is slightly less than a half wavelength. When the operating frequency deviates from f_r, the electrical length βl changes and the two large reflections no longer cancel, thereby producing a significant transmission loss. Thus a bandpass characteristic can be created by simply spacing a pair of series reactances along a transmission line.

The above circuit can be analyzed by using the normalized ABCD matrix method. Referring to part b of Fig. 9-34 with $R_G = R_L = Z_0$,

$$
\begin{bmatrix} \bar{A} & \bar{B} \\ \bar{C} & \bar{D} \end{bmatrix} = \begin{bmatrix} 1 & -j\bar{X}_C \\ 0 & 1 \end{bmatrix} \begin{bmatrix} \cos\phi & j\sin\phi \\ j\sin\phi & \cos\phi \end{bmatrix} \begin{bmatrix} 1 & -j\bar{X}_C \\ 0 & 1 \end{bmatrix}
$$

$$
= \begin{bmatrix} (\cos\phi + \bar{X}_C \sin\phi) & j\{(1 - \bar{X}_C^2)\sin\phi - 2\bar{X}_C\cos\phi\} \\ j\sin\phi & (\cos\phi + \bar{X}_C \sin\phi) \end{bmatrix}
$$

where $\bar{X}_C \equiv X_C/Z_0$, $\phi = \beta l$ and it is assumed that $Z_{01} = Z_0$. By utilizing Eq. (6–31), the following loss equation is obtained.

$$
L_I = L_T = 10\log\left[1 + \bar{X}_C^2(\cos\phi + \frac{\bar{X}_C}{2}\sin\phi)^2\right] \tag{9–96}
$$

The condition for zero loss is

$$
2\cos\phi_r = -\bar{X}_{C_r}\sin\phi_r \tag{9–97}
$$

where ϕ_r is the electrical length of the Z_{01} line at $f = f_r$ and $\bar{X}_{C_r}$ is the value of $\bar{X}_C$ at f_r. This resonance condition can be rewritten as

$$\phi_r = \frac{2\pi l}{\lambda_r} = \pi - \arctan\left(\frac{2}{\bar{X}_{C_r}}\right) \tag{9-98}$$

where $\lambda_r = v/f_r$ is the wavelength at f_r. Note that for $\bar{X}_{C_r} \gg 1$, l is nearly a half wavelength long at the resonant frequency. Spurious bandpass responses occur near frequencies where l is a multiple of a half wavelength. Thus the first spurious response is at approximately twice f_r.

For narrowband filters, $\bar{X}_C \approx \bar{X}_{C_r}$ over the frequency range of interest. By using this approximation and Eq. (9-98), the above loss equation may be rewritten as

$$L_I = L_T = 10 \log\left[1 + \bar{X}_{C_r}^2 \frac{\bar{X}_{C_r}^2 + 4}{4} \sin^2(\phi - \phi_r)\right] \tag{9-99}$$

$$L_I = L_T \approx 10 \log\left[1 + \phi_r^2 \bar{X}_{C_r}^2 \frac{\bar{X}_{C_r}^2 + 4}{16}\left(2\frac{\phi - \phi_r}{\phi_r}\right)^2\right]$$

where the approximate form is useful for $|\phi - \phi_r| < 0.5$ rad. For TEM lines, velocity is independent of frequency and therefore

$$\phi/\phi_r = f/f_r \qquad \text{and} \qquad (\phi - \phi_r)/\phi_r = (f - f_r)/f_r.$$

Equation (9-99) represents the mismatch loss of the filter. Comparing the approximate form with the second term of Eq. (9-15) yields the following expression for the loaded Q of the two-port capacitively coupled resonator.

$$Q_L = \phi_r \frac{\bar{X}_{C_r}}{4}\sqrt{\bar{X}_{C_r}^2 + 4} = \frac{\bar{X}_{C_r}}{4}\sqrt{\bar{X}_{C_r}^2 + 4}\left(\pi - \arctan\frac{2}{\bar{X}_{C_r}}\right) \tag{9-100}$$

where $\phi_r \approx \pi$ for large values of $\bar{X}_{C_r}$. When dissipative losses are included, the overall loss of the bandpass filter is the same as Eq. (9-15). Namely,

$$L_I = L_T = 10 \log\left(1 - \frac{Q_L}{Q_U}\right)^{-2} + 10 \log[1 + (Q_L \epsilon_f)^2] \tag{9-101}$$

where $\epsilon_f = 2(f - f_r)/f_r$, Q_L is given above and Q_U may be approximated by Eq. (9-23). Since the fractional 3 dB bandwidth is $1/Q_L$, Eq. (9-100) verifies our earlier statement that the larger the value of $\bar{X}_{C_r}$, the narrower the bandwidth. A plot of Q_L as a function of $\bar{X}_{C_r}$ is given in Fig. 9-35.

The design procedure for the bandpass filter shown in Fig. 9-34 proceeds in the following manner. Given f_r and the 3 dB bandwidth, calculate Q_L and use Fig. 9-35 to obtain the required value of $\bar{X}_{C_r}$. With Z_0 known, determine X_{C_r} and hence the capacitance value C. Finally, use Eq. (9-98) to obtain the exact resonator length l. The gap spacing d is chosen to produce the required capacitance value. Gap capacitance data may be found in Refs. 9-8 and 9-12. In cases where the value of gap spacing becomes impractically small, overlay capacitors are sometimes used. For wider bandwidth designs, the coupled-line open-circuited resonator arrangement is preferred. This and other wideband techniques are described in Sec. 9-5.

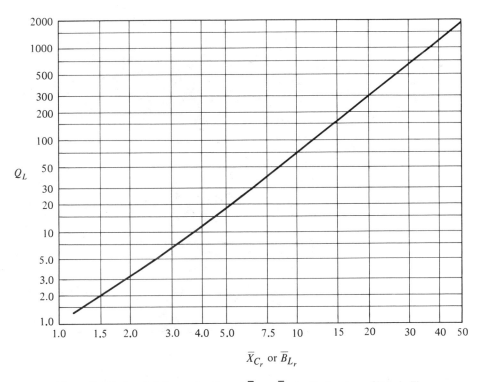

Figure 9–35 Loaded Q as a function of $\overline{X}_{C_r}$ or $\overline{B}_{L_r}$ for the bandpass filters in Figs. 9–34 and 9–36.

Waveguide bandpass filter.   The waveguide bandpass filter shown in Fig. 9–36 utilizes a two-port inductively coupled cavity resonator. As in the previous case, the spacing between the reflecting obstacles is chosen so that their reflections cancel at f_r, the center of the passband. In most designs, the transverse dimensions of the resonator (a and b) are the same as the connecting guides and therefore $Z_{01} = Z_0$. The ABCD matrix analysis is similar to that of the stripline bandpass filter except that the series capacitors are replaced by shunt inductors and λ is replaced by λ_g. For this case, the resonant condition is given by

$$\phi_r = \frac{2\pi l}{\lambda_{g_r}} = \pi - \arctan\left(\frac{2}{\overline{B}_{L_r}}\right) \qquad (9\text{--}102)$$

where λ_{g_r} is the guide wavelength and $\overline{B}_{L_r}$ the normalized inductive susceptance at f_r. An expression that relates $\overline{B}_L$ to the iris opening d is given in Fig. 7–15a. Note that for large values of $\overline{B}_{L_r}$ ($d \ll a$), l is approximately half the guide wavelength. The first spurious response occurs near the frequency where $\phi_r \approx 2\pi$ rad. Because of the waveguide cutoff effect, this frequency is roughly one and a half times f_r.

The loss equation for the waveguide bandpass filter is

$$L_l = L_T = 10 \log\left(1 - \frac{Q_L}{Q_U}\right)^{-2} + 10 \log\left[1 + (Q_L\,\epsilon_\lambda)^2\right] \qquad (9\text{--}103)$$

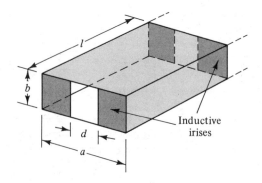

(a) Air-filled waveguide resonator

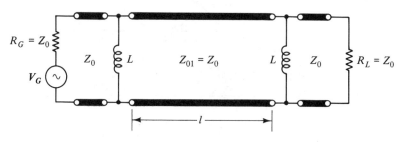

(b) Equivalent transmission-line circuit

Figure 9–36 An inductively coupled waveguide bandpass filter.

where ϵ_λ is defined in Eq. (9–90) and the relation between Q_L and the 3 dB bandwidth is given by Eq. (9–91). For the two-port inductively coupled resonator,

$$Q_L = \phi_r \frac{\bar{B}_{L_r}}{4}\sqrt{\bar{B}_{L_r}^2 + 4} = \frac{\bar{B}_{L_r}}{4}\sqrt{\bar{B}_{L_r}^2 + 4}\left(\pi - \arctan\frac{2}{\bar{B}_{L_r}}\right) \quad (9\text{–}104)$$

This equation is plotted in Fig. 9–35. Values of *unloaded Q* for the TE_{101} cavity mode can be determined from Eq. (9–64).

The design procedure for the waveguide bandpass filter parallels that of the capacitively coupled stripline version described earlier.

Tapped-line bandpass filter.   Two stripline versions of this filter are shown in Fig. 9–37. The one in part *a* uses the open-circuited resonator described earlier in Sec. 9–2a and Fig. 9–12. As explained, the resonator behaves like a shunt-connected parallel resonant circuit in the vicinity of f_r, the frequency at which $l_1 + l_2$ equals one-half wavelength. Since both R_G and R_L equal Z_0, the external Q of the filter is *half* the value obtained from Eq. (9–33). A sketch of the filter's loss characteristic is given in Fig. 9–38, where f_1 and f_2 represent the 3 dB loss points, f_L and f_H the 1 dB loss points and f_r the center frequency of the passband. The center frequency is determined by the length $l_1 + l_2$ (namely, $l_1 + l_2 = \lambda_r/2$). Since l_1 is related to Q_E, it can be adjusted to produce the desired 3 dB bandwidth.

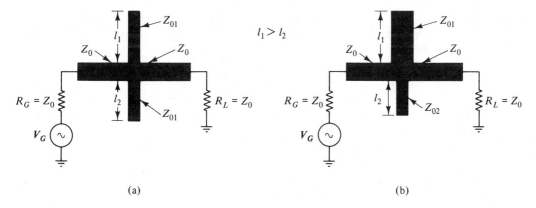

$l_1 > l_2$

Figure 9–37   Center conductor patterns for two tapped-line bandpass filters.

Note that unlike the usual bandpass response, this filter exhibits two infinite loss points ($f_{\infty 1}$ and $f_{\infty 2}$). As a result, the loss slope at the edges of the passband (that is, the skirt selectivity) is greater than that of a simple L-C bandpass filter. Assuming $l_1 > l_2$, $f_{\infty 1}$ is the frequency at which l_1 is a quarter wavelength long. This transforms the open circuit at the end of the Z_{01} line into a short circuit shunting the Z_0 line, thereby producing infinite loss (assuming infinite unloaded Q). Similarly $f_{\infty 2}$ is the frequency at which l_2 is a quarter wavelength long.

A variation of the tapped-line filter is shown in part b of Fig. 9–37. In this case, $f_{\infty 1}$, $f_{\infty 2}$, f_L, and f_H can be independently set by proper choice of l_1, l_2, Z_{01}, and Z_{02}. The following analysis develops the required design equations.

The total normalized admittance of the two open-circuited lines is

$$\bar{Y}_p = j\frac{B}{Y_0} = j[\bar{Y}_{01}\tan\beta l_1 + \bar{Y}_{02}\tan\beta l_2] \qquad (9\text{–}105)$$

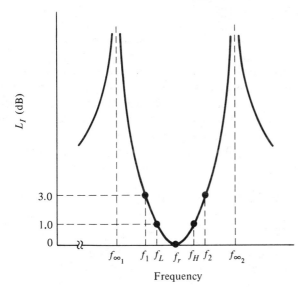

Figure 9–38   Loss characteristic of the tapped-line filter.

where $\overline{Y}_{01} = Z_0/Z_{01}$, $\overline{Y}_{02} = Z_0/Z_{02}$, and l_1 and l_2 are a quarter wavelength long at $f_{\infty 1}$ and $f_{\infty 2}$, respectively. Assuming $R_G = R_L = Z_0$, the insertion and transducer loss due to $\overline{Y}_p$ is given by

$$L_I = L_T = 10 \log \left[1 + \left(\frac{B}{2Y_0} \right)^2 \right] \qquad (9\text{–}106)$$

Therefore, the 1 dB points (f_L and f_H) occur when $B/Y_0 = \pm 1$. Substituting these two conditions into Eq. (9–105) and noting that for TEM lines $\beta l_1 = (f/f_{\infty 1})\pi/2$ and $\beta l_2 = (f/f_{\infty 2})\pi/2$, results in the following design equations.

$$Z_{01} = Z_0 \frac{NP - MR}{N + R} \qquad \text{and} \qquad Z_{02} = Z_0 \frac{MR - NP}{M + P} \qquad (9\text{–}107)$$

where

$$M \equiv \tan \left(\frac{\pi f_L}{2 f_{\infty 1}} \right), \; N \equiv \tan \left(\frac{\pi f_L}{2 f_{\infty 2}} \right), \; P \equiv \tan \left(\frac{\pi f_H}{2 f_{\infty 1}} \right), \; \text{and} \; R \equiv \tan \left(\frac{\pi f_H}{2 f_{\infty 2}} \right).$$

For example, suppose the desired 1 dB and infinite loss frequencies are $f_L = 2.8$ GHz, $f_H = 3.0$ GHz, $f_{\infty 1} = 2.1$ GHz, and $f_{\infty 2} = 3.5$ GHz. Then for, say, $Z_0 = 50$ ohms and $\lambda = 0.7\lambda_0$, the design values for the filter are

$$l_1 = 2.50 \text{ cm}, \quad l_2 = 1.50 \text{ cm}, \quad Z_{01} = 25 \text{ ohms}, \quad \text{and} \quad Z_{02} = 62.4 \text{ ohms}.$$

Multisection bandpass filters. The skirt selectivity of a bandpass filter can be substantially improved by using a multisection design. The reason for this improvement is similar to that given in the discussion on bandstop filters. From a wave-reflection point of view, the operation of a multisection bandpass filter is such that within the passband, the reflections due to the individual resonators tend to cancel, while outside the passband they tend to add. Thus, compared to a single-section design, the multisection filter has lower reflection loss in the passband and much greater loss outside the passband.[13]

Multisection filters generally use the same type resonator for each of the sections. As a result, impedance inverters are needed to simulate the usual ladder network design. Figure 9–39 shows how a bandpass ladder network (part *a*) can be replaced by a cascade of shunt-connected parallel resonant circuits separated by quarter-wave inverters. Note that each section is specified by its resonant frequency and external Q rather than its element values. Q_E of an individual section is defined as its external Q when that section is connected directly to a Z_0 generator and a Z_0 load. The two circuits shown are equivalent under the assumption that the inverter lines are approximately $\lambda/4$ at all frequencies of interest, which is reasonable for narrowband designs.[14]

Most bandpass filters are either of the maximally flat (Butterworth) or equal-

---

[13] The passband *dissipative* loss, however, is greater since the individual dissipative losses are additive.

[14] Mumford (Ref. 9–26) shows that the frequency sensitivity of the quarter-wave lines can be accounted for by a slight adjustment in the external Q of the individual sections.

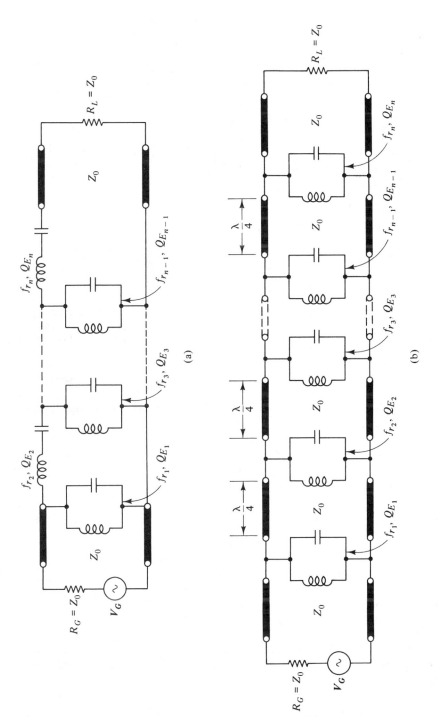

Figure 9–39 A prototype network for multisection bandpass filters.

469

ripple (Tchebyscheff) design. The loss characteristics for three-section versions are shown in Fig. 9–40. The maximally flat, quarter-wave coupled type will be described here. For the Tchebyscheff design procedure, the reader is referred to Ref. 9–2 and Sec. 8.10 of Ref. 9–8.

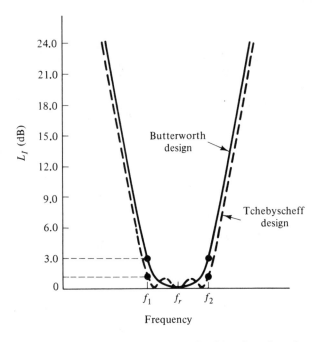

Figure 9–40 A comparison of loss characteristics for three-section Butterworth and Tchebyscheff bandpass designs. (Note that for a given bandwidth, the Tchebyscheff design has *lower* maximum loss in the passband.)

With $R_G = R_L = Z_0$, the loss function for an n-section maximally flat bandpass filter is given by

$$L_I = L_T = 10 \log [1 + (Q_T \epsilon_f)^{2n}] \qquad (9\text{–}108)$$

where $Q_T \equiv f_r/(3 \text{ dB bandwidth})$ and ϵ_f is defined in Eq. (9–12). For waveguide filters Q_T is defined by Eq. (9–91) and ϵ_f is replaced by ϵ_λ which is defined by Eq. (9–90). The above loss equation assumes all resonators have infinite unloaded Q. Losses associated with finite values of Q_U are accounted for by adding the following term to Eq. (9–108).

$$L_D = {}_r\sum_1^n 10 \log \left\{ 1 - \frac{Q_{L_r}}{Q_{U_r}} \right\}^{-2} \qquad (9\text{–}109)$$

where Q_{U_r} and Q_{L_r} are respectively the unloaded and loaded Q of the r th section and n is the total number of resonators. For $Q_U \gg Q_L$, practically all of this loss is dissipative.

Mumford (Ref. 9–26) explains that in order to achieve a maximally flat response, the external Q of the individual resonators must have a sinusoidal distribution. That is,

$$Q_{E_r} = Q_T \sin \left(\frac{2r - 1}{2n} \right) \pi \qquad \text{where } r = 1, 2, \ldots, n \qquad (9\text{–}110)$$

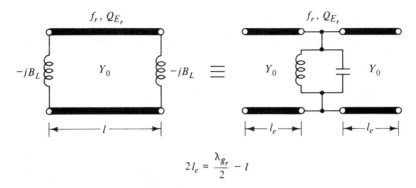

$$2l_e = \frac{\lambda_{g_r}}{2} - l$$

Figure 9–41 Equivalence between a two-port inductively coupled cavity and a shunt-connected parallel-resonant circuit.

Waveguide versions of the bandpass filter usually employ the two-port inductively coupled cavity resonator shown in Fig. 9–36. The equivalence between it and a shunt-connected parallel resonant circuit is given in Fig. 9–41, where $2l_e$ is known as the *excess phase length* of the resonator. Comparing the expression in the figure with Eq. (9–102) yields

$$l_e = \frac{\lambda_{g_r}}{4\pi} \arctan\left(\frac{2}{\bar{B}_{L_r}}\right) \tag{9–111}$$

In designing a multisection filter, the l_e values are incorporated into the quarter-wave inverter sections, thereby making them slightly less than a quarter wavelength long. The following example illustrates the design procedure.

Example 9–3:

Design a three-section, quarter-wave coupled, maximally flat bandpass filter in air-filled rectangular guide having the following specifications.
  Center frequency = 8000 MHz   and   *3dB bandwidth* = 100 MHz.
The width and height of the cavity resonators are the same as those of the input, output, and inverter lines, namely, $a = 3.00$ cm and $b = 1.50$ cm.

Solution: At 7950 and 8050 MHz, λ_g equals 4.855 cm and 4.756 cm, respectively. Therefore from Eq. (9–91),

$$\lambda_{g_r} = 4.805 \text{ cm} \qquad \text{and} \qquad Q_T = 48.5$$

From Eq. (9–110),

$$Q_{E_1} = 24.3, \qquad Q_{E_2} = 48.5, \qquad \text{and} \qquad Q_{E_3} = 24.3$$

and therefore from Fig. 9–35,

$$\bar{B}_{L_1} = 5.7, \qquad \bar{B}_{L_2} = 8.0, \qquad \text{and} \qquad \bar{B}_{L_3} = 5.7$$

From Eq. (9–102),

$$l_1 = 2.145 \text{ cm}, \qquad l_2 = 2.215 \text{ cm}, \qquad \text{and} \qquad l_3 = 2.145 \text{ cm}$$

and from Eq. (9–111),

$$l_{e1} = 0.129 \text{ cm,} \qquad l_{e2} = 0.094 \text{ cm,} \qquad \text{and} \qquad l_{e3} = 0.129 \text{ cm.}$$

The quarter-wave inverting lines must be reduced by the excess phase length. Thus,

$$l_{12} = \frac{\lambda_{g_r}}{4} - l_{e1} - l_{e2} = 0.978 \text{ cm}$$

and

$$l_{23} = \frac{\lambda_{g_r}}{4} - l_{e2} - l_{e3} = 0.978 \text{ cm}$$

The transmission-line circuit of the three-section filter is shown in Fig. 9–42, where the Z_0 line sections represent rectangular waveguide. Any of the inductive iris types shown in Fig. 7–15 may be used to realize the required values of $\bar{B}_L$.

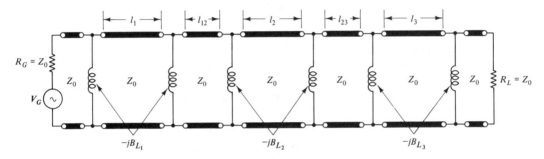

Figure 9–42 Transmission-line representation of the three-section quarter-wave coupled filter described in Ex. 9–3.

This same design procedure can be applied to the capacitively coupled stripline resonators shown in Fig. 9–34. In this case, the resonator can be represented by a series-connected series resonant circuit with line lengths l_e on either side. Equation (9–111) may be used to calculate l_e by simply replacing $\bar{B}_{L_r}$ with $\bar{X}_{C_r}$. Also, since TEM lines are involved, λ_g is replaced by λ in all pertinent equations. The resonator lengths l and the values of $\bar{X}_{C_r}$ are obtained from Eq. (9–98) and Fig. 9–35.

The direct-coupled filter represents an alternate design approach for multisection bandpass filters. These filters have been analyzed by Riblet (Ref. 9–27), Cohn (Ref. 9–28), and Levy (Ref. 9–29). Additional information may be found in Ref. 9–8. Since a direct-coupled filter does not use quarter-wave inverters, it is shorter in length than a corresponding quarter-wave-coupled filter. The following qualitative discussion follows the design approach described by Cohn.

Formulas for the direct-coupled resonator filter may be deduced from the low-pass filter prototype shown in Fig. 9–43a. Element values for maximally flat (Butterworth) and equal-ripple (Tchebyscheff) designs are given by Eqs. (9–117) and (9–118). Values up to $n = 5$ are listed in Table 9–2. Since the generator impedance is unity, all element values represent *normalized* impedances and admittances. Also shown in part a are the loss characteristics for both Butterworth and Tchebyscheff responses. The corresponding loss equations are given by Eqs. (9–123) and (9–124), where $\bar{\Omega}$ is replaced by ω', the *normalized* frequency variable of the prototype filter. The frequency ω' is normalized since the edge of the passband has been defined as $\omega' = 1$ (as shown in the figure). Also indicated in the figure is ω'_s,

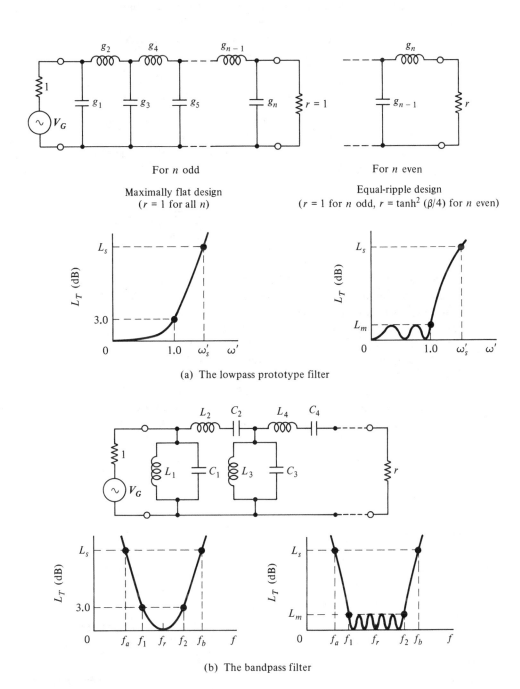

Figure 9–43   The lowpass prototype filter and its corresponding bandpass configuration.

the edge of the *usable* stopband (defined as $L_T = L_s$, where L_s is specified by the designer).

The lowpass response (and circuit) can be converted to a bandpass response by means of the following frequency transformation.

$$\omega' = \frac{f_r}{f_2 - f_1}\left(\frac{f}{f_r} - \frac{f_r}{f}\right) \tag{9-112}$$

where $f_r = \sqrt{f_1 f_2}$ and the edges of the passband region (f_1 and f_2) are indicated in part *b* of Fig. 9–43. The frequencies f_2 and f_1 correspond to $\omega' = \pm 1$ in the lowpass prototype. Similarly, f_b and f_a correspond to $\omega' = \pm\omega_s'$. The element values for the bandpass circuit in part *b* are given by

$$L_k = \frac{w}{2\pi f_r g_k} \quad \text{and} \quad C_k = \frac{g_k}{2\pi f_r w} \quad \text{for } k \text{ odd}$$

$$\tag{9-113}$$

$$L_k = \frac{g_k}{2\pi f_r w} \quad \text{and} \quad C_k = \frac{w}{2\pi f_r g_k} \quad \text{for } k \text{ even}$$

where $w = (f_2 - f_1)/f_r$ is the fractional bandwidth of the bandpass filter.

As explained earlier, it is desirable at microwave frequencies to convert the filter circuit into one that contains either all series-resonant elements or all parallel-resonant elements. This can be done by utilizing impedance and admittance inverters as indicated in Fig. 9–44. These inverters are useful in transforming the properties of a series-connected, series-resonant circuit into that of a shunt-connected, parallel-resonant circuit and vice versa. The properties of ideal J and K inverters are given in part *a* of Fig. 9–45. Note that the J inverter is simply the dual representation of the K inverter. The quarter-wave line represents one type of inverter circuit. Many other types are described in the literature (Ref. 9–8). Part *b* of Fig. 9–45

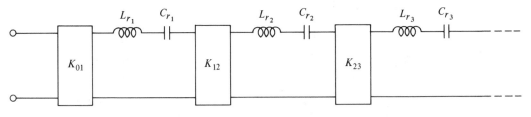

(a) Impedance inverters (K type)

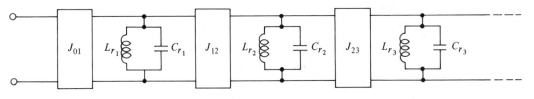

(b) Admittance inverters (J type)

Figure 9–44   The use of impedance and admittance inverters in bandpass filter design.

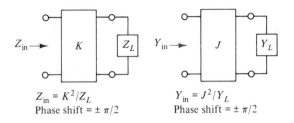

(a) Inverter properties

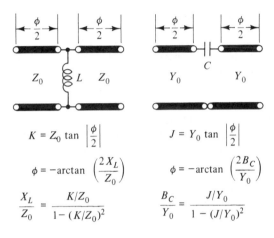

(b) Two forms of inverters

Figure 9–45 Properties of K and J inverters. Two types are shown in part b.

shows two that are particularly useful with transmission-line resonators. Note that in both cases, the reactance values are realizable but the line lengths are *negative*. This does not pose a problem since the negative line lengths are incorporated into the resonator structure itself.

The design procedure, as outlined by Cohn, makes use of these inverters and the transmission-line equivalence indicated in Fig. 9–46. This equivalence is easily verified by equating the ABCD matrices of the two networks (Prob. 9–36). In the vicinity of $l \approx \lambda/2$ ($\beta l \approx 180°$), the series element behaves like a series resonant circuit and the shunt elements like parallel resonant circuits. The ideal 1 to -1 transformer may be ignored since it does not affect the loss characteristics of the filter.

To understand the design procedure, consider a two-section inductively coupled bandpass filter. For this filter, the circuit shown in Fig. 9–47a is used as a starting point. The K inverters are of the type shown in Fig. 9–45. The series resonant circuits can be approximated by half-wavelength lines since, as explained by Cohn, the shunt elements in Fig. 9–46 are negligible compared to the large shunt susceptances of the K inverters. Figure 9–47 describes the development of the final filter form. The two series resonant circuits and K inverters are shown in part a. Part b indicates their approximate realization using the transmission-line circuits in Figs. 9–45b and 9–46, where l equals a half wavelength at the center frequency of the

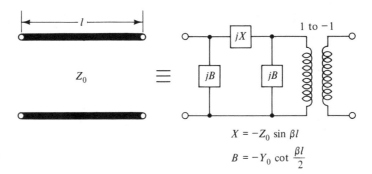

$$X = -Z_0 \sin \beta l$$

$$B = -Y_0 \cot \frac{\beta l}{2}$$

Figure 9–46   The lumped-element equivalent of a lossless transmission-line section.

passband. The final form of the filter is pictured in part c, where the negative line lengths of the inverters have been combined with the half-wave lines. Thus, $l_{12} = l + l_1 + l_2$ and $l_{23} = l + l_2 + l_3$, where l_1, l_2, and l_3 are negative. The negative line lengths at the input and output terminals of the filter can be ignored since they are incorporated into the input and output connecting lines. Waveguide versions of this filter are usually constructed with inductively coupled cavity resonators. A similar procedure utilizing J inverters and capacitively coupled stripline resonators may be used to realize direct-coupled bandpass filters in TEM structures.

The explanation given here is qualitative in nature since it is merely intended to show how transmission-line techniques can be used to realize direct-coupled filters at microwave frequencies. Detailed design procedures with examples may be found in Chapter 8 of Ref. 9–8.

Some of the more widely used narrowband filters have been discussed in this section. Many other types have been developed over the years. For example, the dual-mode waveguide filter has found considerable use in satellite systems. It consists of cylindrical cavity resonators that support orthogonal TE_{11p} modes. These modes are coupled by means of metal screws oriented at 45° with respect to the transverse electric field. This technique is described in Refs. 9–31 and 9–32.

One of the best sources of design information on microwave filters is Ref. 9–8. The reader will also find the bibliographies in Refs. 9–12, 9–16, and 9–23 to be particularly valuable.

9–5 WIDEBAND MICROWAVE FILTERS

This section covers some of the design principles for filters having moderate and wide bandwidths. Many of these filters utilize the frequency sensitive characteristics of open and shorted quarter-wave lines. As explained earlier, the loaded Q of directly coupled, quarter-wave lines are quite low [see Eq. (9–25) and the related discussion]. As a result, they are ideally suited for use in wideband filter design. With multisection versions, both wide bandwidths and high skirt selectivity can be realized. All the filters described here are of the commensurate-line TEM type. This means that for a given design, the electrical length (βl) of all transmission-line

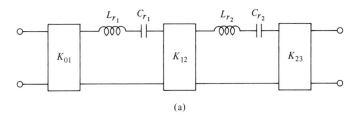

(a)

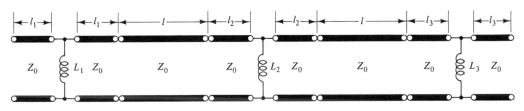

(Note: l_1, l_2, and l_3 are negative)

(b)

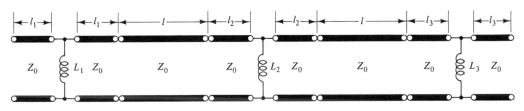

(c)

Figure 9–47   Development of a two-section direct-coupled filter that utilizes inductively coupled resonators.

sections are equal and directly proportional to frequency. With suitable modifications, some of the design techniques can be applied to waveguide configurations.

9–5a Wideband Filters with Quarter-Wavelength Stubs

The design of narrowband microwave filters usually involves the modeling of transmission-line resonators as lumped-element L-C networks. As explained in the previous section, this approximate model is inaccurate for bandwidths in excess of about 10 percent. For wider bandwidths, improved modeling and more exact synthesis procedures are required. In most cases the procedures are based upon Richards' transformation and Kuroda's identities.

Richards' transformation.   The exact design of lumped-element filters on an insertion loss basis is well established. Synthesis procedures and the required element values are tabulated in many texts and articles (for example, Refs. 9–8, 9–14,

9–15, and 9–33). It would, of course, be very desirable if this body of information could be utilized in the design of microwave filters. By use of a frequency transformation developed by Richards (Ref. 9–34), this can indeed be done for TEM transmission-line filters with commensurate line lengths.

The following relation defines Richards' frequency transformation.

$$S = j\Omega = j\tan\left(\frac{\pi}{2}\frac{f}{f_r}\right) \tag{9-114}$$

where f_r represents the frequency at which all transmission lines are one quarter wavelength long (that is, $\beta l = \pi/2$).[15] Note that the new variable Ω ranges from zero to infinity when the frequency is varied from zero to f_r. Thus by replacing the usual frequency variable ω by Ω, the entire frequency response of a lumped-element prototype is reproduced over a finite portion of the frequency band. This is indicated in Fig. 9–48. Part *a* shows the loss characteristics of a typical lumped-element L-C filter. The lowpass response shows the loss becoming infinite as the frequency becomes infinite. Part *b* of the figure shows the response of a corresponding transmission-line filter. The transmission-line version is realized by simply replacing all inductors in the lumped-element filter by shorted lines, all capacitors by open-circuited lines, and setting all line lengths equal to a quarter wavelength at f_r. Note that the entire frequency response of the lumped-element prototype filter is reproduced between zero and f_r for the transmission-line filter. Also, the response repeats periodically for frequency increments of $2f_r$, which corresponds to repeated increments of one-half wavelength in the electrical length of the transmission-line sections. The correspondence between the lumped elements and the quarter-wave lines is given in Fig. 9–49. They derive directly from Richards' transformation and the impedance properties of open and shorted lines (Sec. 3–6). The input impedance of a shorted line having a characteristic impedance Z_{0s} is

$$Z_{in} = jZ_{0s}\tan\beta l = jZ_{0s}\tan\left(\frac{\pi f}{2f_r}\right) = j\Omega Z_{0s} \tag{9-115}$$

where $l = \lambda/4$ at f_r. Similarly, the input admittance of an open-circuited line having a characteristic admittance Y_{0s} is

$$Y_{in} = jY_{0s}\tan\beta l = jY_{0s}\tan\left(\frac{\pi f}{2f_r}\right) = j\Omega Y_{0s} \tag{9-116}$$

Since the impedance of an inductor is $j\omega L$ and the admittance of a capacitor is $j\omega C$, the following equivalences can be established.

1. In the Ω domain, a shorted line of characteristic impedance Z_{0s} and length $l = \lambda/4$ at f_r behaves like an inductor of value Z_{0s} henries.
2. In the Ω domain, an open-circuited line of characteristic admittance Y_{0s} and length $l = \lambda/4$ at f_r behaves like a capacitor of value Y_{0s} farads.

[15] Most articles indicate this frequency as f_0 rather than f_r. Our choice of the subscript r is consistent with its use in previous sections. For example, λ_r (rather than λ_0) denotes the wavelength at the center frequency of bandpass and bandstop filters. Thus, confusion with free-space wavelength (λ_0) is avoided.

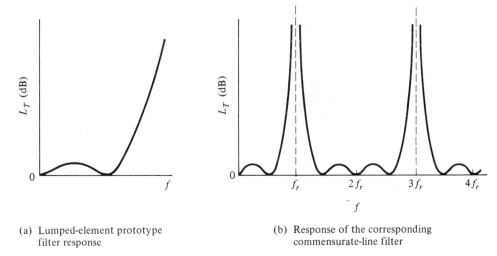

(a) Lumped-element prototype
    filter response

(b) Response of the corresponding
    commensurate-line filter

Figure 9–48 Effect of Richards' transformation on the frequency response of a filter network.

Resistors are unaffected by Richards' transformation since their impedance R is independent of frequency.

The above equivalences suggest a technique for realizing transmission-line filters that utilize commensurate line sections. Since the equivalences are exact, the resultant filter designs are also exact.

Commensurate-line bandstop filter design.   The transducer loss versus frequency response for Butterworth and Tchebyscheff bandstop filters is shown in Fig. 9–50, where f_r denotes the center of the stopband. In the Butterworth design, f_c and $2f_r - f_c$ represent the 3 dB loss points, while for the Tschebyscheff design, they

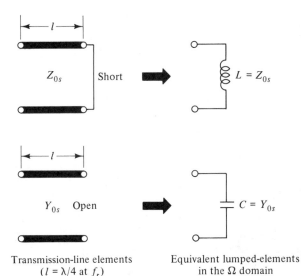

Transmission-line elements
($l = \lambda/4$ at f_r)

Equivalent lumped-elements
in the Ω domain

Figure 9–49 Correspondence between shorted and open-circuited transmission-line sections and reactive elements in the Ω domain.

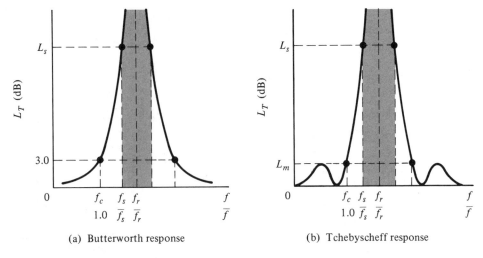

(a) Butterworth response   (b) Tchebyscheff response

Figure 9–50 Bandstop characteristics for Butterworth and Tchebyscheff designs.

represent the maximum passband loss points (L_m) as indicated in the figure. In both cases, the frequency range between f_s and $2f_r - f_s$ represents the usable stopband (defined as $L_T \geq L_s$, where L_s is specified by the designer). Normalized frequency scales ($\bar{f}$) are also included, where $\bar{f} \equiv f/f_c$ and hence $\bar{f}_c = 1.00$.

At microwave frequencies, the desired bandstop characteristic can be obtained by designing a comparable lowpass prototype filter and applying Richards' transformation to convert it into a bandstop filter. As indicated earlier, the bandstop response of the resultant commensurate-line filter repeats itself in increments of $2f_r$.

A lumped-element lowpass filter and its response for Butterworth and Tchebyscheff designs is shown in Fig. 9–51. The element values are tabulated in various texts (for example, Refs. 9–8, 9–14, and 9–15). Closed form expressions for the two designs are given below.

Butterworth (maximally flat) response:

$$g_k = 2 \sin \left[\frac{(2k-1)\pi}{2n} \right], \qquad \text{where } k = 1, 2, \ldots, n \qquad (9\text{--}117)$$

For example, if $n = 3$, $g_1 = g_3 = 1.00$, and $g_2 = 2.00$.

Tchebyscheff (equal-ripple) response:

$$g_1 = 2a_1/\gamma \qquad (9\text{--}118)$$

$$g_k = \frac{4a_{k-1}a_k}{b_{k-1}g_{k-1}}, \qquad k = 2, 3, \ldots, n$$

where

$$a_k = \sin \left[\frac{(2k-1)\pi}{2n} \right], \qquad k = 1, 2, \ldots, n$$

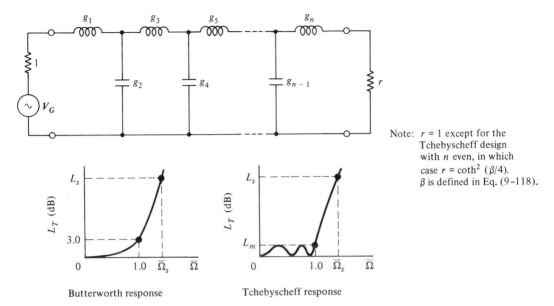

Figure 9–51 A lumped-element prototype lowpass network. Transducer loss characteristics for Butterworth and Tchebyscheff designs are indicated. The dual network (Fig. 9–43a) may also be used as a prototype. For the dual, $r = \tanh^2 (\beta/4)$ for Tchebyscheff designs with n even. (Note: $L_T = L_I$ only when $r = 1$.)

$$b_k = \gamma^2 + \sin^2 (k\pi/n), \qquad k = 1, 2, \ldots , n$$

$$\gamma = \sinh (\beta/2n)$$

and

$$\beta = \ln \coth \left(\frac{L_m}{17.37}\right) \qquad \text{with } L_m \text{ in dB}$$

(Note: In the above equations, β is *not* a phase constant and γ is *not* a propagation constant. They simply represent the indicated quantities.) For example, if $n = 3$ and $L_m = 1.0$ dB, $\beta = 2.855$, $\gamma = 0.494$, and therefore $g_1 = g_3 = 2.024$ and $g_2 = 0.994$. Table 9–2 lists element values up to $n = 5$ for both the Butterworth design and the Tchebyscheff design with 0.20 dB loss ripple. These values represent normalized impedances and admittances since the generator impedance (and admittance) is unity. Since the g values for inductive elements are normalized impedances and those for capacitive elements are normalized *admittances*, the unnormalized values become

$$G_k = g_k Z_0 \quad \text{ohms} \qquad \text{for inductances}$$

and   (9–119)

$$G_k = g_k / Z_0 \quad \text{mhos} \qquad \text{for capacitances}$$

where a matched generator ($R_G = Z_0$) has been assumed. Z_0 denotes the characteristic

TABLE 9–2   Normalized element values for the prototype lowpass filters in Figs. 9–43a and 9–51. More extensive tables may be found in Refs. 9–8 and 9–14.

ELEMENT VALUES FOR BUTTERWORTH DESIGN					
n	g_1	g_2	g_3	g_4	g_5
1	2.000				
2	1.414	1.414			
3	1.000	2.000	1.000		
4	0.765	1.848	1.848	0.765	
5	0.618	1.618	2.000	1.618	0.618

ELEMENT VALUES FOR TCHEBYSCHEFF DESIGN with $L_m = 0.20$ dB					
n	g_1	g_2	g_3	g_4	g_5
1	0.434				
2	1.038	0.675			
3	1.228	1.153	1.228		
4	1.303	1.284	1.976	0.847	
5	1.339	1.337	2.166	1.337	1.339

impedance of the input line. It should be noted that the frequency scale in Fig. 9–51 has been normalized to Ω_c, the value of Ω at $f = f_c$. That is,

$$\overline{\Omega} \equiv \frac{\Omega}{\Omega_c} \tag{9–120}$$

Thus the inductance and capacitance values for the lowpass filter are

$$L_k = G_k/\Omega_c \qquad \text{and} \qquad C_k = G_k/\Omega_c \tag{9–121}$$

If one were designing a *lumped-element* filter, Ω would simply be replaced by ω, the usual frequency variable, and hence $\omega_c = 2\pi f_c$ would replace Ω_c. However, for the equivalent commensurate-line filter (that is, inductors replaced by short-circuited line sections and capacitors by open-circuited line sections), Ω is related to frequency by Richards' transformation. Namely,

$$\Omega = \tan\left(\frac{\pi f}{2 f_r}\right) \qquad \text{and} \qquad \Omega_c = \tan\left(\frac{\pi f_c}{2 f_r}\right) \tag{9–122}$$

The following example illustrates how the lowpass prototype, Richards' transformation, and its associated equivalences (Fig. 9–49) are used to design a bandstop transmission-line filter.

Example 9–4:

Design a three-element, T-type bandstop filter with a 3 dB bandwidth of 800 MHz centered at 2000 MHz. Use a Butterworth design with commensurate length open- and shorted-line sections. Assume $R_G = R_L = Z_0 = 50$ ohms.

Solution: Since the bandwidth is 800 MHz and $f_r = 2000$ MHz, $f_c = 2000 - 0.5(800) = 1600$ MHz and therefore $\Omega_c = \tan 72° = 3.078$. The g values for a three-element Butterworth prototype are given in Table 9–2, namely, $g_1 = g_3 = 1.00$ and $g_2 = 2.00$. For the T-type lowpass filter, g_1 and g_3 represent series inductors, while g_2 represents a shunt capacitor. Thus for a 50 ohm system, the unnormalized values become

$$G_1 = G_3 = 50 \text{ ohms} \qquad \text{and} \qquad G_2 = 0.04 \text{ mhos.}$$

Substitution into Eq. (9–121) yields

$$L_1 = L_3 = 16.24 \text{ H} \qquad \text{and} \qquad C_2 = 0.013 \text{ F}$$

These values represent the elements of the lowpass prototype network. The equivalent commensurate-line filter is shown in part a of Fig. 9–52, where $l = \lambda/4$ at 2000 MHz. From the equivalences in Fig. 9–49,

$$Z_{01} = Z_{03} = 16.24 \text{ ohms} \qquad \text{and} \qquad Y_{02} = 1/Z_{02} = 0.013 \text{ mhos.}$$

This completes the design of the bandstop transmission-line filter. A coaxial version is pictured in part b of Fig. 9–52, where the dielectric-filled coaxial sections are the series-connected shorted lines.

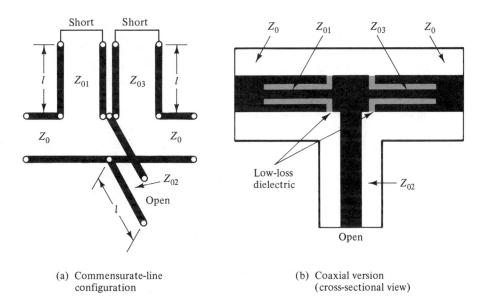

(a) Commensurate-line
   configuration

(b) Coaxial version
   (cross-sectional view)

Figure 9–52 A three-element commensurate-line bandstop filter. The design procedure is illustrated in Ex. 9–4.

The loss characteristic of the bandstop filter is indicated in part a of Fig. 9–50. Note that this same filter can be viewed (and used) as a lowpass filter with a cutoff frequency of 1600 MHz and an infinite loss point at 2000 MHz. However, unlike a *true* lowpass filter, the commensurate-line version has spurious responses at $2f_r$, $4f_r$, etc.

The loss expressions for both Butterworth and Tchebyscheff lowpass filters are given in the literature (for example, Chapter 4 in Ref. 9–8). By replacing the normalized frequency variable with $\overline{\Omega}$, the same expressions may be used for the commensurate-line filters.

For the Butterworth design,

$$L_T = 10 \log (1 + \overline{\Omega}^{2n}) \qquad (9\text{–}123)$$

where $\overline{\Omega}$ is defined by Eqs. (9–120) and (9–122), and n is the number of elements in the filter.[16]

For the Tchebyscheff design,

$$L_T = 10 \log [1 + (\kappa - 1)T_n^2(\overline{\Omega})] \qquad (9\text{–}124)$$

where

$$T_n(\overline{\Omega}) = \cos (n \cos^{-1} \overline{\Omega}) \qquad \text{for } |\overline{\Omega}| \le 1$$

$$T_n(\overline{\Omega}) = \cosh (n \cosh^{-1} \overline{\Omega}) \qquad \text{for } |\overline{\Omega}| \ge 1$$

and $\kappa \equiv$ antilog $(L_m/10)$. L_m is the maximum passband loss as indicated in Figs. 9–50b and 9–51.

The number of elements (n) needed in a particular filter design is generally determined by the required skirt selectivity. With f_c, f_s, f_r, and L_s specified, Eq. (9–123) is used to determine n for a Butterworth design. For a Tchebyscheff design, the required value of n is obtained from Eq. (9–124), where f_c, f_s, f_r, L_s, and L_m are specified. In either case, the value obtained must be increased to the next higher integer value since n must be a positive integer.

If one were to construct the bandstop filter described in Ex. 9–4 and Fig. 9–52, its measured loss response would closely follow that predicted by Eq. (9–123). Thus, from an electrical point of view, it is a good design. From a mechanical point of view, however, it is not! As explained in Sec. 6–6, the implementation of series-connected stubs in TEM structures tends to be awkward and expensive. Therefore, a procedure that would eliminate them without compromising filter performance would be very desirable. The existence of certain network equivalences, known as *Kuroda's identities*, makes such a procedure possible.

Kuroda's identities.   Figure 9–53 shows four network equivalences known as Kuroda's identities of the first and second kind. The equivalences are only valid in the S-plane, where S is defined by Eq. (9–114). Thus, the impedance of an inductor is $j\Omega L$, while the admittance of a capacitor is $j\Omega C$. Kuroda's method has been generalized by Levy (Ref. 9–35) to include any two-port network and unit element. The *unit element* (labeled U.E. in Fig. 9–53) is defined by the following ABCD matrix.

$$\begin{bmatrix} A & B \\ C & D \end{bmatrix} = \frac{1}{\sqrt{1 - S^2}} \begin{bmatrix} 1 & SZ_{0k} \\ S/Z_{0k} & 1 \end{bmatrix} = \frac{1}{\sqrt{1 + \Omega^2}} \begin{bmatrix} 1 & j\Omega Z_{0k} \\ j\Omega/Z_{0k} & 1 \end{bmatrix} \qquad (9\text{–}125)$$

---

[16] For the Butterworth design, we have defined the edge of the passband as the 3 dB loss point. If the edge of the passband is defined as the frequency at which the loss is L_m(dB), the loss equation becomes $10 \log [1 + (\kappa - 1)\overline{\Omega}^{2n}]$, where $\kappa \equiv$ antilog $(L_m/10)$.

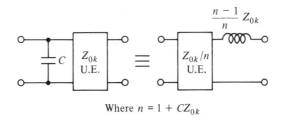

Where $n = 1 + CZ_{0k}$

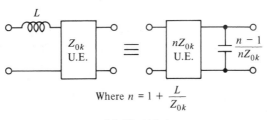

Where $n = 1 + \dfrac{L}{Z_{0k}}$

(a) First kind

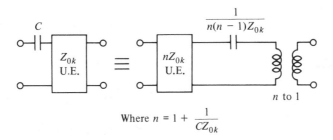

Where $n = 1 + \dfrac{1}{CZ_{0k}}$

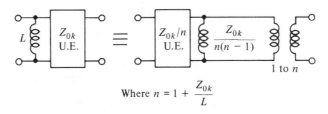

Where $n = 1 + \dfrac{Z_{0k}}{L}$

(b) Second kind

Figure 9–53 Kuroda's identities of the first and second kind. These identities are valid only in the Ω domain. Thus inductors represent shorted line sections and capacitors represent open-circuited line sections. All lines are $\lambda/4$ long at f_r.

where $S = j\Omega$ and from Richards' transformation, $\Omega = \tan{(\pi f/2f_r)}$. A comparison of this expression with the ABCD matrix of a transmission line (Appendix C) reveals that a unit element (U.E.) is simply a lossless transmission line of characteristic impedance Z_{0k} that is one quarter wavelength long at f_r. By utilizing Eq. (9–125), any of the identities in Fig. 9–53 can be verified by comparing the ABCD matrices of the equivalent networks (Prob. 9–42).

Before showing how Kuroda's identities can be used to simplify the design of transmission-line filters, a few general comments are in order. First of all, it should be remembered that in the Ω domain, the inductor L represents a shorted line section, while a capacitor C represents an open-circuited line section. Thus Kuroda's identities of the first kind can be used to replace a series-connected shorted line by a shunt-connected open-circuited line, or vice versa. On the other hand, Kuroda's identities of the second kind can only be used to change the *position* of an element in the circuit. Furthermore, they introduce ideal transformers which can present an obstacle to physically realizing the transmission-line filter. In the case of symmetrical networks, this obstacle is avoided since transformer-pairs can be rearranged (by suitable changes in impedance levels) so that they cancel each other. Finally, it should be noted that if $R_L = Z_0$, any number of unit elements having $Z_{0k} = Z_0$ can be inserted between the filter and the load without altering the input reflection coefficient and the amplitude response of the filter. The phase response, however, is altered. In many filter applications, this change in phase is unimportant. Similarly, if $R_G = Z_0$, the insertion of unit elements with $Z_{0k} = Z_0$ between the filter and the generator does not alter the amplitude response and the *magnitude* of the input reflection coefficient.

The following example illustrates the use of Kuroda's identities in the design of commensurate-line bandstop filters. In this example, all series-connected shorted stubs will be replaced by shunt-connected open-circuited stubs.

Example 9–5:

Design a commensurate-line bandstop filter centered at 3000 MHz having a Tchebyscheff response (Fig. 9–50b). The filter specifications are $L_T \geq 20$ dB between 2750 and 3250 MHz and $L_T \leq 1.0$ dB for $f \leq 2500$ MHz and $f \geq 3500$ MHz. Use only shunt-connected, open-circuited stubs. Assume air-insulated TEM lines and a matched 50 ohm system (that is, $R_G = R_L = Z_0 = 50$ ohms).

Solution: Referring to Fig. 9–50b, $f_r = 3000$ MHz, $f_c = 2500$ MHz, $f_s = 2750$ MHz, $L_m \leq 1.0$ dB and $L_s \geq 20$ dB. From Eq. (9–122), $\Omega_c = 3.73$, $\Omega_s = 7.60$, and therefore $\overline{\Omega}_s = 2.04$.

In order to determine the minimum number of elements required to meet the specifications, Eq. (9–124) is evaluated at $\overline{\Omega} = \overline{\Omega}_s$. Namely,

$$20 \text{ dB} = 10 \log \left[1 + 0.26 \cosh^2 (1.34n) \right]$$

and hence $n = (\cosh^{-1} 19.5)/1.34 = 2.73$. Thus it appears that a three-element filter is needed to meet the specifications. The normalized element values for a three-section Tchebyscheff filter with 1.0 dB ripple are given below Eq. (9–118), namely, $g_1 = g_3 = 2.024$ and $g_2 = 0.994$.

Since we wish to use all shunt-connected, open-circuited stubs, the central element of the prototype filter must be a shunt capacitor. For a three-element filter, this requires a T configuration. Kuroda's identities are then used to convert the two series inductors into shunt capacitors. From Eq. (9–119) with $Z_0 = 50$ ohms,

$$G_1 = G_3 = 101.2 \text{ ohms} \quad \text{and} \quad G_2 = 0.0199 \text{ mhos}.$$

Since $\Omega_c = 3.73$, Eq. (9–121) yields

$$L_1 = L_3 = 27.13 \text{ H} \quad \text{and} \quad C_2 = 5.33 \times 10^{-3} \text{ F}.$$

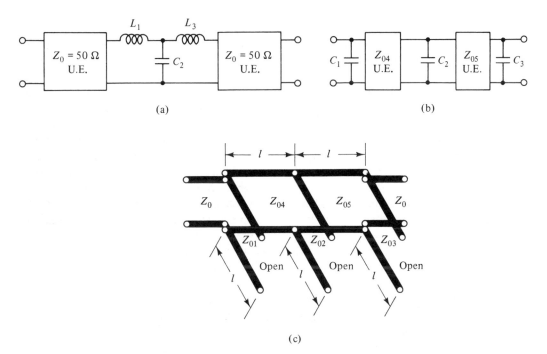

Figure 9–54 Conversion of the prototype filter in Ex. 9–5 into an array of shunt-connected open-circuited line sections.

If the specifications allowed the use of series-connected shorted lines, the commensurate-line version of the filter would be that shown in Fig. 9–52, where in this case $Z_{01} = Z_{03} = 27.13$ ohms, $Z_{02} = 1000/5.33 = 187.6$ ohms, and the stub lines are $\lambda/4$ long at 3000 MHz. However, since all shunt-connected lines are specified, Kuroda's identities must be used to eliminate the series inductors. Figure 9–54 shows the development from the prototype filter to a commensurate-line filter consisting of three open-circuited stubs and two unit elements. Part a shows the prototype filter with 50 ohm unit elements added to the input and output ports. With $R_G = R_L = 50$ ohms, the filter's loss response is unaffected by their presence. Application of Kuroda's identity of the first kind produces the network shown in part b, where $C_1 = C_3 = 7.03 \times 10^{-3}$ F and $Z_{04} = Z_{05} = 77.1$ ohms. By virtue of Richards' transformation, the commensurate-line version of the filter becomes that shown in part c, where

$$Z_{01} = Z_{03} = 1000/7.03 = 142.2 \, \text{ohms} \qquad \text{and} \qquad Z_{02} = 187.6 \, \text{ohms}.$$

Since $\lambda_0 = 10$ cm at 3000 MHz and the lines are air-insulated, $l = 10/4 = 2.50$ cm. This completes the bandstop filter design.

The filter designed in the above example will meet the given loss specifications. In fact since $n = 3$ (rounded off from 2.73), the actual loss at 2750 and 3250 MHz will be slightly greater than 20 dB (see Prob. 9–44). Also, as explained earlier, the commensurate-line filter will have spurious bandstop regions centered at odd multiples of 3000 MHz.

A practical consideration in the design of transmission-line networks is the realizable range of line impedances. For coaxial lines, the usable range of character-istic impedances is from about 5 to 200 ohms. For strip-type transmission lines, the range is even less, typically 10 to 100 ohms. Thus the filter described in the above example is realizable with coaxial lines but not with stripline elements. In some cases, the realizability problem can be overcome by using an alternate design tech-nique. One such technique is the coupled-line filter described in Sec. 9–5b.

Commensurate-line bandpass filter design. The design procedure for commensurate-line bandpass filters is similar to that used for bandstop filters. For the bandpass design, the prototype network is a lumped-element highpass filter. Part a of Fig. 9–55 shows one possible form for the prototype. Like its lowpass counter-part, the alternate form would have a shunt-connected reactance as the first element. Part b of the figure shows the highpass response for a Tchebyscheff design and the resultant bandpass response due to Richards' transformation. The bandpass charac-teristic is achieved by replacing the capacitors by open-circuited line sections and the inductors by short-circuited line sections. As before, all line sections are $\lambda/4$ long at f_r.

The highpass prototype shown in Fig. 9–55a results from a lowpass to high-pass frequency transformation, wherein the frequency variable $\overline{\Omega}$ is replaced by $-1/\overline{\Omega}$ (see, for example, Sec. 11–9 of Ref. 9–14). This transforms the network from a lowpass to a highpass filter since inductors are replaced by capacitors and

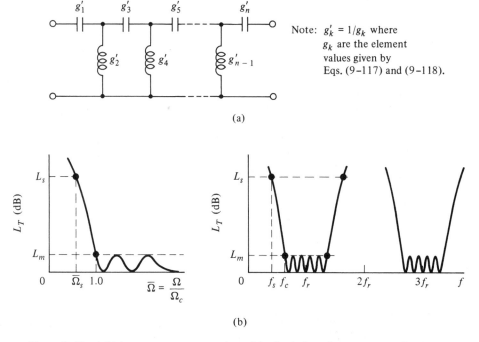

(a)

(b)

Figure 9–55 A highpass prototype network used in the design of commensurate-line band-pass filters. A typical highpass response and its corresponding bandpass response are shown in part b.

capacitors by inductors. As noted in the figure, the normalized element values for the highpass filter (g'_k) are simply the reciprocal of the corresponding element values for the lowpass prototype. By unnormalizing the element values and applying Richards' transformation, a bandpass response centered at f_r is produced. Figure 9–56a shows a three-element bandpass filter based upon this design procedure. A coaxial version is pictured in part b of the figure.

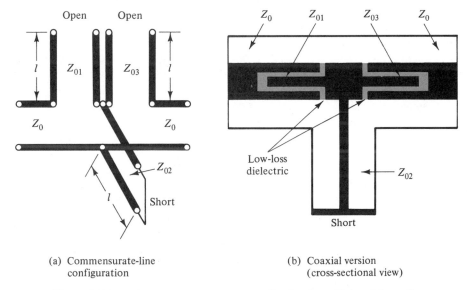

(a) Commensurate-line configuration

(b) Coaxial version (cross-sectional view)

Figure 9–56   A three-element commensurate-line bandpass filter and its realization with coaxial line sections.

To illustrate the procedure, consider the design of a three-element bandpass filter having the same values of f_c, f_r, and L_m as the bandstop filter described in Ex. 9–5. Since the lowpass prototype contained two series inductors and one shunt capacitor, the highpass prototype will consist of two series capacitors with normalized element values $g'_1 = g'_3 = (2.024)^{-1} = 0.494$, and one shunt inductor with $g'_2 = (0.994)^{-1} = 1.006$. For a 50 ohm system, Eq. (9–119) yields $G'_1 = G'_3 = 0.0099$ mhos and $G'_2 = 50.3$ ohms. From Eq. (9–121), $C_1 = C_3 = 2.65 \times 10^{-3}$ F and $L_2 = 13.5$ H, where $\Omega_c = 3.73$. The resultant commensurate-line bandpass filter is shown in Fig. 9–56a, where $Z_{01} = Z_{03} = 1000/2.65 = 377$ ohms and $Z_{02} = 13.5$ ohms.

The loss expressions for Butterworth and Tchebyscheff responses are obtained by replacing $\overline{\Omega}$ with $-1/\overline{\Omega}$ in Eqs. (9–123) and (9–124).

For the Butterworth design,

$$L_T = 10 \log [1 + \overline{\Omega}^{-2n}] \qquad (9\text{–}126)$$

For the Tchebyscheff design,

$$L_T = 10 \log [1 + (\kappa - 1)T_n^2(1/\overline{\Omega})] \qquad (9\text{–}127)$$

where

$$T_n(1/\overline{\Omega}) = \cos\left[n\cos^{-1}(1/\overline{\Omega})\right] \qquad \text{for } |1/\overline{\Omega}| \le 1$$

$$T_n(1/\overline{\Omega}) = \cosh\left[n\cosh^{-1}(1/\overline{\Omega})\right] \qquad \text{for } |1/\overline{\Omega}| \ge 1$$

and $\kappa \equiv$ antilog $(L_m/10)$.

Note that the minus sign disappears when $\overline{\Omega}$ is replaced by $-1/\overline{\Omega}$ since the loss equations are even functions of $\overline{\Omega}$.

The specific bandpass design just described is not practical since the 377 ohm lines cannot be realized in a TEM system. Also, as explained previously, series-connected TEM line sections are difficult and expensive to construct. Therefore, it would be desirable to realize the bandpass filter as a cascade of shunt-connected stubs and unit elements. In the bandstop case, this was achieved by judicious application of Kuroda's identities of the first kind. This technique, however, does not work in the bandpass case since none of Kuroda's identities involve the interchange of series capacitors and shunt inductors. Thus the synthesis of bandpass filters consisting of an array of shunt-connected shorted stubs (or series-connected open-circuited stubs) must follow directly from the prototype networks in Fig. 9–57. In part a, the inductors represent shorted line sections of characteristic impedance $Z_{0k} = L_k$. Both the unit elements and shorted line sections are $\lambda/4$ long at f_r, the center of the passband. The network form in part b is used when the design calls for series-connected open-circuited lines. In this case $Z_{0k} = 1/C_k$. These filter configurations have been analyzed by various workers (Refs. 9–36, 9–37, and 9–39). The procedure developed by Matthaei (Ref. 9–36) is commonly used in the design of both narrowband and wideband filters. Since the filter networks in Fig. 9–57 are equivalent to certain *coupled-line* filters, the design procedure will be outlined in Sec. 9–5b.

Before proceeding to a discussion of coupled-line filters, a few comments are in order regarding the filter configurations in Fig. 9–57. First of all, they are *redun-*

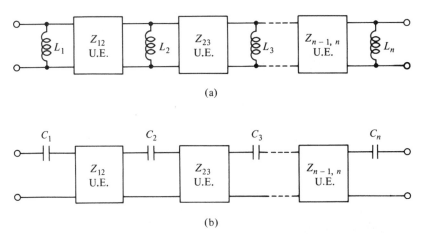

Figure 9–57 Highpass prototype networks used in synthesizing commensurate-line bandpass filters. The networks are converted to bandpass filters by use of Richards' transformation.

dant filters. That is, they contain more degrees of freedom than are needed to realize the required frequency response (see Sec. V in Ref. 9–37).[17] For this reason, one commonly used version sets the characteristic impedance of the unit elements equal to Z_0, the impedance of the input and output lines. A synthesis procedure for this version is described in Appendix I of Ref. 9–37. Tables of normalized admittance values for a maximally flat response are given by Mumford (Ref. 9–41). Figure 9–58 shows a three-stub version, where the impedance of both unit elements is Z_0. As usual, all line sections are $\lambda/4$ long at f_r.

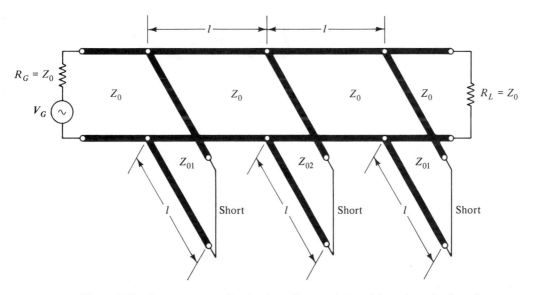

Figure 9–58 A commensurate-line bandpass filter consisting of three shorted stubs and two Z_0 unit elements ($l = \lambda/4$ at f_r).

Another important point regarding the filter networks in Fig. 9–57 is that an n-stub design can be transformed into one that contains anywhere from one to $n - 1$ stubs. This is achieved by judicious use of Kuroda's identities of the second kind. Since the identities are exact, the characteristics of the transformed versions will be identical to those of the original filter. This gives the designer added flexibility in choosing a physically realizable structure that produces the desired loss characteristic. To illustrate the transformation procedure, the three-stub filter in Fig. 9–58 will be reduced to a single shorted stub and two unit elements. The transformation to a two-stub configuration is left to the reader (Prob. 9–48).

Part *a* of Fig. 9–59 shows the three-stub filter of Fig. 9–58, where the S-plane inductors represent shorted stubs. The application of Kuroda's identity (Fig. 9–53b) results in the circuit shown in part *b*. Note that the L_2 inductor (divided by n^2) has been moved to the right of the second transformer. Since the effects of the back-to-back transformers cancel each other, the circuit reduces to that shown in part *c*, where L represents the parallel combination of the three inductors in part *b*. The

---

[17] Nonredundant filters are discussed in Refs. 9–9, 9–23, and 9–40.

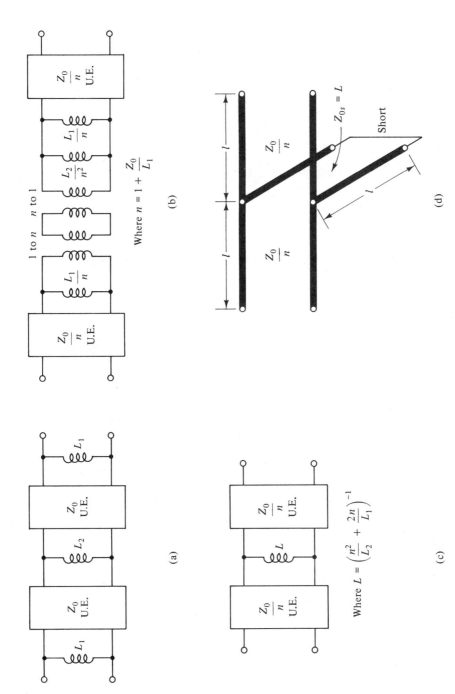

Figure 9–59 The use of Kuroda's identities in converting a three-stub bandpass filter into one containing a single stub.

resultant commensurate-line filter is shown in part *d*. Its filter characteristics are identical to the three-stub version shown in Fig. 9–58.

It is interesting to note that the single-stub version has the same form as the dc return described in Fig. 6–32. Thus, by using Kuroda's identities, broadband dc returns can be designed from the commensurate-line bandpass prototype of Fig. 9–57a. Similarly, dc blocks can be designed from the prototype network in Fig. 9–57b.

9–5b Coupled-Line Filters

The design of capacitively coupled stripline resonators was described in Sec. 9–4b and Fig. 9–34. Another method of capacitive coupling is indicated in part *a* of Fig. 9–60, where $x \ll \lambda$. To achieve wide bandwidths, large capacitance values are required, which means that the gap between conductors must be extremely small. Alternately, the coupling can also be increased by lengthening x, which results in wider, less critical gaps. Part *b* illustrates the case where the coupling length has been increased to a quarter wavelength. Since the length is now an appreciable part of a wavelength, the coupling is *not* simply capacitive. Consequently, the configuration must be analyzed in terms of coupled-line theory (Sec. 8–3b). This configuration represents one example of a large class of filters known as *coupled-line filters*. The procedures used in their design are based on the concepts developed by Jones and Bolljahn (Ref. 9–42) and Ozaki and Ishii (Ref. 9–43). Both bandpass and bandstop filters can be realized using these concepts. Only the bandpass case will be discussed here. The reader is referred to Refs. 9–9 and 9–44 for the design of bandstop coupled-line filters.

Figure 9–61 shows one form of a multisection, coupled-line bandpass filter, wherein the resonator sections are open-circuited at both ends. Both this filter and its dual have been analyzed by Cohn (Ref. 9–45) and Matthaei (Ref. 9–36). The design procedure developed by Matthaei is accurate for both narrowband and wideband versions. The procedure for open-circuited resonators is outlined here.

The first step in deriving the design equations for the coupled-line filter in Fig. 9–61 is to convert the lowpass prototype (Fig. 9–51) into a form that contains impedance inverters and series-connected inductors. This is indicated in Fig. 9–62.

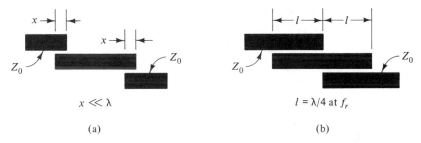

Figure 9–60 The coupled-line filter as an extension of the capacitively coupled bandpass filter.

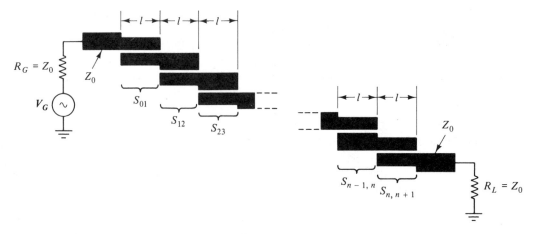

Figure 9–61   Center conductor pattern of a parallel coupled-line stripline filter ($l = \lambda/4$ at f_r).

The properties of an ideal K inverter are summarized in Fig. 9–45a.[18] The correspondence between the lowpass prototype and the network of Fig. 9–62 is obtained by comparing their impedance properties (see Ref. 9–28). This leads to the following relations between the normalized element values and the K inverters.

K-inverter values (end sections)

$$\frac{K_{01}}{Z_0} = \frac{K_{n,n+1}}{Z_0} = \frac{1}{\sqrt{g_1}} \tag{9-128}$$

K-inverter values (interior sections)

$$\frac{K_{k,k+1}}{Z_0} = \frac{1}{\sqrt{g_k g_{k+1}}}, \qquad k = 1, 2, \ldots, n - 1 \tag{9-129}$$

The normalized element values are given by Eqs. (9–117) and (9–118). Some values are tabulated in Table 9–2.

The next step in the derivation is to establish an equivalence between the coupled-line filter in Fig. 9–61 and the prototype network of Fig. 9–62. This is achieved by comparing the image impedance and phase of each section. The image impedance of a parallel-coupled section is derived in Ref. 9–42 and given by

$$Z_I = \frac{\sqrt{(Z_{0e} - Z_{0o})^2 + (Z_{0e} + Z_{0o})^2 \cos^2 \phi}}{2 \sin \phi} \tag{9-130}$$

where $\phi = \pi f / 2 f_r$ and f_r is the frequency at which the coupled-line sections are a quarter wavelength long. Z_{0e} and Z_{0o} are the even- and odd-mode line impedances, respectively. Values for coplanar stripline pairs are given in Fig. 8–28. The image impedance for the interior sections of the prototype in Fig. 9–62 is given by

---

[18] K inverters and unit elements are *not* the same. The unit element behaves like an inverter only at f_r, the frequency at which it is a quarter wavelength long.

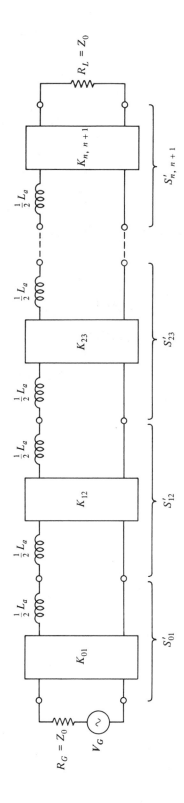

Figure 9–62 Network representation of the coupled-line filter in Fig. 9–61. (After Matthaei, Ref. 9–36.)

Note: $L_a = Z_0$ when $R_G = R_L = Z_0$

$$Z_{k,k+1} = \sqrt{K_{k,k+1}^2 - \left(\frac{\omega' L_a}{2}\right)^2} \tag{9-131}$$

where $\omega' = 2\pi f'$ is the normalized frequency variable for the prototype network.

The lowpass response of the prototype and the bandpass response for the resultant coupled-line filter are shown in Fig. 9–43. By equating the image impedances (within a scale factor) at two frequencies (namely, $f = f_r$ which corresponds to $\omega' = 0$ in the prototype and $f = f_2$ which corresponds to $\omega' = 1$) and the image phases at one frequency (namely, $f = f_r$, which corresponds to $\omega' = 0$), the following design equations can be obtained. (Additional derivation details may be found in Ref. 9–36).

Even- and odd-mode impedances (end sections).

$$(Z_{0e})_{01} = (Z_{0e})_{n,n+1} = Z_0 \left[1 + \left(1 + \frac{g_1}{2}\tan\phi_1\right)^{-1/2} \right]$$

$$(Z_{0o})_{01} = (Z_{0o})_{n,n+1} = Z_0 \left[1 - \left(1 + \frac{g_1}{2}\tan\phi_1\right)^{-1/2} \right] \tag{9-132}$$

where $\phi_1 = \pi f_1/2f_r$ and f_1 is the low end of the passband as defined in Fig. 9–43b.

Even- and odd-mode impedances (interior sections).

$$(k = 1, 2, \ldots, n - 1)$$

$$(Z_{0e})_{k,k+1} = s\left[N_{k,k+1} + \frac{K_{k,k+1}}{Z_0} \right]$$

$$(Z_{0o})_{k,k+1} = s\left[N_{k,k+1} - \frac{K_{k,k+1}}{Z_0} \right] \tag{9-133}$$

where

$$s = \frac{2g_1 Z_0}{2 + g_1 \tan\phi_1} \quad \text{and} \quad N_{k,k+1} = \sqrt{\left(\frac{K_{k,k+1}}{Z_0}\right)^2 + \left(\frac{\tan\phi_1}{2}\right)^2}$$

Equations (9–132), (9–133), (9–128), and (9–129) are the design equations for the coupled-line filter in Fig. 9–61. A similar set of equations for the dual network may be found in Ref. 9–36 and Sec. 10.02 of Ref. 9–8.

The prototype loss equations for Butterworth and Tchebyscheff responses are given by Eqs. (9–123) and (9–124), where $\overline{\Omega}$ is replaced by ω'. The response for the corresponding bandpass filter is obtained by means of an appropriate frequency transformation. For bandwidths up to an octave, the following transformation yields accurate results.

$$\omega' = \frac{2(f - f_r)}{f_2 - f_1} \tag{9-134}$$

where $f_r = (f_1 + f_2)/2$ and f_1 and f_2 are defined in part b of Fig. 9–43. Improved accuracy for wideband designs can be obtained with a more complex transformation. The rationale for its form and use is given in Ref. 9–36.

The design procedure for the coupled-line bandpass filter may now be summarized.

1. Given f_1, f_2, f_b (or f_a), L_s, and L_m, use Eq. (9–134) and the appropriate loss equation [Eq. (9–123) or (9–124)] to determine the number of prototype elements (n) needed to meet the required specifications.
2. Determine the element values (g_k) from Eq. (9–117), Eq. (9–118), or Table 9–2.
3. Use Eqs. (9–128) and (9–129) to obtain the K-inverter values.
4. Use Eqs. (9–132) and (9–133) to determine the even-and odd-mode impedances of the coupled-line sections. Note that the resultant filter contains n resonators and hence $n + 1$ coupled-line sections.
5. Determine the dimensions of the coupled-line sections. The length of all sections are $\lambda/4$ at f_r. The transverse dimensions are related to the values of Z_{0e} and Z_{0o}. For coplanar stripline pairs, the dimensions may be obtained from Fig. 8–28. Data for rectangular bars, circular rods, and overlapping stripline are given in Sec. 5.05 of Ref. 9–8. These three configurations are particularly useful for wideband designs.

The following brief example illustrates the design procedure for the coupled-line filter (Fig. 9–62).

Example 9–6:

Design a Butterworth response, coupled-line, bandpass filter that meets the following specifications. $L_T = 30$ dB at 5500 MHz and 3 dB at 4000 and 5000 MHz. Assume $R_G = R_L = 50$ ohms.

Solution: From Eq. (9–134),

$$\omega_s' = \frac{2(5500 - 4500)}{5000 - 4000} = 2.0$$

since $f_r = (4000 + 5000)/2 = 4500$ MHz. Substituting into Eq. (9–123) yields

$$30 = 10 \log [1 + (2)^{2n}] \qquad \text{or} \qquad n = 4.98.$$

From Table 9–2 for $n = 5$,

$$g_1 = g_5 = 0.618, \qquad g_2 = g_4 = 1.618 \qquad \text{and} \qquad g_3 = 2.000.$$

From Eqs. (9–128) and (9–129),

$$\frac{K_{01}}{Z_0} = \frac{K_{56}}{Z_0} = 1.272, \qquad \frac{K_{12}}{Z_0} = \frac{K_{45}}{Z_0} = 1.000 \qquad \text{and} \qquad \frac{K_{23}}{Z_0} = \frac{K_{34}}{Z_0} = 0.556.$$

From Eqs. (9–132) and (9–133) with $\phi_1 = 4\pi/9$ rad and $Z_0 = 50$ ohms,

$$(Z_{0e})_{01} = (Z_{0e})_{56} = 80.14 \text{ ohms}, \qquad (Z_{0o})_{01} = (Z_{0o})_{56} = 19.86 \text{ ohms}$$

$$(Z_{0e})_{12} = (Z_{0e})_{45} = 44.97 \text{ ohms}, \qquad (Z_{0o})_{12} = (Z_{0o})_{45} = 22.53 \text{ ohms}$$

$$(Z_{0e})_{23} = (Z_{0e})_{34} = 38.68 \text{ ohms}, \qquad (Z_{0o})_{23} = (Z_{0o})_{34} = 26.19 \text{ ohms}$$

The physical realization of parallel coupled-line filters is difficult for bandwidths in excess of 50 percent. This is because the spacing between coupled lines becomes small and hence very critical. For wideband designs, the stub-line filter forms of Fig. 9–57 lead to more easily realizable structures. Any parallel coupled-line filter can be converted to a stub-line configuration by using the equivalences shown in Fig. 9–63. Other useful equivalences are tabulated in Refs. 9–37 and 9–43. Thus a multisection, coupled-line filter with open-circuited resonators can be converted to an array of series-connected, open-circuited stubs, while one with short-circuited resonators can be converted to an array of shunt-connected, short-circuited stubs. Matthaei (Table II in Ref. 9–36) describes a procedure that eliminates the end sections of the coupled-line filter, thus reducing by two the number of sec-

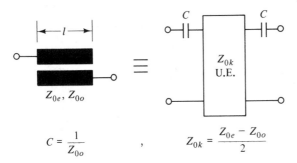

$$C = \frac{1}{Z_{0o}} \qquad , \qquad Z_{0k} = \frac{Z_{0e} - Z_{0o}}{2}$$

C represents an open-circuited
stub of characteristic impedance
$1/C = Z_{0o}$ and length $l = \lambda/4$ at f_r

(a)

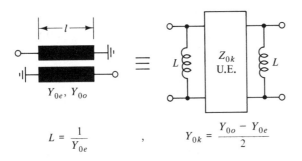

$$L = \frac{1}{Y_{0e}} \qquad , \qquad Y_{0k} = \frac{Y_{0o} - Y_{0e}}{2}$$

L represents a short-circuited
stub of characteristic admittance
$1/L = Y_{0e}$ and length $l = \lambda/4$ at f_r

(b)

Figure 9–63 Coupled-line sections and their Ω-domain equivalences.

tions in the stub-line version. As explained in part *a* of Sec. 9–5, the number of stubs can be reduced further by judicious use of Kuroda's identities. An exact design theory for these and other filter configurations has been developed by Wenzel. For additional information, the reader is referred to Refs. 9–46 and 9–38.

Another form of coupled-line filter is the interdigital filter developed by Bolljahn and Matthaei (Ref. 9–47). Conceptually, it can be viewed as a parallel, coupled-line filter that has been folded back on itself. This is pictured in Fig. 9–64 for a two-resonator configuration. Part *a* shows the parallel-coupled version, where the dashed lines represent voltage nodes. In the interdigital filter (part *b* of the figure), these points are shorted to ground. For an even number of resonators, the input and output ports are on opposite sides of the structure. The dual of the network shown is obtained by interchanging the short circuits and open circuits. Both networks have been analyzed by several workers. Considerable design information is available in Refs. 9–46, 9–47, 9–48, and 9–49.

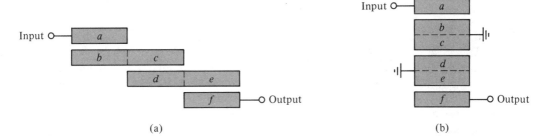

(a)   (b)

Figure 9–64   Development of an interdigital filter from a parallel coupled-line filter.

The coupled-line sections in the interdigital filter are all a quarter wavelength long at the design frequency. By capacitively loading the open-circuited ends, the lengths of the sections can be reduced. This has the advantage that the first spurious response is moved further away from the desired passband. Another version, known as the *comb-line* filter, has all the capacitors on the same side of the resonators. The opposite sides are all grounded. Design details for these configurations may be found in Refs. 9–50, 9–51, and 9–52.

All the filters discussed in this section have been of the Butterworth or Tchebyscheff design. Another type that is widely used at microwave frequencies is the elliptic-function filter. The loss characteristics for a bandpass version is shown in Fig. 9–65. This design is optimum in the sense that for a given minimum loss in the stopband (L_s) and maximum loss in the passband (L_m), the elliptic filter provides the steepest skirt selectivity. Both bandpass and bandstop versions have been developed by several workers. Design procedures as well as tables of element values for the lowpass prototype are given in Refs. 9–9, 9–53, 9–54, and 9–55.

The discussion in this chapter has focused on the loss characteristics of filters, the usual requirements being low loss in the passband and sharp cutoff at the band edges. In some applications, however, the phase characteristics are also very important. For example, filters used in microwave communication systems are required to pass the desired signal with a minimum of distortion. To avoid this distortion, both

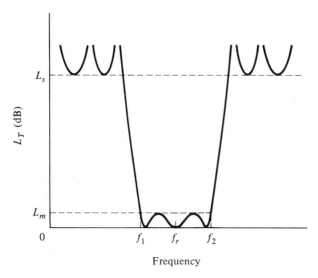

Figure 9–65 Typical response of an elliptic-function bandpass filter.

the insertion loss and group delay of the filter must be constant across the passband. This ideal cannot be achieved perfectly, but is approached remarkably well in some modern filter designs. The group delay will be constant if the transmission phase changes linearly with frequency. For this reason, filters of this type are often called *linear-phase* filters.

All the filters discussed thus far have been essentially dissipationless and of the *minimum-phase* variety. As a result, their phase and loss responses are uniquely related via a pair of Hilbert transforms. Thus the conditions for constant loss and linear phase are not really different, one implies the other. This means that the design techniques that have been described for minimizing passband loss and its variation also favor linear phase across most of the passband. At the band edges, however, the requirement for sharp cutoffs is incompatible with linear phase. To partially circumvent this incompatibility, one must employ more sophisticated circuit topologies. In particular, it has been found useful to introduce a certain amount of cross-coupling between nonadjacent stages in cascaded structures. For additional information on the design and characteristics of linear-phase filters, the reader is referred to Chapter 8 of Ref. 9–9. Also included are several references to the excellent work of J. D. Rhodes, J. H. Cloete, and others.

In concluding this discussion on filters, it should be mentioned that several types have been developed that utilize solid-state microwave technology. These include YIG filters (Refs. 9–21 and 9–22), SAW filters (Ref. 9–56), varactor-tuned filters (Ref. 9–57), and dielectric filters (Refs. 9–19 and 9–20). At millimeter wavelengths, quasi-optical techniques have been widely used in the design of bandpass and bandstop filters (Refs. 9–58 and 9–59).

9–6 SOME FILTER AND RESONATOR APPLICATIONS

Filters and resonators are used to perform a variety of functions in microwave systems. Some of these are described in this section. Most of the explanations given are at an elementary level. References are provided for those interested in more detailed treatments of the various topics.

9–6a Frequency Multiplexers

A frequency multiplexer is a multiport network that separates a wide band of frequencies into a number of narrower bands. This function is indicated in Fig. 9–66 for the case of three frequency channels. The power-frequency spectrum entering the input port is split into three distinct channels centered at f_1, f_2, and f_3. The channels may be contiguous (that is, adjacent to each other) as indicated in the figure, or they may be separated by guard bands. A multiplexer can be realized in a variety of ways. For example, it may consist of an array of bandpass filters properly connected to a multiport junction. This and other multiplexing arrangements will be described briefly.

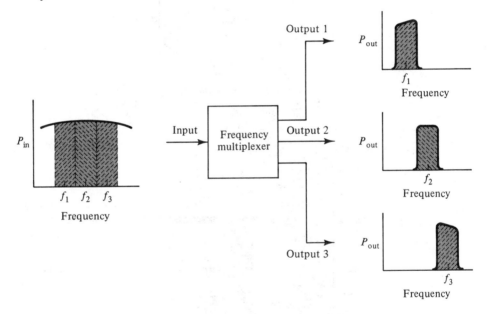

Figure 9–66   A three-channel frequency multiplexer.

Junction-type diplexer.   Two junction-type diplexers are shown in Fig. 9–67. In part *a*, the input line and the two filters are connected in series, while in part *b* they are all connected in shunt. Although the configurations can be expanded to accommodate more outputs, it is seldom done since the connection of three or more filters at a junction is difficult at microwave frequencies. As a result, microwave multiplexing is usually achieved by cascading diplexer networks. Any combination of the four filter types (lowpass, highpass, bandstop, and bandpass) may be used to satisfy the diplexer requirements. The series connection is preferred in waveguide systems, where the junction may be either a bifrucated guide or an *E*-plane tee. For TEM systems, the parallel connection is preferred since the shunt tee junction is easy to construct. In order to insure low loss in the two passbands, the

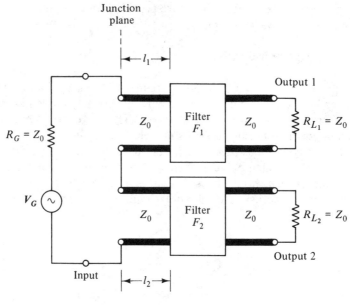

(a) Series connection

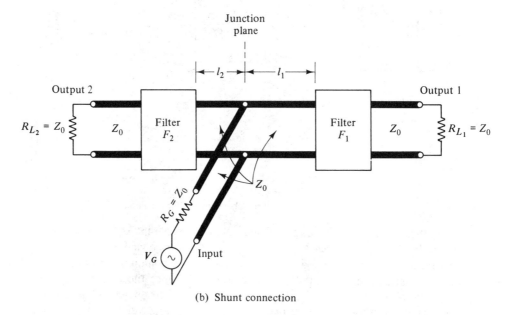

(b) Shunt connection

Figure 9–67. Series- and shunt-connected frequency diplexers.

diplexer input must be well matched across *both* frequency bands. For the series connection, this requires that $Z_1 + Z_2 = Z_0$, where Z_1 and Z_2 are the input impedances of filters F_1 and F_2 *at the junction plane*. For the shunt connection, the match requirement is $Y_1 + Y_2 = Y_0$, where Y_1 and Y_2 are the input admittances of filters F_1 and F_2 *at the plane of the junction*.

For diplexers with guard bands, the usual approach is to use filters that are well matched in their respective passbands. This means that $Z_1 = Z_0$ and $Y_1 = Y_0$ in the frequency band centered at f_1, while $Z_2 = Z_0$ and $Y_2 = Y_0$ in the frequency band centered at f_2. Since the characteristic impedance of the l_1 line is Z_0, Z_1, and Y_1 are independent of l_1. Similarly, Z_2 and Y_2 are independent of l_2. In their respective stopbands, however, the filters are highly reflective and therefore, in the f_2 band, Z_1 and Y_1 are functions of l_1, while in the f_1 band, Z_2 and Y_2 are functions of l_2. With the individual filters well matched in their respective passbands, the input match condition is satisfied by adjusting the line lengths in the following manner.

For the *series* connection (Fig. 9–67a):

1. Set l_1 so that $Z_1 = 0$ in the f_2 band.
2. Set l_2 so that $Z_2 = 0$ in the f_1 band.

For the *shunt* connection (Fig. 9–67b):

1. Set l_1 so that $Y_1 = 0$ in the f_2 band.
2. Set l_2 so that $Y_2 = 0$ in the f_1 band.

Since the line lengths can only be optimized at the center frequencies of the two passbands, this technique is particularly useful in narrowband applications.

Design procedures for diplexers with contiguous channels are described in Refs. 9–11, 9–60, and 9–61. One method involves the use of maximally flat complementary filters. Figure 9–68a shows a shunt-connected version that utilizes three-element filters. In the Ω domain, the capacitors represent open-circuited line stubs, while the inductors represent short-circuited lines. Therefore, the filter on the left has a bandpass response and the one on the right has a bandstop response. Part *b* of Fig. 9–68 shows the frequency response at the two output ports. The input admittance of the diplexer is $Y_p + Y_s$, where Y_p and Y_s represent the input admittances of the bandpass and bandstop filters, respectively. A perfect match at all frequencies requires $Y_p + Y_s = Y_0$. Filter pairs that satisfy this condition are said to be complementary. For the series connection, π filters are used and the match condition requires $Z_p + Z_s = Z_0$. As indicated in the figure, the complementary filters must have the same center frequency and 3 dB bandwidth. For a given Butterworth lowpass prototype, the normalized element values for the complementary highpass filter are simply the reciprocal of the corresponding lowpass values. The proof of this statement is given in several sources (for example, Ref. 9–60). Thus for the three-element filters in Fig. 9–68, the unnormalized values are related by the following expressions.

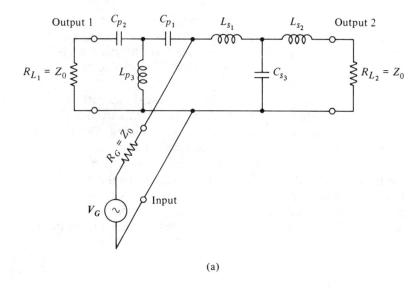

(a)

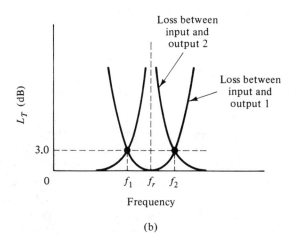

(b)

Figure 9–68 A diplexer network with shunt-connected maximally flat complementary filters.

$$\Omega_c C_{p_1} = \frac{1}{\Omega_c L_{s_1}} \qquad \Omega_c C_{p_2} = \frac{1}{\Omega_c L_{s_2}} \qquad \text{and} \qquad \Omega_c L_{p_3} = \frac{1}{\Omega_c C_{s_3}} \qquad (9\text{–}135)$$

where use has been made of Eq. (9–121).

The g values for a *singly terminated,*[19] three-element, lowpass Butterworth prototype are $g_1 = 3/2$, $g_2 = 4/3$, and $g_3 = 1/2$. Suppose, for example, $Z_0 = 50$ ohms, $f_r = 2000$ MHz, and $f_c = 1600$ MHz. Since from Eq. (9–122), $\Omega_c = 3.078$,

---

[19] The complementary condition requires that the sum of the squares of the normalized transfer function magnitudes equal unity. Since the transfer function assumes an *ideal* source at the input, singly terminated filter values must be used.

the lowpass element values become

$$L_{s_1} = 24.37 \text{ H}, \qquad L_{s_2} = 8.12 \text{ H}, \qquad \text{and} \qquad C_{s_3} = 8.66 \times 10^{-3} \text{F}$$

where use has been made of Eqs. (9–119) and (9–121). From Eq. (9–135), the highpass elements are

$$C_{p_1} = 4.33 \times 10^{-3} \text{F}, \qquad C_{p_2} = 13.0 \times 10^{-3} \text{F}, \qquad \text{and} \qquad L_{p_3} = 12.18 \text{ H}.$$

As before, the capacitors represent quarter-wave, open-circuited lines with $Y_{0s} = C$, while the inductors represent quarter-wave shorted lines with $Z_{0s} = L$. Figure 9–68b shows the loss characteristics of the resultant bandpass-bandstop diplexer. For the above values, both filters will have 3 dB bandwidths of 800 MHz centered at 2000 MHz. Since the filters are complementary, the input match is unity at all frequencies including the crossover regions.

The technique described here cannot be applied directly to Tchebyscheff filter pairs since their impedance functions are not exactly complementary. However, a modified set of element values have been derived by Veltrop and Wilds (Ref. 9–61) that result in quasi-complementary, Tchebyscheff-response diplexers having low input SWR values over a wide frequency range.

Two other diplexer arrangements that produce a well-matched input throughout both frequency bands are shown in Fig. 9–69. The one in part a consists of a pair of 0–90° hybrids and two identical filters (F_1). To understand its operation, assume the filters have bandpass responses centered at f_1 and the hybrids are broadband. For an input signal $a_1/\underline{0}$, the waves arriving at the top and bottom of plane A are $a_1/\sqrt{2}/\underline{-90°}$ and $a_1/\sqrt{2}/\underline{0}$, respectively. Signals within the passband of the filters arrive at plane B with negligible loss. Therefore, the waves at the top and bottom of plane B are $a_1/\sqrt{2}/\underline{-90° - \phi}$ and $a_1/\sqrt{2}/\underline{-\phi}$, where ϕ represents the phase delay through the filters. The action of the second hybrid yields an output at port 1 equal to $a_1/\underline{-90° - \phi}$. Thus, input signals within the filter passband arrive at output 1 with negligible loss. The reflected signals at the top and bottom of plane A are $\Gamma a_1/\sqrt{2}/\underline{-90°}$ and $\Gamma a_1/\sqrt{2}/\underline{0}$, where Γ is the input reflection coefficient of the filters. Since the filters are highly reflective in their stopband ($|\Gamma| \approx 1$), the resultant signal at output 2 is

$$\{\Gamma a_1/\sqrt{2}/\underline{-90°} + \Gamma a_1/\sqrt{2}/\underline{-90°}\}/\sqrt{2} \approx a_1/\underline{-90°}$$

Thus input signals *outside* the filter passband arrive at output 2 with negligible loss. For ideal hybrids and identical filters, the diplexer input is perfectly matched at all frequencies since the net reflected signal at the input is

$$\{\Gamma a_1/\sqrt{2}/\underline{0} + \Gamma a_1/\sqrt{2}/\underline{-180°}\}/\sqrt{2} = 0$$

The diplexer arrangement shown in part b of Fig. 9–69 produces similar results. It is known as a *traveling-wave directional filter* and is described in Refs. 9–62 and 9–63. It consists of a pair of broadband, highly directive couplers connected as shown. The configuration is similar to the traveling-wave resonator (Fig. 9–20b) except that it contains *two* directional couplers. As explained in Sec. 9–4 [below Eq. (9–45)], resonance occurs when the electrical length of the loop is a multiple of a

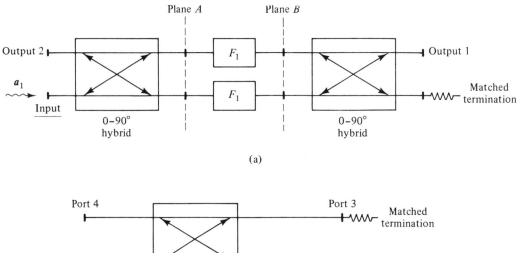

(a)

(b)

Figure 9–69 Diplexer configurations that utilize directional coupler pairs.

wavelength. Coale (Ref. 9–62) shows that at resonance the input signal is delivered to port 4 while port 2 is decoupled. At frequencies far removed from resonance, the input signal is delivered to port 2 while port 4 is decoupled. Since the couplers are highly directive, port 3 is decoupled at all frequencies. Thus for input signals at port 1, the output at port 4 exhibits a bandpass response, while that at port 2 yields the complementary bandstop response. The input is well matched at all frequencies and hence n traveling-wave diplexers can be cascaded to produce a multiplexer with $n + 1$ channels.

Other diplexer arrangements have been developed by many workers. These include cavity-type directional filters (Refs. 9–64 and 9–65) and circulator-coupled multiplexers (Refs. 9–66 and 9–67).

9–6b Resonator Applications

The applications of microwave resonators are also many and varied. Their use in the design of narrowband filters was described in Sec. 9–4. Resonators also form an integral part of microwave oscillators and narrowband amplifiers (Ref. 9–5). Other applications include frequency meters, echo boxes, frequency discriminators and stabilization circuits as well as high-power testing and the evaluation of microwave materials. Some of these are described in this section.

High-power testing. The testing of microwave components under high peak power conditions often requires electric and magnetic fields greater than can be produced by available sources. Resonance phenomena can be used to create the required field values. As an example, consider the two-port inductively coupled resonator in Fig. 9–36. The net power flow toward the load is $a^2 - b^2$, where a and b are the rms values of the forward traveling wave (**a**) and the reflected wave (**b**) within the cavity. Under the assumption of infinite unloaded Q, the net power flow must be the same at all planes including the input and output ports. Therefore,

$$P_{in} = P_L = a^2 - b^2 \qquad (9\text{--}136)$$

Within the cavity, $b = |\Gamma|a$, where Γ is the reflection coefficient due to the inductive iris nearest the load. Thus,

$$P_{in} = P_L = a^2(1 + |\Gamma|)(1 - |\Gamma|) \qquad (9\text{--}137)$$

At resonance, a voltage maximum occurs at the midpoint of the cavity. Therefore, the rms value at that point is

$$V_{max} = V^+ + V^- = \sqrt{Z_0}(a + b) = a\sqrt{Z_0}(1 + |\Gamma|) \qquad (9\text{--}138)$$

where use has been made of Eq. (4–66). Also, since the input SWR is unity at resonance (assuming infinite unloaded Q), $P_{in} = P_A$, the available source power. Hence, $P_{in} = V_{in}^2/Z_0$, where V_{in}, the voltage at the input port, is half the open-circuit voltage of the source. By combining Eqs. (9–137) and (9–138), the following result is obtained.

$$\frac{V_{max}}{V_{in}} = \sqrt{\frac{1 + |\Gamma|}{1 - |\Gamma|}} = \sqrt{\text{SWR in the cavity}} \qquad (9\text{--}139)$$

The SWR due to a shunt susceptance can be deduced from Eq. (7–10). Namely,

$$SWR = \left(\frac{\bar{B} + \sqrt{\bar{B}^2 + 4}}{2}\right)^2 \qquad (9\text{--}140)$$

where $\bar{B}$ is the normalized magnitude of the shunt susceptance. Thus, the ratio of maximum voltage in the cavity to the input (or load) voltage is

$$\frac{V_{max}}{V_{in}} = \frac{\bar{B} + \sqrt{\bar{B}^2 + 4}}{2} \qquad (9\text{--}141)$$

Note that the ratio is greater than unity for all values of $\bar{B}$. In fact for $\bar{B} > 3$, $V_{max}/V_{in} \approx \bar{B}$.

One disadvantage of this technique is that the length of the device under test must be small compared to a wavelength since the voltage maximum only exists at the midpoint of the cavity.

An alternate method of high-power testing utilizes the traveling-wave resonator (Fig. 9–20b). Since its operation involves traveling waves rather than standing waves, there is no restriction on the length of the device being tested. From Eq.

(9–45) with $A_t = 0$, the ratio of rms values at resonance ($\phi = 2n\pi$) is

$$\frac{b_4}{a_1} = \frac{k_c}{1 - \sqrt{1 - k_c^2}} \tag{9–142}$$

where k_c is the coupling coefficient of the directional coupler. For example, if $k_c = 1/\sqrt{2}$ (a 3 dB coupler), the fields within the resonator are 2.41 times greater than the input values.

Microwave frequency discriminators. A frequency discriminator produces an output voltage or current that is proportional to the frequency deviation $(f - f_r)$ of the input microwave signal, where f_r denotes the reference frequency. When the incoming signal is a frequency-modulated wave with a carrier frequency f_r, the output of an ideal discriminator will faithfully reproduce the modulating signal. Three of the more common arrangements are the Pound discriminator, the dual-mode cavity circuit, and the bridge-type configuration. The first two are analyzed in Sec. 8.2 of Ref. 9–4. One version of the bridge discriminator is shown in Fig. 9–70a. The microwave portion consists of a 0–180° hybrid (for example, a magic tee), a 0–90° hybrid (for example, a short-slot coupler), a two-port transmission cavity, and a pair of amplitude detectors (D_1 and D_2). The differential amplifier is used to amplify the difference of the detected voltages. A typical transfer characteristic is shown in part b of Fig. 9–70, where the frequency deviation scale has been normalized to the reference frequency f_r.

The following analysis of the bridge discriminator assumes identical detectors and an infinite unloaded Q for the cavity. Let the incoming microwave signal be denoted by the phasor $a/\underline{0}$, where a^2 equals the input power. Therefore, the signals at planes x_1 and x_2 are both $a/\sqrt{2}\,/\underline{0}$. At planes y_1 and y_2, the signals become $a/\sqrt{2}\,/\underline{0}$ and $\sqrt{T}a/\sqrt{2}\,/\underline{-\phi}$, where T is the transmission loss ratio of the cavity and ϕ is its phase delay.[20] From Eqs. (9–15) and (9–16) with Q_U infinite,

$$T = \frac{1}{1 + (Q_L\epsilon_f)^2} \quad \text{and} \quad \phi = \arctan Q_L\epsilon_f \tag{9–143}$$

where $\epsilon_f \approx 2(f - f_r)/f_r = 2\Delta f/f_r$. The action of the 0–90° hybrid produces the following signals at planes z_1 and z_2, the input to the detectors.

$$\boldsymbol{a}_{z_1} = \frac{a}{2}[1 - \sqrt{T}\sin\phi - j\sqrt{T}\cos\phi]$$

and $\hspace{9cm}$ (9–144)

$$\boldsymbol{a}_{z_2} = \frac{a}{2}[\sqrt{T}\cos\phi - j(1 + \sqrt{T}\sin\phi)]$$

For small signals, the amplitude detectors have a square-law characteristic. This

---

[20] The phase delay from plane x_1 to plane y_1 is actually βl, where l is the distance between the planes. Likewise, the phase delay between x_2 and y_2 is $\beta l + \phi$. Since the operation of the discriminator depends upon the phase difference between the two paths, βl can be set equal to zero without any loss in generality.

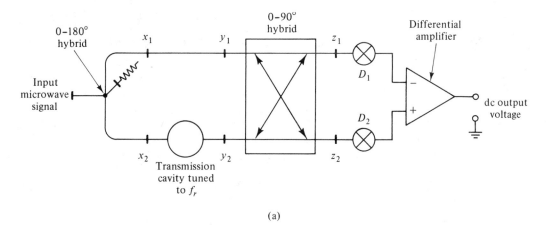

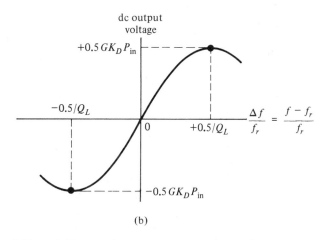

(b)

Figure 9–70   A bridge-type frequency discriminator and its transfer characteristic.

means that the output dc voltage is proportional to the incident microwave power. Thus, $V_{D_1} = K_D a_{z_1}^2$ and $V_{D_2} = K_D a_{z_2}^2$, where a_{z_1} and a_{z_2} represent rms values and K_D is the conversion factor of the detectors. For semiconductor-type detectors K_D usually ranges from 10 to 1000 mV/mW. Assuming square-law operation,

$$V_{D_1} = \frac{a^2}{4} K_D[1 + T - 2\sqrt{T}\sin\phi]$$

and   (9–145)

$$V_{D_2} = \frac{a^2}{4} K_D[1 + T + 2\sqrt{T}\sin\phi]$$

Denoting the voltage gain of the differential amplifier as G and noting that a^2 equals P_{in}, the power level of the incoming signal, the output of the discriminator circuit is

$$V_{out} = G(V_{D_2} - V_{D_1}) = GK_D P_{in} \sqrt{T} \sin \phi \qquad (9\text{-}146)$$

Making use of Eq. (9–143) and noting that $\sin (\arctan Q_L \epsilon_f) = \sqrt{T} Q_L \epsilon_f$,

$$V_{out} = \frac{GK_D P_{in}}{1 + (Q_L \epsilon_f)^2} Q_L \epsilon_f \qquad (9\text{-}147)$$

where $\epsilon_f = 2(f - f_r)/f_r = 2\Delta f/f_r$. This transfer function is plotted in part b of Fig. 9–70. It can be shown (Prob. 9–55) that the peaks occur at $\Delta f/f_r = \pm 0.5/Q_L$ and have values of $\pm 0.5 GK_D P_{in}$.

The discriminator slope is approximately linear for $|Q_L \epsilon_f| < 0.25$. Its value is

$$\frac{d V_{out}}{df} = \frac{2GK_D P_{in} Q_L}{f_r} = \frac{2GK_D P_{in}}{3 \text{ dB } BW} \qquad (9\text{-}148)$$

For example, if $P_{in} = 0.5$ mW, $K_D = 400$ mV/mW, $G = 100$, and the 3 dB cavity bandwidth is 40 MHz, the slope equals 1000 mV/MHz or 1.0 mV/KHz.

In order for the discriminator to accurately reproduce the modulating signal of an incoming FM wave, the carrier and all significant sidebands must satisfy the condition $|Q_L \epsilon_f| < 0.25$ or $|f - f_r| < (3 \text{ dB Bandwidth})/8$. For an FM wave with carrier frequency f_r, a modulating frequency f_m and a maximum frequency deviation ΔF, the significant sidebands lie in the range $|f - f_r| \leq (f_m + \Delta F)$. Therefore the minimum cavity bandwidth required for linear demodulation is $8(f_m + \Delta F)$. A variation of this discriminator circuit is described in Ref. 9–68.

The frequency discriminator can also be used to stabilize the frequency of an oscillator and reduce its FM noise spectra. A sample of the oscillator output is fed to the discriminator input via a directional coupler. The output of the discriminator is fed back to the power supply of the oscillator as a control voltage. (The frequency of most microwave oscillators is voltage controllable.) When the loop gain of the negative feedback circuit is sufficient, a significant reduction in frequency drift and FM noise will result. The loaded Q of the cavity affects both the loop gain and the capture range of the stabilization circuit. Loop gain is directly proportional to Q_L. Capture range is defined as the frequency range over which the slope of the discriminator curve (Fig. 9–70b) is positive (namely, $|\Delta f| < 0.5 f_r/Q_L$). Outside this range, the slope is reversed causing the feedback to become positive, which makes the circuit unstable.

Another method of frequency stabilization is shown in Fig. 9–71, wherein a high-Q cavity is connected to one arm of a tee junction and the oscillator to another arm. In waveguide systems an E-plane tee is commonly used. The circuit makes use of the fact that the output power and frequency of an oscillator is a function of its terminating impedance. By using measured data, contours of constant output power and constant frequency can be plotted on the impedance coordinates of a Smith chart. The resultant chart is known as a *Rieke diagram* (Ref. 9–7). By proper choice of l_1, the constant power contours can be set roughly parallel to the constant resistance circles and the constant frequency contours roughly parallel to the constant reactance lines.

The impedance to the right of plane A (Z_A) is frequency sensitive (due to the cavity) and a function of l_2. In order to frequency stabilize the oscillator, the length

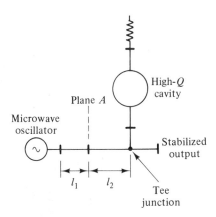

Figure 9–71 A frequency stabilization circuit.

l_2 is adjusted so that the reactance versus frequency slope of Z_A is the negative of the oscillator's frequency versus reactance characteristic. Suppose, for example, l_1 is adjusted so that the frequency of the oscillator decreases when the imaginary part of Z_A increases and the length l_2 is adjusted so that the reactance slope of Z_A is positive. If now the oscillator frequency tends to increase, the reactive portion of Z_A increases. By virtue of the oscillator's frequency-reactance characteristic, this tends to *decrease* the oscillator frequency. Thus the circuit produces negative feedback which stabilizes the frequency of the oscillator. This arrangement is particularly useful in improving short-term frequency stability and reducing FM oscillator noise. As in the previous case, increasing Q_L increases the degree of stabilization but reduces the capture range.

9–6c Periodic Structures

Transmission lines that are periodically loaded with identical obstacles are referred to as *periodic structures*. In most cases, the obstacles are reactive elements such as inductive or capacitive irises and shorted-line sections. These structures are an integral part of many microwave oscillators and broadband amplifiers since their phase velocities can be made comparable to the velocity of an electron beam. This is a necessary condition for significant energy transfer between the beam and the electromagnetic wave. Microwave devices that make use of this interaction include magnetrons, traveling-wave amplifiers, and backward-wave oscillators.

Periodic structures are usually analyzed from both a circuit and field point of view. These concepts are described briefly in the following paragraphs.

Circuit analysis of a reactively loaded line.   Two examples of periodic structures with reactive discontinuities are pictured in Fig. 9–72. Also shown are the equivalent circuits for a unit cell of each structure. In both cases, the line sections between reactances are parallel-plate transmission lines operating in the TEM mode with the electric field direction as shown. The analysis that follows is equally valid for coaxial lines or any other TEM transmission line. Periodically loaded waveguides are analyzed in Refs. 9–4 and 9–7. In part *a* of Fig. 9–72, the line is

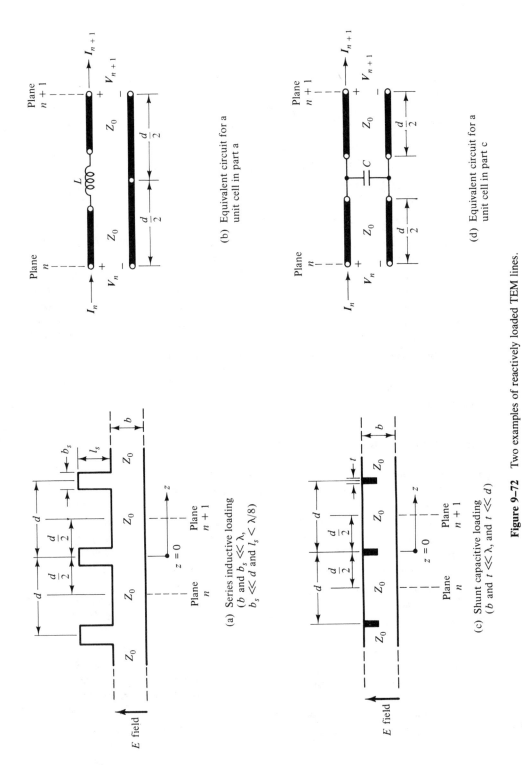

(a) Series inductive loading
(b and $b_s \ll \lambda$,
$b_s \ll d$ and $l_s < \lambda/8$)

(b) Equivalent circuit for a
unit cell in part a

(c) Shunt capacitive loading
(b and $t \ll \lambda$, and $t \ll d$)

(d) Equivalent circuit for a
unit cell in part c

Figure 9–72 Two examples of reactively loaded TEM lines.

periodically loaded by series-connected shorted stubs. For $b_s \ll \lambda$ and $l_s < \lambda/8$, each stub can be approximated by an inductance L as indicated in part b. The structure in part c is periodically loaded with capacitive-type septums. For $t \ll \lambda$, each septum can be approximated by a shunt capacitance C as indicated in part d. Both structures may be analyzed using the ABCD matrix method. The inductively loaded case will be presented here. Since the analysis of the capacitively loaded case is similar, it is left to the reader.

Because of the periodic nature of the structure, its propagation characteristics can be determined by analyzing the unit cell in part b of Fig. 9–72. Assuming loss-less line sections, the ABCD matrix of the circuit is

$$
\begin{bmatrix} A & B \\ C & D \end{bmatrix} = \begin{bmatrix} \cos \dfrac{kd}{2} & jZ_0 \sin \dfrac{kd}{2} \\ \dfrac{j}{Z_0} \sin \dfrac{kd}{2} & \cos \dfrac{kd}{2} \end{bmatrix} \begin{bmatrix} 1 & jX \\ 0 & 1 \end{bmatrix} \begin{bmatrix} \cos \dfrac{kd}{2} & jZ_0 \sin \dfrac{kd}{2} \\ \dfrac{j}{Z_0} \sin \dfrac{kd}{2} & \cos \dfrac{kd}{2} \end{bmatrix}
$$

$$
= \begin{bmatrix} \left(\cos kd - \dfrac{\overline{X}}{2} \sin kd \right) & jZ_0 \left\{ \sin kd + \dfrac{\overline{X}}{2}(1 + \cos kd) \right\} \\ \dfrac{j}{Z_0} \left\{ \sin kd - \dfrac{\overline{X}}{2}(1 - \cos kd) \right\} & \left(\cos kd - \dfrac{\overline{X}}{2} \sin kd \right) \end{bmatrix}
$$

$$(9\text{–}149)$$

where $\overline{X} \equiv \omega L/Z_0$, Z_0 is the characteristic impedance of the TEM line sections and $k \equiv \omega\sqrt{\mu\epsilon}$ is the phase constant. Wave propagation along the z axis of the periodic structure requires that the voltage and current at plane $(n + 1)$ be the same as their values at plane n, *except for a propagation factor*. That is,

$$V_{n+1} = V_n e^{-\gamma d} \quad \text{and} \quad I_{n+1} = I_n e^{-\gamma d} \tag{9–150}$$

where $\gamma = \alpha + j\beta$ is the propagation constant of the periodic structure. As will be shown shortly, the periodically loaded line has alternating passbands and stopbands. In the passbands, $\alpha = 0$ and hence $\gamma = j\beta$, while in the stopbands $\gamma = \alpha$. Equation (9–150) implies that the unit cell may be viewed as a transmission line of length d and propagation constant γ. Furthermore, one can define a characteristic impedance (Z_{0p}) for the unit cell and hence for the periodic structure. Namely,

$$Z_{0p} \equiv \frac{V_n}{I_n} = \frac{V_{n+1}}{I_{n+1}} \tag{9–151}$$

The ABCD matrix elements for a transmission-line section are given in Table C–1. With l replaced by d and Z_{0k} by Z_{0p}, the matrix becomes

$$
\begin{bmatrix} A & B \\ C & D \end{bmatrix} = \begin{bmatrix} \cosh \gamma d & Z_{0p} \sinh \gamma d \\ \dfrac{\sinh \gamma d}{Z_{0p}} & \cosh \gamma d \end{bmatrix} \tag{9–152}
$$

By comparing the matrix elements in Eqs. (9–149) and (9–152), the following relations are obtained.

$$\cosh \gamma d = \cos kd - \frac{\bar{X}}{2} \sin kd \qquad (9\text{-}153)$$

and

$$Z_{0p} = Z_0 \left[\frac{2 \sin kd + \bar{X}(1 + \cos kd)}{2 \sin kd - \bar{X}(1 - \cos kd)} \right]^{1/2} \qquad (9\text{-}154)$$

Equation (9-154) is derived by equating the ratio B/C for the two matrices. It should be noted that the characteristic impedance of a periodic structure is not unique but depends upon the choice of terminal planes for the unit cell. The propagation constant, on the other hand, is unique (Prob. 9-57).

Equation (9-153) states that the propagation constant of the periodic structure is a function of frequency since $k = \omega \sqrt{\mu \epsilon}$ and $\bar{X} = \omega L/Z_0$. At frequencies where the right-hand side of the equation is greater than unity, γ is real (that is, $\gamma = \alpha$). This defines the stopband regions of the structure. Conversely, when $\cosh \gamma d < 1$, γ is imaginary (that is, $\gamma = j\beta$), which defines the passband regions. The passband regions are indicated in the k-β diagram of Fig. 9-73. This diagram is also referred to as an ω-β diagram since k merely represents a normalized form of the frequency variable ω. With γ imaginary, $\cosh j\beta d = \cos \beta d$ and hence the k-β relation in the passband regions is given by

$$\cos \beta d = \cos kd - \frac{\bar{X}}{2} \sin kd \qquad (9\text{-}155)$$

The edges of the passbands occur when the magnitude of the right-hand side equals unity. The lower edges are $k = 0$, π/d, $2\pi/d$, $3\pi/d$, etc., which corresponds to $\beta = 0$, π/d, 0, π/d, etc. The upper edges of the passband regions ($\beta = \pi/d$, 0,

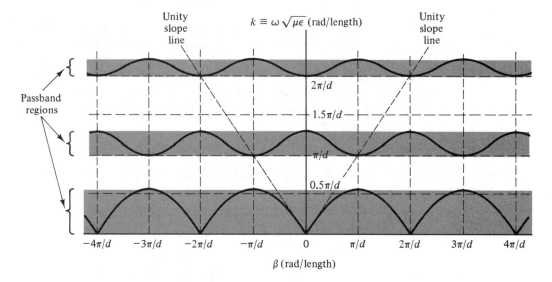

Figure 9-73 The k-β diagram for the inductively loaded TEM line shown in Fig. 9-72a.

π/d, 0, etc.) are obtained by setting Eq. (9–155) to -1 for $\beta d = \pi$ and to $+1$ for $\beta d = 0$. This leads to the following equations for computing their values.

$$\tan \frac{kd}{2} = \frac{2}{\overline{X}} \quad \text{and} \quad \tan \frac{kd}{2} = -\frac{\overline{X}}{2} \quad (9\text{--}156)$$

For example if $\overline{X} = kd$ (that is, $L = Z_0 d \sqrt{\mu\epsilon}$), the upper edges of the first and second passbands are $k = 1.72/d$ and $4.05/d$ rad/length, respectively. If the loading is increased (L larger), the passbands become narrower in frequency. Conversely, reducing the loading widens the passbands. In fact, for $L = 0$, the TEM structure becomes an all-pass network, and since $\beta = \pm\omega\sqrt{\mu\epsilon}$ for TEM transmission, the k-β characteristic reduces to a pair of straight lines with unity slopes. These are indicated in Fig. 9–73 as dashed lines. For non-TEM transmission (for example, waveguides), $k^2 = \beta^2 + \beta_c^2$, where $\beta_c = 2\pi/\lambda_c$ and λ_c is the cutoff wavelength. In this case, the k-β characteristic is a hyperbola with the k-axis intercept at $2\pi/\lambda_c$.

The alternating passband and stopband regions of the periodic structure can be explained in terms of wave reflections from the reactive obstacles. This explanation is similar to that used to describe the multisection bandstop filter in Sec. 9–4. In the passband, the reflections from the obstacles tend to cancel resulting in low-loss transmission through the structure. In the stopband, these reflections are additive. Since the infinite line has an infinite number of obstacles, the reflection is total and hence wave transmission cannot occur.

Group velocity (v_g) and phase velocity (v_p) are defined in Sec. 5–5. For the periodic structure,

$$v_g \equiv \frac{d\omega}{d\beta} = v\frac{dk}{d\beta} \quad \text{and} \quad v_p \equiv \frac{\omega}{\beta} = v\frac{k}{\beta} \quad (9\text{--}157)$$

where $v = 1/\sqrt{\mu\epsilon} = c/\sqrt{\mu_R \epsilon_R}$. Thus for a given operating point on the k-β characteristic, the group velocity is proportional to the slope of the k-β curve at that point. The phase velocity, on the other hand, is proportional to the slope of a line drawn from the origin to the operating point. Note that for the first passband in Fig. 9–73, both v_g and v_p are positive and less than v for $\beta \leq \pi/d$. This is because both slopes are less than that of the unity slope line. For the second passband, however, v_g is negative and v_p is positive when $\beta \leq \pi/d$. Waves in which the group and phase velocities are in opposite directions are known as *backward waves*. For the periodic structures pictured in Fig. 9–72, the primary and all odd-numbered passbands exhibit forward wave propagation, while the even-numbered passbands exhibit backward-wave propagation. For $\beta \leq \pi/d$, the dual configurations (that is, series C and shunt L loading), support backward waves for the primary and odd-numbered passbands and forward waves for the even-numbered ones. Electronically tunable microwave oscillators utilize backward type waves.

In its passbands, a periodic structure has the properties of a low-loss transmission line with phase constant β and characteristic impedance Z_{0p}. In order to properly match the structure to a load termination and/or a generator, Z_L and Z_G must equal Z_{0p}. This is not an easy task since, as indicated by Eq. (9–154), Z_{0p} is frequency dependent. Despite this, matching techniques similar to those described in

Chapter 4 are sometimes used. These will not be discussed here. For additional information, the reader is referred to Secs. 8.4 and 8.5 of Ref. 9–2 and Sec. 7.6 of Ref. 9–4.

Space harmonics in periodic structures. Figure 9–73 indicates that the k-β characteristic for each passband is a periodic function of β. This follows from the fact that, in addition to the primary solution, Eq. (9–155) has solutions given by $\beta_n d = \beta_0 d + 2n\pi$, where β_0 is the primary solution $(0 \leq \beta_0 \leq 2\pi/d)$ and n is an integer. Thus for a given frequency, (that is, k fixed), the periodic structure has an infinite set of phase constants, each one corresponding to a specific field variation in the z direction. These field patterns are known as *space* (or *Hartree*) *harmonics* and their phase constants are denoted by β_n. Space harmonics provide a useful way of analyzing the field characteristics of a periodic structure.

Consider the periodically loaded TEM lines pictured in Fig. 9–72. In the absence of reactive loading, the transverse electric and magnetic field components are $E_y = \hat{E}_y e^{-j\beta z}$ and $H_x = \hat{H}_x e^{-j\beta z}$, where $\hat{E}_y$ and $\hat{H}_x$ are functions of x and y only. (The y axis has been chosen parallel to the electric field vector shown.) Thus, the value of $\hat{E}_y$ (and $\hat{H}_x$) is the same at all points along the line and hence is constant over the length of a unit cell. When the line is reactively loaded, however, this is no longer true since standing waves now exist along the line sections. Furthermore, the field in the vicinity of the loading is significantly different than in the connecting lines. As a result, the field pattern is a periodic function of z and hence can be expressed in terms of an infinite Fourier series. Thus $\hat{E}_y$ and $\hat{H}_x$ can be written in the following forms.

$$\hat{E}_y(x, y, z) = \sum_{-\infty}^{\infty} E_n(x, y)e^{-j2n\pi z/d}$$

and   (9–158)

$$\hat{H}_x(x, y, z) = \sum_{-\infty}^{\infty} H_n(x, y)e^{-j2n\pi z/d}$$

where $\hat{E}_y$ and $\hat{H}_x$ are now functions of z. The Fourier coefficients E_n and H_n are obtained from a knowledge of the field patterns [namely, $\hat{E}_y(x, y, z)$ and $\hat{H}_x(x, y, z)$] in the usual manner. Namely,

$$E_n(x, y) = \frac{1}{d} \int_0^d \hat{E}_y(x, y, z)e^{j2n\pi z/d} dz$$

and   (9–159)

$$H_n(x,y) = \frac{1}{d} \int_0^d \hat{H}_x(x, y, z)\, e^{j2n\pi z/d}\, dz$$

Equation (9–150) states that the voltage and current (and hence the fields) at the terminal planes of a unit cell can only differ by the propagation factor $e^{-\gamma d}$, where $\gamma = j\beta$ in the passbands of the periodic structures. Since the choice of terminal planes is arbitrary, the field at *any* two points that are separated by a distance d along the z axis can only differ by the factor $e^{-j\beta d}$. By making use of Eq. (9–158), the complete field expression can thus be written as

$$E_y = e^{-j\beta z}\hat{E}_y(x, y, z) = \sum_{-\infty}^{\infty} E_n(x, y)e^{-j\beta_n z} \qquad (9\text{-}160)$$

and

$$H_x = e^{-j\beta z}\hat{H}_x(x, y, z) = \sum_{-\infty}^{\infty} H_n(x, y)e^{-j\beta_n z}$$

where $\beta_n = \beta + 2n\pi/d$ and E_n and H_n are defined in Eq. (9-159). Hence the effect of reactive loading on the field pattern can be expressed in terms of an infinite set of space harmonics, each with its own phase constant β_n.

The group and phase velocities of the nth space harmonic are given by $d\omega/d\beta_n$ and ω/β_n, respectively. Therefore,

$$v_{g_n} = \left(\frac{d\beta_n}{d\omega}\right)^{-1} = \frac{d\omega}{d\beta} \qquad \text{and} \qquad v_{p_n} = \frac{\omega}{\beta + 2n\pi/d} \qquad (9\text{-}161)$$

Note that for a given frequency, the group velocity is the *same* for all the space harmonics. On the other hand, phase velocity decreases for increasing values of n. As indicated previously, this is useful in situations that require significant electron-wave interaction. Finally, it must be stressed that space harmonics should not be confused with higher-order modes in transmission lines. Higher-order modes can be excited independent of each other. Space harmonics, however, must be simultaneously present and in the correct proportions to satisfy the boundary conditions of the periodic structure.

For a more detailed discussion of periodic structures and other slow-wave transmission lines, the reader is referred to Refs. 9-1, 9-2, 9-4, and 9-7.

REFERENCES

Books

9-1. Ramo, S., J. R. Whinnery, and T. Van Duzer, *Fields and Waves in Communication Electronics,* John Wiley and Sons, Inc., New York, 1965.

9-2. Collin, R. E., *Foundations for Microwave Engineering,* McGraw-Hill Book Co., New York, 1966.

9-3. Ragan, G. L., *Microwave Transmission Circuits,* Rad. Lab. Series, vol. 9, McGraw-Hill Book Co., New York, 1951.

9-4. Altman, J. L., *Microwave Circuits,* D. Van Nostrand Co., New York, 1964.

9-5. Liao, S. Y., *Microwave Devices and Circuits,* Prentice-Hall, Inc. Englewood Cliffs, NJ, 1980.

9-6. Ghose, R. N., *Microwave Circuit Theory and Analysis,* McGraw-Hill Book Co., New York, 1963.

9-7. Slater, J. C., *Microwave Electronics,* D. Van Nostrand Co., Princeton, NJ, 1950.

9-8. Matthaei, G. L., L. Young, and E.M.T. Jones, *Microwave Filters, Impedance-Matching Networks and Coupling Structures,* McGraw-Hill Book Co., New York, 1964.

9–9. Malherbe, J.A.G., *Microwave Transmission Line Filters,* Artech House, Dedham, MA, 1979.

9–10. Matsumoto, A., Microwave Filters and Circuits, in *Advances in Microwaves,* Sup. 1, L. Young, ed., Academic Press, New York, 1970.

9–11. Matthaei, G. L. and E. G. Cristal, Theory and Design of Diplexers and Multiplexers, in *Advances in Microwaves,* vol. 2, L. Young, ed., Academic Press, New York, 1967.

9–12. Saad, T. S., *Microwave Engineers' Handbook,* vol. 1, Artech House, Dedham, MA, 1971.

9–13. Howe, Jr. H., *Stripline Circuit Design,* Artech House, Dedham, MA, 1974.

9–14. Weinberg, L., *Network Analysis and Synthesis,* McGraw-Hill Book Co., New York, 1962.

9–15. Balabanian, N., *Network Synthesis,* Prentice-Hall, Inc., Englewood Cliffs, NJ, 1958.

Articles

9–16. A 28-year index of the *IEEE Trans. Microwave Theory and Techniques,* MTT-29, June 1981.

9–17. Hansen, W. W., A Type of Electrical Resonator. *J. Applied Phys.,* 9, October 1938, pp. 654–663.

9–18. Sethares, J. C. and S. J. Naumann, Design of Microwave Dielectric Resonators. *IEEE Trans. Microwave Theory and Techniques,* MTT-14, January 1966, pp. 2–7.

9–19. Cohn, S. B., Microwave Bandpass Filters Containing High-Q Dielectric Resonators. *IEEE Trans. Microwave Theory and Techniques,* MTT-16, April 1968, pp. 218–227.

9–20. Gerdine, M. A., A Frequency-Stabilized Microwave Band-Rejection Filter Using High Dielectric Constant Resonators. *IEEE Trans. Microwave Theory and Techniques,* MTT-17, July 1969, pp. 354–359.

9–21. Carter, P. S., Magnetically-Tunable Microwave Filters Using Single-Crystal Yttrium-Iron-Garnet Resonators. *IRE Trans. Microwave Theory and Techniques,* MTT-9, May 1961, pp. 252–260.

9–22. Matthaei, G. L., Magnetically Tunable Band-Stop Filters. *IEEE Trans. Microwave Theory and Techniques,* MTT-13, March 1965, pp. 203–212.

9–23. Young, L., Microwave Filters—1965. *IEEE Trans. Microwave Theory and Techniques,* MTT-13, September 1965, pp. 489–508.

9–24. Rizzi, P. A., Microwave Filters Utilizing the Cutoff Effect. *IRE Trans. Microwave Theory and Techniques,* MTT-4, January 1956, pp. 36–40.

9–25. Torgow, E. N., Hybrid Junction—Cutoff Waveguide Filters. *IRE Trans. Microwave Theory and Techniques,* MTT-7, January 1959, pp. 163–167.

9–26. Mumford, W. W., Maximally-Flat Filters in Waveguide. *Bell System Tech. J.,* 27, October 1948, pp. 684–713.

9–27. Riblet, H. J., Synthesis of Narrow-band Direct-Coupled Filters. *Proc. IRE,* 40, October 1952, pp. 1219–1223.

9–28. Cohn, S. B., Direct-Coupled-Resonator Filters. *Proc. IRE,* 45, February 1957, pp. 187–196.

9–29. Levy, R., Theory of Direct-Coupled-Cavity Filters. *IEEE Trans. Microwave Theory and Techniques,* MTT-15, June 1967, pp. 340–348.

9–30. Riblet, H. J., A Unified Discussion of High-Q Waveguide Filter Design Theory. *IRE Trans. Microwave Theory and Techniques,* MTT-6, October 1958, pp. 359–368.

9–31. Williams, A. E., A Four-Cavity Elliptic Waveguide Filter. *IEEE Trans. Microwave Theory and Techniques,* MTT-18, December 1970, pp. 1109–1114.

9–32. Atia, A. E. and A. E. Williams, Narrow-Bandpass Waveguide Filters. *IEEE Trans. Microwave Theory and Techniques,* MTT-20, April 1972, pp. 258–265.

9–33. Saal, R. and E. Ulbrich, On the Design of Filters by Synthesis. *IRE Trans. Circuit Theory,* CT-5, December 1958, pp. 284–327.

9–34. Richards, P. I., Resistor-Transmission-Line Circuits. *Proc. IRE,* 36, February 1948, pp. 217–220.

9–35. Levy, R., A General Equivalent Circuit Transformation for Distributed Networks. *IEEE Trans. Circuit Theory,* CT-12, September 1965, pp. 457–458.

9–36. Matthaei, G. L., Design of Wide-Band (and Narrow-Band) Band-Pass Microwave Filters on the Insertion Loss Basis. *IRE Trans. Microwave Theory and Techniques,* MTT-8, November 1960, pp. 580–593.

9–37. Wenzel, R. J., Exact Design of TEM Microwave Networks Using Quarter-Wave Lines. *IEEE Trans. Microwave Theory and Techniques,* MTT-12, January 1964, pp. 94–111.

9–38. Carlin, H. J. and W. Kohler, Direct Synthesis of Band-Pass Transmission Line Structures. *IEEE Trans. Microwave Theory and Techniques,* MTT-13, May 1965, pp. 283–296.

9–39. Mumford, W. W., An Exact Design Technique for a Type of Maximally-Flat Quarter-Wave-Coupled Band Pass Filter. *PTGMTT Symposium Digest,* 1963, pp. 57–61.

9–40. Horton, M. C. and R. J. Wenzel, General Theory and Design of Optimum Quarter-Wave TEM Filters. *IEEE Trans. Microwave Theory and Techniques,* MTT-13, May 1965, pp. 316–327.

9–41. Mumford, W. W. Tables of Stub Admittances for Maximally Flat Filters Using Shorted Quarter-Wave Stubs. *IEEE Trans. Microwave Theory and Techniques,* MTT-13, September 1965, pp. 695–696.

9–42. Jones, E.M.T. and J. T. Bolljahn, Coupled-Strip-Transmission Line Filters and Directional Couplers. *IRE Trans. Microwave Theory and Techniques,* MTT-4, April 1956, pp. 78–81.

9–43. Ozaki, N. and J. Ishii, Synthesis of a Class of Strip-Line Filters. *IRE Trans. Circuit Theory,* CT-5, June 1958, pp. 104–109.

9–44. Schiffman, B. M. and G. L. Matthaei, Exact Design of Microwave Band-Stop Filters. *IEEE Trans. Microwave Theory and Techniques,* MTT-12, January 1964, pp. 6–15.

9–45. Cohn, S. B., Parallel-Coupled Transmission-Line-Resonator Filters. *IRE Trans. Microwave Theory and Techniques,*, MTT-6, April 1958, pp. 223–231.

9–46. Wenzel, R. J., Exact Theory of Interdigital Band-Pass Filters and Related Coupled Structures. *IEEE Trans. Microwave Theory and Techniques,* MTT-13, September 1965, pp. 559–575.

9–47. Matthaei, G. L., Interdigital Band-Pass Filters. *IRE Trans. Microwave Theory and Techniques,* MTT-10, November 1962, pp. 479–491.

9–48. Pyle, J. R., Design Curves for Interdigital Band-Pass Filters. *IEEE Trans. Microwave Theory and Techniques,* MTT-12, September 1964, pp. 559–567.

9–49. Cristal, E. G., New Design Equations for a Class of Microwave Filters. *IEEE Trans. Microwave Theory and Techniques,* MTT-19, May 1971, pp. 486–490.

9–50. Robinson, L. A., Wideband Interdigital Filters with Capacitively Loaded Resonators. *G-MTT Symposium Digest*, 1965, pp. 33–37.

9–51. Matthaei, G. L., Comb-Line Band-Pass Filters of Narrow or Moderate Bandwidths. *Microwave Journal*, 6, August 1963, pp. 82–91.

9–52. Wenzel, R. J., Synthesis of Combline and Capacitively Loaded Interdigital Bandpass Filters of Arbitrary Bandwidth. *IEEE Trans. Microwave Theory and Techniques*, MTT-19, August 1971, pp. 678–686.

9–53. Levy, R. and I. Whiteley, Synthesis of Distributed Elliptic-Function Filters from Lumped-Constant Prototypes. *IEEE Trans. Microwave Theory and Techniques*, MTT-14, November 1966, pp. 506–517.

9–54. Horton, M. C. and R. J. Wenzel, The Digital Elliptic Filter—A Compact Sharp-Cut-off Design for Wide Bandstop or Bandpass Requirements. *IEEE Trans. Microwave Theory and Techniques*, MTT-15, May 1967, pp. 307–314.

9–55. Schiffman, B. M. and L. Young, Design Tables for an Elliptic-Function Band-Stop Filter ($N = 5$). *IEEE Trans. Microwave Theory and Techniques*, MTT-14, October 1966, pp. 474–482.

9–56. Tancrell, R. H. and M. G. Holland, Acoustic Surface Wave Filters. *Proc. IEEE*, 59, March 1971, pp. 393–409.

9–57. Benguerel, A. P. and N. S. Nahman, A Varactor Tuned UHF Coaxial Filter. *IEEE Trans. Microwave Theory and Techniques*, MTT-12, July 1964, pp. 468–469.

9–58. Cohen, J. and J. J. Taub, Dielectric Slab Interference Filters Having Wide Stop Bands. *Proc. IEEE*, 54, July 1966, pp. 1016–1017.

9–59. Saleh, A.A.M., An Adjustable Quasi-Optical Bandpass Filter—Parts 1 and 2. *IEEE Trans. Microwave Theory and Techniques*, MTT-22, July 1974, pp. 728–739.

9–60. Wenzel, R. J., Application of Exact Synthesis Methods of Multichannel Filter Design. *IEEE Trans. Microwave Theory and Techniques*, MTT-13, January 1965, pp. 5–15.

9–61. Veltrop, R. G. and R. B. Wilds, Modified Tables for the Design of Optimum Diplexers. *Microwave Journal*, 7, June 1964, pp. 76–80.

9–62. Coale, F. S., A Traveling-Wave Directional Filter. *IRE Trans. Microwave Theory and Techniques*, MTT-4, October 1956, pp. 256–260.

9–63. Standley, R. D., Frequency Response of Strip-Line Traveling Wave Directional Filters. *IEEE Trans. Microwave Theory and Techniques*, MTT-11, July 1963, pp. 264–265.

9–64. Nelson, C. E., Circularly Polarized Microwave Cavity Filters. *IRE Trans. Microwave Theory and Techniques*, MTT-5, April 1957, pp. 136–147.

9–65. Williams, R. L., A Three-Cavity Circularly Polarized Waveguide Directional Filter Yielding a Maximally Flat Response. *IRE Trans. Microwave Theory and Techniques*, MTT-10, September 1962, pp. 321–328.

9–66. Fox, A. G., S. E. Miller, and M. T. Weiss, Behavior and Applications of Ferrites in the Microwave Region. *Bell System Tech. J.*, 34, January 1955, pp. 5–103.

9–67. Brown, J. and J. Clark, A Unique Solid-State Diplexer. *IRE Trans. Microwave Theory and Techniques*, MTT-10, July 1962, pg. 298.

9–68. Robinson, S. J., Comment on Broadband Microwave Discriminator. *IEEE Trans. Microwave Theory and Techniques*, MTT-12, March 1964, pp. 255–256.

PROBLEMS

9–1. The expressions for the loaded Q of the series and parallel resonant circuits in Fig. 9–1 are given by Eqs. (9–6) and (9–7). Derive these two expressions from the relationships given in Eqs. (9–8) and (9–9).

9–2. Derive Eqs. (9–10) and (9–11).

9–3. Prove that Eq. (9–23) is also valid for an open-circuited line section at parallel resonance.

9–4. A shorted air-insulated coaxial line is connected to a 75 ohm generator. The line is 2.50 cm long and has a characteristic impedance of 15 ohms.
   (a) Calculate the two lowest parallel resonant frequencies. Also compute the external Q at these two frequencies.
   (b) Determine the unloaded and loaded Q at the lowest resonant frequency if the attenuation constant for the line is 0.20 dB/m.

9–5. A shorted 3.0 cm long, air-insulated 25 ohm coaxial line is shunted by a thin wire having an inductance of 0.18 nH.
   (a) Calculate the primary resonant frequency of the inductively coupled resonator.
   (b) Determine the external Q when the resonator is connected to a matched 50 ohm generator.

9–6. A capacitively coupled stripline resonator is shown in Fig. 9–11. With $C = 0.15$ pF,
   (a) Determine the resonator length l required for series resonance at 5.0 GHz. Assume $Z_{01} = 50$ ohms and $\epsilon_R = 4.0$ for the strip transmission line.
   (b) Use Eq. (9–8) with $R_G = 0$ to calculate the unloaded Q of the resonator if $\alpha = 0.02$ dB/cm for the Z_{01} line. (Note: $R = Re(Z_{in})$ at 5.0 GHz.)

9–7. Derive Eq. (9–33).

9–8. A tapped half-wave resonator (Fig. 9–12) is constructed using air-insulated TEM lines. The input impedance is zero at 3.0 and 3.6 GHz.
   (a) At what frequency (between 3.0 and 3.6 GHz) will the input admittance be zero?
   (b) What is the external Q if $Z_{01} = 20$ ohms and $Z_0 = 50$ ohms?

9–9. A tapped half-wave resonator (Fig. 9–12) has the following values: $Z_{01} = Z_0 = 50$ ohms, $l_1 = 0.26\lambda$ and $l = 0.50\lambda$ at f_r. Given $\alpha l_1 \approx \alpha(l - l_1) = 0.02$ Np, calculate Y_{in}/Y_0 at f_r.

9–10. Consider the coaxial resonator in part a of Fig. 9–13 with $a = 1.0$ cm, $b = 2.0$ cm, and $l = 5.0$ cm. The coaxial line section is air insulated.
   (a) Calculate the resonant frequencies for $n = 1$ and $n = 2$. Assume a TEM coaxial mode.
   (b) Assume a TE_{11} coaxial mode and repeat the calculation. Make use of Eqs. (5–16) and (5–40).

9–11. A rectangular cavity (Fig. 9–14) resonates in the TM_{111} mode at 5.0 GHz. Given $a = 8.0$ cm and $b = 6.0$ cm, calculate the resonant frequencies for the TE_{101}, $TE_{102,}$ and TE_{111} modes. Assume an air-filled cavity.

9–12. A quartz-filled ($\epsilon_R = 3.8$) cylindrical cavity resonates at 8.0 GHz in the TE_{013} mode. Determine the cavity dimensions if its diameter equals its length.

9–13. A cylindrical cavity with $D = l$ (Fig. 9–16) resonates at 6.0 GHz in the TE_{011} mode.
    (a) Calculate the unloaded Q if the cavity is air filled and has copper walls.
    (b) Compare the result in part a with the unloaded Q of a cubic cavity operating in the TE_{101} mode and having the same resonant frequency.

9–14. An adjustable length air-filled cylindrical cavity operating in the TE_{211} mode is used as a tunable resonant circuit. With $D = 4.0$ cm and $l = 3.0$ cm, the resonant frequency is 8.84 GHz. What is the maximum length that insures TE_{211} mode operation without interference from other cavity modes? What is the resonant frequency for this length? (Hint: Use the mode chart in Fig. 9–17).

9–15. The unloaded Q of the TE_{012} mode in a cylindrical cavity may be determined from Eq. (9–84). What value of D/l yields the highest unloaded Q? Calculate its value when $\lambda_r = 5.0$ cm and $\delta_s = 10^{-4}$ cm.

9–16. An inductively coupled rectangular cavity is shown in Fig. 9–19. The cavity is air filled and has the following dimensions: $l = 6.40$ cm, $a_1 = a = 2.50$ cm, and $b_1 = b = 1.0$ cm.
    (a) What value of $\bar{B}_L$ at f_r will cause the cavity to resonate at 7.5 GHz in the TE_{102} mode?
    (b) Calculate Q_L assuming infinite unloaded Q.

9–17. An inductively coupled rectangular cavity is shown in Fig. 9–19. Calculate the resonant frequency of the TE_{101} mode when $l = 4.0$ cm, $a_1 = a = 2.0$ cm, and $b_1 = b = 1.0$ cm. The cavity is air filled and $\bar{B}_L = 5.0$ at f_r.

9–18. Calculate $\bar{Y}_{in} = Y_{in}/Y_0$ at $0.99f_r$ and $1.01f_r$ for the cavity described in the previous problem. The resonator input is at the plane of the inductive iris. Assume $\bar{B}_L$ is proportional to λ_g and Q_U is infinite.

9–19. A traveling-wave resonator is pictured in Fig. 9–20b. The amplitude of the wave in the ring is a function of the coupling coefficient k_c.
    (a) Calculate the required value of k_c for $b_4 = 5a_1$. Assume $A_t = 0.1$ dB and $\phi = 4\pi$ rad.
    (b) Determine b_4/a_1 when $\phi = 5\pi$ rad and k_c is the same as in part a.

9–20. A capacitively coupled bandstop filter is shown in Fig. 9–28. Given $Z_0 = Z_{01} = 50$ ohms, $f_r = 1.0$ GHz, and $Q_E = 20$,
    (a) Determine the values of C and l. Assume $\lambda = 0.8\lambda_0$.
    (b) Calculate the exact insertion loss at 1.1, 1.5, and 1.9 GHz. Compare your results with the loss characteristics in Fig. 9–29.

9–21. A capacitively coupled bandstop filter (Fig. 9–28) has a 3 dB bandwidth of 150 MHz centered at 6000 MHz. Calculate the insertion loss at f_r if the unloaded Q of the resonator is 2000.

9–22. A waveguide bandstop filter (Fig. 9–30b) has a loaded Q of 50 and a center frequency of 8.0 GHz. All guides are air filled.
    (a) Determine the two frequencies at which $L_l = 3.0$ dB. Assume infinite unloaded Q and $a = 2.50$ cm.
    (b) Calculate d and l when $b_1 = 0.8b$ and $p = 1$. Use the expression in Fig. 7–15a to obtain d.

9–23. Given $a = 4.0$ cm, $b = 2.0$ cm, $b_1 = 1.0$ cm, and $p = 1$, design a waveguide bandstop filter (Fig. 9–30b) having 3 dB loss points at 5.7 and 5.8 GHz. Calculate $\bar{B}_{L_r}$ and l. Assume $Q_U \gg Q_L$ and air-filled waveguides.

9–24. A waveguide bandstop filter (Fig. 9–30b) has infinite loss at 10.0 GHz. Given $\bar{B}_{L_r} = 2.57$, $a = 2.0$ cm, $b = b_1 = 1.0$ cm, $l = 2.00$ cm, and $p = 1$, calculate L_l at

10.1, 12.0, and 15.0 GHz using Eq. (9–18) with ϵ_f replaced by ϵ_λ. Assume air-filled guides.

9–25. Compare the loss values in Prob. 9–24 with an exact insertion loss analysis based on the equivalent circuit in Fig. 9–30a. Assume $\bar{B}_L$ is directly proportional to λ_g.

9–26. Derive a loss expression (in terms of $Q_T = \omega_r L/Z_0$ and ϵ_f) for a three-section version of the quarter-wave coupled bandstop filter shown in Fig. 9–33. Assume that the element values of the middle section are $L/2$ and $2C$ and that $\beta l \approx 90°$ at all frequencies of interest.

9–27. Design a stripline bandpass filter (Fig. 9–34) having a center frequency of 3.0 GHz and a 3 dB bandwidth of 100 MHz. Given $Z_0 = 50$ ohms, $\lambda = 0.7\lambda_0$, and an infinite unloaded Q,
(a) Determine l and C.
(b) Calculate the two frequencies at which the transducer loss is 13 dB.

9–28. Prove that the resonance condition for a stripline bandpass filter with $Z_{01} \neq Z_0$ is given by $\phi_r = \pi - \arctan\{2\bar{X}_{C_r}\bar{Z}_{01}/(\bar{X}_{C_r}^2 + 1 - \bar{Z}_{01}^2)\}$, where $\bar{Z}_{01} \equiv Z_{01}/Z_0$ and $\bar{X}_{C_r} \equiv X_{C_r}/Z_0$.

9–29. An inductively coupled cavity resonator is used as a bandpass filter (Fig. 9–36). The insertion loss is 0.2 dB at 9.00 GHz and 3.2 dB at 8.95 and 9.05 GHz. Calculate Q_U assuming an air-filled cavity with $a = 2.25$ cm.

9–30. Design a waveguide bandpass filter (Fig. 9–36) having 3 dB loss points at 5.7 and 5.8 GHz. Determine the iris opening d and cavity length l. Assume an air-filled cavity with infinite unloaded Q and $a = 4.00$ cm.

9–31. The tapped-line filter in part a of Fig. 9–37 has the following parameter values: $Z_0 = 50$ ohms, $Z_{01} = 25$ ohms, $l_1 = 0.28\lambda_r$, $l_2 = 0.22\lambda_r$, $\lambda = 0.7\lambda_0$, and $f_r = 2000$ MHz. Use Eqs. (9–105) and (9–106) to determine the 3 dB loss points f_1 and f_2.

9–32. Consider a two-stage version of the bandpass filter shown in part a of Fig. 9–37. It consists of two identical resonators spaced $\lambda/4$ apart along the Z_0 line. Prove that

$$L_I = L_T = 10 \log [1 + (\bar{B}^4/4)]$$

where $\bar{B} = (\tan \beta l_1 + \tan \beta l_2)Z_0/Z_{01}$. Assume the spacing is approximately $\lambda/4$ at all frequencies of interest.

9–33. A tapped-line bandpass filter is shown in part b of Fig. 9–37. It is desired that the 1.0 dB loss points occur at 4.35 and 4.65 GHz. Given $Z_0 = 50$ ohms, $\lambda = 0.7\lambda_0$, $l_1 = 1.40$ cm, and $l_2 = 1.00$ cm, calculate the required values of Z_{01} and Z_{02}.

9–34. Calculate the insertion loss at 7.5 and 8.5 GHz for the waveguide filter described in Ex. 9–3. Use Eq. (9–108) with ϵ_f replaced by ϵ_λ.

9–35. Calculate the insertion loss at 8.0 GHz for the waveguide filter described in Ex. 9–3. The unloaded Q of each resonator is 4000.

9–36. Verify the transmission-line equivalence in Fig. 9–46 by comparing the ABCD matrices of the two networks.

9–37. Calculate the normalized element values (g_k) for a four-element lowpass filter having a Tchebyscheff response with 0.5 dB loss ripple. Also determine the normalized load value (r) for the filter form in Fig. 9–51.

9–38. (a) Use Eq. (9–123) to determine the transducer loss at 1.2, 1.6, and 2.2 GHz for the filter design in Ex. 9–4.
(b) Derive the normalized ABCD matrix of the filter (Fig. 9–52a) and verify the 1.6 GHz transducer loss value obtained in part a.

9–39. A Butterworth-type bandstop filter has been proposed to satisfy the following requirements: f_r = 5.0 GHz, 3 dB bandwidth = 1.8 GHz, and $L_T \geq$ 26 dB at 4.6 and 5.4 GHz. Can the specifications be met with a three-element filter?

9–40. Design a three-element, T-type bandstop filter having the specification in Prob. 9–39 except that the stopband loss requirement is reduced to 20 dB. Use a Butterworth design with commensurate-length open- and shorted-line sections. Assume $R_G = R_L = Z_0$ = 50 ohms and $\lambda = \lambda_0$.

9–41. A commensurate-line bandstop filter has the response indicated in Fig. 9–50b. Given f_r = 5.0 GHz, f_c = 4.1 GHz, f_s = 4.6 GHz, $L_m \leq$ 1.0 dB, and $L_s \geq$ 26 dB, determine the minimum number of elements n needed to meet the specifications.

9–42. Verify the first equivalence in part a of Fig. 9–53 by comparing the ABCD matrices of the two networks. Repeat for the second equivalence in part b of the figure.

9–43. Use Kuroda's identities to convert the commensurate-line bandstop filter in Ex. 9–4 to the filter form in part c of Fig. 9–54. Assume $\lambda = \lambda_0$.

9–44. (a) Calculate the transducer loss at 2750 and 3250 MHz for the commensurate-line bandstop filter described in Ex. 9–5 and Fig. 9–54c.
(b) Use the ABCD matrix method to verify that L_T = 1.0 dB at 3500 MHz.

9–45. Use the ABCD matrix method to verify that for the circuit in Fig. 9–58, $|\Gamma_{in}|$ = 0.684 at $0.5f_r$ and $1.5f_r$ when $Z_{01} = 2Z_0$ and $Z_{02} = Z_0$.

9–46. Design a maximally flat, three-element bandpass filter having a center frequency of 4.0 GHz and a 3 dB bandwidth of 1.0 GHz. Use the filter form in Fig. 9–56. Assume $R_G = R_L = Z_0$ = 50 ohms and $\lambda = \lambda_0$.

9–47. The transformed version of the three-stub filter in Fig. 9–58 is shown in part d of Fig. 9–59. Given $Z_{01} = 2Z_0$ and $Z_{02} = Z_0$, use the ABCD matrix method to verify that $|\Gamma_{in}|$ = 0.684 at $0.5f_r$ and $1.5f_r$.

9–48. Use Kuroda's identities to transform the bandpass filter in Fig. 9–58 into an equivalent symmetrical filter consisting of two shorted stubs separated by two unit elements.

9–49. Design a three-section (n = 2) coupled-line bandpass filter (Fig. 9–61) with a maximally flat response. The specifications are L_T = 3 dB at 2.5 and 3.5 GHz. Determine the length of the coupled-line sections and their even- and odd-mode impedances. Assume Z_0 = 50 ohms and $\lambda = 0.8\lambda_0$.

9–50. Design a four-section (n = 3) coupled-line bandpass filter (Fig. 9–61) with an equal-ripple response. The specifications are Z_0 = 50 ohms and a maximum loss of 0.2 dB between 7.0 and 8.0 GHz. Calculate the length and the even- and odd-mode impedances of the coupled-line sections. Assume $\lambda = \lambda_0$.

9–51. Calculate the transducer loss at 6.0 and 9.0 GHz for the bandpass filter described in Prob. 9–50.

9–52. A three-section coupled-line bandpass filter (Fig. 9–61) has the following impedance values:

$(Z_{0e})_{01} = (Z_{0e})_{23}$ = 75 ohms, $(Z_{0o})_{01} = (Z_{0o})_{23}$ = 25 ohms,

$(Z_{0e})_{12}$ = 47 ohms and $(Z_{0o})_{12}$ = 27 ohms.

Use the equivalences in Fig. 9–63 to convert it to an equivalent four-stub configuration.

9–53. Verify Eq. (9–145).

9–54. A 1.0 mW, 5.0 GHz microwave signal is frequency modulated with a 9.0 MHz sine wave. Given ΔF = 3.0 MHz, K_D = 75 mV/mW, and G = 50,

(a) Determine the minimum cavity bandwidth that will accurately reproduce the 9.0 MHz modulation signal.

(b) Use the above value of bandwidth to calculate the rms value of the 9.0 MHz output voltage.

9-55. Prove that the maximum and minimum voltage values in Fig. 9–70b occur at $\Delta f / f_r = \pm 0.5/Q_L$ and their values are $\pm 0.5 G K_D P_{in}$.

9-56. Prove that the characteristic impedance of the periodic structure pictured in part c of Fig. 9–72 is

$$Z_{0p} = Z_0 \left[\frac{2 \sin kd - \bar{B}(1 - \cos kd)}{2 \sin kd + \bar{B}(1 + \cos kd)} \right]^{1/2}$$

when the unit cell is that given in part d.

9-57. Replace the unit cell in part b of Fig. 9–72 by one that consists of *two* series inductors of value $L/2$ separated by a Z_0 line of length d. Prove that Eq. (9–153) is still valid but Eq. (9–154) is not.

APPENDIX A
List of Symbols and Units

This appendix lists the symbols used in this text. The order is alphabetical with Roman and Greek symbols listed separately. It does not include some symbols used only in the course of short derivations or standard symbols such as j, π, e, etc. Also listed is the SI unit and its abbreviation for each quantity.

The notational system used throughout is outlined in Sec. 1–4.

SYMBOL	QUANTITY	SI UNIT (ABBREVIATION)
A, B, C, D	Elements of the ABCD Matrix	A and D are dimensionless. B in ohms (Ω), C in mhos ($\mho$)
$\overline{A}, \overline{B}, \overline{C}, \overline{D}$	Normalized ABCD Elements	——
A_C	Capture Area of an Antenna	Meter^2 (m^2)
A_t	Total Attenuation	Neper (Np) [Decibel (dB) commonly used]
$\vec{a}_x, \vec{a}_y, \vec{a}_z$	Unit Vectors in Cartesian Coordinates	——
$\vec{a}_r, \vec{a}_\phi, \vec{a}_z$	Unit Vectors in Cylindrical Coordinates	——
$\mathbf{a}_k$	Incident Power Wave at Port k	Watt$^{1/2}$ (W$^{1/2}$)
$B, \mathscr{B}$	Magnetic Flux Density	Weber/meter^2 (Wb/m^2) or Tesla (T)
B	ac Susceptance	Mho ($\mho$)
$\overline{B}$	Normalized Susceptance	——
$\mathbf{b}_k$	Scattered Power Wave at Port k	Watt$^{1/2}$ (W$^{1/2}$)
C	Capacitance	Farad (F)
C'	Capacitance per Unit Length	Farad/meter (F/m)
c	Velocity of Light in Free Space	Meter/second (m/s)
$D, \mathscr{D}$	Electric Flux Density	Coulomb/meter^2 (C/m^2)
d	Distance from Load End of a Transmission Line	Meter (m)
$E, \mathscr{E}$	Electric Field Intensity	Volt/meter (V/m)
E_d	Dielectric Strength	Volt/meter (V/m)
F	Force	Newton (N)
f	ac Frequency	Hertz (Hz)
f_c	Cutoff Frequency	Hertz (Hz)
f_p	Plasma Frequency	Hertz (Hz)
G	Conductance	Mho ($\mho$)
$\overline{G}$	Normalized Conductance	——
G'	Conductance per Unit Length	Mho/meter ($\mho$/m)
G_T	Antenna Gain	——
G_k	Unnormalized Element Values of the Lowpass Prototype Filter	Ohm (Ω) or Mho ($\mho$)
g_k	Normalized Element Values of the Lowpass Prototype Filter	——
g'_k	Normalized Element Values of the Highpass Prototype Filter	——

SYMBOL	QUANTITY	SI UNIT (ABBREVIATION)
$H, \mathcal{H}$	Magnetic Field Intensity	Ampere/meter (A/m)
$I, \mathcal{I}$	Electric Current	Ampere (A)
I^+, I^-	Current for Forward and Reverse Traveling Waves	Ampere (A)
$J, \mathcal{J}$	Current Density	Ampere/meter^2 (A/m^2)
J_n	nth Order Bessel Function of the First Kind	——
J'_n	First Derivative of J_n	——
$K, \mathcal{H}$	Surface Current Density	Ampere/meter (A/m)
K_c	Coupling Factor	——
K_d	Directivity Factor	——
K_{nm}	Cutoff Constants for Circular Waveguide	——
$K_{k, k+1}$	Inverter Value between Sections k and $k+1$	——
k	Wave Number $(= \omega\sqrt{\mu\epsilon} = 2\pi/\lambda)$	Meter^{-1} (m^{-1})
k_c	$= 2\pi/\lambda_c$	Meter^{-1} (m^{-1})
k_c	Coupling Coefficient of a Directional Coupler	——
k_t	Transmission Coefficient of a Directional Coupler	——
L	Inductance	Henry (H)
L'	Inductance per Unit Length	Henry/meter (H/m)
L_I	Insertion Loss	Decibel (dB)
L_R	Return Loss	Decibel (dB)
L_T	Transducer Loss	Decibel (dB)
L_m	Maximum Passband Transducer Loss of a Filter	Decibel (dB)
L_s	Minimum Stopband Transducer Loss of a Filter	Decibel (dB)
l	Length (usually the length of a transmission line)	meter (m)
ln	Natural Logarithm	——
log	Common Logarithm	——
M	Magnetization	Ampere/meter (A/m)
M_s	Saturation Magnetization	Ampere/meter (A/m)
m	Magnetic Pole	Weber (Wb)
N_n	nth Order Bessel Function of the Second Kind	——
N_1, N_2	Primary and Secondary Turns of a Transformer	——
n_t	Turns Ratio of a Transformer $(= N_1/N_2)$	——
n_i	Index of Refraction	——
$P_1, P_2, \ldots$	Path Value in a Signal Flow Graph	——
P	Power	Watt (W)
P_A	Available Generator Power	Watt (W)
P^+, P^-	Power in Forward and Reverse Traveling Waves	Watt (W)

SYMBOL	QUANTITY	SI UNIT (ABBREVIATION)
$p, \wp$	Power Density	Watt/meter^2 (W/m^2)
p^+, p^-	Power Density of Forward and Reverse Traveling Waves	Watt/meter^2 (W/m^2)
Q	Circuit Quality	——
Q_U, Q_L, Q_E	Unloaded, Loaded and External Q of a Resonant Circuit	——
Q_T	Total Q of a Multisection Filter	——
q	Electric Charge	Coulomb (C)
q_{nm}	mth Root of J_n	——
q'_{nm}	mth Root of J'_n	——
R	Resistance	Ohm (Ω)
$\bar{R}$	Normalized Resistance	——
R'	Resistance per Unit Length	Ohm/meter (Ω/m)
R_s	Surface Resistivity	Ohm/square (Ω/sq)
R_k	Low-Frequency Characteristic Impedance of a Lowpass Filter	Ohm (Ω)
S	Surface Area	Meter^2 (m^2)
S	Complex Frequency Variable in Richards' Transformation	——
SWR	Standing Wave Ratio	——
S_k	SWR of kth Obstacle	——
S_{mn}	Scattering Coefficient between Port m and Port n	——
T	Period of an ac Signal	Second (s)
T	Transmission Coefficient of a Two-Port Network	——
T_n	Tchebyscheff Polynomial of Order n	——
$\mathcal{T}_{11}, \mathcal{T}_{12},$ $\mathcal{T}_{21}, \mathcal{T}_{22}$	Elements of the Wave-Transmission Matrix	——
tan δ	Dielectric Loss Tangent	——
t	Time	Second (s)
U_E	Stored Electric Energy	Joule (J)
U_M	Stored Magnetic Energy	Joule (J)
$V, \mathcal{V}$	Electric Potential or Voltage	Volt (V)
V_G	Open-Circuit Generator Voltage	Volt (V)
V^+, V^-	Voltage for Forward and Reverse Traveling Waves	Volt (V)
V, v	Volume	Meter^3 (m^3)
v	Particle velocity	Meter/second (m/s)
v	Wave Velocity in an Unbounded Medium ($= c/\sqrt{\mu_R \epsilon_R}$)	Meter/second (m/s)
v_g	Group Velocity	Meter/second (m/s)
v_p	Phase Velocity	Meter/second (m/s)
X	ac Reactance	Ohm (Ω)
$\bar{X}$	Normalized Reactance	——
Y	ac Admittance	Mho (℧)
Y'	Admittance per Unit Length	Mho/meter (℧/m)
Y_0, Y_{0k}, Y_{0s}	Characteristic Admittance of a Transmission Line	Mho (℧)

SYMBOL	QUANTITY	SI UNIT (ABBREVIATION)
Z	ac Impedance	Ohm (℧)
Z'	Impedance per Unit Length	Ohm/meter (℧/m)
Z_0, Z_{0k}, Z_{0s}	Characteristic Impedance of a Transmission Line	Ohm (℧)
Z_{0e}	Even-Mode Characteristic Impedance of a Coupled-Line Pair	Ohm (Ω)
Z_{0o}	Odd-Mode Characteristic Impedance of a Coupled-Line Pair	Ohm (Ω)
Z_{0p}	Characteristic Impedance of a Periodic Structure	Ohm (Ω)
Z_{TE}, Z_{TM}	Wave Impedance for TE and TM Modes	Ohm (Ω)
Z_I	Image Impedance of a Two-Port Network	Ohm (Ω)
Z_T	Characteristic Impedance of a T Filter	Ohm (Ω)
Z_π	Characteristic Impedance of a π Filter	Ohm (Ω)
α	Attenuation Constant	Neper/meter (Np/m), [Decibel/meter (dB/m) commonly used]
α_c	Attenuation Constant due to Conductors	Neper/meter (Np/m), [Decibel/meter (dB/m)]
α_d	Dielectric Attenuation Constant	Neper/meter (Np/m), [Decibel/meter (dB/m)]
β	Phase Constant	Radian/meter (rad/m), [Degree/meter (°/m) commonly used]
γ	Propagation Constant	meter^{-1} (m^{-1})
Γ	Reflection Coefficient	——
Γ_G	Generator Reflection Coefficient	——
Γ_L	Load Reflection Coefficient	——
Γ_e	Even-Mode Reflection Coefficient of a Coupled-Line Pair	——
Γ_o	Odd-Mode Reflection Coefficient of a Coupled-Line Pair	——
δ_s	Skin Depth	meter (m)
ϵ_0	Permittivity of Free Space	Farad/meter (F/m)
ϵ_R	Relative Permittivity or Dielectric Constant	——
ϵ	Permittivity ($= \epsilon_0 \epsilon_R$)	Farad/meter (F/m)
ϵ_f	Normalized Frequency Deviation	——
ϵ_λ	Normalized Guide Wavelength Deviation	——
η	Intrinsic Impedance	Ohm (Ω)
η_s	Skin-Effect Wave Impedance	Ohm (Ω)
θ_I	Insertion Phase	Radian (rad), [Degree (°) commonly used]
θ_i	Angle of Incidence	Radian (rad), [Degree (°)]

SYMBOL	QUANTITY	SI UNIT (ABBREVIATION)
θ_t	Angle of Refraction	Radian (rad), [Degree (°)]
θ_c	Critical Angle	Radian (rad), [Degree (°)]
θ_{pf}	Power Factor Angle	Radian (rad), [Degree (°)]
λ	Wavelength in an Unbounded Medium ($= \lambda_0/\sqrt{\mu_R \epsilon_R}$)	Meter (m)
λ_0	Free Space Wavelength	Meter (m)
λ_c	Cutoff Wavelength	Meter (m)
λ_g	Guide Wavelength	Meter (m)
μ_0	Permeability of Free Space	Henry/meter (H/m)
μ_R	Relative Permeability	——
μ	Permeability ($= \mu_0 \mu_R$)	Henry/meter (H/m)
ρ	Resistivity	Ohm-meter (Ω-m)
ρ_s	Surface Charge Density	Coulomb/meter^2 (C/m^2)
ρ_v	Volume Charge Density	Coulomb/meter^3 (C/m^3)
σ	Conductivity	Mho/meter ($\mho$/m)
τ	Time Constant	Second (s)
Φ	Magnetic Flux	Weber (Wb)
ϕ	Angle of Reflection Coefficient	Radian (rad), [Degree (°) commonly used]
ϕ	Phase Delay or Electrical Length ($= \beta l$)	Radian (rad), [Degree (°)]
ϕ	Angular Variable in Cylindrical Coordinates	Radian (rad), [Degree (°)]
χ_e	Electric Susceptibility	——
χ_m	Magnetic Susceptibility	——
Ω	Frequency Variable in Richards' Transformation	——
Ω_c	Value of Ω at $f = f_c$	——
$\overline{\Omega}$	Normalized Value of Ω ($= \Omega/\Omega_c$)	——
ω	Radian Frequency ($= 2\pi f$)	Radian/second (rad/s)
ω'	Normalized Frequency	——

APPENDIX B
Material Constants

This appendix provides some of the electric and magnetic characteristics for materials discussed in this text.

Table B–1 lists the dc conductivity at room temperature (20° C) for several metals. The materials are tabulated in order of decreasing conductivity. The source for this data is *Reference Data for Radio Engineers* (Ref. 2–2).

TABLE B–1 The conductivity (σ) of some commonly used metals.

MATERIAL	σ (Mhos/m)
SILVER	6.12×10^7
COPPER	5.80×10^7
GOLD	4.10×10^7
ALUMINUM	3.54×10^7
BRASS	1.50×10^7
NICKEL	1.15×10^7
IRON	1.04×10^7
BRONZE	1.00×10^7
STAINLESS STEEL	0.11×10^7
NICHROME	0.09×10^7

Table B–2 lists the relative dielectric constant (ϵ_R) and loss tangent (tan δ) for several insulators commonly used at microwave frequencies. They represent room temperature (20° C) values measured at a frequency of 3.0 GHz. The relative permeability (μ_R) is approximately unity for all the materials listed. The source for this

TABLE B–2 The dielectric constant (ϵ_R) and loss tangent (tan δ) at 3.0 GHz for typical microwave insulators. For all the materials listed, $\mu_R \approx 1.00$.

MATERIAL	ϵ_R	tan δ
AIR	1.0006	Negligible
ALUMINA (AL_2O_3)	9.6	0.0001
GLASS	4 to 7	0.001 to 0.006
MICA (Ruby)	5.4	0.0003
POLYSTYRENE	2.55	0.0003
QUARTZ (Fused)	3.8	0.00006
REXOLITE–1422	2.54	0.0005
STYROFOAM	1.03	0.0001
TEFLON®	2.1	0.00015
TITANIUM DIOXIDE (RUTILE)	96	0.001
*WATER (DISTILLED)	77	0.157
*WOOD (BALSA)	1.22	0.10

*These two materials are very poor insulators at microwave frequencies. Because of their high values of loss tangent, they are occasionally used as absorbers of microwave energy.

data is *Reference Data for Radio Engineers* (Ref. 2–2). For most good microwave dielectrics, the loss tangent is so small that it is very difficult to measure. Tabulated values sometimes represent upper limits rather than accurate determinations.

For the problems given in this text, it is assumed that the above values are valid at all microwave frequencies. For more accurate values of ϵ_R and tan δ at frequencies other than 3.0 GHz, the reader is referred to Refs. 2–2 and 2–3.

There are literally hundreds of ferrite and garnet materials available to the microwave designer. Table B–3 lists some of the important characteristics for three such materials manufactured by Trans-Tech, Inc. of Adamstown, Maryland. The composition of the materials are

TT-1-390:   A magnesium-manganese ferrite.

TT-2-125:   A nickel-aluminate ferrite.

G-610:   An aluminum doped, yttrium-iron garnet.

TABLE B–3 A summary of characteristics for three typical microwave ferrites and garnets. All quantities except M_s and T_c are measured at 9.4 GHz.

MATERIAL	TT-1-390	TT-2-125	G-610
Saturation Magnetization (M_s)* in gauss	2150	2100	680
g-effective (g_{eff})	2.04	2.30	2.00
Line Width (ΔH) in oersteds	540	460	40
Spin Wave Line Width (ΔH_k) in oersteds	2.5	6.1	1.5
Relative Dielectric Constant (ϵ_R)	12.7	12.6	14.5
Dielectric Loss Tangent (tan δ)	< 0.00025	< 0.001	< 0.0002
Curie Temperature (T_c) in °C	320	560	185

*Some texts and data sheets indicate this quantity as $4\pi M_s$.

CGS units (gauss and oersteds) are commonly used to describe the properties of magnetic materials. The conversion to their MKS counterparts are

$$10{,}000 \text{ gauss} = 1.0 \text{ Wb/m}^2 \qquad \text{and} \qquad 0.0126 \text{ oersted} = 1.0 \text{ A/m}$$

In the CGS system, $\mu_0 = 1$, while in the MKS system, $\mu_0 = 4\pi \times 10^{-7} \, H/m$.

APPENDIX C
Transmission Matrices

The characteristics of a linear two-port network may be described in a variety of ways. Two methods that relate output quantities to input quantities are presented in this appendix. In both cases, the properties of the network are described by a square matrix. One is known as the *ABCD matrix* and the other as the *wave-transmission matrix*. These transmission matrices are especially useful in analyzing a cascade of two-port networks since the matrix for the overall network is simply the product of the individual matrices.

The ABCD matrix. The terminal voltages and currents for a two-port are shown in Fig. C–1. These phasor quantities may be related via the following equations

$$V_1 = AV_2 + BI_2$$
$$I_1 = CV_2 + DI_2$$

(C–1)

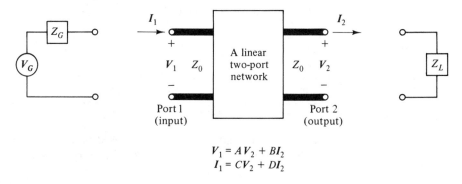

Figure C–1   A linear two-port network and its *ABCD* representation.

where A, B, C, and D are complex quantities that completely describe the two-port network. When the linear network is passive and reciprocal,

$$AD - BC = 1$$

(C–2)

Furthermore, if the network is symmetrical, $A = D$.

Equations (C–1) may be written in matrix form. That is,

$$\begin{Bmatrix} V_1 \\ I_1 \end{Bmatrix} = \begin{bmatrix} A & B \\ C & D \end{bmatrix} \begin{Bmatrix} V_2 \\ I_2 \end{Bmatrix}$$

(C–3)

This is known as the *ABCD matrix representation* of a two-port network. The over-all ABCD matrix for a cascade of two or more networks may be obtained by multiplying the individual ABCD matrices. For example, the relation between V_1, I_1, and V_3, I_3 for the cascade connection shown in Fig. C–2 is

$$\begin{Bmatrix} V_1 \\ I_1 \end{Bmatrix} = \begin{bmatrix} A & B \\ C & D \end{bmatrix} \begin{Bmatrix} V_3 \\ I_3 \end{Bmatrix} \tag{C–4}$$

where

$$\begin{bmatrix} A & B \\ C & D \end{bmatrix} = \begin{bmatrix} A_1 & B_1 \\ C_1 & D_1 \end{bmatrix} \times \begin{bmatrix} A_2 & B_2 \\ C_2 & D_2 \end{bmatrix} = \begin{bmatrix} A_1 A_2 + B_1 C_2 & A_1 B_2 + B_1 D_2 \\ C_1 A_2 + D_1 C_2 & C_1 B_2 + D_1 D_2 \end{bmatrix}$$

Many problems in this text make use of the matrix multiplication described above. For convenience, the ABCD matrix elements of some elementary two-port networks are listed in Table C–1. Note that in all cases, A and D are dimensionless quantities, B is an impedance (ohms), and C is an admittance (mhos).

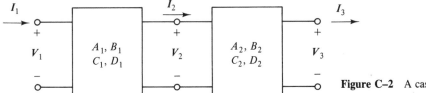

Figure C–2 below:

I_1 $\rightarrow$ + V_1 — | A_1, B_1 C_1, D_1 | I_2 $\rightarrow$ + V_2 — | A_2, B_2 C_2, D_2 | I_3 $\rightarrow$ + V_3 —

Figure C–2 A cascade connection of two-port networks.

All of the elements in Table C–1 can be deduced by writing the voltage and current equations for the particular network. For example, in the case of a series element,

$$V_1 = V_2 + ZI_2 \qquad \text{and} \qquad I_1 = I_2$$

Therefore, $A = D = 1$, $B = Z$, and $C = 0$. The matrix for a section of transmission line is obtained from Eqs. (3–14) and (3–16) by letting $z = 0$ (V_1, I_1) and $z = l$ (V_2, I_2) and solving for V_1 and I_1 in terms of V_2 and I_2.

By making use of matrix multiplication and the list of ABCD matrices in Table C–1, one can determine the ABCD matrix of more elaborate two-port networks. For example, the ABCD matrix for the network shown in Fig. C–3 is obtained by treating it as a cascade of a series element Z_s and a shunt element Z_p. Therefore,

$$\begin{bmatrix} A & B \\ C & D \end{bmatrix} = \begin{bmatrix} 1 & Z_s \\ 0 & 1 \end{bmatrix} \times \begin{bmatrix} 1 & 0 \\ 1/Z_p & 1 \end{bmatrix} = \begin{bmatrix} 1 + Z_s/Z_p & Z_s \\ 1/Z_p & 1 \end{bmatrix}$$

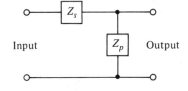

Input Z_p Output

Figure C–3 A two-port network formed by the cascade connection of a series element and a shunt element.

TABLE C–1   *ABCD* matrix elements for some basic two-port networks

Description	Two-port Network Port 1 (Input)	Port 2 (Output)	A	B	C	D
Series Element	Z		1	Z	0	1
Shunt Element	Y		1	0	Y	1
Ideal Transformer	N_1, N_2 $n_t \equiv \dfrac{\text{Turns}}{\text{ratio}} = N_1/N_2$		n_t	0	0	$\dfrac{1}{n_t}$
Lossless Transmission Line	Z_{0k}, β $\alpha = 0$		$\cos \beta l$	$jZ_{0k} \sin \beta l$	$j\dfrac{\sin \beta l}{Z_{0k}}$	$\cos \beta l$
Lossy Transmission Line	Z_{0k}, α, β $\gamma \equiv \alpha + j\beta$		$\cosh \gamma l$	$Z_{0k} \sinh \gamma l$	$\dfrac{\sinh \gamma l}{Z_{0k}}$	$\cosh \gamma l$

An expression for the input impedance of a two-port in terms of the ABCD elements is useful in solving impedance matching problems. With the output port of the network (Fig. C–1) terminated by a load impedance Z_L, Eqs. (C–1) become

$$V_1 = (AZ_L + B)I_2 \qquad \text{and} \qquad I_1 = (CZ_L + D)I_2$$

since $V_2 = Z_L I_2$. Therefore, the input impedance is

$$Z_{\text{in}} = \frac{V_1}{I_1} = \frac{AZ_L + B}{CZ_L + D} \tag{C–5}$$

Figure C–4a shows a two-port network with a generator connected to the input. Its Thevenin equivalent is shown in part *b* of the figure. V_G' and Z_G' may be expressed in terms of the ABCD matrix elements. With the output open circuited, $Z_L \to \infty$ and hence $Z_{in} = A/C$. The open-circuit voltage V_G' is obtained by realizing that $V_1 = V_G Z_{in}/(Z_{in} + Z_G)$ and since $I_2 = 0$, $V_G' = V_1/A$. Thus the Thevenin voltage becomes

$$V_G' = \frac{V_G}{A + CZ_G} \tag{C-6}$$

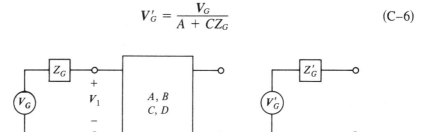

(a)

(b)

Figure C–4   A generator connected to a linear two-port network and its equivalent Thevenin representation.

The Thevenin impedance Z_G' is obtained by shorting V_G and determining the impedance looking into the output port. With V_G replaced by a short circuit, Eqs. (C–1) become

$$V_1 = -Z_G I_1 = AV_2 + BI_2 \qquad \text{and} \qquad I_1 = CV_2 + DI_2$$

Multiplying the second equation by Z_G and adding it to the first yields

$$0 = (A + CZ_G)V_2 + (B + DZ_G)I_2$$

Thus the output impedance is

$$Z_G' \equiv -\frac{V_2}{I_2} = \frac{B + DZ_G}{A + CZ_G} \tag{C-7}$$

in accordance with the polarities defined in Fig. C–1.

It is often useful to normalize the ABCD matrix elements to Z_0, the characteristic impedance of the input and output connecting lines. The normalized matrix is defined as

$$\begin{bmatrix} \bar{A} & \bar{B} \\ \bar{C} & \bar{D} \end{bmatrix} \equiv \begin{bmatrix} A & B/Z_0 \\ CZ_0 & D \end{bmatrix} \tag{C-8}$$

Thus $\bar{A} = A$, $\bar{B} = B/Z_0$, $\bar{C} = CZ_0 = C/Y_0$ and $\bar{D} = D$. When the output of the network is match terminated ($Z_L = Z_0$), Eq. (C–5) reduces to

$$\bar{Z}_{in} \equiv \frac{Z_{in}}{Z_0} = \frac{\bar{A} + \bar{B}}{\bar{C} + \bar{D}} \tag{C-9}$$

and the input reflection coefficient (referenced to Z_0) is given by

$$\Gamma_{in} = \frac{\overline{A} + \overline{B} - \overline{C} - \overline{D}}{\overline{A} + \overline{B} + \overline{C} + \overline{D}} \tag{C–10}$$

Similarly, when $Z_G = Z_0$, Eq. (C–7) reduces to

$$\overline{Z}'_G \equiv \frac{Z'_G}{Z_0} = \frac{\overline{B} + \overline{D}}{\overline{A} + \overline{C}} \tag{C–11}$$

and the reflection coefficient looking into the output port becomes

$$\Gamma_{out} = \frac{\overline{B} + \overline{D} - \overline{A} - \overline{C}}{\overline{A} + \overline{B} + \overline{C} + \overline{D}} \tag{C–12}$$

One can also define a transmission coefficient for the two-port network and express it in terms of the normalized ABCD matrix elements. Consider the set of input and output waves shown in Fig. C–5. The entering waves at ports 1 and 2 are denoted by a_1 and a_2 and the exiting waves by b_1 and b_2. These are known as *power waves* and are precisely defined in Appendix D [Eqs. (D–1), (D–2), and (D–7)]. With port 1 as the input and port 2 terminated in a load impedance Z_L, the transmission coefficient is defined as

$$T \equiv \frac{b_2}{a_1} = \frac{V_2^-}{V_1^+} = \frac{I_2^-}{I_1^+} \tag{C–13}$$

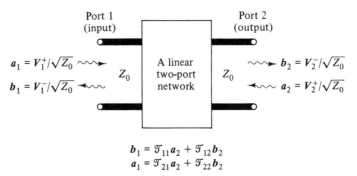

$$b_1 = \mathcal{T}_{11}a_2 + \mathcal{T}_{12}b_2$$
$$a_1 = \mathcal{T}_{21}a_2 + \mathcal{T}_{22}b_2$$

Figure C–5   The wave-transmission representation of a linear two-port network.

where it is assumed that the characteristic impedance of the connecting lines are identical (namely, Z_0). The phase angle of T is the phase difference between the input and output waves, while $|T|^2$ is the ratio of power in the output wave to that in the input wave. T may be expressed in terms of the normalized ABCD matrix elements and the load impedance. Namely,

$$T = \frac{\overline{Z}_L + 1}{\overline{A}\overline{Z}_L + \overline{B} + \overline{C}\overline{Z}_L + \overline{D}} \tag{C–14}$$

The derivation of this expression is as follows:

From Eq. (D–7),

$$2\sqrt{Z_0}\,a_1 = V_1 + Z_0 I_1 \qquad \text{and} \qquad 2\sqrt{Z_0}\,b_2 = V_2 + Z_0 I_2$$

Note that the minus sign in the second equation has been changed so that I_2 is now the current out of port 2. Making use of Eqs. (C–1),

$$\frac{b_2}{a_1} = \frac{V_2 + Z_0 I_2}{V_1 + Z_0 I_1} = \frac{V_2 + Z_0 I_2}{AV_2 + BI_2 + Z_0 CV_2 + Z_0 DI_2}$$

Since $Z_L = V_2/I_2$, the above expression becomes Eq. (C–14), where $\bar{Z}_L \equiv Z_L/Z_0$. When the output is match terminated ($Z_L = Z_0$), the transmission coefficient reduces to

$$T = \frac{2}{\bar{A} + \bar{B} + \bar{C} + \bar{D}} \tag{C–15}$$

For a dissipationless network, $P_{\text{out}} = P_{\text{in}}$, where $P_{\text{out}} = b_2^2 - a_2^2$ and $P_{\text{in}} = a_1^2 - b_1^2$. With the output match terminated, $a_2 = 0$ and therefore $b_2^2 = a_1^2 - b_1^2$ or $(b_2/a_1)^2 + (b_1/a_1)^2 = 1$. Since $b_1/a_1 = |\Gamma_{\text{in}}|$, the magnitude of the input reflection coefficient when $Z_L = Z_0$, and $b_2/a_1 = |T|$, the following useful expression is obtained for a dissipationless two-port network.

$$|\Gamma_{\text{in}}|^2 + |T|^2 = 1 \tag{C–16}$$

The wave-transmission matrix.   Another method of characterizing a two-port network is to relate the waves at the input (b_1, a_1) to similar waves at the output (a_2, b_2). These waves are shown in Fig. C–5 and defined by Eqs. (D–1) and (D–2). For the two-port,

$$b_1 = \mathcal{T}_{11}a_2 + \mathcal{T}_{12}b_2$$
$$a_1 = \mathcal{T}_{21}a_2 + \mathcal{T}_{22}b_2 \tag{C–17}$$

In matrix form,

$$\begin{Bmatrix} b_1 \\ a_1 \end{Bmatrix} = \begin{bmatrix} \mathcal{T}_{11} & \mathcal{T}_{12} \\ \mathcal{T}_{21} & \mathcal{T}_{22} \end{bmatrix} \begin{Bmatrix} a_2 \\ b_2 \end{Bmatrix} \tag{C–18}$$

When the output is match terminated, $a_2 = 0$ and hence

$$\Gamma_{\text{in}} \equiv \left.\frac{b_1}{a_1}\right|_{a_2=0} = \frac{\mathcal{T}_{12}}{\mathcal{T}_{22}} \qquad \text{and} \qquad T \equiv \left.\frac{b_2}{a_1}\right|_{a_2=0} = \frac{1}{\mathcal{T}_{22}} \tag{C–19}$$

Like the ABCD matrix, the $\mathcal{T}$ matrix for a cascade of two-port networks is the product of the individual $\mathcal{T}$ matrices. The following relations are useful for converting between the $\mathcal{T}$ and normalized ABCD matrix forms.

$$\begin{bmatrix} \mathcal{T}_{11} & \mathcal{T}_{12} \\ \mathcal{T}_{21} & \mathcal{T}_{22} \end{bmatrix} = \begin{bmatrix} \dfrac{\bar{A} - \bar{B} - \bar{C} + \bar{D}}{2} & \dfrac{\bar{A} + \bar{B} - \bar{C} - \bar{D}}{2} \\ \dfrac{\bar{A} - \bar{B} + \bar{C} - \bar{D}}{2} & \dfrac{\bar{A} + \bar{B} + \bar{C} + \bar{D}}{2} \end{bmatrix} \tag{C–20}$$

$$\begin{bmatrix} \bar{A} & \bar{B} \\ \bar{C} & \bar{D} \end{bmatrix} = \begin{bmatrix} \dfrac{\mathcal{T}_{11} + \mathcal{T}_{12} + \mathcal{T}_{21} + \mathcal{T}_{22}}{2} & \dfrac{-\mathcal{T}_{11} + \mathcal{T}_{12} - \mathcal{T}_{21} + \mathcal{T}_{22}}{2} \\ \dfrac{-\mathcal{T}_{11} - \mathcal{T}_{12} + \mathcal{T}_{21} + \mathcal{T}_{22}}{2} & \dfrac{\mathcal{T}_{11} - \mathcal{T}_{12} - \mathcal{T}_{21} + \mathcal{T}_{22}}{2} \end{bmatrix} \quad \text{(C--21)}$$

The conversion relations between the $\mathcal{T}$ matrix and the scattering matrix are given in Appendix D.

Transmission matrices provide an orderly way of determining the characteristics of two-port networks. The calculations may be numerous, but the procedures are straightforward and hence easily programmed on a computer. Many standard programs for microwave circuits are based upon these matrix methods.

APPENDIX D
The Scattering Matrix

The scattering matrix is a useful analytical technique for studying multiport microwave networks. Its elements relate forward and reverse traveling waves at the various ports of the network. The technique is a logical extension of the interpretation of transmission-line phenomena in terms of incident and reflected waves. This appendix reviews the scattering matrix and the concept of power waves. Although the analysis that follows is based on a three-port, the results may be generalized to any N-port network.

Figure D–1 shows a linear three-port network with transmission lines connected to each port. Their characteristic impedances are denoted by Z_{01}, Z_{02}, and Z_{03} and are assumed to be real in all subsequent discussions. Also shown are the terminal voltages and currents, V_k and I_k, where k indicates the port number. In classical circuit theory, the network is usually characterized by an impedance or admittance matrix that relates these terminal voltages and currents. This approach is presented in most texts on network theory and will not be repeated here. An alternate method

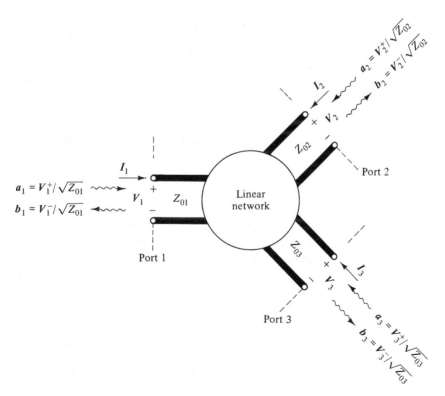

Figure D–1 A linear three-port network and its associated incident and scattered waves.

that is quite useful in microwave analysis is to describe the network behavior in terms of incident and scattered waves. These are shown in Fig. D–1 and are denoted by a_k and b_k, respectively. Note that the outgoing waves (b_k) are *not* labeled *reflected waves* since they are also associated with transmission from other ports. These phasor quantities are defined by the following equations.[1]

$$a_k = \frac{V_k^+}{\sqrt{Z_{0k}}} = I_k^+ \sqrt{Z_{0k}} \tag{D–1}$$

and

$$b_k = \frac{V_k^-}{\sqrt{Z_{0k}}} = I_k^- \sqrt{Z_{0k}} \tag{D–2}$$

where V_k^+, V_k^-, I_k^+, and I_k^- are the incident and scattered voltage and current waves and Z_{0k} is the characteristic impedance of the connecting line at port k. Since Z_{0k} is real, the phase angle of a_k is the same as V_k^+ and I_k^+. Similarly, the phase of b_k represents the phase of both the V_k^- and I_k^- waves. The square of their rms values equal the power flow associated with the incident and scattered waves. That is,

$$P_k^+ = a_k^2 \qquad \text{and} \qquad P_k^- = b_k^2 \tag{D–3}$$

Although a_k and b_k are usually referred to as *power* waves, they are in reality *normalized* voltage (and current) waves.

The net input power at port k is given by

$$P_{k_{in}} = a_k^2 - b_k^2 \tag{D–4}$$

On the other hand, the power dissipated in a termination at port k is

$$P_{k_L} = b_k^2 - a_k^2 = b_k^2[1 - |\Gamma_k|^2] \tag{D–5}$$

where $|\Gamma_k|$ is the magnitude of the load reflection coefficient at port k. Note that $\Gamma_k = a_k/b_k$.

The voltage and current at any point along a transmission line can be expressed in terms of forward and reverse traveling waves [Eqs. (3–14) and (3–16)]. Designating the input terminals of the lines in Fig. D–1 as $z = 0$, we have

$$V_k = V_k^+ + V_k^- = \sqrt{Z_{0k}}\,(a_k + b_k)$$

and $\hfill$ (D–6)

$$I_k = I_k^+ - I_k^- = \frac{1}{\sqrt{Z_{0k}}}(a_k - b_k)$$

where V_k and I_k represent the phasor voltage and current at the input to the kth port.

---

[1] A more general treatment of this subject is found in "Power Waves and the Scattering Matrix" by K. Kurokawa, *IEEE Transactions on Microwave Theory and Techniques*, March 1965, pp. 194–202. This article defines a_k and b_k in terms of source and load impedances. Complex impedances with negative real parts are included in the analysis.

Solving the above equations for a_k and b_k in terms of the terminal voltage and current yields

$$a_k = \frac{1}{2}\left(\frac{V_k}{\sqrt{Z_{0k}}} + I_k\sqrt{Z_{0k}}\right)$$

and   (D–7)

$$b_k = \frac{1}{2}\left(\frac{V_k}{\sqrt{Z_{0k}}} - I_k\sqrt{Z_{0k}}\right)$$

These equations describe the relationship between the power waves and the terminal voltages and currents.

Referring again to the three-port network in Fig. D–1, the incident and scattered waves are related to the network characteristics by the following equations.

$$b_1 = S_{11}a_1 + S_{12}a_2 + S_{13}a_3$$
$$b_2 = S_{21}a_1 + S_{22}a_2 + S_{23}a_3 \qquad (D\text{–}8)$$
$$b_3 = S_{31}a_1 + S_{32}a_2 + S_{33}a_3$$

where the scattering coefficients (S_{11}, S_{12}, etc.) define the characteristics of the linear network.[2] Written in matrix form,

$$\{b\} = [S]\{a\}$$

where $\{a\}$ and $\{b\}$ are column matrices that represent the incident and scattered waves. The scattering matrix for the three-port is given by

$$[S] = \begin{bmatrix} S_{11} & S_{12} & S_{13} \\ S_{21} & S_{22} & S_{23} \\ S_{31} & S_{32} & S_{33} \end{bmatrix} \qquad (D\text{–}9)$$

Since a_k and b_k are phasors, the elements of the scattering matrix are generally complex. These elements are easily measured for an actual network. For example, S_{11}, S_{22}, and S_{33} represent the input reflection coefficients at ports 1, 2, and 3, respectively, when all output ports are match terminated. Thus, S_{11} may be determined by connecting a generator to port 1 and reflectionless loads to ports 2 and 3. Since a_2 and a_3 are zero, measuring the ratio of reflected to incident voltage at port 1 yields S_{11}, where from Eqs. (D–1), (D–2), and (D–8)

$$S_{11} = \frac{b_1}{a_1} = \frac{V_1^-}{V_1^+}$$

S_{22} and S_{33} can be similarly determined. The off-diagonal elements of the scattering matrix represent transmission coefficients. For instance, $S_{21}a_1$ is the wave that emerges from port 2 when a generator is connected to port 1 and all other ports are terminated in reflectionless loads (that is, $a_2 = a_3 = 0$). Under these conditions S_{21}

---

[2] Since the a_k and b_k waves are defined in terms of Z_{0k}, the scattering coefficients are referenced to the characteristic impedance of the connecting lines.

is the transmission coefficient from port 1 to port 2. From Eqs. (D–1), (D–2), and (D–8),

$$S_{21} = \frac{b_2}{a_1} = \sqrt{\frac{Z_{01}}{Z_{02}}} \frac{V_2^-}{V_1^+} = \sqrt{\frac{Z_{02}}{Z_{01}}} \frac{I_2^-}{I_1^+}$$

Note that when $Z_{01} = Z_{02}$, S_{21} equals the transmission coefficient T described in Appendix C when $Z_L = Z_0$ [(Eq. (C–15)]. Similar conclusions may be drawn for the other off-diagonal scattering coefficients.

For reciprocal[3] networks, the scattering matrix is symmetrical. For the three-port, this means that $S_{12} = S_{21}$, $S_{13} = S_{31}$, and $S_{23} = S_{32}$. Since the scattering coefficients are complex, these equalities apply to both their magnitude and phase. Thus for a reciprocal three-port network, there are six independent coefficients and therefore twelve independent parameters in the scattering matrix. When the network is dissipationless, a set of conditions exist which further reduces the number of independent parameters. These conditions are helpful in analyzing multiport networks and hence are derived here.

Necessary conditions for a dissipationless network. Assume that a generator is connected to port 1 and reflectionless loads to ports 2 and 3 of the network shown in Fig. D–1. This means that $a_1 \neq 0$ and $a_2 = a_3 = 0$. If the network is dissipationless, the total power out of the network must equal the total input power. Thus from Eq. (D–3),

$$P_1^+ = a_1^2 = b_1^2 + b_2^2 + b_3^2$$

Taking the magnitude squared of both sides of Eqs. (D–8) with a_2 and a_3 equal to zero and adding yields

$$b_1^2 + b_2^2 + b_3^2 = \{|S_{11}|^2 + |S_{21}|^2 + |S_{31}|^2\}a_1^2$$

Since the left-hand side equals a_1^2,

$$|S_{11}|^2 + |S_{21}|^2 + |S_{31}|^2 = 1 \qquad\qquad\qquad\qquad \text{(D–10)}$$

This equation states that the sum of the squares of the *magnitude* of the scattering coefficients in the first column equals unity. This statement holds true for all columns of the matrix. The following equation generalizes the result to an N-port dissipationless network. Namely,

$$\sum_{n=1}^{N} |S_{np}|^2 = \sum_{n=1}^{N} S_{np} S_{np}^* = 1 \qquad\qquad\qquad \text{(D–11)}$$

for any column p from 1 to N. The asterisk (*) denotes the complex conjugate. This condition, as well as the one that follows, is valid for both reciprocal and nonreciprocal networks.

---

[3] The definition of reciprocity may be found in most texts on network theory. Essentially it means that interchanging the positions of a zero impedance source and a zero impedance ammeter in a network does not affect the ammeter reading.

The scattering coefficients of a dissipationless network are also constrained by the following relationship. Namely,

$$\sum_{n=1}^{N} S_{np} S_{nq}^* = 0 \qquad \text{for } p \neq q \qquad (D-12)$$

where p and q represent two *different* columns of the matrix.

Let us now verify this equation for the first and second columns of a three-port network. Suppose a_1 and a_2 are finite, but $a_3 = 0$. Taking the magnitude squared of both sides of Eqs. (D–8) and adding yields

$$b_1^2 + b_2^2 + b_3^2 = |S_{11}a_1 + S_{12}a_2|^2 + |S_{21}a_1 + S_{22}a_2|^2 + |S_{31}a_1 + S_{32}a_2|^2$$

Since the left-hand side is the power out of the network, it must equal the input power $(a_1^2 + a_2^2)$. Recognizing that for any complex quantity Q, $|Q|^2 = QQ^*$, the above expression may be multiplied out to give the following result.

$$a_1^2 + a_2^2 = \{|S_{11}|^2 + |S_{21}|^2 + |S_{31}|^2\}a_1^2 + \{|S_{12}|^2 + |S_{22}|^2 + |S_{32}|^2\}a_2^2$$
$$+ \{S_{11}S_{12}^* + S_{21}S_{22}^* + S_{31}S_{32}^*\}a_1 a_2^* + \{S_{12}S_{11}^* + S_{22}S_{21}^* + S_{32}S_{31}^*\}a_2 a_1^*$$

Using Eq. (D–11) reduces the above to

$$\{S_{11}S_{12}^* + S_{21}S_{22}^* + S_{31}S_{32}^*\}a_1 a_2^* + \{S_{12}S_{11}^* + S_{22}S_{21}^* + S_{32}S_{31}^*\}a_2 a_1^* = 0$$

Since a_1 and a_2 are independent signals, they may be chosen in any convenient manner. Letting $a_1 = a_2$ and finite yields

$$\{S_{11}S_{12}^* + S_{21}S_{22}^* + S_{31}S_{32}^*\} + \{S_{12}S_{11}^* + S_{22}S_{21}^* + S_{32}S_{31}^*\} = 0$$

Next, letting $a_1 = ja_2$ and finite yields

$$\{S_{11}S_{12}^* + S_{21}S_{22}^* + S_{31}S_{32}^*\} - \{S_{12}S_{11}^* + S_{22}S_{21}^* + S_{32}S_{31}^*\} = 0$$

since $a_1^* = (ja_2)^* = -ja_2^*$.
Combining the above expressions results in

$$S_{11}S_{12}^* + S_{21}S_{22}^* + S_{31}S_{32}^* = 0$$

and   (D–13)

$$S_{12}S_{11}^* + S_{22}S_{21}^* + S_{32}S_{31}^* = 0$$

The first equation is exactly Eq. (D–12) when $p = 1$, $q = 2$, and $N = 3$. The second equation is the case when $p = 2$, $q = 1$, and $N = 3$. Thus for the first two columns of the scattering matrix, the sum of the products of the elements of one column with the complex conjugate of adjacent elements in the other column is zero. Equation (D–12) represents the generalization of this statement to any two columns of an N-port dissipationless network.

It should be noted that if the network is reciprocal, Eqs. (D–11) and (D–12) also apply to the *rows* of the scattering matrix since $S_{np} = S_{pn}$ and $S_{nq} = S_{qn}$.

Effect of a shift in terminal plane on the scattering coefficients. A shift in the terminal plane of any port of the network produces a change in the values of a_k and b_k. This, in turn, affects the various scattering coefficients. When the lines are lossless, only their phase angles are affected. Suppose the scattering coefficients in Eqs. (D–8) and (D–9) are associated with the three terminal planes shown in Fig. D–1. A shift in the terminal planes of the three ports results in the following transformed scattering coefficients ($\tilde{S}$).

Diagonal components:

$$\tilde{S}_{nn} = S_{nn} e^{-j2\beta_n l_n}, \qquad\qquad n = 1, 2, \text{ or } 3 \qquad\qquad \text{(D–14)}$$

where β_n is the phase constant of the transmission line connected to port n and l_n is the amount that the terminal plane is shifted *away* from the network. For a shift *toward* the network, l_n is negative. Since the diagonal components represent reflection coefficients, Eq. (D–14) is merely the reflection-coefficient transformation equation for a lossless line.

Off-diagonal components:

$$\tilde{S}_{mn} = S_{mn} e^{-j(\beta_m l_m + \beta_n l_n)} \qquad\qquad\qquad \text{(D–15)}$$

where $m = 1$, 2, or 3, $n = 1$, 2, or 3, but $m \neq n$.
β_m and β_n are the phase constants and l_m and l_n are the amount that the terminal planes are shifted *away* from the network for ports m and n. For example, if $m = 2$ and $n = 3$, the $e^{-j(\beta_2 l_2 + \beta_3 l_3)}$ term accounts for the additional phase delay experienced by a wave traveling from port 3 to port 2.

To better understand the utility of the scattering matrix, consider the following illustrative example.

Example D–1:

Prove that for a reciprocal three-port network, the ports cannot all be perfectly matched when the network is dissipationless.

Solution: If all three ports are matched, then $S_{11} = S_{22} = S_{33} = 0$ and the scattering matrix reduces to

$$[S] = \begin{bmatrix} 0 & S_{21} & S_{31} \\ S_{21} & 0 & S_{32} \\ S_{31} & S_{32} & 0 \end{bmatrix}$$

where for a reciprocal network $S_{12} = S_{21}$, $S_{13} = S_{31}$, and $S_{23} = S_{32}$. Applying Eq. (D–12) to the first and second columns requires that $S_{31} S_{32}^* = 0$. Thus either S_{32} or S_{31} must be zero. The scattering matrix for these two possibilities are

$$\begin{bmatrix} 0 & S_{21} & S_{31} \\ S_{21} & 0 & 0 \\ S_{31} & 0 & 0 \end{bmatrix} \qquad \text{and} \qquad \begin{bmatrix} 0 & S_{21} & 0 \\ S_{21} & 0 & S_{32} \\ 0 & S_{32} & 0 \end{bmatrix}$$

For the first case, the application of Eq. (D–11) to the second and third columns results in

$$|S_{21}| = 1 \qquad \text{and} \qquad |S_{31}| = 1$$

This clearly contradicts Eq. (D–11) since the sum of the squares of the element magnitudes in the first column equals 2. Similarly, one can show that the second case represents an impossible situation. The conclusion is that all three ports of a reciprocal network *cannot* be simultaneously matched.

Two-port networks are quite common in microwave work. For this case, Eq. (D–8) reduces to

$$\boldsymbol{b_1} = S_{11}\boldsymbol{a_1} + S_{12}\boldsymbol{a_2}$$

$$\boldsymbol{b_2} = S_{21}\boldsymbol{a_1} + S_{22}\boldsymbol{a_2}$$

$$(D-16)$$

The scattering coefficients for some elementary two-ports are listed in Table D–1.

TABLE D–1   Scattering coefficients for some basic two-port networks. (Note: All impedances are normalized to Z_0, the characteristic impedance of the connecting lines.)

Description	Two-port Network with Z_0 connecting lines	S_{11}	S_{12}
Series Element	Z between Z_0 and Z_0; $\bar{Z} \equiv Z/Z_0$	$\dfrac{\bar{Z}}{\bar{Z}+2}$ $(S_{22} = S_{11})$	$\dfrac{2}{\bar{Z}+2}$ $(S_{21} = S_{12})$
Shunt Element	Y between Z_0 and Z_0; $\bar{Y} \equiv Y/Y_0 = YZ_0$	$\dfrac{-\bar{Y}}{\bar{Y}+2}$ $(S_{22} = S_{11})$	$\dfrac{2}{\bar{Y}+2}$ $(S_{21} = S_{12})$
Ideal Transformer	N_1, N_2 between Z_0 and Z_0; $n_t \equiv \dfrac{\text{Turns}}{\text{ratio}} = N_1/N_2$	$\dfrac{n_t^2 - 1}{n_t^2 + 1}$ $(S_{22} = -S_{11})$	$\dfrac{2n_t}{n_t^2 + 1}$ $(S_{21} = S_{12})$
Lossless Transmission Line	Z_0 Z_{0k}, β Z_0; $\alpha = 0$	$\dfrac{j(\bar{Z}_{0k}^2 - 1)\sin\beta l}{2\bar{Z}_{0k}\cos\beta l + j(\bar{Z}_{0k}^2 + 1)\sin\beta l}$ $(S_{22} = S_{11})$	$\dfrac{2\bar{Z}_{0k}}{2\bar{Z}_{0k}\cos\beta l + j(\bar{Z}_{0k}^2 + 1)\sin\beta l}$ $(S_{21} = S_{12})$
Lossy Transmission Line	Z_0 Z_{0k}, α, β Z_0; $\gamma \equiv \alpha + j\beta$	$\dfrac{(\bar{Z}_{0k}^2 - 1)\sinh\gamma l}{2\bar{Z}_{0k}\cosh\gamma l + (\bar{Z}_{0k}^2 + 1)\sinh\gamma l}$ $(S_{22} = S_{11})$	$\dfrac{2\bar{Z}_{0k}}{2\bar{Z}_{0k}\cosh\gamma l + (\bar{Z}_{0k}^2 + 1)\sinh\gamma l}$ $(S_{21} = S_{12})$

It has been assumed that the characteristic impedance of the two connecting lines are identical (that is, $Z_{01} = Z_{02} = Z_0$).

The conversion relations between the scattering matrix and the $\mathcal{T}$ matrix described in Appendix C are as follows.

$$\begin{bmatrix} \mathcal{T}_{11} & \mathcal{T}_{12} \\ \mathcal{T}_{21} & \mathcal{T}_{22} \end{bmatrix} = \begin{bmatrix} \dfrac{S_{12}S_{21} - S_{11}S_{22}}{S_{21}} & \dfrac{S_{11}}{S_{21}} \\ -\dfrac{S_{22}}{S_{21}} & \dfrac{1}{S_{21}} \end{bmatrix} \qquad (D-17)$$

$$\begin{bmatrix} S_{11} & S_{12} \\ S_{21} & S_{22} \end{bmatrix} = \begin{bmatrix} \dfrac{\mathcal{T}_{12}}{\mathcal{T}_{22}} & \dfrac{\mathcal{T}_{11}\mathcal{T}_{22} - \mathcal{T}_{12}\mathcal{T}_{21}}{\mathcal{T}_{22}} \\ \dfrac{1}{\mathcal{T}_{22}} & -\dfrac{\mathcal{T}_{21}}{\mathcal{T}_{22}} \end{bmatrix} \qquad (D-18)$$

With $Z_{01} = Z_{02}$, the condition for reciprocity is $S_{12} = S_{21}$, which requires that $\mathcal{T}_{11}\mathcal{T}_{22} - \mathcal{T}_{12}\mathcal{T}_{21} = 1$.

——————— APPENDIX E ———————
Hyperbolic Functions

This appendix reviews various useful relationships for hyperbolic functions with both real and complex arguments.

Euler's identity relates trigonometric and exponential functions. Namely,

$$e^{j\theta} = \cos\theta + j\sin\theta \qquad \text{and} \qquad e^{-j\theta} = \cos\theta - j\sin\theta$$

where θ is in radians. Conversely,

$$\sin\theta = \frac{e^{j\theta} - e^{-j\theta}}{2j} \qquad \text{and} \qquad \cos\theta = \frac{e^{j\theta} + e^{-j\theta}}{2}$$

Values for these circular trigonometric functions are readily available from tables and calculators. Cosine is an even-order function $[f(x) = f(-x)]$, while sine is an odd-order function $[f(x) = -f(-x)]$. The following useful approximations should be noted. For $\theta < 0.5$ rad ($\approx 30°$),

$$\begin{aligned} &\sin\theta \approx \theta \text{ (in radians)} && \text{within 4 percent} \\ &\tan\theta \approx \theta \text{ (in radians)} && \text{within 9 percent} \\ &\cos\theta \approx 1 && \text{within 12 percent} \end{aligned} \qquad \text{(E–1)}$$

Hyperbolic functions are related to exponential functions in the following manner.

$$\sinh z = \frac{e^z - e^{-z}}{2} \qquad \text{and} \qquad \cosh z = \frac{e^z + e^{-z}}{2} \qquad \text{(E–2)}$$

Also,

$$\tanh z \equiv \frac{\sinh z}{\cosh z} = \frac{e^z - e^{-z}}{e^z + e^{-z}} \qquad \text{(E–3)}$$

Hyperbolic functions for real values of z are tabulated and readily available. Sinh z is an odd-order function with values ranging from $-\infty$ to $+\infty$. Cosh z is an even function with values between $+1$ and $+\infty$. Tanh z is an odd-order function with values between -1 and $+1$. The following useful approximations should be noted. For $z < 0.5$ Np,

$$\begin{aligned} &\sinh z \approx z && \text{within 4 percent} \\ &\tanh z \approx z && \text{within 8 percent} \\ &\cosh z \approx 1 && \text{within 13 percent} \end{aligned} \qquad \text{(E–4)}$$

Hyperbolic functions with complex arguments appear in the impedance transformation equation. The reader will find the following identities useful.

$$\begin{aligned} \sinh z &= -j\sin jz & \sinh jz &= j\sin z \\ \cosh z &= \cos jz \qquad \text{and} & \cosh jz &= \cos z \\ \tanh z &= -j\tan jz & \tanh jz &= j\tan z \end{aligned} \qquad \text{(E–5)}$$

The following relations are useful in computing hyperbolic functions when z is complex. Setting $z = x + jy$,

$$\sinh (x + jy) = \sinh x \cos y + j \cosh x \sin y$$
$$\cosh (x + jy) = \cosh x \cos y + j \sinh x \sin y \qquad \text{(E-6)}$$
$$\tanh (x + jy) = \frac{\tanh x + j \tan y}{1 + j \tanh x \tan y}$$

The following brief example shows a typical calculation.

Example E–1:

Calculate $\tanh (\alpha l + j\beta l)$ for $\alpha l = 0.3$ Np and $\beta l = 3\pi/4$ rad.

Solution: With $\alpha l < 0.5$ Np, $\tanh \alpha l \approx \alpha l$ and hence from Eq. (E–6),

$$\tanh (\alpha l + j\beta l) \approx \frac{\alpha l + j \tan \beta l}{1 + j\alpha l \tan \beta l} = \frac{0.3 - j1}{1 - j0.3} = 0.55 - j0.83.$$

Point-To-Point Transmission

There are two ways in which electrical energy and information can be transmitted from one point to another. One method is via a transmission line between the two points. With a generator connected to one end, the power delivered to a matched load at the other end is $P_L = P_T e^{-2\alpha r}$, where P_T is the power into the line at the generator end. The length of the line equals the distance between the two points and is denoted by r. The transmission loss, defined as $10 \log (P_T/P_L)$, is a function of the attenuation constant α of the line, which increases in value as the frequency increases. The two dashed curves in Fig. F–2 show the transmission loss as a function of distance for a typical coaxial cable. Data is given at two frequencies, 0.5 and 5.0 GHz. Note that in both cases, the loss increases rapidly as the distance between the two points is increased.

The other method of transmitting electrical energy between two points involves a pair of directive antennas aimed at each other. The antenna system approach is described in Fig. F–1. The power source P_T is connected to a transmitting antenna which radiates the electromagnetic energy. A portion of this energy is intercepted by the receiving antenna and delivered to the load. The transmission efficiency is maximized by carefully aiming the antennas at each other. The following analysis shows that P_L/P_T is inversely proportional to r^2.

Let us assume, for the moment, that the transmitting antenna is isotropic, that is, it radiates equally well in all directions. Assuming that air is a lossless medium, the power density (p_T) at a distance r will equal $P_T/4\pi r^2$, since the surface area of a sphere of radius r is $4\pi r^2$. In practice, antennas are not isotropic but directive. Therefore, the actual power density at the receiving antenna is

$$p_T = G_T \frac{P_T}{4\pi r^2} \tag{F–1}$$

G_T is the directive gain of the transmitting antenna and is defined as

$$G_T \equiv \frac{p_T}{p_o} \tag{F–2}$$

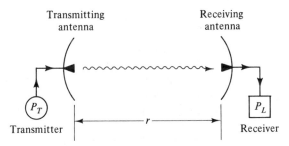

Figure F–1 An antenna system for point-to-point communication.

where p_T is the power density at a distance r for the actual antenna and p_o is the power density at the same distance for an isotropic antenna radiating the same amount of power as the actual antenna. Thus, directive gain is a measure of the *focusing* ability of an antenna. Its value is often expressed in decibels ($10 \log G_T$). For example, an antenna with 6 dB gain is capable of producing in a specified direction, a power density that is four times greater than an isotropic antenna. It is assumed that the antenna has been aimed so as to maximize the power density in the desired direction. Therefore, as used in this discussion, G_T is actually the *maximum* directive gain.

Since the receiving antenna is also directive, it must be aimed at the transmitting antenna to maximize the received signal. Under these conditions, the received power is

$$P_L = p_T A_C = \frac{P_T G_T A_C}{4 \pi r^2} \tag{F-3}$$

where A_C is the *capture area* of the receiving antenna[1] and is defined as

$$A_C \equiv \frac{\text{Power available at the receiving antenna terminals}}{\text{Power density at the receiving antenna}} \tag{F-4}$$

For a well-designed microwave antenna, G_T is proportional to its cross-sectional area A_T. That is,

$$G_T = k \frac{4 \pi A_T}{\lambda_o^2} \tag{F-5}$$

where λ_o is the operating wavelength and k is an efficiency factor (generally, 0.4 to 0.7). Substitution into Eq. (F-3) yields

$$\frac{P_L}{P_T} = \frac{k A_T A_C}{\lambda_o^2 r^2} \tag{F-6}$$

This equation is very significant in that is shows that for a well-designed antenna system, the transmission efficiency increases with increasing frequency (decreasing λ_o). It is this property, more than any other, that makes microwaves so important in communication and radar systems. The derivation of the above expressions assumes that the spacing between the two antennas is sufficient to satisfy the *far-field* criteria, namely,

$$r > 2D^2 / \lambda_o \tag{F-7}$$

where D is the largest transverse dimension of the antenna.

The solid curves in Fig. F-2 show the transmission loss ($10 \log P_T/P_L$) for a typical antenna system at 0.5 and 5.0 GHz. Note that for fixed values of k, A_T, and A_C, the loss at 5.0 GHz is 20 dB less than the loss at 0.5 GHz. This shows the advantage of microwave frequencies in point-to-point transmission.

---

[1] By virtue of reciprocity, the capture area of the receiving antenna and its gain when used as a transmitting antenna are related. That is, $A_C = G_T \lambda_o^2 / 4 \pi$. This relationship holds for all antennas, provided, of course, that corresponding polarized waves are considered for transmission and reception. See, for example, Sec. 1.15 of *Antenna Theory and Design*, by R. S. Elliot (Prentice-Hall, Inc., 1981).

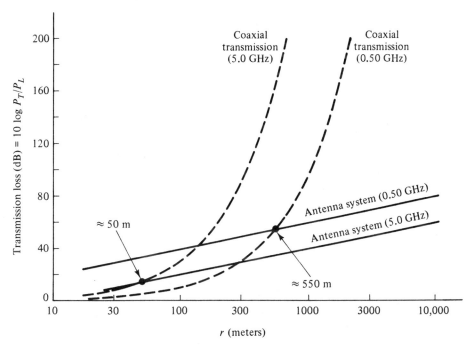

Figure F–2   A comparison between transmission-line and antenna systems for point-to-point communication.

Before concluding this discussion, let us compare the two methods of transmission. At very short distances, it is apparent that transmission lines are superior. On the other hand, at very long distances, the antenna system is clearly more efficient. The actual *crossover* distance is a function of frequency. At 0.5 GHz, for the values given in the figure, the antenna system has lower loss than the coaxial line when r is greater than 550 m. At 5.0 GHz, the antenna system is superior when r is greater than 50 m. At higher frequencies, the crossover distance is even shorter. In fact, for millimeter waves, antennas and lenses are often used for distances of less than a few meters (the length of a typical lab bench). The actual crossover distance in a particular application is not only a function of frequency, but of antenna size and the loss properties of the transmission line.

ANSWERS TO PROBLEMS

CHAPTER 2

2–1. $U_E = 1.285 \times 10^{-10}$ J

2–2. (a) $r = 5.33 \times 10^{-3}$ m
  (b) $G = 6.79 \times 10^{-6}$ mho

2–3. $N = 24$ turns, $R = 0.016$ ohm

2–4. $\epsilon_R = 6.25$

2–5. $\mu_R = 1.15$, $\epsilon_R = 2.42$

2–6. ——

2–7. $\theta = 70.5°$

2–8. (a) $\mathbf{E} = 99.9\underline{/-4 \text{ rad}}$ V/m
  (b) $A_t = 0.082$ dB

2–9. $\delta_s = 1.7 \times 10^{-4}$ cm $= 1.7$ microns

2–10. (a) 6.1×10^{-6} m thick
  (b) $R_{ac} = 3.9$ ohms, compared to 29.0 ohms before plating

2–11. $f = 1250$ MHz

2–12. $\epsilon_R = 6.25$

2–13. $\mathbf{E}_{x_1} = 22.4\underline{/18.4°}$ V/m and
  $\mathbf{H}_{y_1} = 0.179\underline{/71.6°}$ A/m at $d = 1$ cm,
  while $\mathbf{E}_{x_1} = 10\underline{/90°}$ V/m and
  $\mathbf{H}_{y_1} = 0.24\underline{/90°}$ A/m at $d = 2$ cm.
  The first magnetic field maximum occurs at $d = 2$ cm.

2–14. $\Gamma = 0.172$, SWR $= 1.42$

2–15. $\mu_{R_1} = 1.07$, $\epsilon_{R_1} = 2.74$

2–16. $p_t = 4.52$ W/m²

2–17. SWR $= 3.0$ and 5.83

2–18. $n_1 = 1.74$ minimum

CHAPTER 3

3–1. $v = 2.22 \times 10^8$ m/s, $Z_0 = 300$ ohms

3–2. $Z_0 = 25$ or 100 ohms

3–3. ——

3–4. V and $I = 0$ at $t = 5$ ns, $V = 10$ V and $I = 0.10$ A at $t = 15$ ns, and $V = 16$ V and $I = 0.04$ A at $t = 35$ ns.

3–5. $I_L = 0$ for $0 \le t \le 20$ ns, 0.48 A for $20 \le t \le 60$ ns, 0.384 A for $60 \le t \le 100$ ns, and 0.4032 A for $100 \le t \le 140$ ns.

3–6. (a) $V = 0$ for $0 \le t \le 1.5$ ns, 90 V for $1.5 \le t \le 4.5$ ns. Thereafter, the voltage oscillates between 0 and 90 V in 3.0 ns intervals.
  (b) $f = 166.7$ MHz. Either increase l or decrease v.

3–7. 0.28 percent

3–8. (a) $Z_0 = 65.8$ ohms
  (b) $Z_0 = 45.4$ ohms and $\lambda = 6.90$ cm.

3–9. $P = 35.8$ W

3–10. $I = 0.38\underline{/-126.4°}$ A, $P = 7.22$ W

3–11. $\Gamma_L = 0.678\underline{/274.5°}$, $\Gamma_{in} = 0.678\underline{/166.5°}$ and SWR $= 5.21$

3–12. $\Gamma_{in} = 0.509\underline{/14.2°}$

3–13. A capacitor of value 4.97 pF.

3–14. SWR $= 2.33$

3–15. $V_{max} = 48$ V, $V_{min} = 12$ V, $I_{max} = 0.64$ A, and $I_{min} = 0.16$ A

3–16. $L = 68.2$ nH

3–17. ——

3–18. $d = 1.25$ cm. No!

3–19. $R_L = 66.7$ or 600 ohms

3–20. $\Delta P = 0.71$ W and $\Delta P = 0$

3–21. (a) $\mathbf{V}_L = 13.29\underline{/-64.77°}$ V and
  $\mathbf{I}_L = 0.157\underline{/-109.77°}$ A
  (b) $P_{in} = P_L = 1.47$ W

3–22. $P_{in} = 1.63$ W, $P_L = 1.24$ W, and 0.39 W is dissipated by the line.

3–23. $P_{in} = 1.97$ W, $P_L = 1.54$ W, and 0.43 W is dissipated by the line.

3–24. SWR $= 3.0$ and 1.22

3–25. None. All.

3–26. $L_R = 25$ dB at the input.

3–27. Ref. Loss $= 1.25$ dB and 0.044 dB.

3–28. $Z_{in} = 29.3 - j28.6$ ohms

3–29. $Y_{in} = (0.50 + j0.60)/75$ mhos

3–30. $Z_L = 25 + j4.5$ ohms

3–31. Re $(Z_{in}) = 60$ ohms

3–32. $l/\lambda = 0.161$

3–33. $P = 0.47$ W for the one on the left, $P = 0.12$ W for the one on the right.

3–34. (a) $l = 10$ cm and $b/a = 1.95$.
(b) $\Delta P = 153.4$ mW

3–35. $Z_{in} = 55.4 + j83.1$ ohms

3–36. $l = 19.2$ cm

3–37. (a) $X_C = 143.2$ ohms, 15.6 percent
(b) 3.1 cm shorter

3–38. $Z_0 = 60$ ohms, $v = 2.18 \times 10^8$ m/s

3–39. $l \geq 13.33$ m

3–40. SWR $= 7.5$

3–41. $Q = 78.5$

3–42. $l = \lambda/8$, $X_L = 100$ ohms, and $Q = 50$

3–43. SWR $= 1.50$, $\Gamma_L = 0.20\underline{/-53°}$

3–44. (a) $Y_L = (8.4 - j10)10^{-3}$ mhos
(b) $Y = (7.2 + j4.8)10^{-3}$ mhos

3–45. (a) $Z_L = 50 + j80$ ohms
(b) $Z_L = 36 + j92$ ohms

3–46. (a) $l = 1.044$ cm
(b) None

3–47. $l = 0.55$ cm. Yes! $l = 0.38$ cm.

3–48. (a) $l = 0.75$ cm
(b) $l = 1.4$ cm

3–49. (a) $Z_{in} = 41 + j36$ ohms
(b) $Z_{in} = 45 + j108$ ohms

3–50. (a) $Z_{in} = 18.3 - j11.7$ ohms and
$\Gamma_{in} = 0.33\underline{/-121°}$
(b) SWR $= 1.50$ on the 50 ohm line and
SWR $= 2.0$ on the 30 ohm line.

3–51. Yes! A loop is necessary to avoid a counterclockwise trend in the $\overline{Y}_{in}$ plot.

3–52. $d = 3.38$ cm to the first current minimum.
$d = 7.13$ cm to the first current maximum.

CHAPTER 4

4–1. (a) $l = 0.98$ cm and $C = 1.45$ pF.
(b) $\overline{Z}_{in} = 0.92 - j0.17$ and
SWR $= 1.20$ at 1800 MHz.
$\overline{Z}_{in} = 1.1 + j0.15$ and SWR $= 1.18$
at 2200 MHz.

4–2. SWR $= 2.50$, 1.00, and 2.30 at 1800,
2000, and 2200 MHz, respectively.

4–3. (a) $l = 0.90$ cm and $C = 1.01$ pF.
(b) $\overline{Z}_{in} = 0.80 + j0.16$ and SWR $=$
1.32 at 2500 MHz.

$\overline{Z}_{in} = 1.1 - j0.30$ and SWR $= 1.35$
at 3500 MHz.

4–4. (a) $l_s = 1.21$ cm for $Z_{0s} = 50$ ohms and
$l_s = 0.58$ cm for $Z_{0s} = 20$ ohms
(b) For $Z_{0s} = 50$ ohms, SWR $= 1.50$ at
3500 MHz, while for $Z_{0s} = 20$ ohms,
SWR $= 1.37$.

4–5. (a) $l_s = 3.75$ cm and $Z_{0s} = 32.1$ ohms
(b) SWR $= 1.55$ at both 1600 and 2400
MHz

4–6. $l = 1.20$ cm and $Z_{in} = 28 - j21$ ohms

4–7. $Z_{01} = 125$ ohms. SWR $= 1.20$ at 6000
MHz and SWR $= 1.33$ at 5500 and
6500 MHz.

4–8. ——

4–9. (a) $Z_{01} = 25$ ohms and $l = 0.5$ cm
(b) SWR $= 1.25$

4–10. $Z_{01} = 9$ ohms, $Z_{02} = 90$ ohms, $l_1 =$
0.23 cm and $l_2 = 0.32$ cm. From Eqs.
(4–14) and (4–15), $l_1 = 0.22$ cm and
$l_2 = 0.33$ cm.

4–11. ——

4–12. ——

4–13. (a) $Z_{01} = 283$ ohms, $Z_{02} = 35.4$ ohms,
and $l_1 = l_2 = 0.57$ cm
(b) SWR $= 1.70$ and 1.95 at 2000 and
3000 MHz, respectively

4–14. $l_1 = 0.163\lambda$ and $l_2 = 0.358\lambda$

4–15. $\overline{B} = 1.0$

4–16. $Z_L = 26 - j14$ ohms

4–17. $l_1 = 0.189\lambda$, $l_2 = 0.375\lambda$ and
$l_1 = 0.074\lambda$, $l_2 = 0.051\lambda$

4–18. ——

4–19. ——

4–20. (a) $l = 1.57$ cm and $d_1 = 0.283$ cm
(b) SWR $= 1.50$ at both 2700 and 3300
MHz

4–21. (a) $l = 1.57$ cm, $d_1 = 0.169$ cm, and
$d_2 = 0.384$ cm
(b) SWR $= 1.08$ at both 2700 and 3300
MHz

4–22. (a) $l = 1.57$ cm, $d_1 = 0.116$ cm,
$d_2 = 0.283$ cm, and $d_3 = 0.425$ cm
(b) SWR $= 1.02$ at both 2700 and 3300
MHz

4-23. (a) $l = 1.57$ cm, $d_1 = 0.171$ cm, and $d_2 = 0.382$ cm

    (b) SWR = 1.04 at 2700, 3000, and 3300 MHz

4-24. The minimum number is three.

4-25. (a) $S_{max} = 7.0$ (b) $S_{min} = 1.0$

4-26. ——

4-27. ——

4-28. $L_T = 41.7$ dB and the dissipative loss is 36.9 dB

4-29. $L_T = 40.1$ dB and the dissipative loss is 0 dB

4-30. $L_T = 0.07$ dB

4-31. $P_L = 0.124$ W and $P_d = 0.376$ W

4-32. $L_T = 1.45$ dB and $P_L = 0.715$ W

4-33. (a) $L_I = 0.15$ dB at 1.0 GHz and 4.77 dB at 7.5 GHz

    (b) $L_I = 0.20$ dB at 1.0 GHz and 3.23 dB at 7.5 GHz

    (c) $L_I = 0.04$ dB at 1.0 GHz and 11.6 dB at 7.5 GHz

4-34. (a) $L_T = 1.94$ dB of which 0.18 dB is mismatch loss.

    (b) $n_t = \sqrt{2}$ and $L_T = 3.0$ dB, which is all dissipative loss.

4-35. $\theta_I = \arctan \{0.50(\bar{Z}_{01} + \bar{Z}_{01}^{-1}) \tan \beta l\}$ The three conditions are $Z_{01} = Z_0$, $l = \lambda/4$, and $l = \lambda/2$.

4-36. $L_I = 2.1, 3.3, 9.6, 21.5,$ and 40.3 dB at $f = 0.1, 0.3, 1.0, 3.0,$ and 10.0 GHz, respectively. $\theta_I = 13.6°, 37.8°, 85.2°, 130.7°,$ and $163.1°$ at $f = 0.1, 0.3, 1.0, 3.0,$ and 10.0 GHz, respectively.

4-37. ——

4-38. For the overall $\mathcal{T}$ matrix, $\mathcal{T}_{11} = \mathcal{T}_{21} = -j/\sqrt{2}$, $\mathcal{T}_{12} = 0$, and $\mathcal{T}_{22} = j\sqrt{2}$. $\Gamma_{in} = 0$ and $L_T = 3.0$ dB.

4-39. ——

4-40. ——

4-41. (a) $\Gamma_{in} = 0.305\underline{/-31.6°}$ and $L_T = 7.54$ dB

    (b) $\Gamma_{in} = 0.40\underline{/-28.5°}$ and $L_T = 7.6$ dB

4-42. $\mathcal{T}_{11} = 0.50 + j0.65$, $\mathcal{T}_{12} = 0.78 - j0.27$, $\mathcal{T}_{21} = -0.59 + j2.34$, and $\mathcal{T}_{22} = 2.77 + j0.32$. $S_{11} = 0.30\underline{/-26°}$, $S_{12} = S_{21} = 0.36\underline{/-7°}$ and $S_{22} = 0.87\underline{/-82°}$.

4-43. (a) $\Gamma_1 = S_{11} + \dfrac{S_{12} S_{21} S_{11}}{1 - S_{22} S_{11}}$

    (b) $T = \dfrac{S_{21}^2}{1 - S_{11} S_{22}}$

CHAPTER 5

5-1. $l = 4.6$ cm

5-2. $\alpha l = 0.234$ dB

5-3. $b/a = 2.30$. Increased.

5-4. $\alpha_c = 0.093$ dB/m, $\alpha_d = 0.059$ dB/m, and $\alpha = 0.152$ dB/m.

5-5. ——

5-6. $P_{max} = 196$ kW

5-7. $f_c = 7.58$ GHz, $\Delta f_c = 3.69$ GHz

5-8. $d = 2a = 0.16$ cm, $D = 2b = 0.536$ cm

5-9. (a) $\alpha = 0.091$ dB/m

    (b) $\alpha = 0.116$ dB/m

5-10. 1.1 percent

5-11. Keep b fixed and increase w by $2\frac{1}{2}$ times.

5-12. $\alpha = 1.50$ dB/m at 2 GHz and 4.80 dB/m at 8 GHz.

5-13. $b = 0.59$ cm maximum

5-14. $w = 0.205$ inch = 0.521 cm.

5-15. $\alpha_c = 0.33$ dB/m, $\alpha_d = 0.10$ dB/m, and $\alpha = 0.43$ dB/m

5-16. $w = 0.23$ cm and $l = 0.95$ cm

5-17. $f_c = 6.56, 13.12, 14.76,$ and 16.16 GHz, respectively.

5-18. $f_c = 3.28, 6.56, 7.38,$ and 8.08 GHz, respectively.

5-19. ——

5-20. $\lambda_g = 12.29$ cm at 7 GHz and 2.99 cm at 12 GHz

5-21. $\lambda_g = 2.51$ cm at 7 GHz and 1.34 cm at 12 GHz

5-22. $a = 1.07$ cm and $l = 4.23$ cm, minimum

5-23. $b = 1.41$ cm and $l = 2.27$ cm

5-24. SWR = 1.30

5-25. $a = 2.35$ cm and $l = 1.18$ cm

5-26. ——

5-27. 10 percent

5-28. $\alpha = 0.041$ dB/m, which is half that of the coaxial line

5–29. $a = 0.476$ cm and $b = 0.238$ cm

5–30. $f_c = 4.78$ GHz, $f_{max} = 5.93$ GHz, and $f_{max} = 7.52$ GHz

5–31. $\lambda_g = 5.71$ cm at 6 GHz and 3.23 cm at 8 GHz. $Z_0 = 436$ ohms at 6 GHz and 329 ohms at 8 GHz

5–32. (a) $C_p' = 0.097$ pF/cm
(b) $f_c = 6.0$ GHz for the TE_{20} mode and 6.9 GHz for the TE_{30} mode

5–33. $a = 3.26$ cm, $b = 1.47$ cm, and $d = 0.37$ cm

5–34. $\cot\{\pi(a - s)/\lambda_c\} = (b/d) \tan(\pi s/\lambda_c)$

5–35. ——

5–36. $f_c = 2.74$ GHz for the TE_{10} mode and 7.47 GHz for the TE_{20} mode.

CHAPTER 6

6–1. SWR = 1.13, SWR = 1.10

6–2. SWR = 1.16

6–3. $A_t = 27$ dB

6–4. $dX_{in}/d\omega = 12.76 \times 10^{-9}$ ohms/rad/s for the shorted line and 7.16×10^{-9} ohms/rad/s for the lumped inductor.

6–5. ——

6–6. $l = 1.50$ cm, SWR = 1.45, and $L_l = 0.15$ dB

6–7. ——

6–8. $l = 0.74$ cm, $a_1 = 0.12$ cm

6–9. SWR = 1.17 at 5 GHz and 1.06 at 8 GHz

6–10. $R_1 = 36.3$ ohms, $R_2 = 16.3$ ohms, and $\Delta L_l = 1.4$ dB

6–11. ——

6–12. $f_{min} \approx 2.86$ GHz

6–13. $A_t = 10.9$ dB

6–14. $D = 4.0$ cm

6–15. $\Delta\phi = 45°$ in both cases

6–16. $\Delta\theta = 44.6°$ and $\Delta\theta = 45.6°$

6–17. ——

6–18. SWR = 1.00 at f_r and 1.20 at $1.3f_r$

6–19. Minimum $\epsilon_R = 7.86$

6–20. $Z_{01} = 16.67$ ohms. SWR = 1.43 at both frequencies.

6–21. ——

6–22. $L_1/2 = 4.77$ nH and $C_2 = 1.70$ pF.
$L_l = 1.0$, 12.5, and 30.3 dB at 2, 4, and 8 GHz. SWR = 2.68, 69.1, and 4300 at 2, 4, and 8 GHz. $A_t = 18.2$ and 31.8 dB at 4 and 8 GHz.

6–23. (a) $Z_{in} = 2.8 + j93.0$ ohms
(b) $L_1 = 9.55$ nH and $C_2/2 = 0.85$ pF.
$Z_{in} = 1.7 - j60.4$ ohms.

6–24. $L_l = 0.36$ dB at 0.5 GHz and 0.43 dB at 0.75 GHz

6–25. $l_1 = 1.25$ cm, $l_2 = 0.86$ cm, $D = 1.04$ cm, $d_0 = 0.297$ cm, $d_1 = 0.154$ cm, and $d_2 = 0.575$ cm

6–26. $L_l = 26.9$ dB, a 3.0 dB increase

6–27. $l_1 = 1.50$ cm, $l_2 = 1.04$ cm, $D = 1.25$ cm, $d_0 = 0.54$ cm, $d_1 = 0.0035$ cm, and $d_2 = 0.484$ cm

6–28. (a) $f_c = 1.974$ GHz, $R_k = 77.5$ ohms
(b) $Z_{in} = 3.3 - j119.1$ ohms, $L_l = 14.0$ dB

6–29. $l_1 = l_2 = 1.36$ cm, $l_3 = 2.90$ cm, $w_1 = 0.033$ cm, $w_2 = 1.20$ cm, and $w_3 = 0.108$ cm

6–30. $Z_{03} = 106$ ohms, $l_3 = 4.28$ cm

CHAPTER 7

7–1. (a) $l_s = 0.646$ cm, $b = 0.103$ cm
(b) SWR = 1.21 at 12.4 GHz and 1.13 at 18.0 GHz

7–2. $l = 6.89$ cm

7–3. $b = 3.49$ cm

7–4. $L_l = 10 \log\{1 + (\bar{B}^4/4)\}$, where $\bar{B} = \bar{B}_C$ or $-\bar{B}_L$

7–5. $\bar{B}_C = 0.484$, $\bar{B}_L = 0.968$

7–6. A horizontally polarized wave

7–7. (a) SWR = 1.64
(b) $L_l = 0.26$ dB

7–8. $A_t = 173$ dB

7–9. ——

7–10. SWR = 3.29

7–11. $L_l = 2.50$ dB, $\theta_l = 254.1°$

7–12. (a) ——

(b) $L_l = 0.06$ dB for the perpendicular component and 0.85 dB for the parallel component.

7–13. (a) 5417 MHz

(b) $b' = 1.24$ cm. The iris in part a has the higher Q.

7–14. $\bar{B}_L = 4.25$

7–15. (a) $\bar{B}_C = 1.50$

(b) ± 0.05

7–16. (a) SWR = 1.77

(b) $b_1 = 0.576$ cm

(c) SWR = 1.77

7–17. (a) $x = 1.536$ cm

(b) $\bar{Z}_{\text{in}} = -j7.57$

7–18. $L_R = 0$ dB for $\bar{R} = 0$ and 0.0023 dB for $\bar{R} = 0.01$. $L_R = 0$ dB for $\bar{R} = 0$ and 0.14 dB for $\bar{R} = 0.01$.

7–19. (a) $\theta_m = 44.93°$

(b) within $\pm 0.33°$

7–20. (a) Zero by the input card, 110.66 mW by the rotatable card, and 49.43 mW by the output card.

(b) 1.09 mW by the input card and 0.88 mW reflected to the source. Input SWR = 1.14

7–21. $P = 46.9$ mW

7–22. ——

7–23. ——

7–24. (a) $\Delta\theta = 1.054$ rad

(b) $\Delta L_l = 6.13$ dB

7–25. (a) $a = 4.01$ cm

(b) Directivity = 26.13 dB minimum

7–26. (a) $\bar{B}_L = 0.828$

(b) $l = 2.00$ cm and $d = 2.41$ cm

7–27. ——

7–28. (a) SWR = 1.16   **(b)** SWR = 1.40

CHAPTER 8

8–1. P_3 varies from 0.225 W to 0.625 W.

8–2. $P_3 = 16.0$ W for $l_2 = 0.25\lambda$ and 73.5 W for $l_2 = 0.50\lambda$

8–3. ——

8–4. (a) $Z_{01} = Z_{02} = 59.46$ ohms, $Z_{03} = 42.04$ ohms

(b) SWR = 1.42

8–5. $Z_{01} = 45.18$ ohms, $Z_{02} = 78.25$ ohms, $Z_{03} = 39.13$ ohms, SWR = 1.38, and $P_1/P_2 = 1.90$

8–6. SWR = 1.213, $P_1/P_2 = 0.348$

8–7. (a) SWR = 3.13

(b) $L_l = 4.35$ dB

8–8. $P_2 = 3.95$ W, $P_3 = 0.988$ W

8–9. $l_3 = 1.20$ cm

8–10. SWR = 1.42 at ports 1 and 2. SWR = 5.62 at port 3.

8–11. (a) $n_t = 1.414$

(b) $P_{L_1} = 0.20$ W, $P_{L_2} = 0.80$ W

8–12. ——

8–13. $P_1 = 0.416$ W, $P_2 = 0.469$ W, $P_3 = 0.006$ W, $P_4^- = 0.109$ W, and SWR = 1.99

8–14. $P_{1_{\max}} = 0.995$ W, $P_{1_{\min}} = 0.582$ W

8–15. SWR = 1.025

8–16. $P_1 = 5.37$ W, $P_r = 0.39$ W

8–17. $P_1 = 0.05$ mW, $P_4 = 12.5$ mW

8–18. ——

8–19. (a) $f = 6.25$ GHz

(b) $f = 6.17$ to 6.33 GHz

8–20. ——

8–21. Directivity = 53.4 dB at 8.2 GHz, 284.4 dB at 10.0 GHz, and 54.0 dB at 12.4 GHz

8–22. $BW = 2906$ MHz for the Tchebyscheff design and 1835 MHz for the binomial design.

8–23. ——

8–24. $l = 2.0$ cm, $Z_{01} = 43.3$ ohms, and $Z_{02} = 86.3$ ohms.

8–25. For a perfect match, $\bar{Z}_{02} = \bar{Z}_{03} + \sqrt{\bar{Z}_{03}^2 - 1}$. Coupling (dB) = $10 \log \bar{Z}_{03}^2$.

8–26. ——

8–27. Coupling = 3.02 dB, Directivity = 11.9 dB.

8–28. ——

8–29. $s = 0.15$ cm, $w = 0.47$ cm, and $l = 2.00$ cm

8–30. Coupling = 13.6 dB at 4.0 and 6.0 GHz, and 13.2 dB at 5.0 GHz

8–31. Coupling = 15.1 dB, Directivity = 26.8 dB

8–32. Directivity = 23.2 dB

8–33. (a) $l = 2.339$ cm
(b) Coupling = 4.3 dB

8–34. $l = 4.678$ cm

8–35. $P_2 = 0.268$ W, $P_3 = 2.40$ W, and
$P_4 = 2.275$ W

8–36. (a) Coupling = 24.0 dB
(b) $P_2 = 0.89$ mW
(c) $\Delta P = +0.28$ dB and -0.29 dB

8–37. 5.93 W $\leq P_T \leq 8.38$ W

8–38. $P_3 = 7.38$ mW

8–39. $P_2 = 3.6$ mW

8–40. Directivity = 46.0 dB minimum

8–41. $M_{max} = 0.607$

8–42. $-2.29° \leq \theta \leq +2.29°$

8–43. SWR = 1.05 maximum

8–44. Arc loss = 0.22 dB, $P = 2.5$ W
absorbed by each TR tube

8–45. (a) zero leakage
(b) leakage = 2.36 W

CHAPTER 9:

9–1. ——

9–2. ——

9–3. ——

9–4. (a) $f_r = 3.0$ and 9.0 GHz. $Q_E = 3.93$ at
3.0 GHz and 11.8 at 9.0 GHz.
(b) $Q_U = 1364$, $Q_L = 3.92$.

9–5. (a) $f_r = 4.668$ GHz
(b) $Q_E = 73.6$

9–6. (a) $l = 1.39$ cm
(b) $Q_U = 483$

9–7. ——

9–8. (a) $f = 3.27$ GHz
(b) $Q_E = 191$

9–9. $Y_{in}/Y_0 = 9.23$

9–10. (a) $f = 3.0$ and 6.0 GHz
(b) $f = 4.38$ and 6.79 GHz

9–11. $f_r = 4.33$ GHz for the TE_{101} mode,
8.04 GHz for the TE_{102} mode, and
5.00 GHz for the TE_{111} mode.

9–12. $D = l = 3.72$ cm

9–13. (a) $Q_U = 38,700$
(b) $Q_U = 13,900$

9–14. $l_{max} = 4.44$ cm, $f_r = 8.03$ GHz

9–15. $D/l = 1.00$, $Q_U = 39,430$

9–16. (a) $\bar{B}_L = 3.88$
(b) $Q_L = 50.4$

9–17. $f_r = 8.283$ GHz

9–18. $\bar{Y}_{in} = -j2.66$ at $0.99 f_r$ and $+j22.7$ at
$1.01 f_r$.

9–19. (a) $k_c = 0.327$
(b) $b_4/a_1 = 0.17$

9–20. (a) $C = 1.36$ pF, $l = 10.46$ cm
(b) $L_I = 0.03$ dB at 1.1 GHz, 0.21 dB at
1.5 GHz, and 1.81 dB at 1.9 GHz

9–21. $L_I = 34.0$ dB

9–22. (a) $f = 7964$ and 8034 MHz
(b) $d = 1.088$ cm, $l = 2.581$ cm

9–23. $\bar{B}_{L_r} = 2.09$, $l = 2.952$ cm

9–24. $L_I = 2.70$ dB at 10.1 GHz, 0.10 dB at
12.0 GHz, and 0.002 dB at 15.0 GHz

9–25. $L_I = 2.11$ dB at 10.1 GHz, 0.13 dB at
12.0 GHz, and 3.90 dB at 15.0 GHz

9–26. $L_I = L_T = 10 \log \{1 + (Q_T \epsilon_f)^{-6}\}$

9–27. (a) $l = 3.163$ cm, $C = 0.166$ pF
(b) $f = 2.782$ and 3.218 GHz

9–28. ——

9–29. $Q_U = 1785$

9–30. $d = 1.194$ cm, $l = 3.124$ cm

9–31. $f_1 = 1978$ MHz, $f_2 = 2022$ MHz

9–32. ——

9–33. $Z_{01} = 67.4$ ohms, $Z_{02} = 95.9$ ohms

9–34. $L_I = 60.5$ dB at 7.5 GHz and 59.5 dB at
8.5 GHz

9–35. $L_I = 0.21$ dB

9–36. ——

9–37. $g_1 = 1.670$, $g_2 = 1.193$, $g_3 = 2.366$,
$g_4 = 0.842$, and $r = 1.984$

9–38. (a) $L_T = 0.04$ dB at 1.2 GHz, 3.0 dB at
1.6 GHz, and 18.8 dB at 2.2 GHz
(b) ——

9–39. No! Four sections are required.

9–40. $Z_{01} = Z_{03} = 14.5$ ohms,
$Z_{02} = 86.2$ ohms,
and $l = 1.50$ cm

9–41. $n = 3$

9–42. ——

9–43. $Z_{01} = Z_{03} = 204$ ohms, $Z_{02} = 76.9$ ohms,
$Z_{04} = Z_{05} = 66.2$ ohms, and $l = 3.75$ cm

9–44. (a) $L_T = 23.1$ dB at both 2750 and
3250 MHz

    (b) ——

9–45. ——

9–46. $Z_{01} = Z_{03} = 251.3$ ohms, $Z_{02} =$
4.97 ohms, and $l = 1.875$ cm

9–47. ——

9–48. $Z_{0s} = L =$
$Z_{01}(2Z_{02} + Z_0)/(Z_{01} + 2Z_{02} + Z_0)$ and
the characteristic impedance of the unit
elements is $Z_0(2Z_{02} + Z_0)/2Z_{02}$.

9–49. $l = 2.0$ cm for all sections.
$(Z_{0e})_{01} = (Z_{0e})_{23} = 76.2$ ohms, $(Z_{0o})_{01} =$
$(Z_{0o})_{23} = 23.8$ ohms, $(Z_{0e})_{12} =$
52.5 ohms, and $(Z_{0o})_{12} = 25.0$ ohms

9–50. $l = 1.0$ cm for all sections.

$(Z_{0e})_{01} = (Z_{0e})_{34} = 69.1$ ohms, $(Z_{0o})_{01} =$
$(Z_{0o})_{34} = 30.9$ ohms, $(Z_{0e})_{12} =$
$(Z_{0e})_{23} =$
50.9 ohms, and $(Z_{0o})_{12} = (Z_{0o})_{23} =$
35.8 ohms.

9–51. $L_T = 27.4$ dB at both frequencies.

9–52. $Z_{0s} = 25$ ohms for the end stubs and
52 ohms for the center stubs. Z_0 for the
outer unit elements is 25 ohms, while for
the center one it is 10 ohms.

9–53. ——

9–54. (a) $BW = 96$ MHz minimum

    (b) $V_{out} = 165.1$ V rms

9–55. ——

9–56. ——

9–57. ——

INDEX